INTERNATIONAL SERIES OF MONOGRAPHS ON PURE AND APPLIED BIOLOGY

Division: **MODERN TRENDS IN PHYSIOLOGICAL SCIENCES**

General Editors: P. Alexander and Z. M. Bacq

Volume 22

PHOTOCHEMISTRY OF PROTEINS AND NUCLEIC ACIDS

AF873858

OTHER TITLES IN THE MODERN TRENDS IN PHYSIOLOGICAL SCIENCES DIVISION

General Editors: P. ALEXANDER and Z. M. BACQ

Vol. 1. FLORKIN—*Unity and Diversity in Biochemistry*

Vol. 2. BRACHET—*The Biochemistry of Development*

Vol. 3. GEREBTZOFF—*Cholinesterases*

Vol. 4. BROUHA—*Physiology in Industry*

Vol. 5. BACQ and ALEXANDER—*Fundamentals of Radiobiology*

Vol. 6. FLORKIN (Ed.)—*Aspects of the Origin of Life*

Vol. 7. HOLLAENDER (Ed.)—*Radiation Protection and Recovery*

Vol. 8. KAYSER—*The Physiology of Natural Hibernation*

Vol. 9. FRANÇON—*Progress in Microscopy*

Vol. 10. CHARLIER—*Coronary Vasodilators*

Vol. 11. GROSS—*Oncogenic Viruses*

Vol. 12. MERCER—*Keratin and Keratinization*

Vol. 13. HEATH—*Organophosphorous Poisons*

Vol. 14. CHANTRENNE—*The Biosynthesis of Proteins*

Vol. 15. RIVERA—*Cilia, Ciliated Epithelium and Ciliary Activity*

Vol. 16. ENSELME—*Unsaturated Fatty Acids in Atherosclerosis*

Vol. 17. BALABUKHA—*Chemical Protection of the Body against Ionizing Radiation*

Vol. 18. PETERS—*Biochemical Lesions and Lethal Synthesis*

Vol. 19. THOMSON—*Biological Effects of Deuterium*

Vol. 20. DE ROBERTIS—*Histophysiology of Synapses and Neurosecretion*

Vol. 21. KOVACS—*The Biochemistry of Poliomyelitis Viruses*

OTHER DIVISIONS IN THE SERIES ON

PURE AND APPLIED BIOLOGY

BIOCHEMISTRY

BOTANY

PLANT PHYSIOLOGY

ZOOLOGY

PHOTOCHEMISTRY OF PROTEINS AND NUCLEIC ACIDS

BY

A. D. McLAREN

Professor, College of Agriculture,
University of California, Berkeley

AND

D. SHUGAR

Professor, Institute of Biochemistry and
Biophysics, and State Institute of Hygiene,
Academy of Sciences, Warsaw

PERGAMON PRESS

OXFORD · LONDON · EDINBURGH · NEW YORK
PARIS · FRANKFURT

1964

PERGAMON PRESS LTD.
Headington Hill Hall, Oxford
4 & 5 Fitzroy Square, London W.1

PERGAMON PRESS (SCOTLAND) LTD.
2 & 3 Teviot Place, Edinburgh 1

PERGAMON PRESS INC.
122 East 55th Street, New York 22, N.Y.

GAUTHIER-VILLARS ED.
55 Quai des Grands-Augustins, Paris 6

PERGAMON PRESS G.m.b.H.
Kaiserstrasse 75, Frankfurt am Main

Distributed in the Western Hemisphere by
THE MACMILLAN COMPANY · NEW YORK
pursuant to a special arrangement with
Pergamon Press Incorporated

Copyright © 1964
Pergamon Press Inc.

Library of Congress Catalog Card Number 63-16864

Set in Monotype Baskerville 10 on 12 pt.
and printed in Great Britain by the Pitman Press, Bath

Contents

PAGE

PREFACE xi

I. SOME PRINCIPLES OF PHOTOCHEMISTRY 1

A. Introduction 1

1. The laws of photochemistry 1
2. Absorption spectroscopy 4
3. Transfer mechanisms of excitation energy 11

B. Factors of influence in photochemical reactions in solution 15

1. The absorption of light and the quantum yield 15
2. General nature of photochemical reactions in solution 18
   (a) Some reactions of small molecules 18
   (b) Detection and identification of free radicals 24
   (c) Oxygen and photosensitization 26

II. ABSORPTION AND LUMINESCENCE SPECTRA OF NUCLEOPROTEINS AND THEIR COMPONENTS

A. Proteins 29

1. Principal absorbing groups in proteins 29
2. Free and bound phenolic hydroxyl groups in proteins 37
3. Peptide absorption in the range 2400–3200 Å 39
4. Absorption by fibrous proteins 43
5. Apparent absorption of light attributable to scattering 45
6. Extraneous absorption in proteins 46
7. Peptide absorption below 2400 Å 48
8. Excited states of proteins 51

B. Nucleic acids 58

1. Purines and pyrimidines 59
2. Nucleosides and nucleotides 64
3. Nucleic acids, polynucleotides and oligonucleotides 66

C. Nucleoproteins 79

D. Viruses 83

III. Action of Ultraviolet Light on Amino-acids, Peptides and Related Substances

A. Direct Action 88
1. Photochemistry of the end groups 89
2. Photochemistry of the peptide bond 90
3. Photochemistry of side groups and of amino-acids 97

B. Photosensitized Destruction of Amino-acids in the Presence of Dyes 107

IV. Action of Ultraviolet Light on Proteins: General

1. Site of action of ultraviolet light on proteins 110
2. Changes in the absorption spectra of proteins with irradiation 113
3. Photolysis and aggregation of irradiated proteins 118
(a) Photolysis of the peptide bond 118
(b) Effect of ultraviolet light on disulfide linkages 121
(c) Aggregation phenomena 122
4. Irradiation of protein monolayers 129

V. Photochemical and Photosensitized Inactivation of Enzymes

1. Kinetics of enzyme inactivation in solution 133
2. Mechanism of inactivation of enzymes by ultraviolet light 137
3. Kinetics of inactivation of dry enzymes 151
4. Photosensitized inactivation of enzymes 156

VI. Photochemistry of Purine and Pyrimidine Derivatives

1. Carbohydrates 163
2. Purine derivatives 165
3. Purine nucleotides 170
4. Nucleotide co-enzymes 171
5. Pyrimidine derivatives 174
6. Reversible photoproducts of uracil derivatives 176
7. Photochemically induced additions to double bonds 182
8. Dimer photoproducts 184
9. Kinetics and mechanism of photochemical transformation of some uracil analogues 191

10. Reversible photoproducts of cytosine derivatives 196
(a) Nature of photoproducts of cytosine glycosides 198
(b) Cytosine and 1-methylcytosine 200
(c) Quantum yields for cytosine derivatives 203
(d) Kinetics of reverse reaction 204
11. Primary photoproducts of substituted pyrimidine derivatives 209
12. Photodegradation products of pyrimidine derivatives 216

VII. NUCLEIC ACIDS AND OLIGO- AND POLYNUCLEOTIDES
A. General 221
B. Photochemical Degradation of Nucleic Acids 222
1. Deoxyribonucleic acids 222
2. Nucleic acid films 227
3. Ribonucleic acids 230
C. Nucleoproteins 231
D. Biological Inactivation 232
1. Infectious RNA 232
2. Transforming deoxyriboneucleic acids (T-DNA) 233
3. RNA core 242
E. Kinetics of Biological Inactivation of Nucleic Acids 242
F. Reversibility of Nucleic Acid Photolysis 246
1. Physico-chemical evidence for reversibility 246
(a) Natural nucleic acid preparations 246
(b) Pyrimidine dimer formation in DNA 249
(c) Natural and synthetic oligo- and polynucleotides 251
(i) Purine oligonucleotides 253
(ii) Oligonucleotides of uridylic acid 253
(iii) Oligonucleotides of thymidylic acid 260
(iv) Oligonucleotides of cytidylic acid 263
(v) Hetero-oligonucleotides 264
(vi) Twin-stranded polynucleotides 267
2. Reversal of biological inactivation 270
G. Summary of Primary Effects of Ultraviolet Light on Nucleic Acids 273
H. Indirect Irradiation Effects 275

VIII. INACTIVATION OF VIRUSES
A. Introduction 279
B. Inactivation Studies with a Plant Virus 280

C. Photochemically Produced Vaccines 291

D. Bacteriophages 295

1. DNA content and UV lethality 295
2. Action spectra 297
3. Quantum yields 298
4. Kinetics of inactivation of infectivity 299
5. Phage-host interactions 301
   (a) Kinetics of survival 303
   (b) Photoreactivation 304
   (c) Multiplicity reactivation 306
   (d) Cross reactivation 309
   (e) Gene-controlled reactivation 310

E. Photosenisitized Inactivation 313

IX. SOME SELECTED PROBLEMS IN PHOTOBIOLOGY

A. Radiation Receptors in Biological Systems: Action Spectra 320

1. Wavelength dependence of quantum yields 324
   (a) Nucleic acids 324
   (b) Proteins 327
2. Examples of action spectra studies 328
3. Partial cell irradiation 333

B. Reversal of Initial Photobiological Effects: Photoreactivation 334

1. Ultimate effects of ultraviolet irradiation 335
2. Characteristics of photoreactivation 336
3. Radiation receptors in photoreactivation 339
4. Physical models as analogies of photoreactivation 340
5. Alternative methods for reversal of ultraviolet effects 342
6. Potential reversible photochemical transformations in nucleic acids 343

C. Radiation Induced Syntheses: Origin of Life 346

APPENDIX. Some Techniques in Photochemistry 358

A. Light Sources 358

1. Mercury arcs 358
   (a) Medium pressure mercury arcs 358
   (b) High pressure mercury arcs 358
2. Carbon arcs 359
3. Xenon arcs 359
4. Low-pressure mercury resonance lamps 360
5. 1849 Å resonance lamp 362

6. Simple laboratory sources of 2537 Å radiation for inactivation studies and preparative photochemistry 362
7. Monochromators 365
8. Flash discharge sources 365
B. Flash-photolysis Flash-spectroscopy 367
C. Spectral Filters 369
1. Solid filters 369
2. Chemical filter combinations 371
(a) Short wavelength cut-off filters 371
(b) Bandpass filters 373
D. Measurement of Light Intensity 376
1. Physical methods 376
2. Chemical actinometry 378
(a) Uranyl oxalate actinometer 378
(b) Malachite green leucocyanide actinometer 380
(c) Ferrioxalate actinometer 381
(d) Other actinometer systems 383
E. Microbeam, or Partial Cell, Irradiation 384
BIBLIOGRAPHY 387
INDEX 435

Preface

THIS book is offered as a contribution to the field of photobiology. The pronounced upsurge of interest in the biological effects of various types of radiation, which followed on the advent of nuclear fission, has undergone further stimulation in recent years. On the one hand the development of numerous new techniques, such as paper and column chromatography, has made it possible to isolate many of the small quantities of radiation products normally encountered in biological systems, while electron spin resonance, flash spectroscopy and nuclear magnetic resonance, amongst others, have furnished powerful tools for the investigation of excited states and photoproduct structures. On the other hand the phenomenal advances recorded in the field of molecular biology, in particular the remarkable progress achieved in the elucidation of the structure of proteins and nucleic acids and the role of the latter in genetics and protein biosynthesis, have made available a solid theoretical basis for an interpretation at the molecular level of the chemical modifications resulting from the action of radiation on these key cellular constituents.

At the time we initially considered the advisability of undertaking the preparation of a short monograph on the effects of ultraviolet and visible radiations on proteins and nucleic acids, we rather naively entertained the hope of being able to provide a reasonably complete coverage of the work done in this domain. We have been to some extent frustrated in the attainment of this goal, owing in large measure to the growing interest in the field and an increased rate of publication. In particular, phage photochemistry has now developed into a highly specialized field of its own, an adequate coverage of which would require a separate small volume. While these developments have made our task more difficult, they simultaneously testify to the need of a new review, insofar as proteins and nucleic acids are concerned, one which would at least partially fill the gap that has arisen since the publication during the last decade of Vols. II and III of Hollaender's comprehensive collective *Radiation Biology*.

An initial chapter, outlining briefly some of the principles of photochemistry, is followed by a survey of the absorption spectra of nucleoproteins and their constituents. Succeeding chapters deal with the physical and chemical effects of ultraviolet and visible radiation on proteins, nucleic acids, viruses and their primary building blocks. Some attention is devoted to such problems as energy transfer in macromolecules, photosensitized reactions, etc. An effort is made, where possible, to interpret these effects in terms of the theory, however spotty, of photochemical reactions.

Throughout, some of the implications of this study for the field of photobiology are incorporated.

We hope that the Appendix on some photochemical techniques will fulfil a practical need for many of those working in the photobiological field. If any apology is required for the section on Radiation Induced Syntheses in Chapter IX, it is only that it is not as comprehensive as might be desired; space research is now well out of the realm of fantasy and the role of ultraviolet and visible radiation both in space research and in the problem of the origin of life has attained such significance that an increasing number of research workers are now devoting their attention in whole or in part to this new domain.

It is a pleasure to acknowledge partial support to one of us (A. D. McL.) of some of the research reported in this book, by the U.S. Atomic Energy Commission.

Berkeley and Warsaw A. D. McLaren and D. Shugar

CHAPTER I

Some Principles of Photochemistry

A. INTRODUCTION

1. *The Laws of Photochemistry*

Photochemistry as a branch of physical chemistry still suffers from a poverty of sweeping general rules which one can use as a guide for prediction. It is still largely an empirical branch. An absolute principle, sometimes called the first law, is that only light which is absorbed by a system can induce a chemical reaction; i.e. the mere passage of light through a system has no catalytic effect (Rollefson and Burton, 1942). It does not follow, however, that absorbed light will always induce a reaction, for the absorbed energy may be dissipated as heat via collisions of a second kind (whereby an excited energy-rich molecule collides with a normal molecule with an increase in translational energy of the molecules) or lost as fluorescence or both as consecutive steps. In considering the absorption of light by a reactant, the photochemist collides head-on with the eternal dilemma in the conceptual scheme of light (Raman, 1922). In keeping with N. Bohr's theory of complementariness in nature, the photochemist uses a *wavelength* λ in calculating relevant extinction coefficients and counts *quanta* in calculating the efficiency of a photochemical reaction!

The absorption of light by a molecule is a quantized event; i.e. as a molecule absorbs light it receives energy in the form of discrete units called photons or quanta. The energy q of a quantum is $h\nu$ ergs where h is Planck's constant (6.61×10^{-27} erg sec) and ν is the frequency of radiation ($\nu = c/\lambda = 3.0 \times 10^{10}$ cm per sec/λ). With wavelength λ expressed in angstrom units Å (one Å $= 0.1$ mμ),

$$1 \text{ quantum} = h\nu = 1.98 \times 10^{-8}/\lambda \text{ ergs}$$

Since one does not ordinarily observe changes in single molecules of reactants, but rather in tangible quantities, another useful unit is the einstein, i.e.

$$\begin{aligned} 1 \text{ einstein} &= Nq = 6.02 \times 10^{23} \times 1.98 \times 10^{-8}/\lambda \text{ ergs} \\ &= 11.9 \times 10^{8}/\lambda \text{ joules} \\ &= 2.85 \times 10^{8}/\lambda \text{ calories} \end{aligned}$$

In other words, if Avogadro's number N of molecules each absorb one quantum at 2537 Å, the energy absorbed would be 112,000 calories per mole. The einstein is formally analogous to the faraday. At 6000 Å (red light) an einstein is 47,400 calories and at 2000 Å (for ultraviolet light) it is 142,000 calories. These energies may be compared with thermochemical bond strengths of typical bonds found in proteins, Fig. 1-1. Bond strengths vary

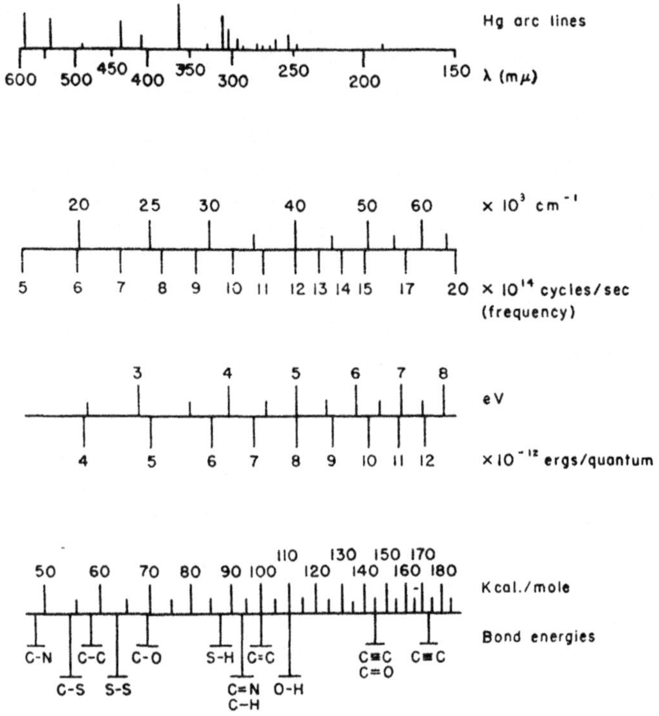

FIG. 1-1. Energy conversion chart (adapted from Bowen, 1946) and bond energies (from Pauling, 1948).

somewhat from one kind of molecule to another. For a photochemical reaction to occur, the quantum absorbed must be equal to or greater than the energy required to dissociate or sufficiently activate the molecule, depending on the nature of the effect considered.

The Stark–Einstein law of photochemical equivalence, sometimes called the second law, states that in the primary process one quantum of light is absorbed per molecule of absorbing and reacting substance. By primary process is meant the immediate effect of the light on the absorbing molecule (Noyes and Leighton, 1941). It is possible for a single molecule to absorb more than one quantum in successive steps; however, the concentration of molecules which have absorbed one quantum is so low in a population that the absorption of further quanta by the excited molecule is highly improbable (the lifetime of excitation being short), and there is no positive evidence that such steps are ever of importance in photochemistry. An attempt to demonstrate this possibility with a very intense light source and a polychromophoric molecule (pepsin) failed (McLaren *et al.*, 1952).

A molecule which has absorbed a quantum is said to be in an excited state. Aside from the afore-mentioned fluorescence (or phosphorescence) or degradation of excitation energy to heat, the energy-rich molecule may dissociate to yield atoms or free radicals, or rearrange, eventually or immediately, into final products. These atoms, radicals, or isomers may be regarded as having been formed in the *primary process*. The time interval between activation and dissociation or rearrangement may be shorter than the time for an intramolecular vibration (*ca.* 10^{-13} sec) or as long as some minutes (as seems to be the case with some rearranged forms of irradiated chymotrypsin molecules). Via secondary reactions, excited molecules may lose energy by reactive collision with other molecules, thereby initiating chemical reactions. Such molecules are said to perform the role of photosensitizers. Thus excited acriflavine can inactivate tobacco mosaic virus and excited methylene blue can inactivate chymotrypsin.

A consideration of the photochemical equivalence law has led to a definition of great practical importance. The quantum yield is defined as

$$\Phi = \frac{\text{number of molecules reacting per unit volume}}{\text{number of quanta absorbed by the reactant per unit volume}} \tag{1-1a}$$

or

$$\Phi = \frac{\text{number of moles reacted per unit volume}}{\text{number of einsteins absorbed by the reactant per unit volume}} \tag{1-1b}$$

Sometimes the experimentally determined quantity is the number of moles of product formed, and one can also usefully define a yield Φ_p as

$$\Phi_p = \frac{\text{moles product formed}}{\text{einsteins absorbed}} \tag{1-2}$$

Φ_p can be more or less than Φ, depending on whether more than one mole of product is formed per mole of reactant changed, whether polymerization is initiated by excited molecules, and depending on whether all products are identified and summed.

The observed rate of reaction may be expressed as ΦI_{abs}, in which Φ may be a constant or a function of concentration of reactants in a system and I_{abs} is the energy absorbed as einsteins per unit volume and time, i.e. the intensity.

2. *Absorption Spectroscopy*

Light of intensity I_0 impinging on a solution with a path-length l is partly absorbed (I_{abs}) and partly transmitted with an intensity of I. For any given substance, pure or in an optically transparent solvent with a molar concentration C, these quantities are related by the Beer–Lambert equation, $I = I_0 10^{-\varepsilon Cl}$. The proportionality constant, ε, is called the molecular extinction coefficient. The variation in the extinction coefficient with wavelength is called an absorption spectrum. A complete theoretical understanding of absorption spectra is available for only the most simple molecules; we shall summarize some of the aspects which apply to these molecules and point out a few of the complications which render difficult their detailed application to complex molecules.

Just as with atoms, the absorption of energy $q = \Delta E$ by a molecule leads to an increase in energy $\Delta E = h\nu = E_1 - E_0$, where E_0 is the energy of the entity before, and E_1 that following absorption of a light quantum, viz.

$$\mathrm{A} + h\nu \underset{k''}{\overset{k'}{\rightleftharpoons}} \mathrm{A}^*$$

where A* is an "electronic isomer" (Bowen, 1947) of the normal molecule, i.e. the energy-rich, excited state of a molecule A, and k' is the specific reaction rate of the forward reaction—the fraction converted per quantum absorbed—(Rollefson and Burton, 1942). The extinction coefficient is dependent on λ because the excitations possible are peculiar to the kinds of chemical linkages found in a molecule, such as individual bonds:

e.g. —CONH—, —S—S— etc.

or groups of bonds,

e.g. 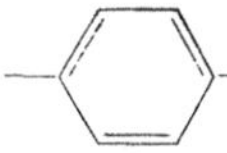—OH, —C=C—C=O, etc.

and are related by the Bohr–Planck relationship† $\Delta E = hc/\lambda$. ΔE must not only be sufficient for the transition but the latter must also be permissible.

An upper value for ε may be obtained by the following consideration (Braude, 1945). Consider an intensity I falling on a slice of thickness dl of a cell of unit area filled with a dilute absorbing solute in a transparent solvent, at a concentration C moles/l. Let the average effective absorbing area ("chromophore area") of an absorbing molecule in the plane perpendicular to the incident light be σ. Then the illumination absorbed is

$$-\mathrm{d}I = \frac{\text{area of absorbing molecules}}{\text{total area}} \times I \times F = \frac{FCN\,\mathrm{d}l\,I\sigma}{1000}$$

† A diagram showing the approximate position of the absorption bands of some commonly occurring chromophores has been published by Scott (1955).

where F represents the fraction of the light falling on the absorbing molecules which is absorbed. Upon integration

$$\varepsilon = \frac{1}{Cl} \log \frac{I_0}{I} = \frac{FN\sigma}{2300}$$

If $F = 1$ and $\sigma = [\sigma]$, the cross section of an entire molecule, taken as 10 Å^2 for illustration, $\varepsilon \cong 10^5$ as a maximum value for a simple molecule. Clearly ε is proportional to the absorption probability k' (Rollefson and Burton, 1942). ε is generally less than 10^5 because $\sigma < [\sigma]$, because orientation of a molecule with respect to the beam can influence absorption, and because $F < 1$ since the transition between energy states E_1 and E_0 does not always occur when a quantum, conceivably of the right magnitude, falls on the molecule in the optimum orientation.

Transitions are governed by certain selection rules and the Franck–Condon principle (Livingston, 1955). These principles are best discussed by reference to Fig. 1-2, a potential energy diagram for a diatomic molecule.

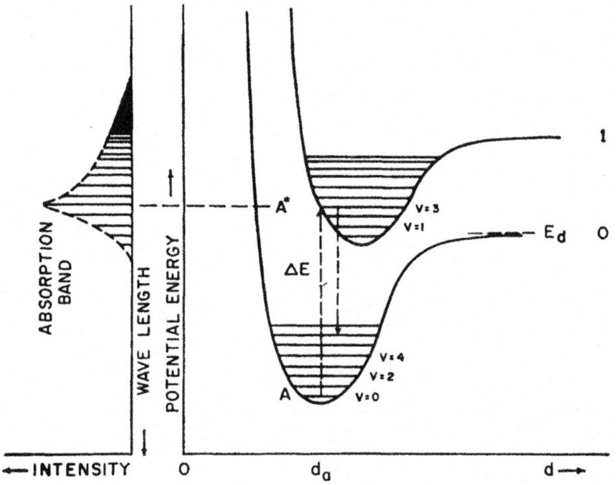

FIG. 1-2. Potential energy diagram for a diatomic molecule. One atom is considered fixed at the vertical axis and vibration due to thermal energy can be represented by movement of the other atom through a distance d along the horizontal axis. E_d is the energy of dissociation for the ground state, 0. Note the similarity of these curves to those describing linear harmonic oscillators. The curves for molecules are quantized and the lowest vibrational energy level extant is $\frac{1}{2}\nu_0 = E_v$. Only the first few vibrational energy levels are identified by quantum numbers and rotational energy levels are not shown (after Bowen, 1947).

Curve 0 represents the parabolic potential energy curve which describes the energy of attraction as a function of distance d between two neutral atoms which have been brought together to form a molecule. The equilibrium distance d_a is the distance at which the magnetic attraction of a single valence pair of unlike spinning electrons just balances the repulsion of the two atomic nuclei. Curve 1, the upper curve, is a similar vibration curve for the molecule A*, one of whose valency electrons has "jumped" by light absorption to a higher level, causing the molecular link to lengthen. The horizontal lines inside each parabola represent the possible discrete, vibrational energy levels of both the excited state and the unexcited or ground state.

Whereas electron jumps in atoms correspond to light absorption only at certain narrow lines in the spectrum, such as the familiar 2537 Å line of mercury, in molecules broad regions of line absorption, or bands, are observed because simultaneous changes in both electronic and vibrational, and usually rotational energies take place. For each line of a band the Bohr frequency condition obtains

$$\nu_{v,r,e} = \frac{\Delta E}{h} = (\Delta E_{\text{electronic}} + \Delta E_{\text{vibrational}} + \Delta E_{\text{rotational}})/h$$

Let us consider each of these changes of energy. Changes in rotational energy alone are small in magnitude and give rise to lines in the long wavelength region of a spectrum. For molecules with a dipole moment

$$\nu_{J+1,J} = \frac{2(J+1)h}{8\pi^2 i}$$

where i is the moment of inertia and $J = 0, 1, 2, 3, \ldots$ For absorption, ν corresponds to a change from J to $J + 1$ or to $J - 1$, (a selection rule), and the lines compose a set of equal spacing. These lines are of no importance photochemically. The next simplest spectra are those corresponding to a simultaneous change of vibrational and rotational energies. Each horizontal line of the curves of Fig. 1-2 represents to a first approximation the energy of a harmonic oscillator

$$E_v = (v + \tfrac{1}{2})h\nu_0$$

where ν_0 is the fundamental frequency of the lowest energy state and v is the vibrational quantum number. In a strictly harmonic oscillator Δv is limited to one (a selection rule) and

$$\nu_{v,r} = (v'' - v')\nu_0 \pm \frac{h}{4\pi^2 i}(J' + 1)$$

The double primes refer to the higher energy states, the single primes to the lower. Since the energy contribution of the first term is much greater than the second, it determines where on the spectrum ν will lie. The second term, allowing for changes in rotational energies along with a change in vibrational energy, gives the fine structure about $(v'' - v')\nu_0$, thereby forming a band in the near infrared.

In visible and ultraviolet portions of the spectrum, in the region of interest to the photochemist,

$$\nu_{e,v,r} = \nu_e + \nu_v + \nu_r$$

and the corresponding energy change is indicated in Fig. 1-2 by the vertical line connecting the two oscillator curves. Since ν_0 is characteristic of a particular electronic state, a change to a new state involves a new set of vibrational states; there is no selection rule which limits the change in vibrational quantum number in transitions of this type. There are, however, other selection rules which apply, based on the following spectral designations.

Spectral designations for molecular states. It is convenient for many purposes to have a method of defining the electronic state of a molecule and one similar to that describing atoms is employed. It will be recalled that the motion of extranuclear electrons in Bohr orbits may be described in terms of four quantum numbers. For a single electron, one particular orbit corresponds to each set of numbers. In terms of wave mechanics, to each set of quantum numbers there corresponds a particular probability of finding the electron at any given point in space (and in Fig. 1-2, the potential energy curves would be represented by broad smears of probable values of energy for a given separation of nuclei). The quantum number $\underline{n}$ represents the *size* of the orbit, the number $\underline{l}$ represents the angular momentum or *shape* of the orbit, and the number $\underline{s}$ stands for the *spin* of the electron. Associated with the orbital motion of the electron there is a magnetic moment designated by a number $\underline{m}$. According to Pauli, for a given atom no two electrons can have identical sets of the four numbers. Heitler and London found with hydrogen that when the electrons are in their lowest states, two hydrogen atoms will attract each other as pictured in Fig. 1-2 provided the spins of the electrons are opposite. In other words a stable molecule will form if the two electrons are paired off. Such a pair is equivalent to a valence of one.

The resultant azimuthal quantum number for the molecule, which is analogous to $\underline{l}$ for an atom, is represented by $\underline{\lambda}$, and the symbols Σ, π, and $\underline{\Delta}$ are used as term symbols for $\underline{\lambda} = 0, 1, 2$ respectively. $\underline{\lambda}$ specifies the component of the total resultant angular momentum of the electrons in the direction of the axis of the molecule. A selection rule states that $\underline{\lambda}$ may change by $0, \pm 1$, i.e. transitions between two Σ or two π or between a Σ and a π or between a π and a $\underline{\Delta}$ term are permissible. By further analogy to atomic spectra the multiplicity of the Σ, π, $\underline{\Delta}$ energy terms are designated by superscripts, such as $^3\Sigma$. The multiplicity is equal to twice the vector sum of the electron spins plus one, where each electron is spinning with an angular momentum represented by $\frac{1}{2}$. The multiplicity is therefore odd for an even number of electrons, typical of most normal molecules. The inter-combination selection rule "forbids" transitions between states of different multiplicities. This is one of the few selection rules which also applies to complex molecules.

Of greater importance in determining the photochemical behavior of molecules is the principle of Franck and Condon which states that electronic transitions which involve appreciable changes in the positions or momenta of constituent nuclei are of low probability (Livingston, 1955). Hence the transition line A → A* is drawn vertically in Fig. 1-2 to be probable. In other words, an electronic change is very fast compared to vibrations. At room temperature most molecules are in their lowest vibrational states of the ground electronic state; transitions from the ground state being most probable, one finds an absorption maximum, Fig. 1-2.† Since the ground state of most molecules is singlet (no unpaired electrons) this means that the easily excitable states will also be singlets, i.e. in which the spins are still paired, although the excited electron is now in a new orbit (Reid, 1957).

In the gas state, at pressures so low that collisions of excited molecules are highly infrequent, an excited molecule of the kind pictured in Fig. 1-2 would emit one fluorescent quantum with generally different and reduced frequency and with the specific rate k'', for each one absorbed. This bathochromic displacement is a generality known as Stokes' rule and accounts for the visibility of many substances exposed to ultraviolet light in a darkened room. In a condensed phase or in solution, an excited molecule can also undergo a loss of vibrational energy by collisions of the "second kind" with other molecules whereupon most of the fluorescence transitions start from the lowest vibrational level of the excited state and the fluorescence spectrum is at longer wavelengths than the absorption spectrum.

An excited singlet molecule of the kind pictured in Fig. 1-2 can also lose energy by another process. Vibrational energy may be lost to the surroundings until the molecule exists in a state where, at some instant, it has the same potential energy as that of a molecule with a triplet excited state. Under the perturbing influence of other atoms or molecules, a change of spin occurs and a radiationless transition from a singlet to a triplet state (unpaired spins) takes place. A transition from the triplet state to the ground state is "forbidden," so the state is long lived and phosphorescence may be observed. In other words, the triplet level is an energy level lower than the lowest excited singlet level.

The absorption of a quantum by a molecule of the type indicated in Fig. 1-3, will result in the formation of an excited molecule which is very unstable; the vibrational energy reached is greater than that required to

† Energy absorbed by a solution, in einsteins per mole, is proportional to the extinction coefficient, and it is the latter which is plotted versus wavelength in practice (cf. equation 9, below).

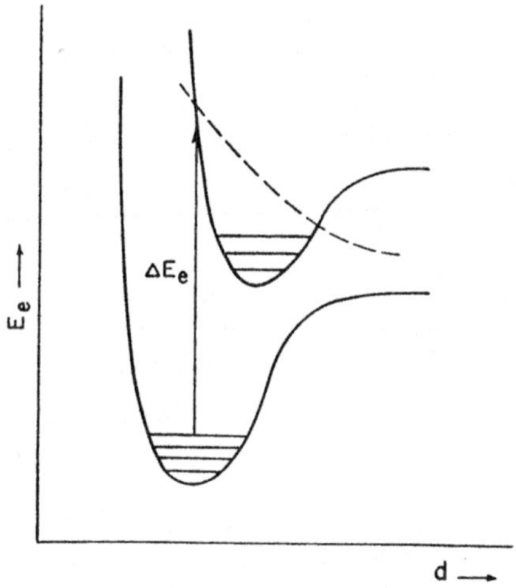

FIG. 1-3. Potential energy diagram for a diatomic molecule showing possibility of dissociation by electronic excitation.

dissociate the excited molecule into a normal and an excited atom. "Optical dissociation" will occur within the period of one molecular vibration, instead of fluorescence. In this figure, transitions to the excited electronic state need not be quantized, since the upper level possesses more energy than the minimum required for dissociation. The spectrum loses its line character in the corresponding region and is said to be continuous, cf. Fig. 1-2.

In addition to the position and intensity of the absorption band in a spectrum, one wants to know about its width and shape. If the minimum of of the upper curve is well to the right of that in the lower curve, Fig. 1-2, point A* lies on the steep part of the upper curve and the transitions grouped near the line A → A* cut the upper curve over a wide energy range and the band will be wide, as for iodine in chloroform. Molecules structurally more complicated than iodine, e.g. benzene, can have a narrower band. Aromatic systems have excitable electrons which are "non-localized" in particular bands and the excitation of such an electron will not weaken any individual bond very much, nor is the distance between atoms altered appreciably. Referring to Fig. 1-2 again, we can say that the minimum of the upper curve would be almost above that of the lower one.

Transitions occurring to the right of the line AA* cut the upper curve at a flatter slope than transitions to the left. The shape of the absorption curve describing the band is much steeper on the long wave than on the short wave side; i.e. the "cut-off" of the curve will be sharper on the long wavelength side.

With more complex molecules the number of dimensions required for graphical representation (one for potential energy plus distance coordinates) becomes too great for a clear physical picture. Since selection rules for diatomic molecules break down in condensed systems due to molecular electromagnetic fields, it seems probable that related rules for polyatomic molecules would not be rigid owing to the presence of such intra-molecular fields. The study of the spectra of most polyatomic molecules has not progressed to a point which permits of the designation of electronic states. For example, although benzene vapor yields a well-defined absorption spectrum, a satisfactory interpretation is not at hand (Reid, 1957). Nevertheless, it is frequently fruitful to draw qualitative analogies with diatomic systems. Polyatomic molecules, other than saturated hydrocarbons, contain one or more resonating systems, each consisting of a group of atoms which absorb together as a *chromophore* in the visible and photographically available (near) region of the ultraviolet. These include —COOH, —$CONH_2$, and —C_6H_4OH for example. If these chromophores are separated by insulating groups, such as —CH_2— or better —CH_2CH_2—, the extinction coefficient of a molecule at a given wavelength will be approximately the sum of the characteristic coefficients of the individual chromophores (Brode, 1939). Just how well this generality holds will be illustrated under a discussion of the absorption spectra of proteins and nucleic acids.

Influence of liquid condensed state on absorption spectra. The absorption bands of molecules are broad compared to the spectra of atoms because the molecules have a choice, in part temperature dependent, of many initial and final vibrational—rotational states belonging to a given electronic transition. The distribution of individual molecules among the various possible (vibrational) energy states is given by

$$n_i/n_0 = e^{-\Delta E_v/RT} = e^{-(E_i - E_0)/RT}$$

where n_i and n_0 are the number of molecules in the ith and zero states respectively, R is the gas constant and T the absolute temperature. As ΔE_v increases, the number of molecules in the ith state decreases rapidly (Rollefson and Burton, 1942). Such an equation fails to describe the initial and final states of an electronic transition in a liquid or solution state, largely due to the modifying and dampening effects of solvent molecules. Because of the proximity of molecules which prevent unhindered vibration, average positions of vibration are recorded in spectra, with a decrease in absorption on each side of the positions. For example, the fine line spectrum of benzene vapor is changed to continuous absorption bands which appear as "band-envelopes" in the extinction curve (cf. Fig. 1-2) (Brode, 1939). Solute–solvent interaction is not large in hexane, and spectra of hexane solutions usually most nearly resemble gas phase spectra. A change of solvents affects λ_{max} more than ε_{max} and wavelength displacements can sometimes be correlated with dipole moments of the solvent (a measure of solvation tendency). The polar effects of solvent molecules on a solute can, however, lead to alterations of the transition rules and thereby alter the intensity of an absorption band (Bowen, 1947).

At sufficiently low temperatures solute–solvent interaction is sufficiently reduced for a re-appearance of some vibrational structure. The resolution is even more marked in spectra of thin films at low temperature (Fig. 1-4)

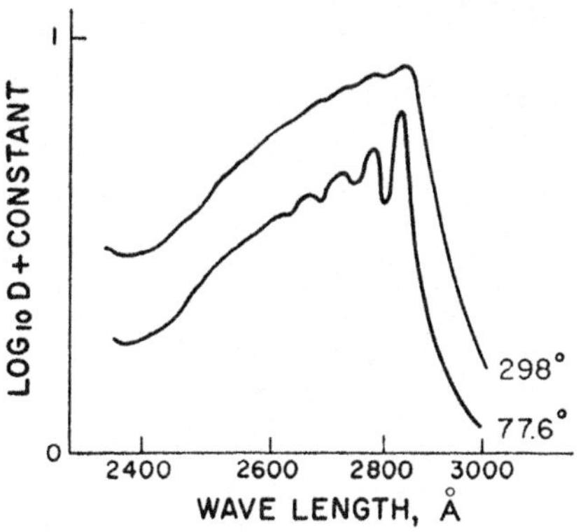

FIG. 1-4. Absorption spectra of sublimed film of thymine at 298°K and 77.6°K.

(Scott *et al.*, 1948). At liquid air temperature all molecules commence an electronic transition from the lowest vibrational energy level. At higher temperatures the molecular population in the higher vibrational levels increases, which increases the number of possible transitions. Details of the low temperature spectrum become replaced by what appears to be a single band of considerable breadth.

The extinction coefficients of broad bands can be additive, provided the chromophores are separated by insulators, Fig. 1-5.

3. *Transfer Mechanisms of Excitation Energy*

An important question which immediately arises after electronic excitation of organic molecules, be they as small as methyl ethyl ketone or as large as a protein, is whether the absorbed energy can be transferred from the site of absorption (the chromophore) to another chemical bond and whether the latter bond can either radiate or undergo chemical reaction. This problem has been discussed a great deal in recent years and some experiments point to the reality of this notion. It has been evaluated most recently by Förster (1960) who states that "resonance transfer of excitation is able to

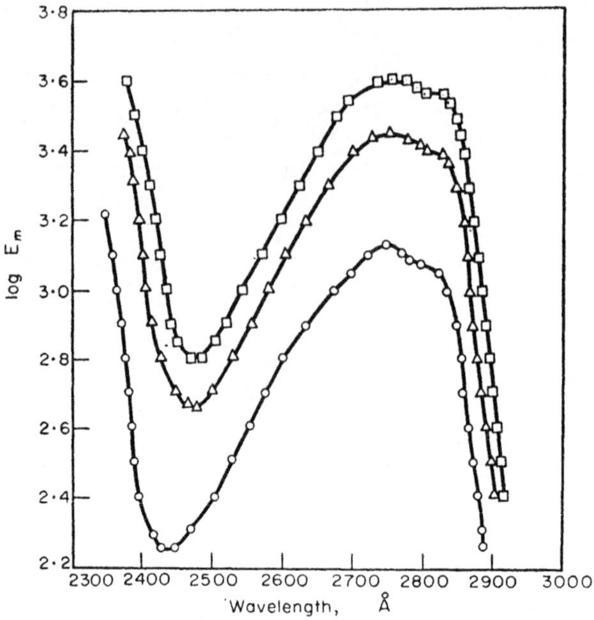

FIG. 1-5. Ultraviolet absorption spectra of tyrosyl-tyrosyltyrosine (1), tyrosyltyrosine (2) and tyrosine (3) (Bardoll and Ross, 1944). The insulator group is

$$\text{—CH}_2\text{—}\underset{|}{\text{CH}}\text{—CONH—}\underset{|}{\text{CH}}\text{—CH}_2\text{—}$$

in this wavelength region, pH 4.5.

explain, in principle, the different photochemical and fluorescence experiments with protein and photosynthetic systems." This may be over optimistic, however.

Two kinds of experiments have been carried out with input quanta having energies less than those required for ionization or dissociation of chemical bonds. One of these is exemplified by Bücher and Kaspers (1947). Photochemical decomposition of carbon monoxide–myoglobin occurs with the same high quantum yield, regardless of whether the active light is absorbed in the porphyrin component or in one of the aromatic amino-acids of the protein. In the other, Bannister (1954) found, for example, that phycocyanin, a chromoprotein, will fluoresce following absorption of a quantum by either the amino-acid residues or by the porphyrin moiety and with a quantum yield roughly the same for either absorption site. Such results show that energy is transferred as electronic excitation energy to the final acceptor; the excited acceptor can either react photochemically or release the energy by fluorescence emission of high efficiency, depending on the nature of the acceptor.

The problem is to discover how energy absorbed by such chromophores as tryptophanyl, phenyl, tyrosyl, etc., with easily excitable π (double bond)-electrons, separated by insulating groups of atoms with σ (single bond)-electrons requiring higher excitation energies, can be transferred from one to another. We know that σ-electron systems are good insulators among these chromophores since absorption spectra of proteins in solution of neutral pH are roughly the sums of the spectra of constituent amino-acids.

Förster rules out the trivial mechanism whereby an excited chromophore emits and the emitted quantum is absorbed by another chromophore on the ground that the fluorescent step is often known in isolated chromophores and is of low efficiency, say 20 per cent, whereas the experiments of Bücher and Kaspers revealed a transfer of unit quantum efficiency. Instead he develops the concept of *resonance transfer* or transfer by *inductive resonance.* This mechanism is in competition with fluorescence emission and with intramolecular non-radiative processes responsible for fluorescence quantum yields which are lower than one. As a simplified model, let the electronic system of a molecule be represented by a single electron with an electronic ground state and only one excited state. With suitable interaction between the different electronic systems, the excitation can jump from one molecule to another before the first one has time to emit a quantum of fluorescence (see Fig. 1-6). The detailed nature of the resonance condition and the transfer process itself depend on molecular vibrations and relaxation. In the model with an electronic ground state and one excited state there is superimposed a set of vibrational levels with a frequency $\nu = 1/(2\pi t)$, where t is the vibrational time constant of the order of 10^{-14} sec. Such an excited molecule will

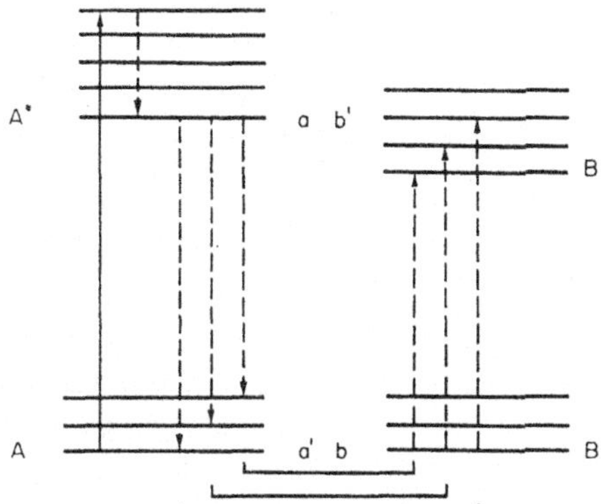

FIG. 1-6. Simplified energy-level diagram of sensitizer (A) and acceptor (B). Radiative transition, ———; non-radiative transition, ↓; coupled-transfer transitions, -------; Resonance transfer can take place from a–a^1 to b–b^1 without fluorescence. The absorption spectrum of the energy acceptor overlaps the emission spectrum of the energy donor (after Förster, 1960).

generally have higher vibrational energy than that found among the solvent molecules, and during a relaxation time t' it will equilibrate thermally with its surroundings. t' is about 10^{-12} sec. The excited molecule, now in its lowest vibrational level, will become deactivated, either by fluorescence, or by a nonradiative intramolecular process in its lifetime, τ; or if another molecule with a low-lying excited state is in the neighborhood of the first one, excitation transfer may take place. If the time of transfer is shorter than t of molecular vibrations, essential differences would be found between the absorption and fluorescence spectra of the free components and those in the intact protein, according to Förster. Since this is not the case (the absorption spectra of different amino-acids and prosthetic groups in protein can be recognized moderately well) the transfer is slower than t. To date we do not know just how slow the transfer is, but Förster estimates that a time between t' and τ is required: only a slow mechanism would be able to account for a single step transfer from tryptophanyl or tyrosyl residues to the prosthetic group in a heme-protein. His extension of earlier work by Perrin, London, and Vavilov leads to the following expression for the rate constant for the transfer process:

$$K_{A\to B} = \frac{1}{\tau_s}\left(\frac{R_0}{R}\right)^6 \tag{1-3}$$

where R is the distance between molecules (or parts of a macromolecule), R_0 is the critical transfer distance for which excitation transfer and spontaneous deactivation of the sensitizer are of equal probability, and τ_s is the actual mean lifetime of the excited sensitizer. τ_s is connected to the radiative lifetime of the excited sensitizer τ'_s and the quantum yield Φ_s of the sensitizer fluorescence (without transfer) by

$$\tau_s = \tau'_s \Phi_s$$

where R_0 is of the order of 10 Å in proteins.

By means of eqn. (1-3) one can answer the question as to whether excitation transfer in proteins occurs in a single step from the sensitizer, A, to the acceptor, B, or via several steps between different aromatic residues. That is, we need to know the ratio $K_{A_1\to A_2}/K_{A\to B}$. Förster calculates a ratio of 1/9 in myoglobin, indicating a negligible amount of multistage transfer. This can follow in spite of the fact that the critical distance for (slow) transfer between amino-acid residues is less than for transfer to the prosthetic group ($R_0 \sim 30$ Å), because in myoglobin there are only nine aromatic residues for one heme. Structural requirements for this kind of transfer are not thought to be critical. The requirement is only that the system needs a moderate concentration of constituents with electronic levels of the appropriate energy (cf. Karreman *et al.*, 1958). With regard to the action of ultraviolet light on proteins, it does not appear that transfer from aromatic groups to disulfide bonds occurs, or at least not enough to contribute significantly to the sum of all processes leading to photochemical inactivation (Förster, 1960). Tyrosine to tryptophan energy migration also seems unlikely (Stryer, 1960). A discussion of the application of this approach to green plant materials and other ordered systems has been published (Tollin, *et al.*, 1958), and a review has just appeared (Steiner and Edelhoch, 1962).

B. FACTORS OF INFLUENCE IN PHOTOCHEMICAL REACTIONS IN SOLUTION

1. *The Absorption of Light and the Quantum Yield*

Before taking up a more detailed consideration of photochemical reactions, we will consider the rate of absorption of ultraviolet light by a reacting substance, namely I_{abs} (Sinsheimer, 1955). As already stated, the absorption of a photon by a molecule is an all-or-none event, i.e. it is a quantized process, and in a transparent solvent the molecules of a single absorbing solute species have an equal probability of the absorption occurring. The fraction of the incident number of quanta falling on a homogeneous solution per unit area per unit time which is transmitted, $\mathrm{d}I/I$, will thus decrease as the concentration increases, i.e. $-\mathrm{d}I/I \propto \mathrm{d}C$ for a given wavelength. It is understood that the path length through which the light traverses, l, is fixed as the concentration is varied so that $-\mathrm{d}I/I \propto l\mathrm{d}C$. This simple relationship is the Beer–Bouguer–Lambert equation which integrates to

$$\int_{I_0}^{I} -\frac{\mathrm{d}I}{I} \propto \int_{0}^{C} l\mathrm{d}C = \ln\frac{I_0}{I} \propto lC = \propto' \log_{10}\frac{I_0}{I} \qquad (1\text{-}4a)$$

The proportionality constant is dependent on the choice of units and the one which has come into most general use is written as ε when C is expressed in moles/l. and the optical path length in cm; ε is then the molecular (or molar) extinction coefficient and

$$\varepsilon Cl = \log_{10}\frac{I_0}{I} = D \qquad (1\text{-}4b)$$

where D is referred to as the optical density (or occasionally "extinction" or "absorbance") and is directly proportional to the concentration. For a given path length, D is an additive quantity for a mixture of solutes, as long as Beer's law is valid, i.e. in the absence of light scattering and molecular association or interaction. Under these circumstances the fraction of monochromatic light absorbed by a constituent x in a mixture of solutes a, b, · · ·, x, · · ·, z may be calculated from the relationship

$$\varepsilon_x C_x/(\varepsilon_a C_a + \varepsilon_b C_b + \cdots + \varepsilon_x C_x + \cdots + \varepsilon_z C_z) = D_x/D \qquad (1\text{-}5)$$

For large molecules some of the incident light may fail to traverse the solution as a result of Rayleigh scattering and an appropriate correction must then be made (Chapter II, Section 1–5). The application of eqn. (4) to associated chromophores such as are encountered in polypeptides and polynucleotides will be taken up in turn.

Some additional expressions for the proportionality constant are summarized in Table 1-1. Where molecular weights are unknown the molar extinction coefficient may be referred to the molar content of some component of the substance, e.g. to phosphorus in the case of the nucleic acids.

TABLE 1-1. SYMBOLS FOR THE PROPORTIONALITY CONSTANT IN BEER'S LAW WITH OPTICAL PATH LENGTH IN CENTIMETERS, AND A LISTING OF RELATED QUANTITIES

log base	C, (molecules/cm³)	C, (moles/l.)	C, (moles/cm³)	C (g/l.)
e	σ	E	β	μ
10	—	ε (molar extinction)	—	k (specific extinction)

$\varepsilon \simeq 1000\,\beta$

I = quantum intensity (mole quanta/cm² per hour = einsteins/cm³ per unit time for unit path length)

l = thickness of solution layer (cm).

In a typical photochemical reaction we may write for the rate of disappearance of a reactant A by a one quantum ("one-hit") plus one molecule process, namely, $A + h\nu \rightarrow P$, the equation

$$-\mathrm{d}[A]/\mathrm{d}t = k \cdot [A] \cdot I_0 \tag{1-6}$$

where k is a proportionality constant. If $\beta_P \approx \beta_A$, the fraction of light absorbed at any instant by unreacted A is $[A]/[A_0]$ where $[A_0]$ is the initial concentration of reactant. From equation (1-1b)

$$\Phi = \frac{[A_0] - [A]}{I_0(1 - 10^{-D}) \cdot t \cdot [\bar{A}]/[A_0]} \tag{1-7}$$

wherein $[\bar{A}]$ is the average amount of reactant present during the time of irradiation t

$$[\bar{A}] = \frac{[A_0]}{t} \int_0^t e^{-kI_0 t}\, \mathrm{d}t$$

from (1-6) and on substituting into (7) we obtain the equation

$$\Phi = \frac{[A_0] \cdot \ln [A_0]/[A]}{I_{abs} \cdot t} \qquad (1\text{-}8a)$$

Equation (1-8a) has been derived in many ways and is very useful for calculating Φ from inactivation data for enzymes and viruses (Claesson, 1956; McLaren, 1949).

For optically dilute solutions

$$I_{abs} = I_0 \left(\beta Cl - \frac{\beta^2 C^2 l^2}{2!} + \cdot \cdot \cdot \right) \approx I_0 \beta Cl \qquad (1\text{-}9)$$

and equation (8a) simplifies to the form first used by Warburg, (1949), namely

$$\Phi = \frac{[A_0] \ln [A_0]/[A]}{I_0 \beta [A_0] lt} = \frac{\ln [A_0]/[A]}{I_0 \beta lt} \qquad (1\text{-}8b)$$

This equation for evaluating Φ has the advantage of not requiring a knowledge of the concentration of the reactant. If an enzyme is being studied, only a knowledge of the fraction of the activity remaining after an irradiation time is needed, along with the thickness of the layer irradiated, the characteristic extinction coefficient β and the incident intensity (for dilute solutions we may consider a uniform intensity throughout). The experimental difficulty may be that of measuring $[A]/[A_0]$. Also, such conditions do not give rise to enough products to allow extensive physicochemical evaluation of the nature of the reaction undergone by the species A.

It is frequently convenient to utilize absorption cells of 1 cm path length and to express I_0 and I_{abs} as einsteins/cm³ per min, falling on or absorbed by the solution respectively. Another form of equation (1-8b) is the following:

$$[A]/[A_0] = e^{-\Phi I_0 \beta t} = e^{-\Phi I_q \sigma t} \qquad (1\text{-}8c)$$

where σ is the absorption cross-section in cm²/molecule ($\sigma = 3.83 \times 10^{-18} \beta$) and I_q is the light intensity in quanta/cm² per unit time (Arnold and Oppenheimer, 1950). The product $I_q t$ is called the dose and the product $\Phi\sigma$ the inactivation cross-section (Setlow, 1957). The absorption cross-section represents the probability that an incident quantum will be absorbed, and the quantum yield measures the probability that an absorbed quantum will produce a reaction. The former is considered to bear some, generally unknown, relation to the physical cross-section of a molecule. A plot of inactivation cross-section versus wavelength should parallel the absorption spectrum of a sensitive reactant if the quantum yield for the reaction is a constant, independent of wavelength. Generally with enzymes, β is a more sensitive function of wavelength than is Φ in the actinic region between 2200 and 3000 Å (McLaren, 1949).

Incidentally, equation (1-8c) can be adapted to the irradiation of micro-organisms with photons in thin layers (Morowitz, 1950).

Finally, we introduce the concept of the action spectrum. Combination of equations (1-4a) and (1-8b) gives

$$\log C_0/C = \frac{I_0 \Phi t}{2.3C_0} \log I_0/I = \frac{E_0 \Phi \lambda}{2.3C_0 Nhc} \log I_0/I \tag{1-10}$$

where E_0 is the incident energy at wavelength λ, and C is molar concentration. If Φ is a constant, independent of wavelength, then for a constant dose of quanta, the *action*, $\log C_0/C$, will be proportional to the optical density of the solution, $\log I_0/I$. In other words, the amount of action will vary with wavelength as does the optical density, and at those wavelengths at which light is most efficiently absorbed, the most photochemical action is observed. Alternatively one may plot $1/E_0$ vs. wavelength and normalize the curve to unity at the absorbance maximum. Action spectra constitute one of the important tools in biology, and have been used to identify the chemical nature of some tissue enzymes and of genes (Loofbourow, 1948; Warburg, 1949; Blum, 1950; Setlow, 1957).

2. *General Nature of Photochemical Reactions in Solution*

(a) *Some reactions of small molecules.* Photochemical reactions include decompositions, oxidations, polymerizations, internal rearrangements, and chain reactions. Of these all but the last have been observed with proteins. We shall consider a few "simple" reactions to start with. In ordinary thermal reactions, the course of the reaction is apt to be governed by the lowest energy barrier encountered and the number of products is not large. For example, the lowest activation energy requirement for the hydrolysis of acetylalanine in the presence of hydrochloric acid is that for the production of acetic acid and alanine. By contrast, the photohydrolysis induced by ultraviolet light of 2537 Å leads to the formation of ammonia, acetaldehyde, pyruvic acid, alanine, and one or more highly absorbing and therefore probably unsaturated, unknown substances. At present there are few rules-of-thumb for predicting the likely structures from a given decomposition reaction except for molecules as simple as amines, ketones or aldehydes. The results for some photochemical reaction mechanisms in the vapor state have been fairly well clarified, but we are more concerned here with reactions in the condensed phase, where progress has been slower.

Among simple organic compounds, aldehydes have been given a large share of attention and studies of absorption spectra in the vapor phase and in hexane solutions have been invaluable in an elucidation of reaction mechanisms. Aldehydes (and ketones) have two absorption regions, one with a maximum at 2800–2900 Å and the second which begins below 2400 Å. It is concluded that excitation of an electron takes place in the carbonyl group since: (a) a band at 3500–2500 Å appears in those solvents which do not react with the carbonyl group whereas reaction of the carbonyl group with water to form the hydrate $=C(OH)_2$ results in a shift of the absorption maximum; (b) the molecular extinction coefficients of homologous aldehydes are nearly the same, whereas absorption by C—H linkages would result in increases with increasing size of the molecule; (c) the extinction coefficient is not larger with acetone and therefore —CH linkages are not involved; (d) the extinction coefficient is not larger with acetone and hence

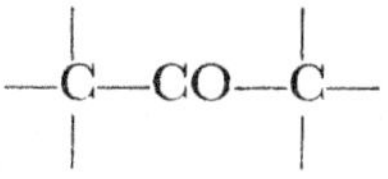

is not the chromophore.

The spectrum of formaldehyde is discrete from 3500 to 2750 Å and diffuse from 2750 to 2500 Å, and the corresponding ranges for acetaldehyde are 3484–3050 Å and 3050–2660 Å (Rollefson and Burton, 1942). A continuous background is observed below 2670 Å. Carbon monoxide and hydrogen are found as a result of absorptions in all three regions of the spectrum, even though the three regions correspond to three different lifetimes of the activated state. The fine line structure between 3570 and 2750 Å indicates a relatively long life and fluorescence is observable.

If no chemical process, such as optical dissociation can take place, the half life of the excited state is greater than 10^{-9} sec and the molecule will lose all or part of its energy of excitation by fluorescence. By the einstein relationship (Livingston, 1955) the half life of the excited state, $\tau_{\frac{1}{2}}$, is inversely proportional to the extinction coefficient,

i.e.

$$\tau_{\frac{1}{2}} = \frac{NC \ln 2}{8\pi\nu^2\varepsilon 10^3}$$

Above $\lambda = 2750$ Å, insufficient energy is available to break a C—H bond to produce hydrogen atoms. Since the quantum yield is nearly unity from 3500 to 2600 Å, the fluorescence efficiency must be low. It has therefore been concluded that an excited molecule exists in a relatively long-lived state during which vibration enters a phase which is identical with the separating half of a collision between a molecule of H_2 and CO. Separation completes the reaction to yield hydrogen molecules.

In homologous aldehydes (and ketones) the C=O bond is not broken but other bonds are. Below 2750 Å (corresponding to the 103 kcal required to break a C—H bond) H and HCO are formed by a process called predissociation. Henri (Rollefson and Burton, 1942) analyzed absorption bands which are neither continuous nor composed of discrete lines but nevertheless exhibit a definite though blurred structure. Such diffuseness is associated with cases in which the absorption of a quantum does not lead at once to dissociation into radicals, as with a continuous spectrum, but rather to a transition from a higher energy state to a state of dissociation by means of an intermediate state. The diffuseness or absence of rotational structure in bands arises from the fact that the life of the excited state is too short for quantization of rotational energy. A picture which can be used to illustrate one kind of predissociation process is represented in Fig. 1-3, for a diatomic molecule. A molecule in the activated state produced by the absorption of a quantum of energy ΔE undergoes a radiationless transfer to another electronic state represented by the dotted curve and separates into two fragments in *ca.* 10^{-13} sec. A transition between the two excited states essentially obeys the Franck–Condon principle as regards separation of the atomic nuclei. There are selection rules for predissociation but a breakdown of rules may occur during collisions with other molecules, resulting in induced predissociations. A similar scheme pertains to polyatomic molecules, and since collisions or magnetic fields can nullify selection rules it would seem that in polyatomic molecules, which are themselves centers of complex fields, the selection rules are not rigorous. Some restrictions do apply, however, since acetaldehyde can absorb more than the minimum amount of energy required for dissociation and yet remain in an excited state long enough to show some fluorescence.

The following steps summarize the most important reactions for formaldehyde vapor.

(1) $HCHO + h\nu \rightarrow HCHO^* \rightarrow CO + H_2$ (primary step at long wavelengths)

(2) $HCHO + h\nu \rightarrow H\cdot + \cdot CHO$ (primary step)

(3) $\cdot CHO \rightarrow H\cdot + CO$ (a secondary step)

(4) $H\cdot + HCHO \rightarrow H_2 + \cdot CHO$ (a secondary step)

(5) $H\cdot + H\cdot + M \rightarrow H_2 + M\nearrow$ (termination step)

In step (5) a third particle M (which may be on the walls of a reaction vessel, HCHO, or solvent) is required in a three-body collision in order to carry off the energy of dissociation of hydrogen with the formation of stable H_2.

In the presence of oxygen other secondary reactions are possible,

(6) $O_2 + \cdot CHO \rightarrow HO_2\cdot + CO$

(7) $2HO_2\cdot \rightarrow H_2O + \frac{3}{2}O_2$

The simple aliphatic aldehydes decompose by four primary processes, two of which are analogous to those for formaldehyde, viz.

(1a) $RCH_2CH_2CHO + h\nu \rightarrow RCH_2CH_3 + CO$

(2a) $\rightarrow RCH_2CH_2\cdot + \cdot CHO$

(8) $\rightarrow RCH{=}CH_2 + HCHO$

(9) $\rightarrow RCH_2\cdot + \cdot CH_2CHO$

All four processes occur and the relative extent of each is dependent on wavelength. Processes (1a) and (8) are relatively more important with shorter wavelengths for n-butyraldehyde, although they appear to require the least energy (Masson *et al.*, 1956). Carbon–oxygen bonds in aldehydes would require energy for breakage corresponding to wavelengths near the transmission limit of quartz.

Summing up the discussion thus far, one can say that the photochemical dissociation of a polyatomic molecule in the vapor state may involve rupture of one bond or the formation of stable molecules. Guesses as to where a split will occur depend on a knowledge of energies required to break a bond (Fig. 1-1) and the site of absorption of light. Analogy with carefully studied cases is most helpful, along with steric considerations. This last point is illustrated by the behavior of ketones. By analogy with acetaldehyde, acetone gives rise to carbon monoxide and ethane. But ketones having a chain of three or more carbon atoms adjacent to the carbonyl group can undergo decomposition by a mechanism different from those illustrated above. Methyl butyl ketone is converted to acetone and propylene at 2200–3000 Å, a region in which the CO is the chromophore, by means of a rearrangement in the primary process. That is, the bond broken is located in a different part of the molecule from that of the chromophore (Masson *et al.*, 1956). A single-step cleavage is probably based on the existence of a hydrogen-bonded, cyclic intermediate:

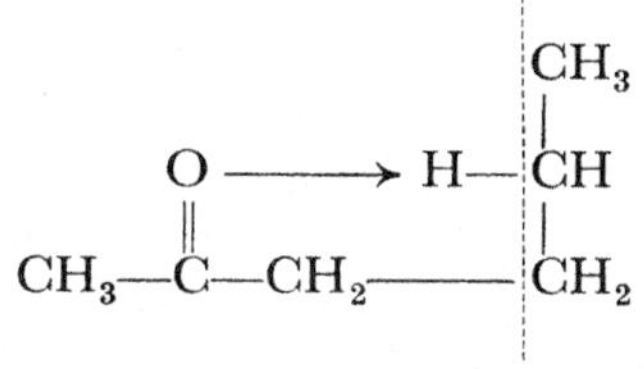

A γ hydrogen is supposed to migrate from the γ carbon. For more complex molecules, predictions can not as yet be made with any degree of certainty.

A pressing question to the biologist is what about the possible influence of solvent on such photolyses in solution? The results of numerous experiments have led to the hypothesis of Franck and Rabinowitch (Rollefson and Burton, 1942); the hypothesis can be illustrated with di-n-propylketone.

Dipropyl ketone decomposes via the following reaction with a quantum yield of about 0.3 in either the vapor phase or in iso-octane solution:

$$CH_3CH_2—CH_2—CO—CH_2CH_2CH_3 + h\nu \rightarrow$$
$$CH_3CH_2CH_2CHO + CH_3CH{=}CH_2$$

It also decomposes simultaneously by another mechanism in the vapor state, namely with the production of free radicals:

$$CH_3CH_2—CH_2CO—CH_2CH_2CH_3 + h\nu \rightarrow$$
$$CH_3CH_2CH_2\cdot + CH_3CH_2CH_2CO\cdot$$

For this reaction the quantum yield is about 0.2. In solution, however, the quantum yield for this step is only *ca.* 0.01. Thus, in solution at room temperature, only that reaction tends to proceed which results in the formation of stable molecules in the primary step. What happens, then, to any free radicals which may be formed? According to the hypothesis, they recombine (or react with solvent if possible) as a result of collision processes. In a simple rupture process the lifetime of the excited state is of the order of 10^{-13} sec. In a gas under standard conditions the mean free time is about 10^{-10} sec so the primary yield of the reacting molecule will not be influenced by other molecules. In a solvent, however, the mean free time is of the order of 10^{-13} sec and the excited molecule can lose its absorbed energy by collision, or free radicals can recombine under the influence of three body collisions with solvent. The increase in Φ with a decrease in wavelength for the photolysis of some alkyl halides in solution has been attributed to the increased translational energy given to the radicals at lower wavelengths. The increase in energy prevents the immediate recombination of primary products since they fly farther apart and get past the "sheath of surrounding molecules."

So far we have assumed that a gaseous molecule in solution does not differ from its identity in the gaseous state. Substances of biological importance, however, are associated more or less as complexes with water molecules in aqueous solution and may also exist in ionic form. In addition, quantum yields and photochemical mechanisms will depend on whether a solvent absorbs light or is affected by the products of the reaction. Correction for light absorption by solvent in computing quantum yields must be taken into consideration. The solvent itself may take part in a given reaction and yet remain unchanged. For example, the quantum yield for the photolysis of alkyl halides is nearly the same in hexane and in benzene, although benzene absorbs an appreciable fraction of the light. It seems that the benzene which absorbs the ultraviolet light transfers the absorbed energy to an alkyl halide, the latter behaving as if it absorbed the radiation (a photosensitized transfer of energy).

With complex organic substances, temperature may have a significant effect on the nature of products produced photochemically and on their rates of formation. If methyl radicals are produced during absorption of light at room temperature, the absorption of a hydrogen atom from aliphatic molecules is less likely to occur with the production of methane than is the case at higher temperatures. At higher temperatures the heat of activation for reactions of the type: $CH_3\cdot + RH \rightarrow CH_4 + R\cdot$ is available. Quantum yields for the inactivation of enzymes are temperature dependent for reasons which are not clear (Chapter V).

Substances capable of ionizing in water show photochemical behavior which can be correlated with more than one form in solution, typical examples of which are glycine and pyrimidine derivatives (Chapters III and VI).

Since, in solution, the solvent always participates to at least the extent of being coupled to the solute in vibrational modes, in a sense solutes are polymeric molecules (Franck and Platzman, 1954). An excited molecule in solvent is efficiently deactivated, without luminescence or dissociation, with the appearance of heat. (Generally luminescence occurs only if the excited chromophore is not coupled with solvent.) It is suggested that initially a radiationless transition from a low vibration level of an excited electronic state to a high vibrational level of a lower electronic state takes place (*internal conversion*), followed by collision with surrounding molecules with a loss of vibrational energy which makes the reverse transition impossible. Internal conversion can also lead to the formation of dissociation products other than radicals (Livingston, 1955).

We include two relatively well understood examples of photochemically induced molecular rearrangements to illustrate a little of the scope in this area. The absorption of ultraviolet light by an ethylenic double bond can lead to an excited state in which free rotation is apparently permissible. The less stable form of the molecule can often be produced in favorable amounts. Fumaric and maleic acids actually achieve a photochemical equilibrium:

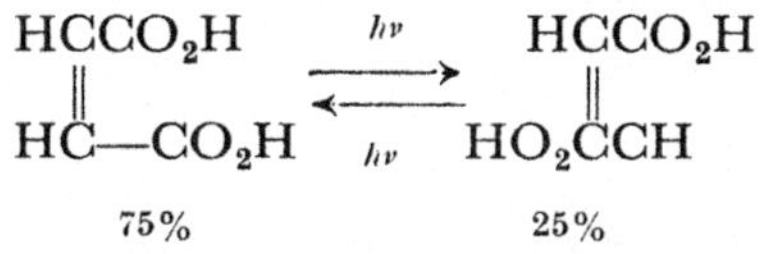

During the conversion of ergosterol to vitamin D in the neighborhood of 2810 Å, several intermediate steps are involved.

Masson *et al.* (1956) list many examples of typical photochemical reactions. In addition to the decomposition, oxidations, and photosensitizations already mentioned, there are polymerizations, halogenations, and addition to olefins. Intramolecular changes and free-radical formation are both probably important in the photochemistry of proteins and nucleic acids. There is evidence that free-radical formation can lead to aggregated protein molecules in solution. Irradiated protein can even initiate the polymerization of acrylonitrile.

(*b*) *Detection and identification of free radicals.* Classical methods for the detection of free radicals include the removal of a metallic mirror from the walls of a vessel containing a reacting gas mixture and the addition polymerization of a vinyl monomer, e.g. acrylonitrile, in solution (Rollefson and Burton, 1942).

One of the newer methods for studying free radical formation and decay is electron paramagnetic spin resonance (Zavoisky, 1945; W. Gordy, 1959; R. Livingston, 1959; Hausser, 1960). Gordy writes that "Electron paramagnetic resonance within solid matter is in its entirety an extremely complex phenomenon capable of baffling the smartest expert, and yet its simpler aspects can be grasped and used to advantage by the reasonably smart novice." Resonance in a radical arises from the interaction of an unpaired electron with an imposed magnetic field. There are two stable orientations of a spinning electron in a magnetic field, one with the magnetic vector pointing in the direction of the field and the other with the vector pointing in opposition to the applied field. These two orientations differ slightly in energy, and the transition from the state of lower energy to the higher can occur by the absorption of a quantum $h\nu = gBH$ where ν is the resonance frequency, g is the spectroscopic splitting factor, B is the Bohr magneton, and H is the field strength. Most organic free radicals have a g value nearly that for the free electron spin, namely 2.0023. Important information about free radicals is obtained from hyperfine structure in the resonance spectrum. The structure arises from the influence of the nuclear field on the applied field; that is, the effective field applied to the electron depends slightly on the nuclear field and this dependence is quantized.

In solids one observes only damaged regions, the broken parts of molecules produced by radiation. The probing radiation is a microwave which does not alter the solid. Results have been obtained with proteins, nucleic acids, etc., which have been irradiated with ultraviolet and ionizing radiation. Radicals present in irradiated alanine, for example, have been described as ethyl ($\cdot CH_2CH_3$).

Various proteins give detectable signals which are indistinguishable, whereas the signals differ among the amino acids and peptides. Perhaps the most interesting feature of the protein work is that the unpaired electron always seems to occur on the cystine or glycine residues. It is presumed, therefore, that the irradiated protein must in some manner be able to transmit either an electron vacancy, or the impulse that produces it, through the molecule until it reaches one of these two residues. There is also evidence, with this technique, of the transfer of radiation damage within nucleic acids and steroids, from one ringed group to another. Gordy assumes that for the nucleic acids as for the proteins, the different unit constituents do not have the same probability of damage when joined together as when separated. Of the various locations in a polymer, the electron density of an odd electron produced by ionizing radiation has a probability of occurring on —C=C—, —C=O, and sulfur in increasing order.

Evidence for free radicals in photosynthesizing systems has been discussed by Sogo *et al.* (1959) and Commoner *et al.* (1957). Although the electron spin resonance signals thus far obtained from isolated oxidation–reduction enzyme systems (e.g. a substrate–dehydrogenase–co-enzyme system) are too small to permit a detailed characterization, the kinetic behavior observed is in keeping with Michealis' hypothesis that biochemical oxidation occurs in successive univalent steps which give rise to free radical intermediates. The detection of a five-component electron spin resonance line at $g = 2.005$ in chloroplasts is evidence for the participation of a specific organic free radical. The amplitude of this resonance in illuminated-chloroplast preparations and in *Chlorella* cells is regarded as the evidence for the participation of a free radical in the electron transport system associated with photosynthesis. Clearly, free radical mechanisms are common to all biochemistry and biology and not just to systems irradiated with photons and ionizing particles.

With the advent of flash photolysis, pioneered by G. Porter and R. G. W. Norrish (Porter, 1959) and expanded by S. Classon, L. I. Grossweiner and others, it has been possible to obtain free radicals in quantities heretofore unattainable in photochemical reactions. In flash photolysis a non-equilibrium situation is created in a reacting system in a short interval of time, e.g. 5 μsec. It is applicable to all the states of matter and at any available temperature range. Its main applications are in studies of the primary photochemical process as a function of wavelength and of the spectral and chemical kinetic behavior of unstable intermediates such as free radicals and molecules in excited states.

A typical flash photolysis arrangement consists of three parts: a bank of condensers, for storing from a few hundred to half a million Joules, which can be discharged through a lamp (e.g. helium or mercury at low pressure), a reaction vessel equipped for constant temperature control, filters and a reflector to condense light from the flash, and some arrangement for detection and estimation of transient intermediates. This last part may consist of an arrangement for producing a second spectroflash, of short duration, at some specified time after the photolysis flash and a spectrograph and recorder. The spectroflash consists of a continuum so that one can record the absorption spectra of reaction intermediates photographically in a single flash of a few microseconds' duration.

The readily available region of the photolysis flash spectrum, down to 2000 Å, covers absorption by aromatic molecules, aldehydes, ketones, and olefins. In this region vapor-phase photolysis results in electronic excitation and dissociation into neutral molecules and free radicals, but not ions. In the liquid state ions may be formed as well. Since the intensity is so high, the number of quanta per flash being of the order of Avogadro's number, the radical concentration may correspond to from 0.002 to 20 mm Hg pressure. The lifetime of a radical can vary from 10^{-8} to 10^{-4} sec between these pressure limits, if recombination is very efficient. Assuming 10^{-6} sec as the limit of the flash method, Porter calculates that a radical concentration of 10^{-5} moles/l. can be obtained, which can give an optical density of the order of 0.1 to 1 per ten cm path length. Physically unstable molecules with excited states are accessible to study by this method only if lifetimes exceed 10^{-6} sec. Metastable triplet states are accessible since they are populated from the higher singlets and have very long radiation lifetimes. In solution, radiationless processes may decrease the lifetime, but it is still usually well above the observable limit.

In condensed phases the benzyl free-radical and many neutral semi-quinone radicals and radical-ions have been found. Triplet absorption spectra have been observed in solution for both aromatic and aliphatic compounds. The disappearance of benzyl radicals in solution has been found to be second order and diffusion controlled (Porter, 1959).

(c) *Oxygen and photosensitization.* Oxygen often plays an important role in governing the course of a reaction and the nature of the products formed. In an open vessel, irradiation with the mercury 1849 Å line can produce ozone. Ozone production can be easily avoided by simply removing from the light source the wavelength region absorbed by oxygen. Suitable filters are a long path of air or dilute solutions of acetic acid.

It seems that oxygen can quench the singlet fluorescent state of some aromatic compounds with the formation of peroxides and can react with energy-rich non-fluorescent triplet states of others to produce these compounds (Reid, 1957). Quenching of fluorescence can also occur with some systems without the formation of detectable products. The importance of these reactions in aqueous solution remains to be clarified.

Photosensitized oxidations have been known for a long time. Many oxidation reactions which occur slowly or not at all in the absence of absorbed light are promoted by the presence of a light-absorbing sensitizer substance which is itself usually not destroyed. For example ergosterol absorbs oxygen under the influence of light and in the presence of a sensitizer, such as eosin, to yield transannular peroxides. With this scheme amino-acids may be altered by visible light in the presence of methylene blue, and tobacco mosaic virus can be inactivated with blue light if absorbed acriflavine is present. The biological counterpart of such reactions is called photodynamic action (Blum, 1941).

An excited molecule may collide with oxygen with quenching of fluorescence. For a molecule M* in a singlet excited state we have these possibilities.

Excitation $M + h\nu \rightarrow M^*$

Emission $M^* \rightarrow M + h\nu'$ (fluorescence)

Solvent and internal quenching

$$M^* \rightarrow M$$

External quenching

$$M^* + O_2 \rightarrow M + O_2$$

The Stern–Volmer equation for this set of reactions is

$$\frac{\Phi_0}{\Phi} - 1 = K(O_2)$$

where Φ_0 is the quantum yield for M^* in the absence of O_2 and Φ is the yield in the presence of this quencher (Reid, 1957). K is a composite constant for the rate constants of this set. Oxygen is a very efficient quencher and very few molecules survive more than one or two collisions without deactivation. Its effect is probably to induce a singlet–triplet conversion by virtue of its paramagnetic unpaired electrons.

Eosin has been a popular sensitizer in photochemical reactions and its chemistry has been described by Schenck as follows, where S is the substrate to be oxidized.

$$M + h\nu \rightarrow M^* \text{ (singlet)}$$
$$M^* + O_2 \rightarrow M^* + O_2 \text{ (triplet)}$$
$$M^* + O_2 \rightarrow MO_2$$
$$MO_2 + S \rightarrow MO_2S$$
$$MO_2S \rightarrow O_2S + M$$

The role of the sensitizer is seen to be that of a carrier of reactive oxygen to the substrate. Many peroxides have been prepared by Schenck and his collaborators in this manner, for example the peroxide of cyclohexadiene (Schenck, 1948).

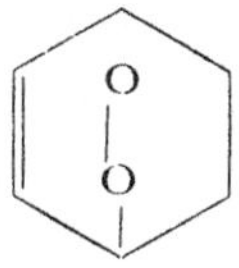

Eosin Y undergoes reduction in the presence of alkyl thiourea without an appreciable drop in fluorescence intensity, and the quantum yield is low (max =0.092). Kinetic studies indicate that the reactive intermediate is the long-lived triplet state. This long-lived excited state has a life time about

5×10^4 times greater than that of the first excited singlet state (Oster and Adelman, 1956). Oster *et al.* (1959) claim to have demonstrated that only those dyes which are capable of being photo-reduced can act as sensitizers for photo-oxidation. Both reactions (photo-reduction of the dye or its action as sensitizer) proceed via metastable long-lived excited states of the dye. We shall revert to the subject of photosensitization in subsequent chapters.

CHAPTER II

Absorption and Luminescence Spectra of Nucleoproteins and their Components

REFERENCE has already been made in the previous chapter to the importance of an adequate knowledge of absorption spectra, and associated absorbing structures, in attempts to interpret the nature of a photochemical reaction. It must now be emphasized that this is all the more important with respect to proteins and nucleic acids inasmuch as the absorption spectra of these substances and their derivatives are very frequently used for following the effects of radiation as well as the nature of the resulting photoproducts. Absorption spectra are also essential in action spectroscopy which, in many instances, is the only procedure available for pin-pointing the receptors of radiation in biological systems.

If we disregard certain prosthetic groups in proteins, such as the heme in hemoglobin, the absorption spectra of proteins and nucleoproteins may be accounted for qualitatively by the absorption characteristics of their constituent components; these exhibit bands with maxima in the wavelength region 200–320 mμ, accessible to measurements with ordinary spectrophotometers and embracing the region in which ultraviolet light exhibits major biological effects. Quantitatively, the absorption spectra are modified to a greater or lesser degree by a variety of secondary factors such as Tyndall scattering (which is of considerable importance for large molecules such as viruses), effects due to covalent and secondary bonds linking the various constituents in the macromolecules, as well as the configuration of the latter, and the nature of the solvent medium. Detailed studies of some of these factors have contributed considerably to our knowledge of the structure of proteins and nucleic acids and have been of some value in the interpretation of photochemical reaction mechanisms involving these macromolecules.

A. PROTEINS

1. *Principal Absorbing Groups in Proteins*

In an examination of protein absorption spectra and their origin, it will be found convenient to base our discussion on two regions in the ultraviolet, that from 2400 to 3200 Å and that below 2400 Å. Except for occasional references the latter will be treated in a separate section below.

The characteristic absorption maxima of proteins in the region above 2400 Å are due primarily to their content of the aromatic amino-acids tyrosine, tryptophan and (to a lesser extent) phenylalanine, the chromophores of which are, respectively, the phenol, indole, and phenyl rings. Although this was first recognized as early as 1883 and repeatedly reaffirmed, it is rather astonishing that up to only a few years ago some disagreement still existed on this question. A small contribution to the extinction coefficient of some proteins is due also to their cystine content. If we exclude for the moment cysteine (see below), the extinction coefficients of other amino acids in this region are negligible by comparison with those already mentioned (Claesson, 1956).

Fig. 2-1 exhibits the spectra of tryptophan, tyrosine, phenylalanine,

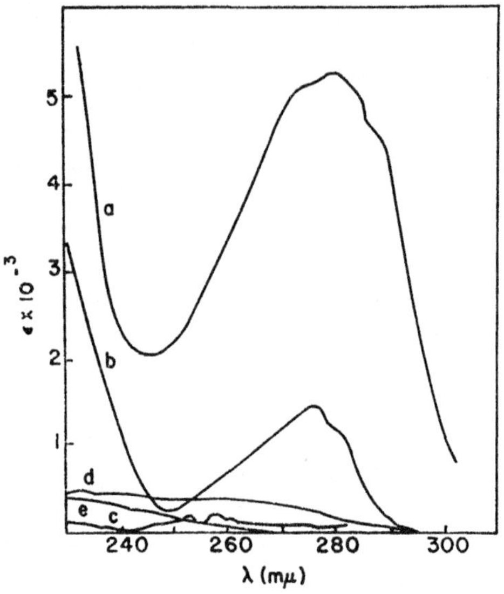

FIG. 2-1. Absorption spectra at neutral pH of tryptophan (a), tyrosine (b), phenylalanine (c), cystine (d) and cysteine (e).

cystine, and cysteine, all at neutral pH, plotted as molar extinction coefficients on the same scale so as to give a representative picture of the relative contribution of each to the spectrum of a given protein.

Cysteine deserves special mention at this point because of the fact that, until very recently, it had been practically ignored, its extinction having been assumed to be negligible on the basis of measurements made in acid medium. The spectrum of this important amino-acid has been investigated in connection with attempts to measure the acid strength of the sulfhydryl anion (Gorin, 1956; Benesch and Benesch, 1955; DeDeken *et al.*, 1956).

That the thiol group absorbs in alkaline medium had been earlier demonstrated by Noda *et al.* (1953) from studies on butyl mercaptan. Figure 2-2 exhibits the absorption spectrum of cysteine at various pH values, from which

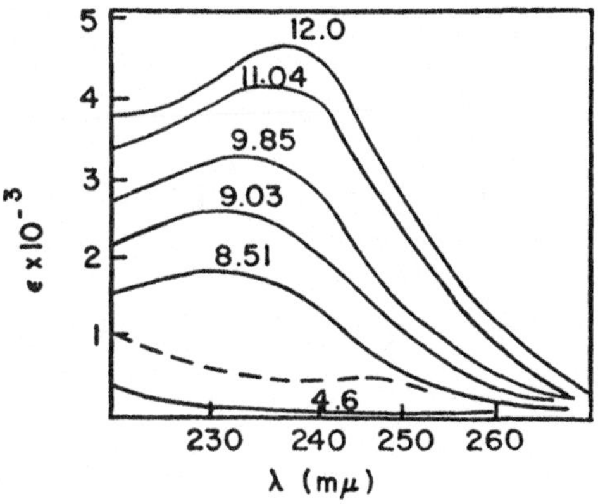

FIG. 2-2. Absorption spectrum of cysteine at pH values indicated (from DeDeken *et al.*, 1956 and Benesch and Benesch, 1955). The dotted lines represents cystine at neutral pH, for purposes of comparison.

it will be observed that the extinction coefficient at alkaline pH is of significance even at wavelengths longer than 2500 Å. The spectrum of cystine at neutral pH is included for comparison purposes. The spectral titration curve of cysteine has been variously interpreted in terms of "mixed p*K*'s" involving the ammonium and sulfhydryl groups (Benesch and Benesch, 1955) and, on the basis of a comparison with other model compounds, in terms of intramolecular hydrogen bonding between the —SH group and C=O and $—NH_2$ acceptors (DeDeken *et al.*, 1956; cf. Evans and Ellman, 1959).

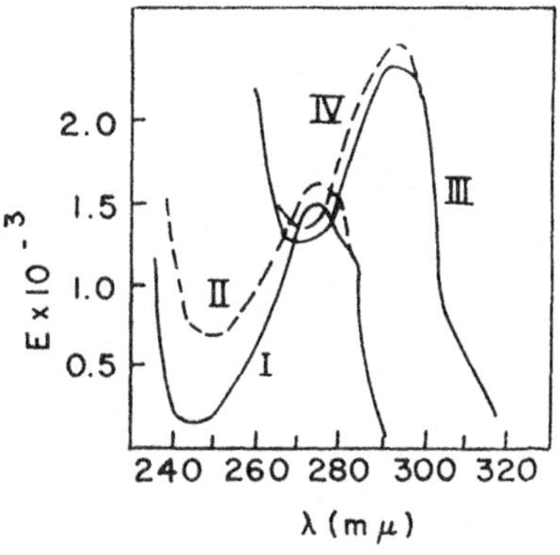

FIG. 2-3. Absorption spectra of leucyltyrosine (I, pH 7.8; III, in 0.1 N NaOH) and tyrosyl-cysteine (II, pH 7.8; IV, in 0.1 N NaOH) (Fromageot and Schnek, 1950).

The potential importance of cysteine in the absorption of polypeptides is illustrated by a comparison of the spectra of leucyltyrosine and tyrosylcysteine at pH 7.8 and 13 (Fig. 2-3). The absorption minimum of the latter is considerably higher at 2500 Å due to the presence of the cysteine component. At acid pH practically no difference would be observed between the two peptides.

Tyrosine is the only amino-acid possessing an ionizable aromatic chromophore, the dissociation of which in alkaline medium is accompanied by a marked bathochromic shift, with a considerable increase in extinction coefficient, at the principal maximum of 2750 Å. This is due to the increased number of low-energy resonance structures possible for the dissociated phenol ring and the resultant increase in stability of its first excited state (West, 1956). The changes resulting in the spectrum of tyrosine with change

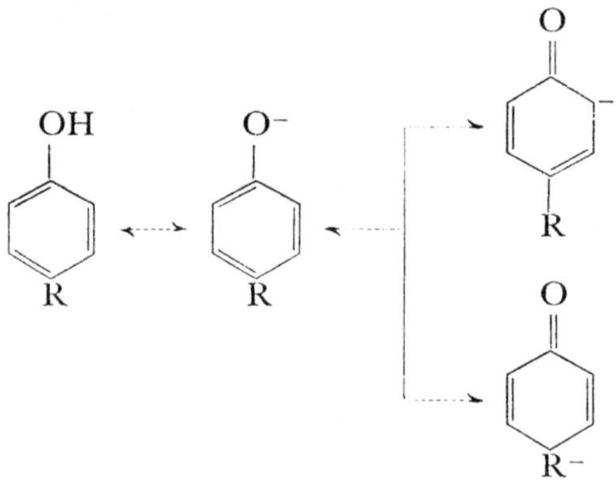

of pH are illustrated in Fig. 2-4.

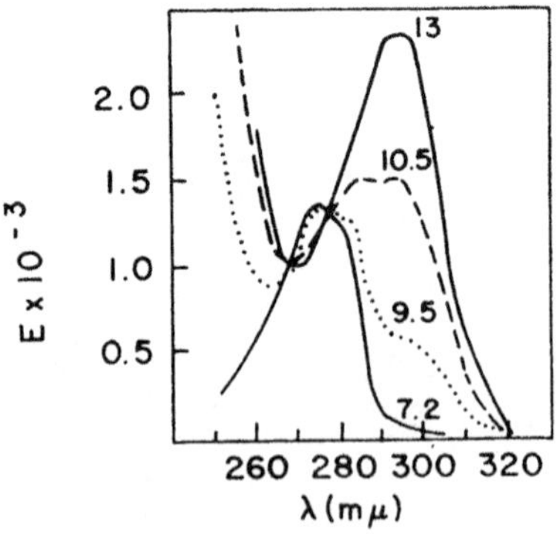

Fig. 2-4. Absorption spectrum of tyrosine at pH values indicated. Isosbestic points are at 267 mμ and at approximately 277.5 mμ (Shugar, 1952a).

For a protein such as ribonuclease, which contains no tryptophan, the behavior is strikingly similar (Fig. 2-5).

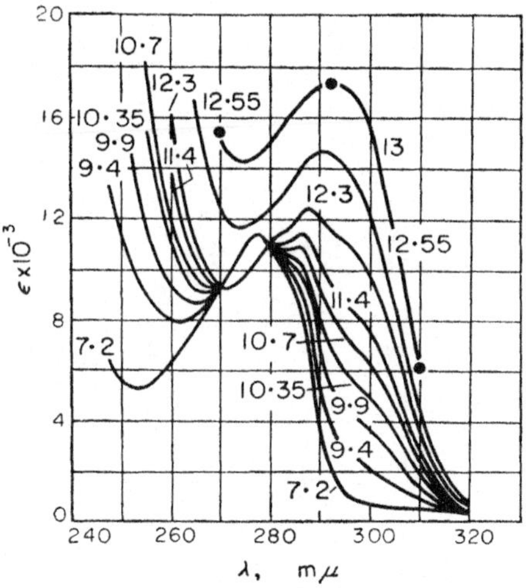

FIG. 2-5. Absorption spectrum of ribonuclease at various pH values indicated. Isosbestic points for curves from pH 7.2 to 12.3 are at 270 mμ and 280 mμ (Shugar, 1952a).

By comparison, the effect of pH on the tryptophan spectrum is small (Fromageot and Schnek, 1950; Beaven and Holiday, 1952; Hermans *et al.*, 1960). Taking advantage of this, as well as the fact that above 2800 Å the major absorbing constituents are tyrosine and tryptophan, Holiday (1936) demonstrated that the contents of these two amino-acids in proteins could be estimated on the intact molecule from the absorption spectrum in alkaline medium by treating the solution as a two-component system. This finding provided an important stimulus to the study of protein absorption spectra at a time when hydrolysis conditions for proteins and chemical estimations of amino-acids were still subject to considerable uncertainty and error, this uncertainty extending amongst others to tryptophan.

Table 2-1 presents the chemically and spectrophotometrically determined values for tyrosine and tryptophan in a number of proteins; it will be seen that the general agreement is reasonably satisfactory. The subsequent development of accurate analytical procedures for amino-acid analyses of proteins has now relegated spectrally determined values of the aromatic amino-acids to the background as analytical quantities; but has simultaneously provided a further impetus to more precise studies of protein absorption spectra since an accurate knowledge of the amino-acid composition of a given protein provides a useful tool for investigations of protein structure through analyses of the minor spectral modifications undergone by the aromatic amino-acids as a result of their incorporation into polypeptide chains. Attempts have been made to improve by empirical means the procedure of Holiday (see, e.g., Goodwin and Morton, 1946) but it is doubtful that the results obtained are more accurate for *both* amino-acids (Beaven

TABLE 2-1. CHEMICAL AND SPECTROPHOTOMETRIC ESTIMATIONS OF TYROSINE AND TRYPTOPHAN IN VARIOUS PROTEINS AND POLYPEPTIDES*

Protein	Per cent tyrosine		Per cent tryptophan	
	Chemical	Spectral	Chemical	Spectral
Horse serum globulin	5.7	6.3	2.8	2.7
Horse serum albumin	5.9	6.4	0.8	0.9
Casein	6.1	6.7	1.1	1.15
Gelatin	variable but low	low	low	low
Gliadin	3.1	3.4	0.84	0.65
Zein	5.9	6.3	0.17	0.0
Insulin	12.9	12.7	0.0	0.0
Edestin	4.3	4.6	1.6	1.2
β-lactoglobulin	4.3	4.9	1.9	2.2
Pepsin	8.5	8.3	2.4	2.4
Tyrocidin	13.0	14.7	16.3	15.8
Gramicidin	0.0	0.0	39.6	39.3
Protamine	0.0	0.0	0.0	0.0
Desoxyribonuclease	7.4 (1)	8.0 (2)	1.6 (1)	2.0 (2)
Trypsin	7.8	7.6	3.6	3.7
Actomyosin	5.5	8.0	1.4	1.3
Lysozyme	3.7	4.7	8.3	8.1
Ribonuclease:† Native	7.2 (3)	8.4 (4)	0.0	0.0
After pepsin digestion	7.2	8.0 (4)	0.0	0.0
at pH 7 after oxidation	7.2	7.0 (5)	0.0	0.0
at pH 12 after oxidation	7.2	7.0 (6)	0.0	0.0
Aldolase	5.3 (7)	5.6 (8)	2.3 (7)	1.1 (8)

* Most of data from Beaven *et al.*, 1950, and Beaven and Holiday, 1952, where original references may be found.

Additional references:

(1) Gehrmann and Okada, 1957
(2) Kunitz, 1950
(3) Hirs *et al.*, 1954
(4) Shugar, 1952a
(5) Harrington and Schellman, 1956
(6) Tramer and Shugar, 1958a
(7) Velick and Ronzoni, 1948
(8) Labeyrie and Shugar, 1951

† Recalculated from published values, using molecular weight of 13,700.

and Holiday, 1952). Furthermore, in the light of the remarkable advances of the past few years in the analytical chemistry of proteins, it is questionable whether further efforts in this direction are worthwhile, although the procedure of Holiday will undoubtedly retain its usefulness for identification and orientation purposes.

Coming back now to the absorption spectrum of tyrosine (Fig. 2-4) it will be clear that the spectra at pH 7.2 and pH 13 represent, respectively, the undissociated phenol ring and its anion; a spectral titration curve of the changes in extinction coefficient at a given wavelength makes possible the calculation of the p*K* of the tyrosine phenol group, which is 10.0. The existence of an ionic equilibrium is demonstrated also by the isosbestic points at 2670 Å and approximately 2775 Å. The latter isosbestic point is, however, not clearly defined until the pH has exceeded 9.5 and the suggestion was advanced that this is due to perturbation of the phenol chromophore by dissociation of the α-amino group which has a p*K* of about 9 (Shugar, 1952a). This is supported by the observation that in ribonuclease, where the amino groups are involved in peptide linkages and therefore not free to dissociate, both isosbestic points are sharply defined (Fig. 2-5). More recently the effect of amino group dissociation on the tyrosine spectrum has been demonstrated in a rather elegant manner by Wetlauffer *et al.*, (1958) by means of differential spectrophotometry of *o*-methyltyrosine.

The effect of dissociation of the amino group is small, as might be expected, since the phenol ring is separated from the amino group by two single bonds

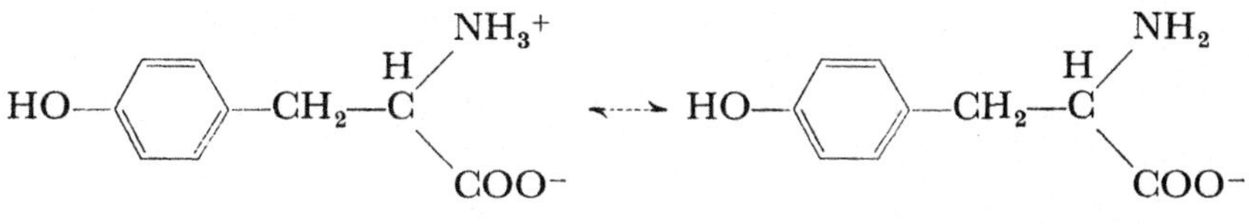

which are usually expected to insulate a chromophore. Braude (1949) has, however, shown that a second-order effect is to be expected and this is likewise evident in the case of phenylalanine and tryptophan, both of which exhibit distinct differences in their absorption spectra as between acid, neutral, and alkaline pH (Fromageot and Schnek, 1950; Beaven *et al.*, 1950; Beaven and Holiday, 1952; Hermans *et al.*, 1960). Dissociation of the carboxyl group also induces minor spectral modifications, the best illustration of this being again the work of Wetlauffer *et al.* (1958) on the difference spectrum of *o*-methyltyrosine in acid medium and that of Hermans *et al.* (1960) on tryptophan.

It might be expected, on the basis of these secondary effects due to amino and carboxyl group dissociations, that peptide bond formation would exert some influence on the absorption spectra of the aromatic amino-acids and such has indeed proved to be the case (Beaven *et al.*, 1950), although the effect is small. Starting with the investigations of Coulter *et al.* (1935), a number of workers (for review see Beaven and Holiday, 1952) have shown that the absorption maxima of aromatic amino-acids in proteins do not coincide exactly with values predicted from the summation spectra of the amino-acid constituents. A comparison of Fig. 2-4 for tyrosine and Fig. 2-5 for ribonuclease, taking into account the position of the isosbestic points, demonstrates that it is the entire absorption band that has been shifted and not the maximum alone. A tabulation of available data by Beaven and Holiday (1952) indicates shifts of the order of about 4 Å for the tyrosine maximum and 9 Å for the tryptophan maximum as a result of peptide bonding.† This shift, the so-called weighting effect, is inadequate to account for the red shifts of 10–40 Å observed in the absorption maxima of proteins and the foregoing authors therefore conclude that the protein macromolecular structure itself constitutes an environment which also influences the fine structure features. Thus, by the action of various proteolytic enzymes, the protein absorption bands may frequently be caused to revert to those wavelengths at which their constituent amino-acids exhibit their maxima (Haurowitz and Astrup, 1939; Beaven *et al.*, 1950; Beaven and Holiday, 1952; Shugar, 1952), from which one might be led to conclude that peptide linkages are indeed responsible for the effect. However, contrary to the experience of Beaven *et al.* (1950), this shift may be partially abolished in the case of ribonuclease by acidification (Shugar, 1952a), in which case it is reversible; and also by heat denaturation (Sela and Anfinsen, 1957; Tramer and Shugar, 1959) in which case it is irreversible. It is clear, therefore, that secondary bonds in the native structure of the protein molecule do indeed play some role, and we shall revert to this point in several instances below.

A rather unusual absorption spectrum is exhibited by the enzyme triosephosphate dehydrogenase in that the minimum at 2500 Å exhibits a rather high absorption (Fig. 2-6). This has been shown to be due to the fact that the native enzyme binds one molecule of diphosphopyridine dinucleotide. If the latter is partially removed by absorption on charcoal, the spectrum more nearly resembles that of a normal protein (Taylor *et al.*, 1948). The contribution of the prosthetic group in this case to the total absorption of the protein at 2500 Å is of sufficient importance that it must be taken into account in quantitative photochemical studies. Furthermore, reduction of the DPN moiety provides us with a protein, the prosthetic group of which possesses an absorption maximum at 3400 Å, where

† Beaven *et al.* (1950 have stressed the need of extending such measurements to tripeptides and, in particular, to those in which the aromatic amino-acid is "sandwiched" between two aliphatic amino-acids. To date no such studies appear to have been made although it is now possible to prepare such peptides by chemical syntheses.

the protein itself is transparent, and the photochemical behavior of which is of considerable interest.

An illustration of an abnormal absorption spectrum is that of a crystalline protein isolated from plants infected with turnip yellow mosaic virus (Markham *et al.*, 1948); it exhibits a very broad maximum at 2710 Å and

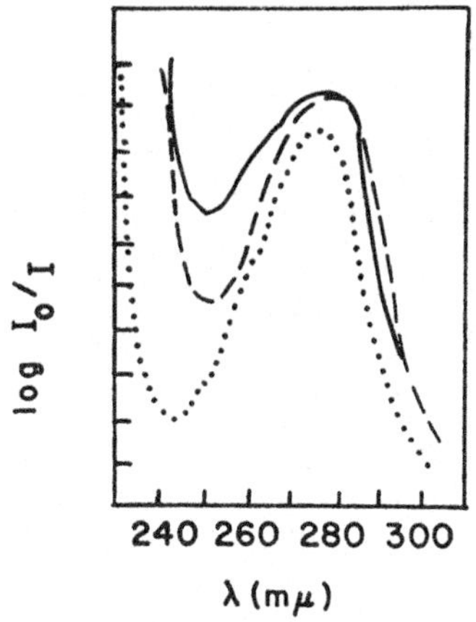

FIG. 2-6. Absorption spectrum of glyceraldehyde phosphate dehydrogenase, 1 mg/ml at pH 7 (———); after removal of 65 per cent of bound DPN (-------); and according to theoretical content of tyrosine and tryptophane (······) (Taylor *et al.*, 1948).

the usual minimum at 2500 Å, but has an exceptionally low ratio of maximum to minimum, namely 1.25. The absence of any data as to the chemical composition and molecular weight makes it difficult to explain this anomaly.

2. *Free and Bound Phenolic Hydroxyl Groups in Proteins*

By taking advantage of the marked changes in absorption exhibited by tyrosine in alkaline medium, Crammer and Neuberger (1943) first demonstrated that it was possible to study the nature of the binding of phenolic hydroxyl groups in native proteins by spectral titration; and showed that, while these groups are "free" in the insulin molecule, most of them are "bound" and unable to dissociate in the ovalbumin molecule until the latter had been denatured by some means such as heat, urea, alkali, etc. The hypothesis advanced by these authors, that the groups not free to dissociate in the native molecule are involved in hydrogen bonding with carboxyl groups, has received wide support, although alternative explanations have more recently been proffered (Sela *et al.*, 1957; Tanford *et al.*, 1956; Wetlauffer *et al.*, 1958; Tramer and Shugar, 1959).

Undoubtedly the most interesting illustration, and that most extensively investigated, of the spectral behavior of free and bound phenolic groups in a protein is that for ribonuclease. In this protein approximately 60 per cent of the phenolic groups are bound; this can be seen from an examination of Fig. 2-5 where, up to pH 12.3, all the curves pass through both isosbestic points; while, if the pH is brought back to 7.0, the original spectrum is obtained. Above pH 12.3 the curves break away from the isosbestic points and the extinction increases with time at a rate dependent on the pH; this is due to alkaline denaturation of the protein. At pH 13 all the phenolic groups are dissociated. Similar results have been obtained with chymotrypsin (Finkelstein and McLaren, 1949). These investigations have been extended to non-aqueous solvents (Singer, 1962).

All of the six tyrosyl residues in ribonuclease dissociate normally in 8 M urea; since the enzyme is fully active in this medium it has been concluded that the detailed structure involving hydrogen bonding of tyrosyl hydroxyls is not essential to ribonuclease activity (Blumenfeld and Levy, 1958). The absorption maximum of pepsin is displaced from 2780 to 2760 Å in 8 M urea and it has been proposed that this involves breakage of

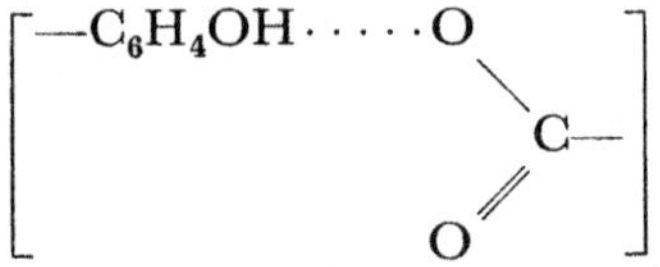

linkages, since both urea and a pH above 6 cause inactivation and tyrosine residues have been postulated as essential for the proteolytic activity of pepsin (McLaren and Pearson, 1949; McLaren and Lewis, 1950; Perlman, 1956). Bound phenolic groups are, however, not always necessary for maintenance of the native structure of a molecule; they are not required in bovine serum albumin (Tanford and Roberts, 1952), nor in insulin (Crammer and Neuberger, 1943) or lysozyme (Fromageot and Schnek, 1950).

It has usually been assumed that the long wavelength shoulder of the principal absorption maximum of proteins at pH 13 is due entirely to dissociated phenolic groups. In those instances where the curves no longer pass through the isosbestic points this may, however, not be true. For serum albumin it has been shown that this is not so (Tanford and Roberts, 1952) and apparently a similar situation prevails to a lesser extent for ribonuclease (Tramer and Shugar, 1959). The origin of this supplementary absorption, which is much more pronounced at shorter wavelengths (see Fig. 2-5) and is associated in some way with the alkaline denaturation of the protein (Shugar, 1952a; Tanford and Roberts, 1952), is unknown.

During the past few years the minor spectral shifts exhibited by most proteins upon acidification to pH 1–2 have been extensively studied by a number of workers who have advanced the hypothesis that this shift, which is usually completely reversible (and is best followed by the technique of difference spectrophotometry), is due to hydrogen bonding to carboxylate acceptors of phenolic (Laskowski *et al.*, 1956; Scherega, 1957) and other groups (Donovan *et al.*, 1958). If specific hydrogen bonding were involved, one would expect such difference spectra to disappear on denaturation of the protein; but tests made on several proteins indicate that such is not the case (Tramer and Shugar, 1959). Furthermore Wetlauffer *et al.*, (1958) have demonstrated that qualitatively similar difference spectra may be obtained, in the case of tyrosine, as a result of dipole–dipole interactions. On the whole, although these spectral shifts in acid medium are indeed due to some type of interaction involving the phenolic ring, it cannot be considered as established that specific hydrogen bonding, rather than local charge and solvation effects (Yanari and Bovey, 1960), is involved (Blumenfeld and Perlmann, 1959; Tramer and Shugar, 1959; Bigelow, 1959; Cha and Scheraga, 1960; Leach and Scheraga, 1960). An as yet unidentified difference spectral peak at 2300–2350 Å also appears during protein denaturation (Glazer and Smith, 1960). Care should be taken that possible light scatter is taken into account in difference spectra (Leach and Scheraga, 1960). For a review see Wetlaufer (1962).

3. *Peptide Absorption in the Range* 2400–3200 Å

The extent to which the amide linkage absorbs in this region of the spectrum is still the subject of some controversy, estimates varying from 1 to 5 per cent of the total extinction for proteins at 2537 Å. Arguments in favor of peptide chromophores as the general source of polypeptide absorption have been effectively disposed of by Beaven and Holiday (1952) and will not be further discussed here. In what follows we shall concern ourselves mainly with values at 2537 Å, since this is the wavelength most frequently used in photochemical studies. At longer wavelengths the contribution due to peptide absorption decreases.

A number of attempts have been made to calculate the absorption of the peptide bond as the difference between the absorption spectrum of a protein and that of a mixture of its constituent amino acids. Rideal and Roberts (1951), for example, have done this with insulin, bovine serum albumin, and gelatin, and arrive at the conclusion that 5 per cent of the absorption at 2537 Å is due to the keto–imino linkages. They assume, however, that gelatin contains no tyrosine whereas this protein is known to contain small but varying proportions of tyrosine which have been quantitatively placed in evidence both chemically and spectroscopically (Beaven *et al.*, 1950); more comprehensive data on the chemical composition of gelatin may be found in a paper by Russell (1958). Rideal and Roberts cite the fact that acetylglycine exhibits a higher absorption at 2537 Å than glycine and conclude that, since the methyl group is transparent at this wavelength, the additional absorption must be due to the peptide linkage;

while undoubtedly true in this case, it should be noted that acetyltryptophan has a lower extinction at 2537 Å than tryptophan itself (Edwards, 1949), although this latter observation merits verification.†

Figure 2-7A shows Rideal and Robert's curves for bovine serum albumin and a mixture of its amino-acids. At 2500 Å the protein absorption is 15

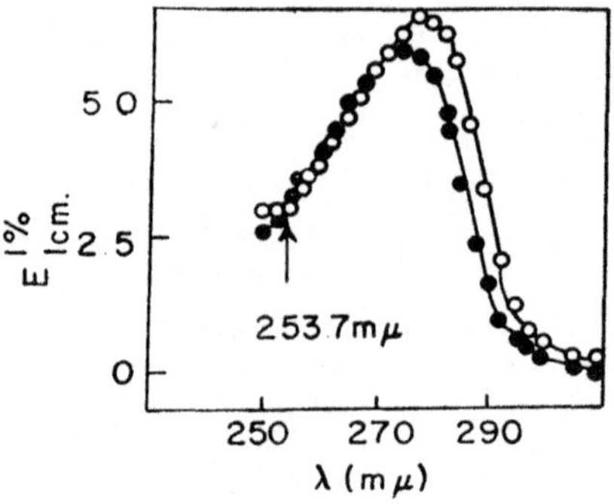

FIG. 2-7A. Absorption spectra of bovine serum albumin (–○–○–○–) and an equivalent mixture of its constituent amino acids (–●–●–●–) (Rideal and Roberts, 1951).

per cent higher than that of its amino-acid mixture. However, an appreciable portion of this is due to the fact that the protein curve shows a higher extinction and a maximum farther to the red by 20–40 Å as a result of the incorporation of the aromatic amino-acids into the polypeptide chain. It is no simple matter to quantitatively correct for this. At 2537 Å the difference between the two curves is much less. It should, however, be noted that, for reasons which are somewhat obscure, different preparations of the same protein do not always exhibit identical spectra, particularly at the shorter wavelengths where the above comparison was made. Attention has been drawn to the importance of the ratio of maximum to minimum absorption as a means of more accurately characterizing the spectra of polypeptides (Shugar, 1952a). The minimum of the absorption spectrum for most proteins is in the neighborhood of 2500 Å; the curve for insulin presented by Rideal and Roberts shows a ratio of maximum to minimum of 2.0 as compared to 2.5 for that of Crammer and Neuberger (1943), 2.1 for McLaren and Waldt (1952), and 1.8 for a curve given by Beaven *et al.* (1950). The differences in absorption between these four samples at 2500 Å are of the same order of magnitude as the difference found by Rideal and Roberts between their preparation and the equivalent mixture of amino-acids.

For chymotrypsin, McLaren and Waldt (1952) found that at 2500 and

† The molar extinction coefficient of tryptophan at 2537 Å is about 2300. This is so large by comparison with the extinction for the peptide linkage, that a small error in concentration would be sufficient to give misleading results.

2537 Å the absorption of the equivalent amino-acid mixture was actually higher than that of the protein, Fig. 2-7B.

It should also be noted that the absorption of peptide bonds in small peptides has been shown to be influenced markedly by the nature of the side chains and the relative position of the amino-acids (Goldfarb *et al.*, 1951) as well as by the length of the peptide chain (Goldfarb, 1953). At shorter wavelengths, where peptide absorption is appreciable, approximate com-

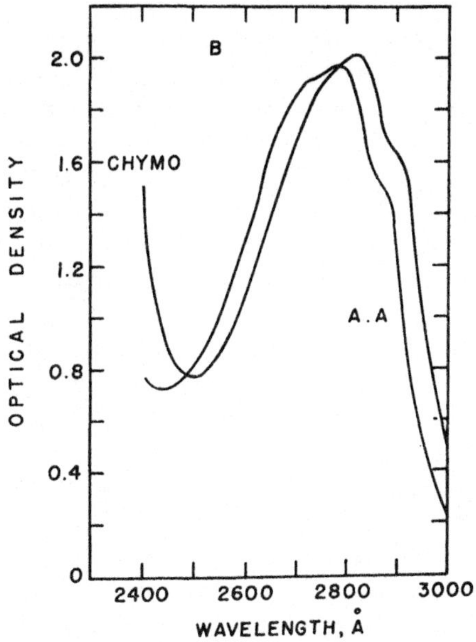

FIG. 2-7B. Absorption spectra of chymotrypsin and the corresponding mixture of amino-acids (A.A.) (McLaren and Waldt, 1952).

parisons such as the above can and have been made; at longer wavelengths considerably more experimental evidence would be required to establish their validity.

Surprisingly enough, attempts to calculate peptide bond absorption from that of simple peptides have not given unequivocal results, perhaps due partly to electrostatic and stereochemical effects which may more than counterbalance the low extinction of the amide linkage, and partly to impurities.

Castille and Ruppol (1928) report that at 2537 Å $\log_e \varepsilon$ is 0.25 for acetylglycine and -0.7 for glycine; from this $\varepsilon_{CONH} = 1.28$. Estermann *et al.* (1956) have compared acetylalanine

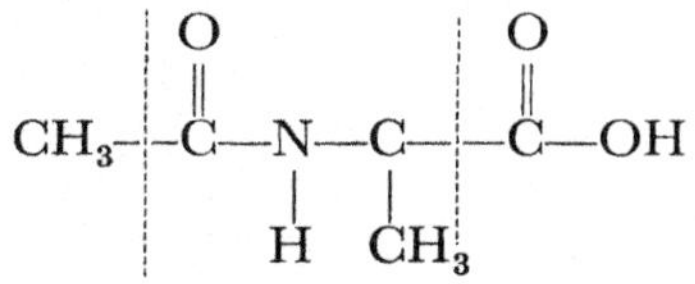

for which ε is 0.226 in 0.2 N acid and 0.146 in 0.2 N base, with acetic acid

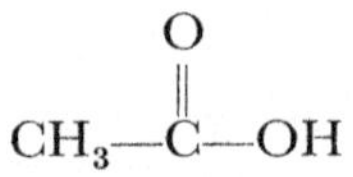

for which ε is 0.02, and calculate ε_{CONH} as equal to 0.2 at 2537 Å. For simple peptides with no side chains, such as diglycine, Claesson (1956) estimates that $\varepsilon_{CONH} \sim 0.1$. By contrast, Setlow and Guild (1951) compared leucine, glycine, and glycyl-leucine; from their figure it appears that $\varepsilon_{CONH} = 3$ at 2537 Å. However, their curve for leucine alone appears to have an extinction of about 3, which is undoubtedly too high. Finally at wavelengths above 2650 Å the extinction coefficient for glycyl-leucine actually increases, a most unlikely possibility, suggestive of the presence of impurities.

Undoubtedly the best estimate for peptide bond absorption would be that obtained by using an aliphatic polypeptide for which the molecular weight is low enough to definitely exclude absorption due to scattering. Such polypeptides are now available by synthetic routes but, oddly enough, have not been used for this purpose, although they have been applied to similar studies in the region 2000–2200 Å (Goldfarb, 1953). An estimate for the peptide bond extinction coefficient may be obtained from the naturally occurring polypeptide clupein (molecular weight 4000), the absorption of which has been measured by McLaren (1949), McLaren and Waldt (1952)- and Claesson (1956). This polypeptide is devoid not only of aromatic amino-acids, but also of cystine. At 2537 Å it exhibits an optical density of about 0.18 for a concentration of 0.8 per cent. It has been suggested by Beaven and Holiday (1952) that this absorption may be ascribed to light scattering; but calculation by means of the scattering equation (p. 45) shows that the absorption due to this source at 2537 Å is equivalent to an optical density of about 0.001 or 0.5 per cent of the measured extinction. From the measured optical density we can calculate $\varepsilon_{CONH} = 1.7$. If we subtract from this the measured absorption of the equivalent mixture of amino-acids which, according to Rideal and Roberts (1951) is slightly more than one-third of the absorption of the polypeptide, we obtain $\varepsilon_{CONH} = 1.2$.

Both Goldfarb and Saidel (Goldfarb, personal communication, 1960) have measured the absorption of synthetic, aliphatic peptides containing from two to six residues and a variety of amino-acids, at concentrations up to near saturation. No band was apparent in the region 2500–3000 Å. At 2500 Å the molar extinction is about 1.0 and either remains at this level or decreases, up to 3000 Å.

Taking all the evidence together it appears not unreasonable to accept a value for $\varepsilon_{CONH} \sim 1$ at 2537 Å, as suggested by Beaven and Holiday (1952). Using this figure we can make some rough calculations of the fractional absorption of a protein due to its peptide linkages and the results for several typical proteins are presented in Table 2-2.

TABLE 2-2. MOLECULAR EXTINCTIONS FOR SOME CHROMOPHORES AT 2537 Å AND THE APPROXIMATE PER CENT ABSORPTION DUE TO PEPTIDE LINKAGES FOR SEVERAL PROTEINS

Chromophore	Model compound	ε_{2537}
—COOH	Acetic acid	0.02
$—NH_2$	Ethylamine	very low
—CONH—	Acetylalanine	0.2
—CONH—	Clupein	1.3–1.7
—S—S—	Cystine	270

Protein	Molecular weight	Approximate no. of peptide bonds	Total ε_{2537}	Per cent CONH absorption
Ribonuclease	13,900	125	8200	2.5
Insulin	12,000	102	6000	1.5
Chymotrypsin	37,000	300	30,500	1.0
Gelatin	100,000	1000	6000	16.0
Serum albumin	69,000	600	20,000	3.0
Desoxyribonuclease	60,000	500	24,000	2.0
Lysozyme	14,700	130	18,000	0.7

We have tacitly assumed in the above that the contribution of each peptide link in a given polypeptide is the same; but, as already mentioned above, the absorption of the peptide bond in small chains is dependent on a variety of factors such as nature of side chains, length of chain, etc. An extensive study of this question has been made by Saidel (1955a, 1955b, 1955c) at wavelengths below 2400 Å where the effects are appreciable and are readily measured. Because of the small absorption of the peptide linkage above 2400 Å, it is by no means certain that such effects will be readily differentiated. From the point of view of photochemistry of proteins, it is undoubtedly of more importance to establish first whether peptide absorption is involved.

4. *Absorption by Fibrous Proteins*

Considerable experimental evidence has been presented by Schauenstein and his collaborators (see Schauenstein, 1949, 1957) for the existence in actomyosin solutions and in orientated films of fibrous proteins, such as collagen and silk fibroin, of a rather broad absorption band in the neighborhood of 2500 Å and with an extinction comparable in magnitude with that exhibited by the aromatic amino-acids. The effect of this band is, in most cases, to flatten out almost completely the minimum exhibited normally by proteins at 2500 Å. The origin of this absorption has been ascribed to an enolized form of the peptide linkage, i.e.

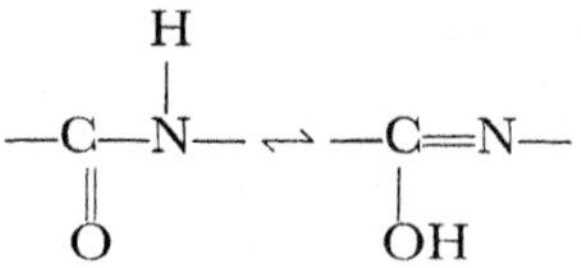

resulting from an exchange of hydrogen atoms between two adjacent polypeptide chains and the formation of interchain hydrogen bonds.

Beaven and Holiday (1952) and Doty and Geidushek (1953) have discussed in great detail the evidence for and against the existence of absorption by "peptenol" chromophores, as they are referred to by Schauenstein, and the reader should consult these articles for further details. It must be pointed out that, in contrast to other theories which have purported to demonstrate the existence of polypeptide fabric absorption, that of Schauenstein *et al.* cannot lightly be dismissed, even though there are strong arguments against it.

On the other hand most of the evidence adduced by Schauenstein *et al.* for peptenol absorption is, as mentioned above, based on studies of oriented fibrous proteins and hence also of the ultraviolet dichroism associated with them. The existence of such chromophores *in vivo* is conceivable and, since their absorption maximum is in the neighborhood of 2500 Å, would be of considerable significance in photochemical studies on living organisms inasmuch as most of such work is done at 2537 Å.

More recently Schauenstein *et al.* (1955) have reported the presence of "peptenol" absorption in heat-denatured γ-globulin and bovine serum albumin, based on the observation that the extinction at about 2500 Å increases as a result of heating, appropriate corrections being made where necessary for turbidity (Bayzer and Schauenstein, 1955). The additional absorption produced by heating in the region 2500 Å is postulated as being due to aggregation resulting from the formation of intermolecular peptide–hydrogen bridges. Some issue may be taken with regard to this interpretation. For example, the spectra of the heated solutions exhibit additional absorption at wavelengths above 3000 Å. Furthermore, the general appearance of the spectra of heated solutions is similar to that of ultraviolet irradiated proteins in some respects as well to that of alkaline denatured protein (cf. Tramer and Shugar, 1958a). More direct evidence on this point is obviously desirable. It should be noted that no claim is made for the existence of peptenol absorption in native γ-globulin or serum albumin.

5. *Apparent Absorption of Light Attributable to Scattering*

The measured absorption of a solution of macromolecules is usually greater than the true (electronic) absorption as a result of Rayleigh scattering. For large molecules scattering may constitute an appreciable fraction of the spectrophotometrically measured optical density, and it is frequently of importance to be able to correct for this. There are also numerous instances where examples of low and apparently "non-specific" absorption of proteins are explained away as due to scattering, without any attempt to verify whether this is actually the case. Since we shall have frequent occasion to refer to this, it is advisable at this point to examine the magnitude of the scattering correction.

If the optical density due to true absorption is D_a and that due to scattering is τ, then the measured extinction will be

$$D = D_a + \tau \qquad (2\text{-}1)$$

With λ_0 the wavelength of the incident light beam in a vacuum, n the refractive index of the solution, n_0 that of the solvent, N Avogadro's number, c the concentration (in g/ml) and M the weight average molecular weight of the solute, then for molecules the longest dimension of which is of the order of 0.1λ the optical density due to scattering is

$$\tau = \frac{32\pi^3 n_0^2}{3N\lambda_0^4}\left(\frac{n - n_0}{c}\right)^2 cM = HcM \qquad (2\text{-}2)$$

An assumption is that there is no molecular interaction; this latter condition can usually be satisfied by extrapolation of measurements to high dilutions (Oster, 1950; Doty and Edsall, 1951; Doty and Geidushek, 1953; Stacey, 1956).

For a molecule with a molecular weight of about 10^5 at a concentration of about 0.1 per cent, such as is normally used in spectrophotometry of proteins, the extinction due to scattering at a wavelength of 3500 Å (which is outside the region in which proteins and nucleic acids absorb), can be calculated from the above formula to give an optical density of about 0.0025. From an experimental point of view this is very small.

The above formula may be applied to dilute solutions even when molecular weights of a few million are involved. However, we are usually more interested in calculating the numerical value of the extinction due to scattering at those wavelengths at which the molecules absorb light; under these conditions the refractive index is anomalous and markedly dependent on wavelength. Frequently we are also concerned with molecules the dimensions of which may be of the order of the magnitude of the wavelength of the light used. Under these circumstances it becomes extremely difficult to calculate the scattering extinction and frequent recourse has been had to extrapolation methods.

If, in the above formula for the scattering extinction, the solute concentration is constant, we have

$$\tau = K/\lambda^4 \tag{2-3}$$

where K is referred to as the scattering constant. More generally the Rayleigh–Mie theory of dependence of τ on λ is

$$\tau = K/\lambda^n \tag{2-4a}$$

We therefore have

$$\log \tau = \log K - n \log \lambda \tag{2-4b}$$

If now we measure τ at several wavelengths in a region in which the molecules do not absorb, a plot of log τ vs. log λ should give a straight line with a slope n. We can then calculate K and use these values of n and K to calculate the extent of scattering at any other wavelength. The maximum value of n is 4 when the particles are several times smaller than the wavelength of the incident light. Treiber and Schauenstein (1949) and Schauenstein and Treiber (1950) have examined in detail scattering corrections by extrapolation and photographic methods for protein absorption spectra (cf. Dannenberg, 1951). It must be emphasized that the extrapolation procedure is merely more convenient than direct calculation and is subject to the same objections cited above for the latter procedure.

6. *Extraneous Absorption in Proteins*

Numerous proteins exhibit some absorption at neutral pH above 3000 Å where neither tyrosine nor tryptophan absorb. The origin of this absorption has been variously ascribed to impurities; and Goodwin and Morton (1946) refer to this as "pigment" absorption, which they consider must be corrected for in order to obtain better estimates for the tyrosine and tryptophan content by the method of Holiday (1936).

Since such end-absorption exists also for proteins of low molecular weight, it cannot be due to light scattering. From Fig. 2-5 it will be seen that ribonuclease exhibits some absorption even above 3200 Å. Recently Setlow and Doyle (1957), in an investigation of the action spectrum for ultraviolet inactivation of ribonuclease, reported a quantum yield for this enzyme at 3130 Å comparable in magnitude to that prevailing at other wavelengths. This is reasonable evidence that ribonuclease absorption above 3100 Å is not "extraneous." It has also been observed by Steele and Szent-Györgyi (1958) that irradiation of proteins at about 3400 Å results in the excitation of fluorescence, suggesting the existence in this region of low-intensity absorption bands. As a sequel to this, concentrated solutions of a variety of proteins were examined in this spectral region; all of the proteins tested were found to exhibit a maximum, but numerical values for extinction coefficients were not given.

If we accept the chemically determined value for the tyrosine content of ribonuclease, it will be seen from Table 2-1 that the spectrally determined value for the *native* protein is 16 per cent higher; following pepsin digestion the discrepancy is reduced, but the spectral value is still about 11 per cent too high. Only following oxidation of the —S—S— bridges does the spectral value, both in neutral and alkaline media exhibit reasonable agreement with the chemically determined value. The inference to be drawn from these results is that the extinction of ribonuclease above 2700 Å is 15 per cent higher than that to be expected on the basis of its tyrosine content! Calculation shows that at this wavelength the extinction due to other absorbing constituents in the molecules does not contribute more than 2 per cent of the total. Such a discrepancy is too serious to be lightly dismissed and is of significance not only in relation to the spectrum and structure of the ribonuclease molecule, but also in attempts to delimit the chromophores responsible for light absorption in the photochemistry of this protein.

Finally, it must be emphasized once more that the ratio of maximum to minimum absorption for a given protein preparation is apparently not readily reproducible in different laboratories. This could be due to impurities. But Fredericq (1955, 1956) reports that this ratio, in the case of highly purified insulin, is markedly dependent on the concentration and also varies following aeration of the solution or following neutralization from acid medium. The suggestion that aeration results in oxidation of tyrosyl residues, thus resulting in modification of the ratio of maximum to minimum absorption, is questionable. It is clear that further work is necessary to completely elucidate the origin of the absorption spectra of proteins.

An additional effect, likely to be encountered in the absorption spectra of synthetic polypeptides of aromatic amino acids, is that of hyperchromicity, which will be discussed in greater detail in connection with nucleic acids (see Sect. B.3). When two aromatic rings in a polymeric chain are adjacent to each other, there may result some interaction between the π-electron orbitals, the net effect being a modification (usually a decrease) in extinction. The probability of two aromatic amino-acids being adjacent to each other in a protein is relatively small; but an examination of the spectral data of Katchalski *et al.* (1953, 1954) for synthetic polypeptides of tyrosine, tryptophan, and phenylalanine demonstrates appreciable decreases in molar extinction coefficients which are most likely due to this effect. Of the natural polypeptides, gramicidin might be expected to exhibit hypochromicity because of its high tryptophan content (Table 2-1), but its solubility in aqueous medium is too low to test this.

7. *Peptide Absorption below* 2400 Å

Measurements in this region of the spectrum are more difficult to carry out, particularly below 2200 Å where errors due to stray light may be serious, depending on the state of the equipment used, and requiring appropriate corrections (Beaven and Holiday, 1950; Saidel *et al.*, 1951). Below 2000 Å vacuum spectrographs must normally be employed and Ley and Arends (1932) in this way examined a number of amino-acids down to 1770 Å, finding only a general rise in absorption with no maxima. Recent technical advances should considerably enhance the possibilities of spectral investigations at shorter wavelengths. Taylor and Jones (1956) have described a procedure for converting commercial spectrophotometers so as to make possible measurements to 1920 Å; while Buell and Hansen (1959) report details for the construction of an instrument suitable for measurements in the range 2300–1900 Å, based on the use of a solar-blind photomultiplier. In addition, commercial spectrophotometers (Beckman, Unicam) are now available with modification which extend the spectral range to 1900 Å, at the same time markedly reducing the amount of scattered light.

The early literature on the spectra of amino-acids is summarized by McLaren (1949b), Saidel *et al.* (1952), Beaven and Holiday (1952), and Claesson (1956). For the region 2000–2400 Å Saidel *et al.* (1952) provide quantitative data for all the commonly occurring amino-acids. A recent study (Preiss and Setlow, 1956) has confirmed and extended earlier amino-acid spectra (Beaven and Holiday, 1952). A broad maximum has been found for phenylalanine at 1900 Å, both in solution and as a film; at pH 7 the absorption spectrum of a solution is similar to that for a dry film. By contrast, the sharp maximum for tryptophan at *ca.* 2160–2180 Å in solution (see Fig. 2-8) is, in the solid state, shifted about 60 Å towards the red, and the extinction coefficient markedly decreased.

At wavelengths below 2400 Å it has been fairly well established that peptide bond absorption is responsible for the steep rise in absorption exhibited by all proteins(see Figs. 2-7A, 2-7B), as well as for the high absorption minimum in nucleo-proteins, notwithstanding that the aromatic amino-acids exhibit additional bands in this region with fairly high extinction coefficients (Fig. 2-8) (e.g. for tryptophan, ε is 38,000). The absorption spectra of diglycine, triglycine, and bovine albumin are exhibited in Fig. 2-9. The maximum at about 1870 Å exhibited by the peptides is about that to be expected from the known absorption characteristics of saturated carboxylic acids and amides. Although the maximum for albumin is not shown, it is at least the same wavelength, according to Goldfarb and Saidel (1951). The long wavelength shoulder of this band corresponds therefore to the steep rise in absorption exhibited by all proteins below the minimum at 2500 Å. The reality of this absorption as being due to the peptide linkage is further testified to by the fact that hydrolysis of simple peptides results in its disappearance. This has been taken advantage of by Saidel and Goldfarb (1951) to follow quantitatively the enzymatic hydrolysis of glycyl–glycine, and by Mitz and Schlueter (1958) for the measurement of acylase activity by using as substrate acetyl-L-methionine, the amide linkage of which exhibits a molar extinction coefficient of 24 at 2380 Å. Tombs *et al.* (1959;

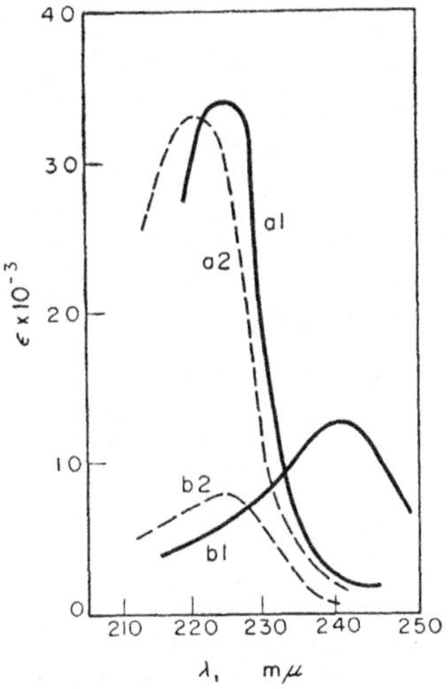

Fig. 2-8. Absorption spectra of tryptophan (a_1 in 0.1 N NaOH, a_2 in 0.1 N HCl) and tyrosine (b_1 in 0.1 N NaOH, b_2 in 0.1 N HCl).

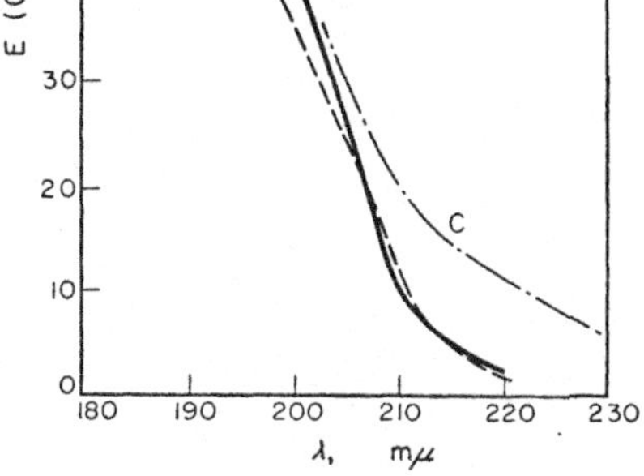

Fig. 2-9. Absorption spectra of diglycine, triglycine and bovine albumin (Goldfarb, Saidel and Mosovitch, 1951; Ham and Platt, 1952).

1961) have described a procedure for analytical estimation of protein in solution, based on peptide bond absorption at 2100 Å.

According to Goldfarb *et al.* (1951) the form of the curve exhibited by bovine albumin (Fig. 2-9) is duplicated by a number of other proteins investigated. Additional work has been confined mainly to the region 2000–2300 Å and it is pointed out that in this region the absorption by aromatic amino-acids is of minor importance by comparison with the peptide linkages. The best evidence for this is the fact that gelatin, which contains only traces of the aromatic amino-acids (Table 2-1), exhibits a spectrum with an absorption coefficient similar to that for other proteins.

Saidel *et al.* (1951) calculate a value for —CONH— of $\varepsilon_{2050} = 2800 \pm 500$, which should be compared with 20,400 for tryptophanyl, 8600 for phenylalanyl and 2200 for cysteinyl R groups in a primary protein structure. At 2050 Å the absorption by peptide bonds is important simply because there are so many of them, and at shorter wavelengths this is even more pronounced. For example, for chymotrypsin Preiss and Setlow (1956) found ε_{1900} to be 900,000. This enzyme contains 220 peptide bonds, 4 tyrosine, 5 phenylalanine, and 7 tryptophan residues per molecule. Taking for ε at 1900 Å the values: —CONH— 3000, phenylalanine 2500, tyrosine 35,000, tryptophan 20,000, the authors calculated a value for the total extinction of 950,000 on the assumption that the individual values were additive. The contribution of the peptide linkages at this wavelength is therefore about 65 per cent of the total. The absorption of the 200 or so aliphatic amino-acid residues was neglected in the above calculation since, e.g. alanine has an ε_{1900} of only 100; effects due to bonding of residues was also ignored (cf. Fig. 2-10).

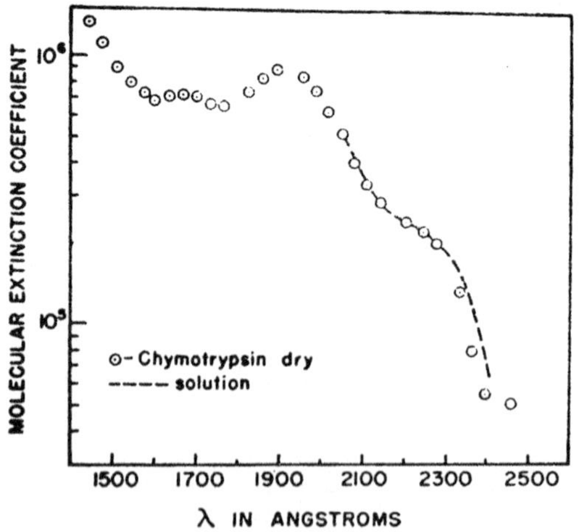

FIG. 2-10. Absorption spectrum of chymotrypsin. The dry film curve is normalized to the solution curve above 2000 Å (Preiss and Setlow, 1956).

Saidel (1955a, 1955b) has considerably extended the analyses of absorption spectra of peptides in the region below 2500 Å. For compounds containing only a single peptide link, the absorption of the latter was found to be affected by such factors as the presence and ionic state of the carboxyl and amino groups on either side of the linkage, as well as by the nature of the amino-acids on either side of the linkage (cf. Magill *et al.*, 1937). There is, however, little or no evidence for hyperconjugation of two or more peptide groups across intervening saturated carbon bonds (Ham and Platt, 1952) so that it may be tentatively concluded that *residue spectra* are approximately additive (cf. Goldfarb, 1953). For full details reference should be made to the original publications (Goldfarb *et al.*, 1951; Saidel *et al.*, 1952; Goldfarb, 1953; Saidel, 1955a, 1955b, 1955c) as well as some more recent ones (Saidel and Lieberman, 1958; Goldfarb *et al.*, 1958).

With poly-L-glutamic acid as a model compound, Imahori and Tanaka (1959) have been able to show that the optical density of peptide absorption increases as the molecule is transformed from a helical to a random coiled configuration. Light-scatter and ionization of carboxyl groups of side-chains were taken into account in reaching this conclusion. A structure containing a large percentage of helix seems to be rather stable until more than one-third of the carboxyl groups of the side-chains are ionized. These results undoubtedly have some bearing on the observations of Goldfarb and Saidel (1951) that the optical density of a protein solution at 2100 Å increases as the protein is denatured.

Apart from the utility of the above studies in connection with polypeptide structure, they underline the potential importance of peptide bond absorption in the photochemistry of proteins and nucleoproteins in the wavelength region 2300–2000 Å where quantum yields are often exceptionally high.

8. *Excited States of Proteins*

Luminescence is any emission of light resulting from absorption; fluorescence is the light emitted as a result of and only during the absorption of radiation from an outside source. In addition to fluorescence, some systems exhibit an "afterglow" or phosphorescence.

Fluorescence–excitation spectra and fluorescence spectra of the three aromatic amino-acids have been established (Teale and Weber, 1957). The quantum yield of the fluorescence of a substance in solution is defined as the ratio of the number of quanta emitted, F, to the number of quanta absorbed, I_{abs}, and is independent of wavelength at > 2000 Å; that is

$$F/I_{abs} = k\Phi$$

where k is a constant depending on the geometry of the system and the angle of observation. From Beer's Law,

$$I_{abs} = I_0(1 - 10^{-D})$$

and by substitution we obtain

$$F = k\Phi I_0(1 - 10^{-D})$$

For an arbitrary wavelength, λ', we have

$$\Phi k = F'/[I_0'(1 - 10^{-D'})]$$

so that

$$F = \frac{F'I_0(1 - 10^{-D})}{I_0'(1 - 10^{-D'})}$$

Teale and Weber then define a quantity Ω as

$$\log \Omega \equiv \log 1\bigg/\left[1 - \frac{FI_0'}{F'I_0}(1 - 10^{-D'})\right] = D$$

where λ' is conveniently chosen as the wavelength of the maximum of the absorption band of least frequency. This equation shows that a plot of log Ω against wavelength should reproduce the absorption spectrum of a pure fluorescent compound. $FI_0'/F'I_0$ is simply the fluorescence at wavelength λ relative to that set up by excitation with light of wavelength λ'—referred to equal photon excitation. With the three aromatic acids a very close correspondence was found between the fluorescence–excitation spectra and the corresponding absorption spectra, when solutions were irradiated with light from a monochromutor at a number of wavelengths (2000–3200 Å).

Fluorescence spectra may be obtained with a single line source such as the mercury 2537 Å line, since these spectra are independent of the exciting wavelength. These spectra continue, in theory, the absorption spectra towards longer wavelengths. Upon excitation, a technical spectrum is obtained. The molecular spectrum is then calculated with corrections for attenuation (emission at wavelengths at which the fluorescence and absorption spectra overlap appreciably appears attenuated) and re-emission (some of the fluorescent light may be reabsorbed to excite further fluorescence at those wavelengths where overlap is negligible).

Maxima and band widths of the fluorescence spectra of the aromatic amino-acids in neutral aqueous solution are given in Table 2-3. As may be seen, the bands are symmetrical about the maximum. In contrast to the absorption and fluorescence–excitation spectra, the emission spectrum of phenylalanine gives no sign of fine structure. This absence of fine structure indicates the increased interaction between excited molecules and permanent water dipoles. About 15 per cent of the emission from tryptophan appears as violet fluorescent light to the unaided eye.

The quantum yields for fluorescence were 0.04 for phenylalanine, 0.2 for tyrosine, and 0.2 for tryptophan. Possible interactions between tyrosine and some of the groups commonly found in proteins have also been investigated (Teale, 1960). The fluorescence is strongly quenched by *ca.* M/10

TABLE 2-3. MAXIMA AND BAND WIDTHS OF THE FLUORESCENCE OF AMINO-ACIDS (TEALE AND WEBER, 1957)

	λ_{max}	$\lambda_{-1/2}$	$\lambda_{+1/2}$
Phenylalanine	2820	2700	2980
Tyrosine	3030	2870	3210
Tryptophan	3480	3230	3830

solutions of glycylglycine and other substances with carboxylate groups. Charged amino and urea amino groups are without effect.

The overall fluorescence quantum yields of a large number of proteins at 2800 Å have also been measured by Teale, a few of which are given in Table 2-4. The low fluorescence of tryptophanyl and tyrosyl residues in proteins is attributable in part to the presence of peptide bonds (Cowgill, 1963).

TABLE 2-4. FLUORESCENCE YIELDS OF AMINO-ACIDS AND PROTEINS, AND PERCENTAGE ABSORPTION BY TRYPTOPHAN AT 2800 Å (TEALE, 1960). WAVELENGTHS OF MAXIMUM EMISSION ARE ALSO SHOWN

Substance	Φ	λ_{max} Å	Phenyl-alanine residues	Tyrosine residues	Tryptophan residues	Per cent absorption by tryptophan
Insulin	0.04	3040	6	8	0	—
Ribonuclease	0.02	3040	3	6	0	—
Trypsin	0.08	3320	4	9	4	65
Chymotrypsin	0.1	3340	6	4	7	90
Pepsin	0.13	3420	13	16	4	51
Phenylalanine	0.04					
Tyrosine	0.2					
Tryptophan	0.2					

None of the proteins tested gave a fluorescence maximum at 2800 Å as does phenylalanine, except gelatin in which phenylalanine is the only aromatic residue. With gelatin some vibrational structure for phenylalanine fluorescence was noted, which probably reflects the lower polarizability of the molecular environment for this residue as compared to that of the amino-acid in water (see above). Proteins containing tyrosine but no tryrophan showed tyrosine fluorescence in low yield or not at all. The fluorescence–excitation spectra of insulin and ribonuclease were similar in band position (λ_{max} 2270 and 2760 Å) to that of pure tyrosine in water (2240 and 2750 Å). On the other hand, "proteins containing tryptophan displayed the fluorescence spectrum characteristic of this aromatic residue to the exclusion of the tyrosine type of spectrum, even where tyrosine was responsible for most of the absorption at the exciting wavelength." Excitation spectra were similar in profile to that of pure tryptophan. Although the absolute quantum yield for fluorescence of tryptophan is independent of wavelength in the range 2537–3000 Å, as found by experiment, the quantum yields increase rapidly with wavelength above 2850 with pepsin and above 2930 Å with chymotrypsin. This information has been interpreted to mean the following.

The relative positions of the absorption and fluorescence spectra of the aromatic residues could permit resonance–energy transfer from phenylalanine and tyrosine to tryptophan, and also between tyrosine groups. Fluorescence–polarization measurements with insulin and ribonuclease had already shown inter-tyrosine energy transfer (G. Weber, 1960). This transfer permits energy migration through those protein molecules in which this residue predominates. The virtual absence of tyrosine (and phenylalanine) fluorescence in those proteins containing tryptophan and showing a tryptophan emission spectrum, does not in itself establish complete energy transfer to the latter, however, for there may be considerable quenching of tyrosine fluorescence (compare ribonuclease and chymotrypsin, Table 2-4). Quenching undoubtedly takes place in ribonuclease wherein half of the tyrosine phenolic groups are "bound," i.e. not free to dissociate, and Φ is only 0.02 compared to insulin with a Φ of 0.04. In insulin the phenolic groups are free. The quenching of tyrosine *per se* by glycine peptides (and formate ion) has already been mentioned. Furthermore, the fluorescence yield of ribonuclease is doubled by denaturation, a process which frees the bound phenolic groups. Denaturation does not cause a large change in molecular shape (McLaren and Peterson, 1961). Above pH 8.5 lysozyme shows a reduction in fluorescence intensity characteristic of tryptophan and indicative of an overlap of the absorption spectrum of ionized tyrosyl with the emission spectrum of tryptophanyl (Steiner and Edelhoch, 1961).

On the other hand, the nearly constant absolute quantum-yield for fluorescence between 2700 and 2850 Å and the increase in yield between 2850 and 2950 Å roughly parallels a plot of the fractional absorption of light by tryptophan in a mixture of tyrosine and tryptophan adjusted in concentration so that half the light is absorbed by each at 2800 Å. This suggests that efficient energy transfer from tyrosine to tryptophan does not take place, for if it did, uniform quantum-yields should have been found in the range 2700–3000 Å, Fig. 2-11. Considerable quenching also occurs with proteins. The similarity between the fluorescence–excitation spectra of the proteins and tryptophan itself indicates that very little contribution is made to fluorescence by cystine, histidine or —CONH—. These results are very important to an understanding of the photochemistry of proteins, for from them one can infer that the site of photochemical reaction in proteins lacking heme chromophores is most probably at the site of absorption of a quantum.

In addition to fluorescence of proteins, a short-lived phenomenon, a few people have studied phosphorescence (Debye and Edwards, 1952; Steel

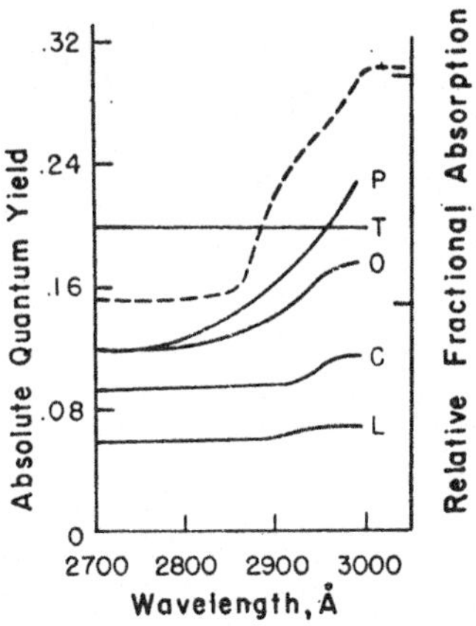

FIG. 2-11. Variation of the fluorescence yield with wavelength of the exciting light in tryptophan (T) and proteins containing tryptophan; lysozyme (L), chymotrypsin (C), ovalbumin (O), and pepsin (P). The fractional absorption by (T) in a mixture having equal absorption by both (T) and tyrosine at 2800 Å is shown by the broken line (after Teale, 1960).

and Szent-Györgyi, 1958; and Vladimirov and Litvir, 1960). Phosphorescence is determined by the aromatic amino-acids. Vladimirov has illuminated frozen protein solutions at −150 to −170°C with monochromatic light from a hydrogen lamp for 5 sec. A shutter was closed for one second and quanta emitted from the previously illuminated specimen were then counted for 5 sec with a photomultiplier arrangement. Solutions of amino-acids and proteins do not display phosphorescence at room temperature. With this procedure the short afterglow of Debye and Edwards (*ca.* 0.1 sec) is barely detectable.

The afterglow spectra of tryptophan, tyrosine and egg albumin are depicted in Fig. 2-12. It may be seen that the spectra of the albumin is apparently due to tryptophan and that the main maximum of the latter in the protein (4360–4380 Å) is shifted relative to the corresponding peak (4300 Å) of amino-acid in solution. The explanation offered by Vladimirov is that the energy absorbed by tyrosine (and phenylalanine) undergoes migration to tryptophan, since zein, which lacks tryptophan, shows a very strong afterglow at low temperature similar in spectra to that of tyrosine. This conclusion is in contradiction to that of Teale, just discussed, but Vladimirov failed to consider seriously the probable quenching of tyrosine phosphorescence by a phenol–carboxylate or some similar interaction in ovalbumin. Denatured albumin, which lacks this interaction, was not studied. In these afterglow spectra of proteins there was no indication of

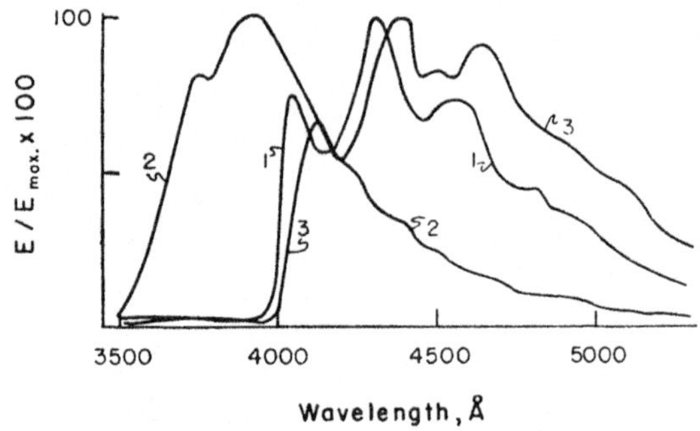

FIG. 2-12. After-glow spectra. 1, tryptophan; 2, tyrosine; 3, egg albumin. E is the intensity of after-glow. Vladimirov and Litvin, 1960.

migration of energy to tryptophan from groups absorbing in the 2000–2500 Å region (SH, SS, CONH).

Fluorescence depolarization studies indicate that some tyrosine–tyrosine and tryptophan–tryptophan energy transfer can take place (Stryer, 1960).

To date a few spectra have been obtained of the excited states of proteins and amino-acids in aqueous medium at room temperature. Spectroscopic techniques have been applied to amino-acids and one protein, ovalbumin, using flash photolysis (see Appendix for experimental details). Tyrosine, glycyltyrosine, and phenol give transient absorption spectra lasting less than 50 μsec, Fig. 2-13.

A characteristic band, in both neutral and alkaline media, at 4120 Å, is ascribed to the phenol radical–cation, produced by photo-ejection of an electron. The primary step in neutral solution is believed to be as follows:

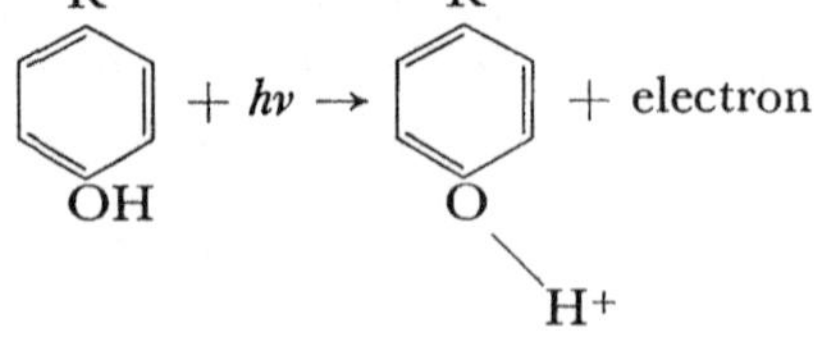

In alkaline medium it may be:

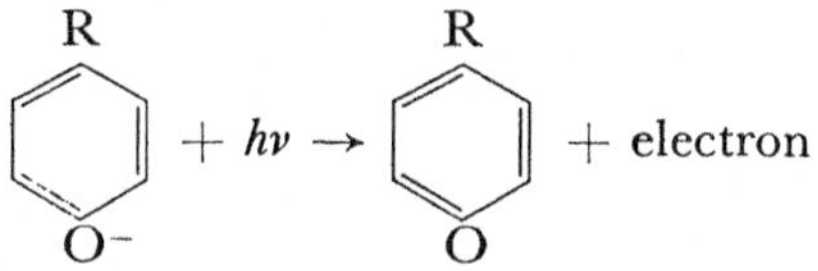

Irradiation of tryptophan likewise gives rise to a transient intermediate with a rather broad band at 5300 Å and another in the near ultraviolet, both due to the excited indole ring, but the structural assignment of which is more difficult to establish. The absorption spectrum of the excited state of ovalbumin includes contributions from both tyrosine and tryptophan but little, if any, from phenylalanine, cystine, histidine, and peptide linkages (Grossweiner, 1956); Grossweiner and Mulac, 1959); this should not be

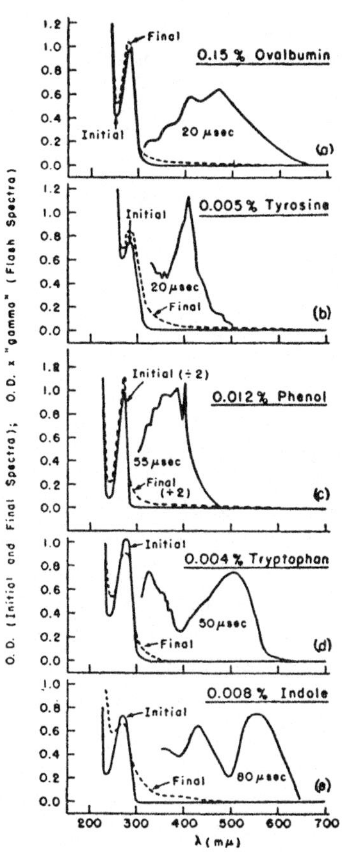

FIG. 2-13. Absorption spectra of metastable states in ovalbumin and simple related compounds (Grossweiner, 1956).

taken to exclude involvement of the latter in the photochemical reactions undergone by proteins, but only that their excited states have not been placed in evidence by flash photolysis techniques.

It has been suggested by the above authors that, in the presence of oxygen, the ejected photoelectrons may reduce water with the formation of HO_2 radicals, the reaction of which with the protein might explain the well-known consumption of oxygen by irradiated ovalbumin (Harris, 1926), possibly via oxidation of tryptophan residues (Deschreider and Renard, 1955).

B. NUCLEIC ACIDS

The selective absorption exhibited by nucleic acids in the neighborhood of 2600 Å, and due primarily to their component purine and pyrimidine bases, embraces the spectral region where irradiation results in major biological effects; most photobiological investigations are conducted near this peak because of the ready availability of 2537 Å mercury resonance lamps. The precise location of the absorption maximum of a given sample, as well as the magnitude of the extinction coefficient, are on the other hand dependent on a number of factors such as the degree of polymerization, secondary bonds in individual polynucleotide chains, and linkages between adjacent chains. These, in turn, are profoundly dependent on the state of the medium in which the nucleic acids sample exists as well as on the procedure used for its isolation. It is therefore not surprising that, although the spectral properties of nucleic acids have received a great deal of attention, it is only during the past two or three years that some order is being brought out of the existing chaos, largely as a result of studies on model oligo- and poly-nucleotides synthesized by enzymatic and chemical procedures, as well as on smaller oligonucleotides prepared by the chemical or enzymatic degradation of natural and synthetic polynucleotides.†

Ultraviolet spectroscopy has, and continues to play, a major role in the investigation of the properties and structure of nucleic acids and their derivatives. Furthermore, the ease and accuracy with which individual derivatives may be characterized and measured by spectral techniques has resulted in these latter becoming the most useful tool in the analytical chemistry of nucleic acids. An enormous literature has therefore been accumulating in this field and we can deal here only with some of the salient points relating as much as is possible to data required in studies on the photochemistry of these important compounds.

† With the advent of enzymatically and chemically prepared oligo- and poly-nucleotides, it becomes necessary perhaps to emphasize that the term "nucleic acids," as used in this text, applies only to those polymers extracted from natural sources.

1. *Purines and Pyrimidines*

Probably the most characteristic feature of purines and pyrimidines, including those derived from nucleic acids, is that most of them contain ionizable chromophores so that pH control is of primary importance in any studies of the optical properties and structure of these compounds (Marshall and Walker, 1951; Shugar and Fox, 1952; Beaven *et al.*, 1955) as well as their photochemical behavior (Shugar and Wierzchowski, 1958a). A typical illustration is the spectrum of cytosine, Fig. 2-14, which exhibits

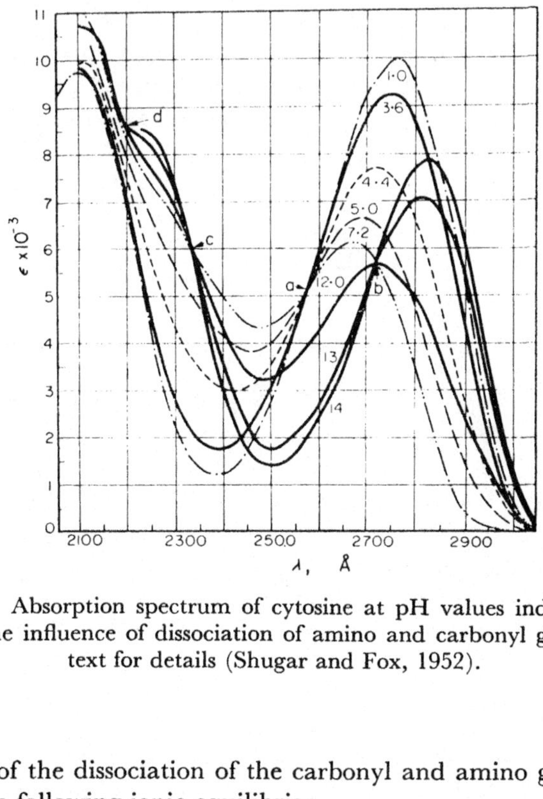

FIG. 2-14. Absorption spectrum of cytosine at pH values indicated and showing the influence of dissociation of amino and carbonyl groups; see text for details (Shugar and Fox, 1952).

the influence of the dissociation of the carbonyl and amino groups, corresponding to the following ionic equilibria:

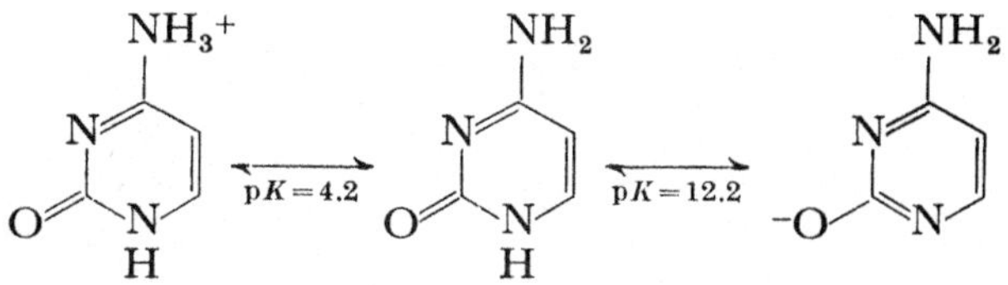

The first is represented by the curves with isosbestic point† at *a* and the second with isosbestic points at *b*, *c*, and *d*. For orotic acid there are three equilibria which may be readily distinguished spectroscopically, with p*K* values of 2.8, 9.45 and > 13, corresponding to the carboxyl, 2- and 4- carbonyl groups respectively (Shugar and Fox, 1952; Fox *et al.*, 1957).

The existence of these equilibria in intact nucleic acids has been demonstrated by a number of observers (e.g. Shack and Thompsett, 1952; Laland *et al.*, 1954; Cavalieri and Stone, 1955; Lawley, 1956; see Beaven *et al.*, 1955, for details) and has proved extremely useful in the interpretation of electrometric titration curves as well as in studies on the structure of nucleic acids. One can therefore readily appreciate the importance of pH in the photochemistry of nucleic acids and their derivatives.

Examination of a variety of purine, pyrimidine and triazolopyrimidine derivatives by Cavalieri *et al.* (1948) and Cavalieri and Bendich (1950) led to the suggestion that the major chromophore in purines and pyrimidines is

$$-\overset{|}{C}=\overset{|}{C}-\overset{|}{C}=\overset{|}{N}-$$

although occasionally it may also be

$$-\overset{|}{C}=\overset{|}{C}-\overset{|}{C}=O$$

A number of subsequent workers have accepted this without reservations although the original authors themselves suggested that only a working hypothesis has been established and that the actual chromophore in any particular case is the resultant of one of the above with the rest of the molecule. To this must be added also the effect of dissociation of functional groups, as illustrated by the ionic equilibrium for cytosine presented above.

Of considerably greater significance in the case of pyrimidines is the 5,6 double bond in the absence of which, as in dihydropyrimidines, the characteristic absorption maximum disappears (Batt *et al.*, 1954). For dihydrouracil and its analogues (e.g. thymine and orotic acid) the resultant spectra in acid and neutral medium exhibit only end-absorption in the ultraviolet; in alkali a new absorption maximum appears at 2300 Å, but this rapidly disappears due to alkaline lability of dihydroderivatives accompanied by opening of the ring at the 3,4 bond; the true alkaline absorption spectrum

† Although protonation of the cytosine ring is shown here to take place on the exocyclic amino group, more recent evidence places the position of protonation on the ring N_3 nitrogen (Dekker, 1960; Katritzky and Waring, 1962; Fikus *et al.*, 1962). Attention is drawn to this point at the present time because of the importance of the position of protonation in 4-aminopyrimidines on the interpretation of their photochemical behavior (Wierzchowski and Shugar, 1963).

must therefore be obtained by extrapolation of the observed spectrum to zero time following alkalization of the solution (Batt *et al.*, 1954).

Figure 2-15(a) exhibits the absorption spectrum of uracil at neutral pH and of dihydrouracil at neutral and alkaline pH. The lack of specific

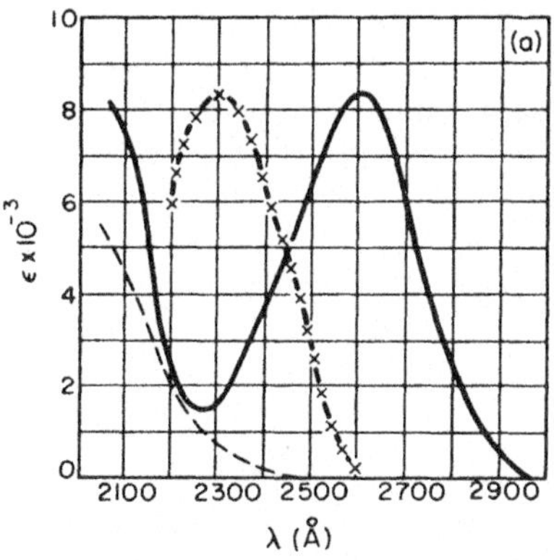

FIG. 2-15. (a) Absorption spectra of uracil at neutral pH (———) and of dihydrouracil at neutral (-------) and alkaline (-×-×-) pH (Batt *et al.*, 1954 and Janion and Shugar, 1960).

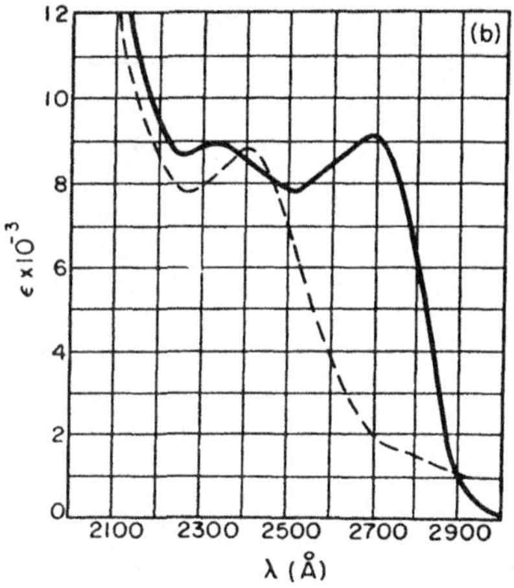

FIG. 2-15. (b) Absorption spectra of cytidine 2′:3′-phosphate (———) and of dihydrocytidine 2′:3′-phosphate (-------) at neutral pH. Dotted curve also represents dihydro-derivatives of cytosine and other cytosine glycosides; but extinction coefficient should not be regarded as accurate because of instability of dihydrocytosine derivatives (Janion and Shugar, 1960).

absorption of dihydrouracil at neutral pH is explicable if it is assumed to exist in the keto form, while the absorption maximum in alkali is due to dissociation, as follows (Janion and Shugar, 1960):

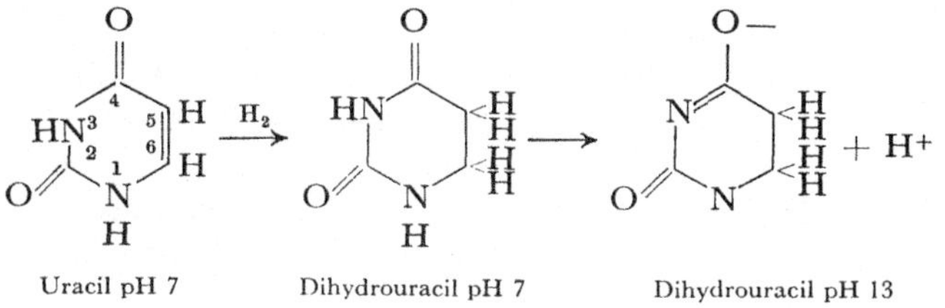

from which it will be seen that dihydrouracil at neutral pH possesses no conjugated double bonds, while at alkaline pH the 2-carbonyl is conjugated with the 3,4 double bond. The position of enolization at alkaline pH was established by a study of *N*-methylated derivatives.

A somewhat different situation prevails for cytosine analogues, as may be seen from Fig. 2-15(b) for cytidine-2′:3′-phosphate and its dihydroderivative. In this case the disappearance of the principal absorption maximum at about 2700 Å, upon saturation of the 5,6 double bond, is accompanied by the appearance of a new maximum at about 2400 Å. The spectrum of dihydrocytosine is similar to that for dihydrocytidine-2′:3′-phosphate, at neutral pH; and is consequently due to the fact that the dihydrocytosine derivatives still possess two conjugated double bonds, as follows,

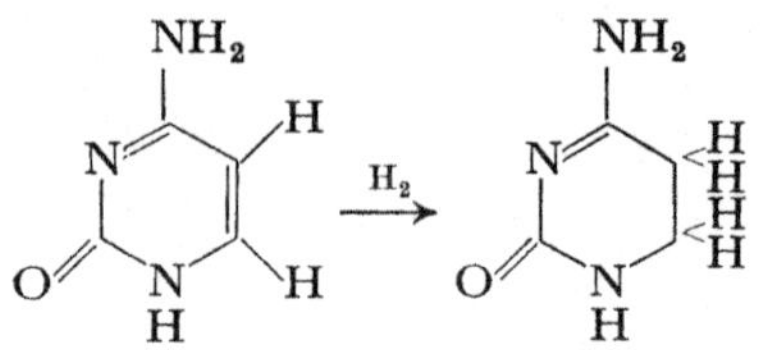

and hence somewhat analogous to dihydrouracil in alkaline medium (see above).

Cytidine-2′:3′-phosphate, instead of cytosine, was selected for purposes of illustration in Fig. 2-15(b) because it exhibits a second, minor, absorption maximum at 2340 Å (Shugar and Wierzchowski, 1959b) which may bear some relationship to the 2400 Å maximum of the dihydroderivative. Cytosine itself and 1-methylcytosine do not exhibit this second maximum; ribosides and deoxyribosides exhibit only a point of inflexion at about 2360 Å; but pyranosylcytosines (Shugar and Fox, 1952; Fox and Shugar, 1952) and isopropylidenecytidine do exhibit this maximum and it has been proposed that its appearance is related to some type of interaction between the cytosine and carbohydrate rings.

The significance of the absorption spectra of the dihydropyrimidines in the interpretation of the photochemical behavior of pyrimidine derivatives will be discussed in Sect. VI-6,10. It should, however, be mentioned that dihydrocytosine derivatives are relatively unstable in aqueous solution and that it is practically impossible to isolate them in undegraded form by any known procedure (Green and Cohen, 1958; Janion and Shugar, 1960).

Somewhat unusual is the fact that 1-methyl-dihydrouracil, in acid or neutral medium, exhibits an absorption maximum at 2140 Å with an extinction coefficient of about 5000. In 1,3-dimethyl-dihydrouracil this maximum also appears, but shifted to 2220 Å, notwithstanding that in the latter compound no possibility exists for keto-enol dissociation or for the existence of conjugated double bonds. A similar situation prevails for barbital, barbituric acid, and their *N*-alkylated derivatives in acid medium where only keto forms are present and conjugation of double bonds excluded; the most striking example amongst these is 1,3-dimethyl barbital,

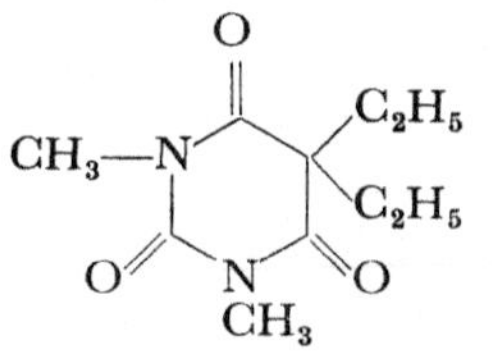

which possesses a maximum at 2280 Å with an extinction of 6.3×10^3. It appears difficult to avoid the conclusion that these maxima are due to carbonyl absorption, despite the fact that $>$C=O bands with extinction coefficients up to about 1000 are normally encountered at, or to the violet of, 1900 Å (Braude, 1955).

The optical properties of purines and pyrimidines as well as their nucleosides and nucleotides, particularly those derived from nucleic acids, are extensively reviewed by Beaven *et al.* (1955), who present also many absorption spectra, including some not previously published elsewhere, and including extensive literature references. The structural significance of the the spectra of these compounds has been critically reviewed by Jordan (1951, 1952, 1955) in a series of articles which are undoubtedly the most informative in this field. The electronic structure of the bases derived from nucleic acids, calculated by means of the molecular orbital method, have been tabulated (Pullman *et al.*, 1956, 1958; Pullman and Pullman, 1958, 1959) in connection with a search for some relationship between electron structure and chemotherapeutic activity.

The nature of the electronic transitions responsible for the absorption bands of purines and pyrimidines in the quartz ultraviolet has been investigated by several observers (Mason, 1959; Rich and Kasha, 1960; Krishna and Goodman, 1961; these also give references to earlier literature). In general the purine bands are due to two independent and mutually perpedicularly polarized $\pi \rightarrow \pi^*$ (excited electron) transitions involving the conjugation electrons of both the pyrimidine and imidazole rings. In addition, for both purines and pyrimidines there is an $n \rightarrow \pi^*$ transition region immediately to the red of the principal $\pi \rightarrow \pi^*$ transition band. These $n \rightarrow \pi^*$ transitions (due to the promotion of a non-bonding, or n-orbital, lone-pair electron on an N or O hetero-atom to an empty π-antibonding molecular orbital) are of relatively low intensity as compared to the $\pi \rightarrow \pi^*$ transitions. Furthermore, the $n \rightarrow \pi^*$ transition moments are polarized out-of-plane as compared to those for the $\pi \rightarrow \pi^*$ transition moments, which are polarized in the planes of the rings. Consequently, in a helical polynucleotide, the $\pi \rightarrow \pi^*$ transitions would be expected to be perpendicular, and the $n \rightarrow \pi^*$

transitions parallel, to the helix axis. This has, in fact, been verified in an elegant manner by Rich and Kasha (1960) by measurements of the ultraviolet dichroic absorption curves of oriented films of DNA and various synthetic polynucleotides.

Practically no concrete data have yet been reported on the excited states of nucleic acid derivatives, although such studies are now under way in several laboratories. In view of the extensive knowledge we now possess regarding the photochemistry of these compounds, as well as their polymers, studies of their photochemically excited states offers an extremely fertile field of research, which can be expected to develop rapidly in the near future, and which should also assist in the interpretation of the radiation chemistry of the nucleic acids.

2. *Nucleosides and Nucleotides*

The naturally occurring nucleosides are *N*-D-ribosides and *N*-D-2-deoxyribosides, the point of attachment of the sugar to the purine or pyrimidine rings being in the 9-position for the former and the 1-position for the latter. Of historical interest is the fact that for the purine nucleosides this was established by ultraviolet absorption spectroscopy from a comparison of the spectrum of the nucleoside with those of the corresponding *N*-alkylated derivatives.

It should perhaps be noted that some naturally occurring nucleosides have been isolated in which the carbohydrate component is neither ribose nor deoxyribose (Neufeld *et al.*, 1957; Bergmann and Burke, 1955), while numerous nucleosides containing other sugars have been prepared by enzymatic and chemical pathways (for review see Fox and Lampen, 1959).

The saturated carbon chains of carbohydrates would not be expected to exhibit specific absorption in the quartz ultraviolet and it was, in fact, demonstrated many years ago that carefully purified sugars possess only extremely low end-absorption below 2200 Å at neutral pH (Kwiecinski *et al.*, 1928; Marchlewski and Urbanczyk, 1933). In alkaline medium at room temperature, however, most sugars develop end-absorption beginning at about 2400 Å and, with time, a well-defined absorption maximum appears at a wavelength of about 2800 Å, the intensity of which increases with time (as well as with increasing pH of the medium) and is accompanied by the production of a yellowish or yellowish-brown color. Spectral studies suggest that the chromogen responsible for this absorption is due to the condensation of compounds produced by the fragmentation of sugars in alkaline solution (Berl and Feazel, 1954).

Exposure of nucleosides (or nucleotides) to alkali is without effect on the carbohydrate components. Hence for all practical purposes the absorption of carbohydrate moieties of nucleic acids or their derivatives may be considered as negligible.

To a first approximation the aborption spectrum of a nucleoside resembles that of the corresponding base methylated in the same position, and it was this principle which Gulland *et al.* (1936) profited from to fix the position of the sugar in purine nucleosides (see above). A closer examination shows that in the case of the pyrimidine nucleosides the nature of the carbohydrate component does influence the spectrum to a measurable degree and makes it possible to differentiate between ribofuranoside and deoxyribofuranoside nucleosides, as well as between glycopyranosides and glycofuranosides (Fox and Shugar, 1952b). In addition, the spectra of pyrimidine nucleosides undergo small shifts in alkaline medium, which have been shown to be due to dissociation of sugar hydroxyl groups. These findings demonstrate quite clearly the existence of some type of interaction between the aromatic and carbohydrate rings manifested, not through the glycosidic linkage but, by some perturbation of the π-electron system of the pyrimidine ring by the carbohydrate hydroxyls. Additional evidence for some kind of interaction is provided in the case of the cytosine nucleosides by the dependence of the pK for dissociation of the amino group on the nature of the carbohydrate moiety.

It will be shown subsequently that the photochemical behavior of pyrimidine nucleosides (and nucleotides) is intimately linked with the nature of the carbohydrate component and is even capable of providing some clues as to the type of interaction involved between this latter and the pyrimidine ring to which it is linked.

Phosphorylation of the sugar hydroxyls of a nucleoside would not be expected to have a marked influence on the absorption spectrum of the resulting nucleotides (as compared to the nucleoside) since the phosphate group is fairly well insulated from the purine or pyrimidine ring by the saturated carbon chain of the carbohydrate moiety; and most nucleotides do in fact resemble the corresponding nucleosides, particularly as concerns the purine derivatives. However, the modifications in absorption spectra of pyrimidine nucleosides produced by different sugar components suggested that the various nucleotide isomers would exhibit measurable spectral differences (Fox and Shugar, 1952b) and this has indeed been found to be the case (Harris *et al.*, 1953; Fox *et al.*, 1953; Cohn, 1955; see also Beaven *et al.*, 1955), to the extent that the different isomers may be identified on the basis of their absorption spectra. This finding has been extensively exploited for analytical purposes (Cohn, 1955; Volkin and Cohn, 1954).

In at least one instance, that of cytidine-2′:3′-phosphate, the resulting spectrum differs from that of the other isomers by the presence of a new maximum at 2360 Å at neutral pH. The resemblance of this spectrum (see Fig. 2-15(b)) to that of pyranosylcytosines (Fox and Shugar, 1952b) is probably more than fortuitous and suggestive of some type of interaction between a phosphoric acid hydroxyl and the pyrimidine ring through hydrogen bonding or some charge effect (cf. Cavalieri, 1953), or as a result of the blocking of one or more of the carbohydrate hydroxyls. The spectral differences between uridine-2′:3′-phosphate and uridine-3′-phosphate are significant enough to make possible the use of the former as a substrate for the measurement of ribonuclease activity by difference spectrophotometry (Richards, 1955; Shugar, 1955). We have found cytidine-2′:3′-phosphate equally suitable for this purpose (cf. Stockx, *et al.*, 1960).

3. *Nucleic Acids, Polynucleotides and Oligonucleotides*

The characterization of nucleic acids by molecular weight is not a simple task, while various preparations are not homogeneous as regards degree of polymerization and base composition. Furthermore the water of hydration of different preparations may vary considerably and cannot always be readily removed without damage to the preparation (Peacocke, 1954). Hence, although absorption spectra were in the past based on optical densities per unit weight, a more suitable method of presentation is that of Chargaff and Zamenhof (1948) which makes use of the fact that polynucleotide chains contain one phosphorus atom for each nitrogenous base. They therefore define the extinction of a given preparation as

$$\varepsilon(\mathrm{P}) = \frac{D}{C}$$

where D is the optical density for a 1 cm light path and C is in g atoms P/1. solution. This procedure is now widely followed, but it must be recognized that ε(P) may not be the same even for samples obtained by similar isolation procedures and its principal advatange is that it makes possible a better comparison of the results of different observers. Published values of ε(P) vary from about 6000 to 9000 for DNA and from 7000 to slightly over 10,000 for RNA, the precise value in each case being a fairly good index of the degree of polymerization of the specimen which, in turn, is dependent not only on the method of isolation employed, but also on the nature of the medium in which it is dissolved.

It has long been recognized that the absorption of a nucleic acid preparation is lower than the additive absorption of its constituent nucleotides. For instance the enzymatic hydrolysis of DNA by DNAase is accompanied by a 30–35 per cent increase in ε(P) at 2600 Å (Kunitz, 1950; Tsuboi, 1950; Laland *et al.*, 1954), a fact which serves as the basis of one method for the measurement of DNAase activity (Kunitz, 1950). It was also noted by Kunitz (1946) that enzymatic hydrolysis of RNA resulted in a small shift in absorption, which he applied to the measurement of RNAase activity. However, the RNA preparation he employed must have been highly degraded since Tsuboi (1950) subsequently demonstrated that a highly polymerized specimen exhibited an increase in absorption of 20 per cent during digestion with RNAase.

A nucleic acid spectrum is therefore *hypochromic* with respect to a mixture of its constituent mononucleotides. If the nucleic acid is subjected to a degradative treatment which produces an increase in absorption, it is said to exhibit *hyperchromicity*. Maximum hyperchromicity will, of course, result from complete degradation to mononucleotides. If a given sample of nucleic acid exhibits 60 per cent of the absorption of its constituent mononucleotides, at a given wavelength, then its hypochromicity is 40 per cent; but its maximum hyperchromicity is 40/60 or 67 per cent.

The precise way in which the degree of polymerization affects the value of $\varepsilon(P)$ was for some time not too clear from studies on nucleic acids alone. For a variety of DNA samples isolated from different sources, digestion with DNAase was reported to increase $\varepsilon(P)$ only after the viscosity had dropped almost to zero (Laland *et al.*, 1954), from which it was concluded that treatment which results in a decrease in intermolecular forces does not affect the absorption spectrum, whereas scission of hydrogen bonds does. This derives some support from the studies of Butler *et al.* (1952) and Pouyet (1952), both of whom demonstrated that at least part of the abnormally high viscosity of DNA may be abolished without change in molecular weight and is therefore due to molecular interaction.

Degradation of RNA to mononucleotides by alkali produces an increase in absorption, or hyperchromicity, of 20–40 per cent (Tsuboi, 1950; Magasanik and Chargaff, 1951). The latter authors associated a major portion of this hyperchromic effect with the non-dialyzable core resulting from the RNAase hydrolysis of RNA, and which is particularly rich in guanylic acid. A similar suggestion was advanced for the RNA from TMV (Reddi, 1958). On the other hand Mihalyi *et al.* (1957) presented evidence for involvement of cytidylic acid residues.

Figure 2-16 shows the effect of acid, alkaline, and enzymatic hydrolysis on the absorption spectrum of DNA. Most studies on hyperchromicity in nucleic acids have been conducted on DNA since it is more readily obtainable in highly polymerized form. On the other hand the total hyperchromicity of an RNA sample is easier to measure since it may be quantitatively hydrolyzed to mononucleotides simply by treatment with acid or alkali. There is one example of a DNA sample the absorption of which has been measured prior to, and following, exhaustive digestion with DNAase and phosphodiesterase, the hyperchromicity thus found being 80 per cent (Sinsheimer and Koerner, 1952); calculation shows that for a highly polymerized specimen with an $\varepsilon(P)$ of 6000 the hyperchromicity should be 70 per cent (Lawley, 1954).

Since the total hyperchromicity of highly polymerized DNA (70–80 per cent) is about twice that prevailing for highly polymerized RNA (up to 40 per cent), it would appear reasonable to assume, from the appreciable variations in absorption spectra of the individual bases accompanying keto–enol tautomerism on dissociation, that one-half of the hyperchromicity of DNA is due to hydrogen bonding of the keto and amino groups of adenine–thymine and guanine–cytosine base pairs in accordance with the Watson–Crick two-stranded structure (Crick and Watson, 1954; Pauling and Corey,

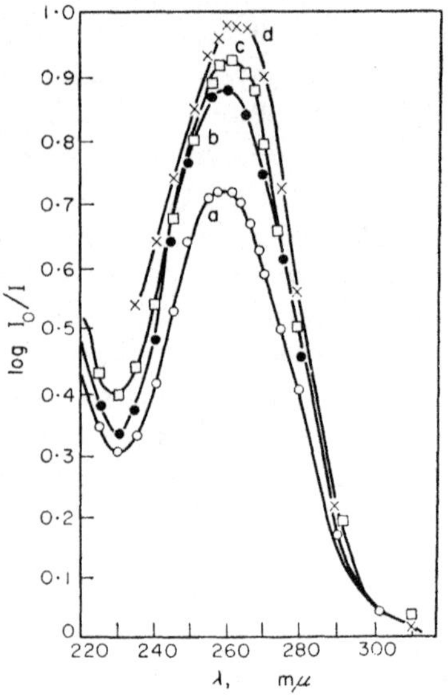

FIG. 2-16. Ultraviolet absorption at pH 7 of desoxyribonucleic acid, 34.4 γ/ml.
(a) non-hydrolyzed
(b) alkaline hydrolyzed
(c) enzymatically hydrolyzed
(d) acid hydrolyzed
(from Tsuboi, 1950).

1956). In support of this is the fact that if a DNA molecule is subjected to ultrasonic irradiation, which results in double chain scission without separation of the two strands, the ultraviolet absorption is unchanged (Doty *et al.*, 1958); whereas subjection to heat, extremes of pH, salts, etc., which cause uncoiling of the strands, result in a *partial* abolition of hyperchromicity (Thomas, 1954; Laland *et al.*, 1954; Lawley, 1956; Doty, 1957; Doty *et al.*, 1959). For an outline of the earlier experimental data reference should be made to the review article of Beaven *et al.* (1955).

Since the publication of the aforementioned review, additional information has been forthcoming at a phenomenal rate from investigations on (a) small oligonucleotides prepared by chemical synthesis and by isolation from nucleic acids degraded under defined conditions, and (b) model oligo- and polynucleotides synthesized by chemical (Michelson, 1958, 1959, 1960) and enzymatic (Grunberg-Manago *et al.*, 1955) procedures. The advantage of the polynucleotides is that they can be prepared as copolymers containing a mixture of the bases normally encountered in nucleic acids or as homopolymers containing a single type of nucleotide. In the latter case a considerable simplification in interpretation might be expected and such has been amply substantiated. In addition the enzymatically synthesized polymers may be degraded by enzymatic methods to yield di-, tri-, and tetranucleotides which may be easily separated and identified by chromatographic and spectral procedures (Heppel *et al.*, 1957a, 1957b).

Figure 2-17 exhibits the magnitude of the hyperchromic effect encountered in enzymatically synthesized homopolymers of high molecular weight containing only uridylic (U) or cytidylic (C) acids. It will be seen that, whereas for poly-U the hyperchromicity is only about 6 per cent, it is of the order of 40 per cent for poly-C (Warner, 1957). For poly-I the value is 60 per cent and for poly-A it is 55 per cent (Felsenfeld and Rich, 1957), all at neutral pH. However, Shapiro and Chargaff (1957) have shown that thymidine diphosphate exhibits an extinction coefficient 8 per cent lower than that for thymidylic acid; if a similar situation prevails for uridine-2′(3′),5′-diphosphate, this alone would account for the 6 per cent hyperchromicity of poly-U. Such an explanation is not applicable to other polymers since, e.g. deoxycytidine diphosphate is also hyperchromic to the extent of 8 per cent with respect to deoxycytidine-5′-phosphate, whereas the hyperchromicity of poly-C is 40 per cent. The increase in absorption, or decrease in hyperchromicity, of such single-stranded polymers under the influence of urea or heat has been regarded as evidence for involvement of intramolecular hydrogen bonding between bases (Warner, 1958) which, in the case of poly-A, may be between amino groups and phosphate oxygens (Beers and Steiner, 1958). Before discussing this in greater detail, we shall first deal with the behavior of small oligonucleotides.

Quantitative observations on small oligonucleotides show that hydrogen bonding is not the main source of hyperchromicity. Contrary to some statements in the literature (Sinsheimer and Koerner, 1952; Smith and Allen, 1953; Beaven *et al.*, 1955; Magasanik, 1955) there are numerous scattered observations on absorption spectra of di- and tri-nucleotides (e.g. Elmore and Todd, 1952; Merrifield and Woolley, 1952; Sinsheimer, 1954) which are explicable only on the basis of the existence of hyperchromicity, although their structural significance has not always been appreciated. Shapiro and Chargaff (1956) present spectral data, as well as absorption spectra, demonstrating hyperchromicity in several dinucleotides; while de Garilhe and Laskowski (1956) measured the hyperchromicity of various combinations of dinucleotides and several oligonucleotides, the data being obtained by

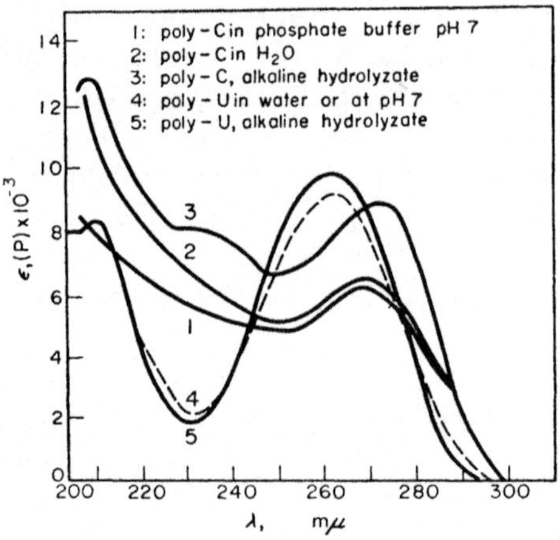

FIG. 2-17. Absorption spectra of polyuridylic acid (poly-U), polycytidylic acid (poly-C) and their alkaline hydrolyzates (from Warner, 1957).

phosphodiesterase digestion to mononucleotides, a technique deserving of wider application. The most organized study of this effect to date is that of Michelson (1958a, 1959a, 1959b, 1959c) on a series of chemically synthesized ribo-oligonucleotides of varying lengths and sequences. Although these compounds differ from the natural ribopolynucleotides in that the internucleotide linkages are a mixture of 2′,5′ and 3′,5′ (instead of only

TABLE 2-5. HYPERCHROMICITY OF SOME ISOMERIC OLIGONUCLEOTIDES, ILLUSTRATING THE INFLUENCE OF INTERNUCLEOTIDE LINKAGES AND OF VARIOUS BASE SEQUENCES (FROM MICHELSON, 1959A)

Compound*	Per cent alkaline hyperchromicity
U2′p5′A	5.2
U3′p5′A	3.2
A2′p5′U	11.1
A3′p5′U	4.1
CpUpGpA*	7.7
CpGpApU*	12.9
CpUpGpA*	6.2
GpCpApU*	12.3
GpCpUpA*	6.6
ApUpGpC*	8.6

* These compounds include mixtures of 2′, 5′, and 3′, 5′ internucleotide linkages. See footnote on p. 197.

3′,5′), this not only does not detract from the value of the results but even enhances their significance since it renders possible a comparison of the effect of the non-natural 2′,5′ internucleotide linkage. This latter fact is admirably illustrated in the upper half of Table 2-5 for the four possible isomers of a dinucleoside monophosphate. The lower half of the same table demonstrates the effect of variations in base sequences. In Table 2-6 we see the effect of chain length on the hyperchromicity of several homopolymers; included for comparison purposes are the values for enzymatically synthesized poly-A and poly-C, the chain lengths of which are very high. It

TABLE 2-6. INFLUENCE OF CHAIN LENGTH ON HYPERCHROMICITY OF OLIGO- AND POLYNUCLEOTIDES*

Polymer	Per cent hyperchromicity in		
	0.1 N HCl	H_2O (pH ~ 7)	0.2 N NaOH
Diadenylic	0	—	15.1
Triadenylic	6.8	—	22.8
Tetra-adenylic	13.2	—	30.7
Penta-adenylic	16.4	—	32.9
Poly-A (13.6)	22.9	~36	36.7
Poly-A†‡	—	55	—
Poly-G (3.2)	9.0	10.3	0
Poly-G (5.4)	14.6	16.5	0
Poly-G (12.5)	18.0	23.0	0
Dicytidylic	—	—	8.7
Tricytidylic	—	—	13.9
Tetracytidylic	—	—	15.3
Poly-C (9.8)	—	~15	15.9
Poly-C†§	—	40	—

* Data from Michelson (1959) except where otherwise indicated.
† Enzymatically synthesized polymers of high molecular weight.
‡ Felsenfeld and Rich, (1957).
§ Warner, (1957).

will be seen that hyperchromicity in small oligonucleotides reaches a limiting value at chain lengths of the order of 10–15; but that these values for poly-A and poly-C are appreciably lower than those for the enzymatically synthesized polymers which are known to possess chain lengths of up to several hundred residues. This "additional" hyperchromicity in high polymers is undoubtedly due to some change in configuration, and we shall revert to this question below. Some of the conclusions emerging from the studies on small oligonucleotides are as follows: (a) the hyperchromic effect is dependent on nucleotide composition as well as on base sequences; (b) the variation of hyperchromicity with pH excludes hydrogen bonding as the principal source of hyperchromicity in *small* oligonucleotides (Michelson, 1959b); (c) the fact that there are appreciable shifts in the p*K*'s of ionizable groups in small oligonucleotides, as well as the influence of these latter on the metachromasy of basic dyes, leads to the conclusion that even small oligonucleotides (not excluding di-nucleotides) possess a definite configuration and that changes in spectral properties may be due to the formation of new chromophores arising from interaction of the π-electron systems of adjacent rings, the resultant electron orbitals extending over two or more adjacent aromatic rings (cf. Laland *et al.*, 1954; Lawley, 1956) with a lower total energy and, from a geometrical point of view, a smaller chromophoric area.

Our discussion of hyperchromicity in small oligonucleotides would be incomplete without at least a brief reference to the fact that this effect exists not only in oligonucleotides but also in some polypeptides (see Sect. A.6) and nucleotide co-enzymes such as DPNH (Weber, 1958), FAD (Whitby, 1953), DPN (Siegel *et al.*, 1959). The high hyperchromicity of DPNH, about 22 per cent, is of interest in that, if it is due in whole or in part to merging of the π-electron orbitals of the nicotinamide and adenine rings, this might be expected to lead to energy transfer of absorbed radiant energy. Such energy transfer has actually been demonstrated for DPNH (Weber, 1958) and undoubtedly occurs also in nucleotide chains (cf. Rhodes, 1961).

Coming back now to the origin of hyperchromicity in highly polymerized polynucleotides, reference should be made once more to Table 2-6, where it is shown that limiting hyperchromicity in small oligo-A chains is reached at a chain length of about 15, for which the hyperchromicity is about 35 per cent; whereas for highly polymerized poly-A it is about 55 per cent. It is therefore of some significance that heating of a solution of poly-A to about 80° results in a reduction of the hypochromicity to about 35 per cent; upon cooling the original hyperchromicity is restored (Fresco and Klemperer, 1959). This effect has been interpreted by Doty and his collaborators (Doty *et al.*, 1959a, 1959b) as due to the existence in the poly-A chain of helical regions stabilized by hydrogen bonds. Heating results in the "melting out" of these helices to a random coil with accompanying loss of hyperchromicity, and on cooling these helical regions are reformed. A similar behavior has been found for single stranded natural polynucleotide chains such as RNA from various sources; and the foregoing interpretation is supported by measurements of optical rotation and chemical reactivity. It is perhaps worth noting that this reversible temperature induced abolition of hyperchromicity in a natural RNA was first observed in connection with investigations on the photochemistry of RNA (Shugar and Wierzchowski, 1958c).

It would clearly be desirable to establish at what chain lengths secondary structure begins to manifest itself, and initial attempts in this direction have been made with adenine oligoribonucleotides, the criterion for appearance of secondary structure being the modification of hypochromicity with increase in chain length. It was found that some degree of secondary structure is already apparent at chain lengths of several residues (Singer *et al.*, 1963). A more extensive investigation has been carried out in the case of a synthetic polypeptide, making use of measurements of hypochromicity in the 1900 Å region. For oligo-methyl-L-glutamate peptides, helix formation was detectable in the nonamer and was almost fully developed (95 per cent) in the tridecamer (Goodman and Listowsky, 1962).

A particularly interesting polynucleotide is poly-U. This polymer exhibits a rather low hyperchromicity, about 6–10 per cent, which is unaffected by heating and other treatments and consequently appears to be devoid of secondary structure. This seems all the more puzzling in that polyribothymidylic acid, which differs from poly-U only in that uracil is replaced by 5-methyluracil, has been reported to exhibit a hyperchromicity of from 25 to 33 per cent (Griffin *et al.*, 1958). Some light is thrown on this apparent discrepancy by the observation of Lipsett (1960) that a considerable decrease in optical density of poly-U takes place with decrease in temperature, the resultant "melting point" T_m being 5.8°C (cf. Szer and Shugar, 1961). It therefore follows that poly-U does exhibit secondary structure, but only at temperatures below about 10°C. Its difference in behavior from poly-ribo-T may consequently be ascribed to the difference in T_m values for the two polymers (Shugar, 1960; Szer and Shugar, 1961). It still remains to be clarified as to why this difference exists. It was found that poly-rT at room temperature exhibits a highly ordered secondary structure. More significant is the fact that the 1:1 twin-stranded complex of poly-rT with poly-A exhibits a T_m about 20°C higher than that of poly-U with poly-A. Two important conclusions emerge from these findings: (a) the common practise of using the A—U base pair in the twin-stranded complex of poly-U with poly-A as a model of the A—T base-pair in DNA is not valid; and, (b) since the hydrogen bonding properties of thymine should not be substantially different from those for uracil, it follows that hydrogen bonding does not contribute significantly to the stability of a polynucleotide helix. This is, of course, not meant to negate the importance of the specificity of the base-pair hydrogen bonds in helix formation (Shugar and Szer, 1962; Szer *et al.*, 1963).

The approach to the origin of hyperchromicity in DNA was considerably simplified by the discovery that, under suitable conditions, synthetic homopolymers may be induced to form complexes with corresponding increases in molecular weight. In fact, complex formation was first observed by spectral methods, which are still the most convenient for this purpose. Such complex formation has now been observed for a variety of combinations, including 1:1 and 2:1 ratios of poly-U: poly-A (Warner, 1957; Felsenfeld and Rich, 1957); for a 1:1 ratio of poly-C: poly-I (Davies and Rich, 1958); for 1:1 and 1:2 ratios of poly-A: poly-I. Figure 2-18 exhibits

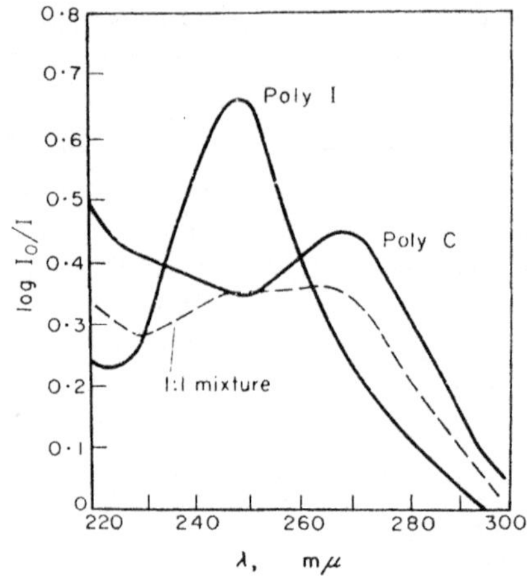

Fig. 2-18. Absorption spectra of equal concentrations of polyinosinic acid (poly-I), polycytidylic acid (poly-C) and a 1:1 mixture of the two, in 0.1 M NaCl and 0.01 M sodium cacodylate pH 6.7 (Davies and Rich, 1958).

the very marked changes in spectral properties associated with the mixing of equal concentrations of poly-C and poly-I in 0.1 M NaCl sodium cacodylate buffer at pH 6.7. Simultaneous titration and spectral studies on these complexes, coupled with molecular weight determinations and X-ray diffraction measurements, show that they resemble Watson–Crick structures for DNA in several respects (Warner, 1957, 1958; Felsenfeld *et al.*, 1957; Davies and Rich, 1958; Beers and Steiner, 1958; Rich, 1958) and that interaction between stacked base pairs is responsible for the major decrease in absorption on complex formation. Furthermore it has been found that some single polymers will form double and triple-stranded complexes (Fresco and Doty, 1957) while poly-I may form a three stranded complex (Rich, 1959; Doty *et al.*, 1959). Reference should be made to the publications of Doty *et al.* (1959a, 1959b) and Rich for other such complexes and to Donohue (1956) for a theoretical discussion of the various combinations of base pairs which may form satisfactorily in terms of hydrogen bond distances.

Even this by no means exhausts all possibilities, and Pauling (1960) has drawn attention to others on the basis of the crystal structure of a 1:1 complex of 1-methylthymine with 9-methyladenine, in which one of the hydrogen bonds is not that to be expected from the Watson–Crick model but involves, instead, the N_7 of the imidazole ring of adenine in place of the N_1 of the six-membered ring (cf. Rich, 1960).

The proposal has been advanced that additional information, with regard to sites suspected of participating in hydrogen bonding in polynucleotides, may be obtained by blocking one or more of them, e.g. by methylation, and thus preventing the formation of helical complexes. It has actually been demonstrated in this way that methylation of the N_3 positions in the uracil residues of poly-U prevents complex formation with poly-A, in agreement with expectations. The extension of this technique to other synthetic polynucleotides is immediately obvious (Szer and Shugar, 1961). It has yet to be demonstrated conclusively that steric factors do not interfere in this approach; but it might equally well be argued that, if the uracil N_3 positions were *not* involved in twin-strand bonding, then methylation of this position would be unlikely to inhibit complex formation with poly-A.

The foregoing findings, extrapolated to the DNA molecule, adequately account for slightly more than half its hyperchromicity in terms of interaction between base pairs in the two-stranded helical Watson–Crick structure. As already mentioned above, this hyperchromicity is lost at elevated temperatures where the DNA molecule is known to "denature" along with a separation of the two-stranded structure. Marmur and Doty (1959) have shown that this partial loss of hyperchromicity (or increase in absorption) occurs over a certain temperature range characteristic for a given DNA, the mid-point of the temperature range being designated by T_m.

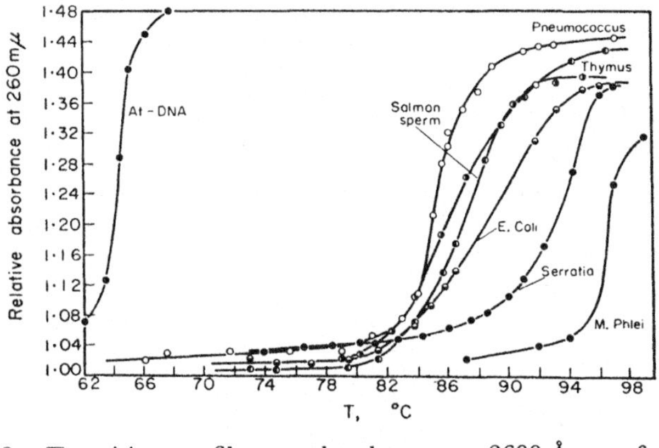

FIG. 2-19. Transition profiles, or absorbances at 2600 Å as a function of temperature of solution, of DNA samples, from various sources. The curve marked AT-DNA is for a synthetic sample containing only adenine and thymine (Marmur and Doty, 1959).

Figure 2-19 demonstrates a number of such curves for DNA from various sources. More significant still is the demonstration that the denaturation temperature, T_m, is related to the guanine–cytosine content of the DNA, as shown in Fig. 2-20. This is explicable on the basis of the fact that guanine-cytosine base pairs may form three hydrogen bonds as compared to two for

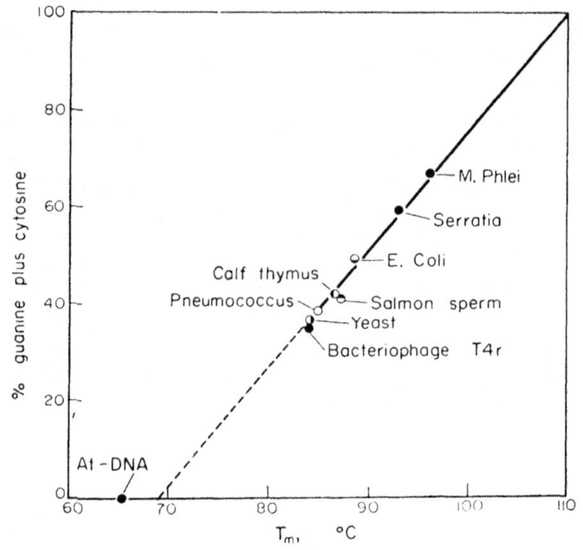

FIG. 2-20. Dependence of the denaturation temperature T_m on the guanine–cytosine content of various samples of DNA (Marmur and Doty, 1959).

the adenine–thymine pair in DNA (Pauling and Corey, 1956). Hence the larger the number of guanine–cytosine base pairs in the DNA double-stranded structure, the more stable it might be expected to be, a fact amply confirmed by Fig. 2-20. Of some interest is the observation that the denaturation temperature, T_m, of a sample of DNA may be sufficiently modified as to be measurable following submission to a radiation dose not exceeding that required to provoke biological inactivation (Marmur and Doty, 1958).

To sum up, therefore, the abnormal absorption spectrum (or hyperchromicity) of a nucleic acid is due (a) to a small extent to the internucleotide linkages in a chain, (b) to interaction between the π-electron orbitals of adjacent aromatic rings in a chain, (c) to the formation of helices, probably via hydrogen bonding, in long single-stranded chains, and (d) to the hydrogen bonding of base pairs in native, double-stranded DNA chains.

It becomes pertinent at this point to examine the extent to which base-pair hydrogen bonding is actually involved in the stability of a twin-stranded polynucleotide helix (see footnote, page 58). Following formulation of the Watson–Crick (1953) structure for DNA as a twin-stranded helix with base pairing via hydrogen bonds, it was perhaps not too surprising that, by analogy with the views prevailing at that time with regard to the role of hydrogen bonding in the stability of polypeptide chains, the implicit assumption arose that the postulated hydrogen bonds between base pairs likewise provided the forces necessary for the maintenance of helical structure. This conception was reinforced by the observation, amongst others, of Marmur and Doty (1959), referred to above, that the thermal stability of a DNA helix is a function of its G—C (or A—T) base-pair content. Since, as pointed out by Pauling and Corey (1956), the G—C base pair can form three hydrogen bonds as against two for the A—T base pair, it appeared reasonable to conclude that this is the reason for the increased stability of both natural and synthetic polynucleotides with increasing G—C content. The view also prevailed for some time that the hypochromicity of polynucleotides was due, in large part, to the effects of hydrogen bonding on the absorption of the individual bases (Beaven *et al.*, 1955). This latter postulate was soon shown to be invalid (Lawley, 1956; Michelson, 1960; Szer and Shugar, 1961, 1963) and, to a considerable extent, studies on the origin of hypochromicity in polynucleotides (see below) simultaneously provided the required impetus for theoretical investigations of the factors responsible for the degree of stability of such polynucleotide helices.

It is perhaps not out of place to recall that the lengths of the base pair hydrogen bonds in DNA postulated by Pauling and Corey (1956) are about 3 Å. These are so long that they must be considerably weaker than the O—H . . . N, O—H . . . O and N—H . . . O bonds formed between the base residues and water in the random coil configuration, and weaker still than the water–water hydrogen bonds in liquid water (cf. Falk *et al.*, 1963). This in itself implies that hydrogen bonding cannot be responsible for the stability of a polynucleotide helix in aqueous medium. Experimental investigations on the denaturation of DNA in mixtures of aqueous and organic solvents strongly underline the importance of the solvent structure on helix stability and to the relatively minor role of base pair hydrogen bonding (Herskovits *et al.*, 1961; Geiduschek and Herskovits, 1961; Herskovits, 1962, 1963). Similar conclusions follow from a study of the effects of hydration on the structure of oriented, hydrated, DNA films (Falk *et al.*, 1963), and from an examination of the relative stabilities of halogenated polyuridylic acids (Szer and Shugar, 1963).

It is consequently understandable why intensive efforts, based on a variety of different experimental approaches, are under way with a view to delineating the factors involved in helix stability. Not the least important of these are theoretical investigations. While the results of one such study have already been published (DeVoe and Tinoco, 1962a), the conclusions arrived at leave much to be desired when they are compared with experimental findings, and further attempts in this direction will have to be awaited.

A number of attempts, using different approaches, are under way with a view to accounting quantitatively from theoretical considerations for the hypochromicity of natural nucleic acids. Ladik (1960) bases his calculations on the assumption that the principal source of hyperchromicity is the mutual interaction between overlapping π-electron orbitals of the aromatic bases stacked parallel to each other in the chains. Tinoco's (1960) initial results were based on calculations of the interactions between the transition dipoles induced in the individual bases by the incident light; this perturbation treatment has been further extended in a subsequent paper (DeVoe and Tinoco, 1962b). Rhodes' (1961) treatment deals with the perturbation of the electron distribution in the individual bases as a result of their mutual interaction due to London dispersion forces. While each of these treatments is reported to account for more or less of the known experimental data on DNA, they are perhaps somewhat too ambitious in scope in attempting to deal at one stroke with complicated natural twin-stranded polymers containing various base sequences. All of the foregoing methods might profitably be applied to calculations of hyperchromicity in model oligonucleotides and synthetic polynucleotides, thus providing considerable simplification and simultaneously making possible an accurate comparison of theoretical calculations with experimental data for a variety of known base sequences (see, e.g. Tables 2-5 and 2-6). The results of such calculations would be extremely useful in assessing the validity of those for the more complicated natural polymers (see Nesbet, 1962).

It should nonetheless be noted that Tinoco's (1960) treatment predicts that the alignment of transition moments parallel to the helical axis should result in a net *increase* in absorption, i.e. hyperchromicity. The $n \rightarrow \pi^*$ transition moments of purines and pyrimidines (see p. 63) would be expected to result in such hyperchromicity at about 2900 Å and this has, in fact, been verified experimentally by Rich and Kasha (1960). Verification at other wavelengths may not be achieved easily because of the large values for hyperchromicity due to the more intense $\pi \rightarrow \pi^*$ transition moment. Rich and Kasha's (1960) findings were based on a study of the ultraviolet dichroism of oriented films of the twin-stranded complex poly-A + poly-U. An analogous approach has been employed by Gellert (1961), using as a model system the alternating twin-stranded poly-dA-dT; while this polymer also exhibits hyperchromicity at 2800 Å, the ultraviolet dichroism at this wavelength was found not to differ appreciably from that at other wavelengths, and it was consequently concluded that $n \rightarrow \pi^*$ transitions are not involved. No immediate explanation for the discrepancy between these two findings has been proferred.

C. NUCLEOPROTEINS

The spectra of nucleoproteins generally exhibit a maximum at about 2600 Å, similar to that for nucleic acids. However, the minimum at about 2300 Å is considerably higher than for free nucleic acids, and below 2300 Å the curves rise very steeply as a result of the contribution of the peptide linkages. The reasons for this are quite clear from an examination of typical nucleic acid and protein spectra, Fig. 2-21. In the figure, curve (*a*) exhibits a

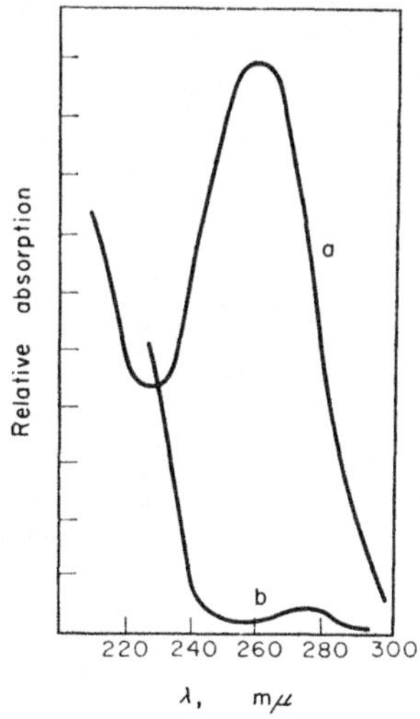

FIG. 2-21. Relative absorption of solutions containing similar concentrations, by weight, of a purified nucleic acid (a) and a protein with a high content of aromatic amino-acids (b).

typical nucleic acid spectrum, curve (*b*) in the same figure is that for a protein with a high content of tyrosine and tryptophan (actually the average for lysozyme, serum albumin and ribonuclease was used) at the same concentration as the nucleic acid. It will be observed that in addition, if the two curves are added together, the resulting spectrum is almost identical with that of NA down to 2400 Å; below this wavelength the entire curve, including the minimum, will be raised considerably.

Fish tropomyosin, which contains about 2.5 per cent ribonucleic acid exhibits a maximum (Fig. 2-22) only slightly to the violet of 2800 Å, but with a rather high minimum at about 2470 Å, typical of nucleoproteins (cf. Fig. 2-6 for triosephosphate dehydrogenase). Hamoir (1951a) has also succeeded in isolating from fish muscle a crystalline preparation of this protein which contains 10–15 per cent nucleic acid; this is the only known crystalline preparation of a nucleoprotein, aside from viruses, although it was subsequently shown that it is not a molecular entity but rather an association of the two components in varying proportions (Hamoir 1951b). From Fig. 2-22 it will be seen that a content of only 10 per cent nucleic acid is sufficient to bring the maximum to 2600 Å, the minimum to 2300 Å, and to give a very high ratio of maximum to minimum.

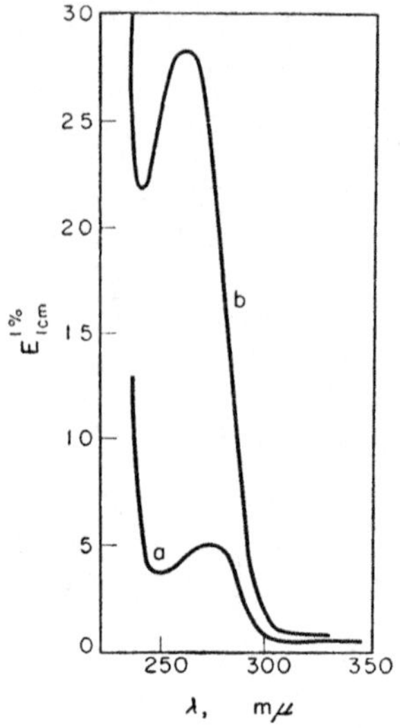

FIG. 2-22. Ultraviolet absorption spectra of tropomyosin (ribonucleic acid content about 2.5 per cent, curve (a) and nucleotropomyosin (ribonucleic acid content about 10 per cent, curve (b) (Hamoir, 1951a).

In many instances the absorption of the protein component above 2600 Å is even less than in the examples cited above. Furthermore, the nucleic acid content varies from about 10–40 per cent, only rarely being as low as 6 per cent (e.g. tobacco mosaic virus). Hence, almost all nucleoproteins exhibit spectra similar to that for nucleotropomyosin, the differences being generally reflected in the ratio of maximum to minimum extinctions. The ratio of maximum to minimum provides a rough estimate of the protein content of a nucleoprotein since at 2300 Å peptide bonds begin to contribute appreciably to protein absorption (Sect. A.6).

In the interpretation of photochemical effects *in vivo*, where nucleic acids are in combination with proteins, it follows that it is the absorption characteristics of the nucleoproteins which must be taken into account. It is consequently of some importance to know what are the nature of the linkages between nucleic acid and protein and to what extent, if any, these may influence the spectrum of the resulting nucleoprotein complex (as compared to simple additivity of the two components), as well as the transfer of absorbed energy between the two.

One might expect spectral studies of nucleoprotein complexes to shed some light on the nature of the bonds between the protein and nucleic acid moieties. Unfortunately, nucleoprotein spectra have hitherto drawn only scattered interest, to a considerable extent because of the experimental difficulties involved in effecting quantitative separation of the components of natural nucleoproteins without in some way modifying one or both of them during the separation procedure (see, however, Sect. D on viruses). Elson (1958, 1959), for example, found that a ribonucleoprotein extracted from *E. coli* may be separated into its component nucleic acid and protein by treatment with urea, from which it was concluded that hydrogen bonding is involved in the original complex. It is, however, quite conceivable that dissociation of the complex resulted from, or was accompanied by, the urea denaturation of the protein component. We have, in fact, found that heat denaturation markedly influences the nature of the artificial complexes formed by proteins with nucleic acids *in vitro*.

In addition, until very recently, considerable doubts existed as to whether nucleoprotein, and particularly deoxyribonucleoprotein, preparations isolated from living cells were not artifacts arising during the extraction procedure, largely because of the earlier belief that DNA occurs only in salt-like combination with histone or protamine in the nucleus. It is indeed only comparatively recently that quantitative studies on the dissociation of nucleohistone in salt solution have been made (Crampton *et al.*, 1954). Sufficient evidence has now accumulated to show that at least some of the nucleoproteins extracted from living cells are definite complexes existing *in vivo*. For some of the earlier references to this subject, the reader is referred to Monty and Dounce (1958) and for a critical discussion of this subject to Zubay and Doty (1959).

Working with carefully purified preparations, Davison and Butler (1956) found that thymus nucleoprotein from different sources may be characterized as a definite molecule on the basis of the existence of a constant ratio of N/P (4.0–4.1) and of arginine/P and lysine/P (cf. Felix, 1953; Vendrely and Vendrely, 1953). Vendrely *et al.*, (1958) have confirmed the finding of Crampton *et al.* (1957) that nucleohistones from different tissues of the same animal are very similar in composition; even more significant are the studies of Crampton (1957), in which it was demonstrated that nucleohistone could be prepared by methods such that dissociation cannot occur during the extraction procedure, and the recent procedure of Zubay and Doty (1959).

Monty and Dounce (1958) have isolated a deoxyribonucleoprotein from liver cell nuclei in which most of the DNA is firmly bound to residual nuclear protein remaining after extraction of globulins and histones. They present considerable evidence, including the isolation of nucleotide amino-acid complexes from enzymatically degraded material, that the complex is a real one. The nature of the bonds between nucleic acid and protein were, however, difficult to characterize because of the large number of functional groups involved. Potter and Dounce (1956) have also obtained from RNA alkali-stable oligonucleotide fractions bound to amino-acids or peptides in a 1 : 1 ratio. Finally, contrary to the earlier conclusions of Riley and Arndt (1953), Feughelman *et al.* (1955) have presented chemical as well as X-ray diffraction data, supplemented with molecular models, in favor of the conception that nucleoprotamine is a definite compound of protein and nucleic acid.

Numerous illustrations exist of the formation *in vitro* of complexes of proteins with nucleic acids by mixing at various pH values. One of these which is especially interesting is that of Alexander (1953), who showed that, whereas protamine forms insoluble complexes of varying composition with DNA in concentration solution, a soluble complex may be formed in dilute solution with a composition, as well as other properties, similar to those of natural nucleoprotamine. Kawade (1954) likewise demonstrated that in strong saline solution DNA-histone complexes similar to thymus nucleohistone may be formed.

Some observations have been recorded on the spectra of such artificial complexes of protein and nucleic acid, but these have not been very systematic, nor have the results always been such as to inspire confidence. There has also been a notable lack of independent evidence of complex formation. Seibert (1953) found a decrease in absorption of mixtures of nucleic acid and tuberculin proteins over a wide range of pH values. Differences from simple addition of up to 10 per cent were tabulated, but are not at all obvious from the accompanying graphs. Nurnberger (1953) presents an extensive series of absorption spectra of mixtures of RNA and DNA with serum proteins, in which absorption was found to run about 5–10 per cent higher than simple additivity and concluded from this that significant interaction occurs but that it is not of sufficient importance to affect cytoplasmic measurements of nucleic acid content by spectral methods. In molar NaCl solution Blout and Assadourian (1954) observed that the decrease in absorption of DNA due to the presence of salt was further accentuated by the addition of bovine plasma albumin or lysozyme. The importance of light scatter in these complexes is not known.

Klamerth (1957) found a non-linear increase in absorption of DNA-histone complexes at 2590 Å with increasing concentrations of histone. Two points of inflexion were noted, the first at an N/P ratio (for the complex) of 3.8, characteristic of normal DNA-histones (cf. Davison and Butler, 1956; see above); and a second at an N/P ratio of 9, which is the limit of histone binding of the complex. These complexes were only incompletely degraded by deoxyribonuclease, in agreement with other observations on susceptibility of nucleoproteins to enzymatic degradation (cf. Shugar, 1955).

Although Beaven *et al.* (1955) regard the question as an open one, there is little room for doubt that the change in absorption resulting from formation of a nucleoprotein complex is due to the nucleic acid component. However, in view of the wide variety of agents which can modify the absorption of a nucleic acid and, particularly, DNA (see Sect. B.3), a modification of nucleic acid absorption resulting from the addition of a protein can hardly be regarded as conclusive evidence of formation of a nucleoprotein-like complex. There are a number of instances where protein-nucleic acid complexes have been formed and demonstrated *in vitro* by light scattering, diffusion, electrophoresis, and sedimentation, but without regard to spectral observations (Shooter *et al.*, 1954; Nath *et al.*, 1955; Kawade, 1957; Zubay and Doty, 1959). There is a very definite need for more precise studies of absorption spectra of natural and artificially formed nucleoproteins under conditions where the physicochemical characteristics of these complexes have been clearly defined by supplementary observations. Additional motivation for such investigations derives from the now reasonably well established role of the histones in the duplication, and coding properties, of DNA; although the mechanism by means of which the histones influence the coding properties of DNA is not yet well understood, complex formation with DNA is apparently involved.

D. VIRUSES

The absorption spectra of many viruses have been reported in the literature. As is to be expected, the spectra of purified viruses closely resemble those of nucleoproteins, and exhibit a high ratio of maximum to the minimum extinction resulting from peptide absorption below 2400 Å. At longer wavelengths the protein component of the absorption is masked by that of the nucleic acid moiety, so that maxima are usually in the neighborhood of 2600 Å. Even for a virus such as tobacco mosaic, where the nucleic acid content is only 6 per cent, careful measurements are required to reveal the tryptophan notch at 2900 Å (Butenandt *et al.*, 1942; Schramm and Dannenberg, 1944). Butenandt *et al.* (1942) have done this rather elegantly by examining an orientated preparation of the virus (obtained by passing a solution through a quartz capillary) with polarized light; under these circumstances the tryptophan maximum at 2900 Å stands out quite sharply when the plane of polarization of the light is perpendicular to the tube axis and suggests that the indole rings lie in a plane perpendicular to the long axis of the virus particles. More recently Seeds and Wilkins (1950) and Perutz *et al.* (1950) have re-investigated this and find the absorption to be greatest when the plane of polarization is parallel to the virus axis.

The much larger dimensions of virus particles result in the scattering of a significant proportion of light, so that the optical densities measured are not only higher than that corresponding to the true absorption, but the shape of the band may also be somewhat modified. Oster (1955) describes an influenza virus preparation in which the characteristic nucleic acid maximum is almost entirely absent. If the virus is suspended in a glycerol solution of the same refractive index as the virus particles, so that scattering is eliminated, the resultant absorption curve is that of the nucleic acid type and is in quantitative agreement with the chemical analysis of the virus. This technique is essentially similar to that proposed by Mitchell (1950) for measuring the nucleic acid content of micro-organisms and undoubtedly merits further study in its application to viruses. Another technique which accomplishes essentially the same purpose, but does not require the use of glycerol in the solution, is that of Shibata *et al.* (1954) involving only a simple modification of the spectrophotometer cuvettes.

Oster and McLaren (1950) calculated the absorption of tobacco mosiac virus from its known content of nucleic acid and aromatic amino-acids, accounting in this way for 50 per cent of the absorption at 2537 Å, from which they conclude that the remaining 50 per cent is due to scattering. Theoretical calculations, using light scattering formulae and the refractive index of the virus for visible light, were stated to account for 80 per cent of the scattered light calculated from the chemical composition of the virus. These figures were later revised (McLaren and Takahashi, 1957) and checked experimentally by a direct method (Chapter VIII).

A procedure more widely used is that involving extrapolation from a region where the virus components are known not to absorb, and already outlined above. Figure 2-23 illustrates the calculation of the scattering

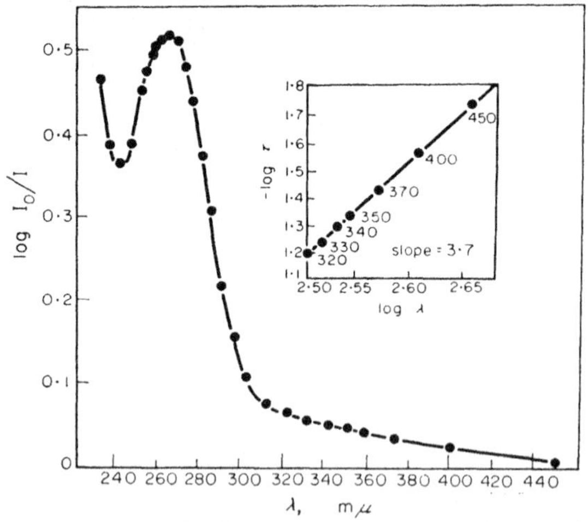

FIG. 2-23. Absorption of purified phage T2, 5.5 × 10^{10} infecting units/ml. in M/15 phosphate buffer pH 7 plus 10^{-3} M $MgSO_4$ and effect of scattering in region where virus components do not absorb (320–450 mμ). Insert shows plot of log (optical density due to scattering) vs. log (wavelength) for pure scattering region (Dulbecco, 1950).

exponent n for a typical case, namely the absorption spectrum of purified phage T2. The inset shows a plot of log (optical density) vs. log λ for the region 3200–4500 Å where the virus constituents do not absorb. The slope of the straight-line plot, n, is 3.7. The scattering constant K may be evaluated from eqn. (2-4B). The scattering curve is extrapolated to shorter wavelengths and the true virus absorption is obtained by simple subtraction. Figure 2-24A

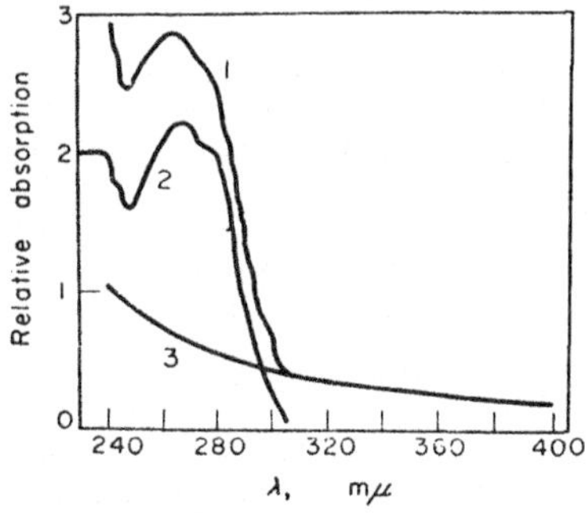

FIG. 2-24A. Absorption spectrum of tobacco mosaic virus; curve 1 is the measured absorption, curve 3 is the calculated absorption due to scattering and curve 2 the absorption of the virus corrected for scattering (i.e. curve 1 minus curve 3) (Schramm and Dannenberg, 1944).

shows the corrected absorption spectrum of tobacco mosaic virus obtained in this way by Schramm and Dannenberg (1944). From the curves it will be observed that the amount of light scattered by the virus is about 25 per cent of the total absorption at 2500–2600 Å. Using an analogous procedure, with special precautions to avoid aggregation, Reddi (1957) obtained curves (Fig. 2-24B) from which it will be seen that only about 16 per cent of the absorption is due to scattered light.

Although the extrapolation method gives some idea of the per cent absorption due to scattering, certain theoretical objections prevail against them. It should be noted that the Rayleigh–Mie theory predicts a scattering exponent n equal to 4 for small particles and <4 as the particle dimensions increase. It is therefore surprising that n is as high as 3.7 for influenza virus (Dulbecco, 1950) and 4 for tobacco mosaic virus (Reddi, 1957), the dimensions of which are comparable to the wavelength of the light used.

A possible source of experimental error in the extrapolation method is the assumption that *all* the absorption above 3200 Å is due to scattering. We have already drawn attention to the fact that, even for proteins, the dimensions of which are small enough to exclude the possibility of scattering, one frequently encounters "extraneous" absorption at 3200 Å, the origin of which is by no means clear. If such absorption exists also in the case of a virus (and most viruses are considerably more difficult to purify than proteins), the extrapolated scattering correction will be too high.

For tobacco mosaic virus, Reddi (1957) has also estimated the absorption of the nucleic acid component by two procedures, (a) measurment of the protein absorption alone and subtraction of this from the corrected virus absorption to give the nucleic acid absorption (Fig. 2-24B), (b) by isolation

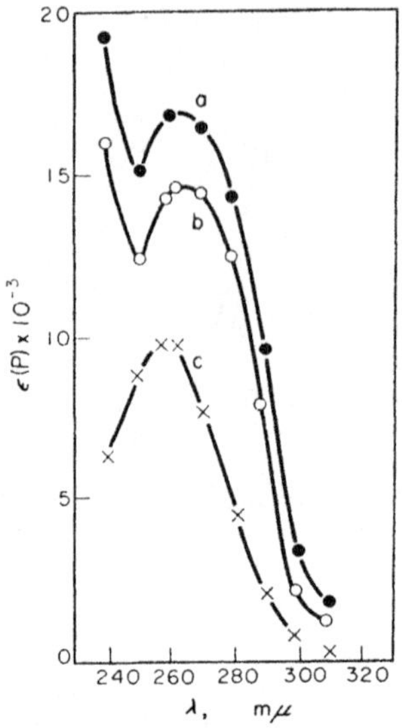

FIG. 2-24B. Absorption spectrum of tobacco mosaic virus; (a) as measured, (b) after deducting for scattering, (c) after deducting for scattering and protein absorption (Reddi, 1957).

of the nucleic acids directly. From Fig. 2-24B it will be seen that nucleic acid makes up about 65 per cent of the corrected absorption of the virus at 2500–2600 Å which is in reasonably good agreement with the value obtained earlier by Schramm and Dannenberg (1944), as calculated from their figure. It should, however, be noted in this connection that the isolated protein component of TMV exhibits spontaneous aggregation as a function of the pH and temperature of the medium (Lauffer *et al.*, 1958). The effect is very pronounced and has been demonstrated not only by viscosity, sedimentation measurements, and electron microscopy, but also by the very pronounced scattering absorption at 3200 Å, which without doubt exists also at shorter wavelengths. This phenomenon will be considered further in Chapter VIII.

On the basis of the observation that the calculated nucleic acid absorption of TMV (obtained by subtraction of the protein absorption from that of the virus corrected for scattering) is similar to the absorption of the isolated nucleic acid, Reddi suggests that the pyrimidine and purine rings do not participate in the formation of the nucleoprotein complex, and that complex formation probably proceeds via the nucleic acid phosphate groups and the protein guanidino groups.

On the whole, it must be admitted that our knowledge of the absorption spectra of viruses, as well as the contribution of each component, is still rather meagre. Such information is of interest not only with respect to virus structure and the nature of the bonds between the two components, but is also necessary in order to distinguish between the effects resulting from the action of ultraviolet light on the different biological properties of a virus. To date action spectra appear to have had more success in this direction. With the increasing availability of a variety of purified viruses, a fertile field of useful research is made available which, it is to be hoped, will be shortly exploited.

CHAPTER III

Action of Ultraviolet Light on Amino-Acids, Peptides and Related Substances

A. DIRECT ACTION

A generalized formula for the *primary* structure of a protein chain is as follows:

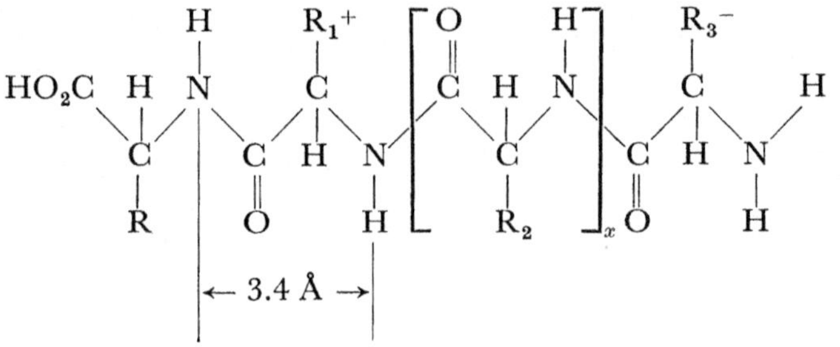

R_1, R_2 and R_3 may or may not be charged or potentially charged side groups of amino-acid residues, x. The degree of polymerization for a typical protein of molecular weight 20,000, $x + 3$ as pictured, is of the order of 170 if we concede an average residue weight of 110. In order to approach the photochemistry of such a polymer we invoke the following conceptual scheme. We assume first of all that the probability that a quantum will be absorbed at a given type of site is proportional to the extinction coefficient of the chromophore making up the site and to the number of such sites in the chain. Secondly, we assume that the probability of a chemical reaction will be roughly the same for the site whether it be part of a macromolecule or a low molecular weight model compound.

Although it is true that excitation energy may be exchanged between different groups within a complex molecule by such mechanisms as "classical resonance" (R. Livingston, 1955), we can only assume, in the absence of suitable criteria, that the net result will be the same as for this scheme. Some supporting data for the second assumption will be found in Chapter V. The status of the subject is without a detailed description at present but may ultimately be aided by a study of model compounds containing the chromophores shown in the primary structure above.

1. *Photochemistry of the End Groups*

Beginning with the carboxyl group, it may be recalled that most of the simple aliphatic acids are transparent down to 2300–2400 Å, Fig. 3-1.†

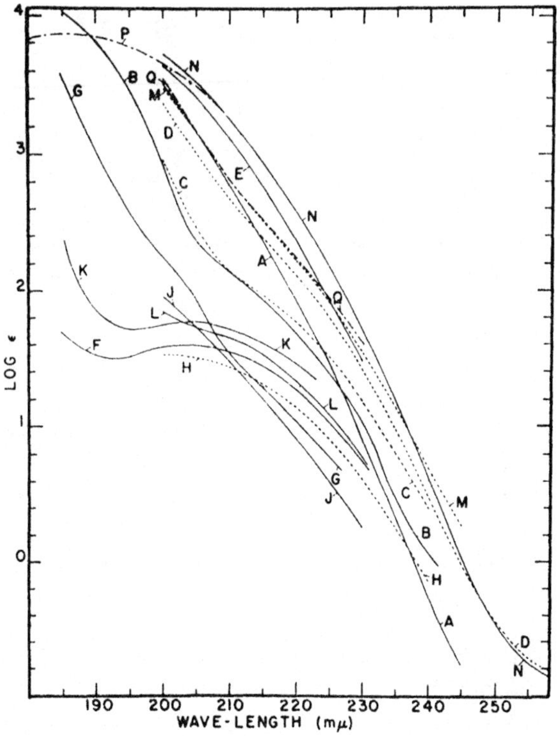

FIG. 3-1. Absorption spectra of some amino-acids, peptides and related compounds (L. J. Saidel, 1955).

A. *N*-methylacetamide, pH 4.3–5.6
B. Acetamide in water
C. Glycineamide, pH 5.8–6.0
D. Acetylglycine, pH 0.9–1.1
E. Acetylglycine, pH 5.8–10.0
F. Acetic acid in water
G. Sodium acetate in water
H. Glycine, pH 1.0: J. glycine in water
K. Ethyl acetate in water
L. Glycine ethylester, pH 3.9–4.5
M. Glycylglycine, pH 1.0; N. same pH 5.1–7.0; P. same, water
Q. Glycylglycine ethylester, pH 5.0

† A dilute solution of acetic acid makes a very fine filter for the 1849 Å line which contaminates resonance lamps designed principally for the output of mercury 2537 Å radiation (Scott and Sinsheimer, 1955).

In aqueous solution, acetic acid in the undissociated form decomposes to give methane, methanol, formic acid and carbon dioxide (Rollefson and Burton, 1939).

$$CH_3CO_2H + h\nu \rightarrow CH_4 + CO_2 \tag{1}$$

$$CH_3CO_2H(H_2O) + h\nu \rightarrow CH_3OH + HCO_2H \tag{2}$$

The Franck–Rabinowitch cage effect tends to prevent free-radical decomposition in the liquid state and manifests itself in a simplicity of products. Rearrangement to yield hydrogen and an unsaturated residue is ruled out on the basis of energy requirement (Burton, 1948).

Amines show absorption at wavelengths beginning between 2100–2500 Å, owing to an electron jump in the NH_2 group. The quantum yields for decomposition are low at room temperatures. A number of products may be obtained from ethylamine in vapor form, namely hydrogen, methane, ethane, nitrogen and polymer.

Since the extinction coefficients are so low for the carboxyl and amino groups in the wavelength range usually employed with proteins, and since α-amino and α-carboxyl groups are only free at the ends of polypeptide chains, the important question really resides in whether peptide bond (—CONH—) splitting by ultraviolet light is an important step and whether R groups are altered. Pertinent model compounds which have been studied are the peptides and amino-acids with the corresponding R groups.

2. *Photochemistry of the Peptide Bond*

The chemical action of ultraviolet light on a few model compounds has been explored in water with the following results.

Acetamide solutions increase in acidity during irradiation at 2537 Å; ammonia is produced with a quantum yield of less than 0.005. Above 58°C the quantum yield becomes appreciable, rising from a value of 0.023. The yields of ammonia are even less with propionamide and butyramide (Volman, 1941). It was reported that the quantum yield decreases with increasing concentration of amide, but no corrections were made in the calculations for the probable change in absorption spectra during irradiation, cf. Fig. 3-2. The increase in Φ with increasing temperature was explained on the basis of the reactions

$$CH_3CONH_2 + h\nu \rightarrow CH_3CO\cdot + NH_2\cdot$$

$$CH_3CO\cdot + NH_2\cdot \rightarrow CH_3CONH_2$$

$$CH_3CO\cdot + H_2O \rightarrow CH_3CO_2H + H\cdot$$

and

$$NH_2\cdot + H\cdot \rightarrow NH_3$$

The non-recombination reactions would be expected to have considerable activation energy, while the recombination reaction would not. Thus there would be an increase in Φ with an increase in temperature. However, CO_2, CO, CH_4 and N_2 were also formed during irradiation and without a simple stoichiometric relationship among them. The CO_2 and CH_4 probably originate from the acetic acid; the CO may arise from $CH_3CO\cdot$ via the thermally aided reaction, where M is any molecule,

$$CH_3CO\cdot + M \rightarrow CH_3\cdot + CO + M^{\nearrow}$$

The source of N_2 as a product was not speculated upon.

In addition to amides there is another group of low molecular weight compounds containing the —CONH— bond, namely the hemipeptides made up of aroyl, acyl and arylacyl derivatives of amino-acids. The hemipeptides lack free α-amino group characteristic of peptides; an example is acetylalanine. Quantum yield data, based on ammonia determinations carried out directly with irradiated acetylalanine solutions by the Conway microdiffusion method at room temperature, involve corrections for the large increase in optical density during irradiation, Fig. 3-2. In calculating

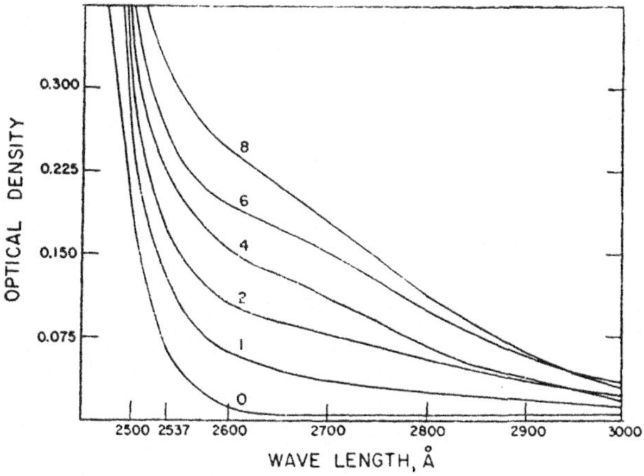

FIG. 3-2. The absorption spectrum of acetylalanine (0.30 M in water) as a function of time of irradiation (intensity 0.2×10^{-4} einstein/ml/hr at 2537 Å) for a 1-cm path length. Times of irradiation in hours are indicated in the figure (Estermann *et al.*, 1956).

Φ, one needs to know the fractional intensity absorbed which is absorbed by the reactant; this is given at any instant by (D_A/D_{A+B}). D_A is equal to $C\ \varepsilon_{2537}$ where C is the concentration of acetylalanine having a molecular extinction of ε_{2537}; D_{A+B} is the optical density of the solution. The subscript B refers to all absorbing products of the reaction. In the experiment cited, after 8 hr of irradiation only 0.34 per cent of the reactant had liberated ammonia, so C was taken as constant. The average fraction of light absorbed during the total time of irradiation is computed from a plot of per cent of incident intensity absorbed by the solution as a function of time of irradiation. The correct working equation is

$$\Phi = \frac{\text{Moles ammonia formed/ml per hr}}{\dfrac{(\text{Intensity, } I_0)(\text{time of irradiation})}{(\text{average fraction of } I_0 \text{ absorbed})(D_A/D_{A+B})}}$$

Thus for a 4 hr irradiation period

$$\Phi = \frac{5.0 \times 10^{-7}}{(2.0 \times 10^{-5})(4.0)(0.315)(0.45)} = 44 \times 10^{-3}$$

This formula is a somewhat better approximation than that published (Estermann *et al.*, 1956). Although it has sometimes been proposed that the weakest bond is broken most efficiently in a compound and that generally only one bond undergoes fission in the primary photolytic process (McLaren, 1949; E. Bergmann and R. Samuel, 1941), the yield of ammonia from acetylalanine is about ten times as great as alanine. If the photolysis of acetylalanine produced only acetic acid and alanine directly, the absorption spectrum of the irradiated solution would decrease (Magill *et al.*, 1937). Instead, an increase is observed, Fig. 3-2 (after 1 hr, 1.3×10^{-2} einsteins had been absorbed per mole). By analogy with the photochemistry of alanine (Weizman *et al.*, 1936; Mandl *et al.*, 1951) lactic acid, acetaldehyde, and pyruvic acid were considered as possible products. Alanine itself is probably deaminated only at wavelengths much shorter than 2537 Å and, if formed from acetylalanine, should not have been lost. No lactic acid was found, but acetaldehyde and pyruvic acid were. The results add up as follows (Table 3-1).

TABLE 3-1. PRODUCTS FROM THE IRRADIATION OF ACETYLALANINE

acetaldehyde:	$\Phi = 0.030$	ammonia:	$\Phi = 0.044$
pyruvate:	$\Phi = 0.017$	alanine:	$\Phi = 0.005$
	0.047		0.049

Thus the ammonia produced directly is equivalent to the production of carbonyl compounds. None of these products, however, account for the observed increase in the spectrum. A dehydrogenation has been suggested (Allen *et al.*, 1937), viz.,

$$CH_3CONHCH(CH_3)CO_2H + h\nu \rightarrow CH_3CON{=}C(CH_3)CO_2H$$

The suggested intermediate, acetyliminopropionic acid, can be hydrolyzed to yield pyruvic acid and ammonia. With irradiated acetylalanine, only one-third as many moles of pyruvate were found as were moles of ammonia. Further, after acid or alkaline hydrolysis of irradiated acetylalanine, a fourfold increase in ammonia was obtained (a total Φ of 0.2), whereas the yield of pyruvate was only slightly increased. A possible intermediate giving rise to additional ammonia during acid hydrolysis is acetamide but this substance would not contribute to the increase in spectrum observed. The highly absorbing product has yet to be identified. Irradiated clupein shows a general increase in spectrum during irradiation, similar to that with acetylalanine (Claesson, 1956).

Ford (1957) has carried out a kinetic analysis of the formation of an absorption band in the spectrum of nylon during and following irradiation with polychromatic light. The new, broad band at 2900 Å increases in intensity in irradiated films stored in the dark at a rate greater than that immediately prior to storage. On re-irradiation the band intensity falls rapidly to a value corresponding to that just prior to storage. Also, the increase in band intensity during storage is a constant fraction of the total intensity produced during the irradiation. He interprets this behavior in terms of the three reactions

$$\mathrm{A} + h\nu_1 \rightarrow \mathrm{B}\cdot + \mathrm{C}\cdot$$

$$\mathrm{B}\cdot + \mathrm{M} \rightarrow \mathrm{BM}\cdot \quad \text{(the new absorbing chromophore)}$$

$$\mathrm{BM}\cdot + h\nu_3 \rightarrow \mathrm{B}\cdot + \mathrm{M}$$

where A is presumably —CONH—; B· and C· are free radicals. The presence of a "dark" reaction, together with a detailed kinetic analysis involving variations in intensities I_1 and I_3 (polychromatic light of frequencies ν_1 and ν_3) supports the view that the new chromophore is a free radical. The species M is dissolved oxygen and BM· is most likely to be —COOO·. In the absence of oxygen no specific peak appears at 2900°, rather only a general increase in absorption from 2400 to 3500 Å. If the absorption center is a free radical, the 2900 Å band should disappear when the film is dissolved; it does, leaving only a general increase in absorption from 2400 to 3500 Å. The overall results thus suggest that in the case of solids, the diffusion processes involved in a chemical reaction may be separated into two classes depending on relative rates. Gaseous diffusion of, say, oxygen is fast, whereas polymer chain or segment diffusion is very much slower. The latter is speeded up enormously if the irradiated polymer is dissolved. These results are probably applicable to the photodegradation of sheets and fabrics made of plastics and wool.

The addition of a phenyl group to compounds containing —CONH— increases the extinction coefficients enormously, of course. If sufficiently insoluble, such compounds can absorb enough light to be photolyzed at measurable rates in the form of monolayers on water. (Rideal and Mitchell, 1937; Carpenter, 1940; cf. McLaren, 1949 for a discussion of these and earlier work on peptides.)

The reaction scheme for stearylamides is

$$C_{17}H_{35}CONH(CH_2)_nC_6H_5 + h\nu + H_2O \rightarrow C_{17}H_{35}COOH + C_6H_5(CH_2)_nNH_2.$$

With these compounds n has been varied from zero to two with comparable results. Carpenter suggested that energy absorbed by a phenyl group could be transmitted through —$(CH_2)_n$— to the peptide bond with a resultant cleavage. Certainly most of the absorbed energy is taken up by the phenyl group initially. Since *N*-methylstearylamide also cleaves during irradiation, the proposed mechanism is not necessary for these qualitative results (cf. Zeelen, 1956).

A group of compounds which is of interest from a spectroscopic point of view and from which some empirical data has been gained on the subject of energy transfer in organic molecules is

$$C_6H_5(CH_2)_n\text{—}CONHCH(CH_3)COOH$$

with n equal to 0, 1, 2, 3, and 4 (McLaren *et al.*, 1953). The spectra of these compounds are shown in Fig. 3-3.

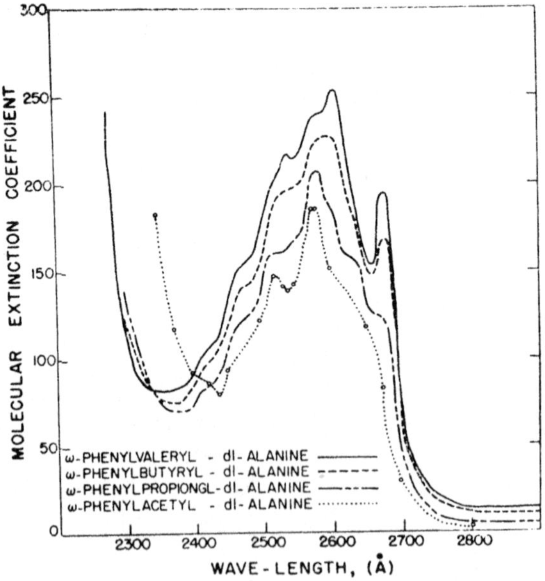

FIG. 3-3. Absorption spectra of phenylacyl-alanines (McLaren *et al.*, 1953).

The chromophore in benzoylalanine C_6H_5CONH— is distinct from the two chromophores in C_6H_5—$(CH_2)_n$—CONH—$CH(CH_3)CO_2H$. By contrast ε_{2537} for benzoyl-DL-alanine is 3000 and the absorption curve rises continuously from $\varepsilon_{2800} = 480$ to a maximum of $\varepsilon_{2300} = 10{,}800$ (not shown). The spectrum of propionylphenylalanine is quite comparable with that of phenylpropionylalanine above 2375 Å but falls between that of phenylpropionylalanine and phenylacetylalanine at shorter wavelengths (Mandl *et al.*, 1950).

Table 3-2 summarizes the analyses for alanine following irradiation of these compounds.

TABLE 3-2. QUANTUM EFFICIENCIES FOR THE LIBERATION OF ALANINE FROM SOME HEMIPEPTIDES (2537 Å)

Compound	$\Phi \times 10^3$	ε_{2537}
Acetylalanine	5	0.2
Benzoylalanine	3	3000
Phenylacetylalanine	4	142
Phenylpropionylalanine	4	163
Phenylbutyrylalanine	55	197
Phenylvalerylalanine	3	218

Except for acetylalanine, the other compounds gave rise to less ammonia than alanine when ammonia was analyzed for by microdiffusion at room temperature. As with acetylalanine, intermediates are produced during irradiation from which additional ammonia is liberated if warm sodium hydroxide is added following irradiation. The presence of a phenyl group so changes the course of the photochemical reactions as to yield alanine, in appreciable amounts, but with $n = 3$ a 10-fold increase of alanine is produced per quantum absorbed. Reference to Fig. 3-3 shows no irregularity in the spectrum of phenylbutyrylalanine. Also the dissociation constant of this compound is only a little lower than that of the next higher and lower members of the series.

Here we encounter a problem of primary importance. How does energy absorbed by a phenyl group influence the photolysis of a —CONH— linkage? If only light absorbed by —CONH— results in photolysis, then the quantum yield for the photolysis of acetylalanine should be about one thousand times higher than for the phenylacyl compounds. Based on liberation of alanine this is not so, and based on ammonia plus alanine (0.04 + 0.005) we still find a difference of over an order of magnitude from expectation. Further, the idea that energy may be transmitted efficiently by —$(CH_2)_n$— is not consistent with the insulator properties exhibited by methylene groups in absorption spectra nor with the unexpectedly relative high yield with phenylbutyrylalanine, Table 3-2. It is true that Fisher–Hirschfelder models of the $n = 3$ and $n = 4$ compounds show the possibility of folding the peptide bond flat against the benzene ring, which is not true with the lower members of the series; but, since the n_3 and n_4 compounds also show a large difference in Φ, photosensitization is not an adequate explanation.

Hentz and Burton (1951) consider that the probability that energy distributed in the π-electron system would transfer to a single bond, in amount sufficient for a decomposition process, should increase as the amount of energy required decreases. For example, in ethylbenzene the weakest bond is that ruptured, as shown

$$C_6H_5CH_2CH_3 + h\nu \rightarrow C_6H_5CH_2\cdot + CH_3\cdot$$

namely β to the ring. The quantum yield is of the order of 10^{-4}. In this case the rearrangement $C_6H_5CH_2CH_3^* \rightarrow C_6H_5CH{=}CH_2 + H_2$ may be of importance. This low quantum yield may be contrasted with the higher value found for the photolysis of the peptide bond in phenylacetylalanine. There also exist fluorescence data indicative of energy transfer in small molecules (Weissmann, 1952) and energy transfer between molecules (West and Miller, 1940). Neither intermolecular energy transfer nor internal conversion seem applicable to these curious results, however.

Compounds containing a phenyl group in addition to —CONH— would expectedly undergo changes in the ring. The large changes in the absorption spectra of such compounds have been interpreted in this way (Mandl *et al.*, 1950). One probable reaction is the addition of a free radical or hydrogen atom to the ring. Since both fluorescence and molecular deactivation of excited phenyl compounds can take place to an important extent, chemical decomposition processes involving primarily the benzene ring may, relatively speaking, make a minor contribution to the total picture (Hintz and Burton, 1951). Obviously only a systematic and exhaustive study of one or two compounds can lead to a reasonable understanding of the various ways in which an absorbed quantum can lead to decomposition. Fortunately, as will be seen below, it is probably only necessary to know the quantum efficiency for the loss in identity of a chemical moiety in the primary structure of an enzyme to account for its inactivation, rather than a detailed knowledge of all the possible ways this moiety can undergo change with the production of all possible products.

Incidentally, the yield of ammonia from propionylphenylalanine is almost the same as that from phenylproprionylalanine (Mandl and McLaren, 1949). No phenylalanine is liberated. In this connection it must be noted that very little or no free amino-acid is liberated during irradiation of acetyltryptophan, glycyltryptophan, glycyltyrosine or leucyltyrosine either. The spectra of all of these aromatic substances change markedly during irradiation, which indicates that the action of ultraviolet light is to modify the aromatic groups. According to Weizmann *et al.* (1938), it is easier to split off a terminal amino group on a peptide than it is to liberate nitrogen from the peptide bond.

3. *Photochemistry of Side Groups and of Amino-Acids*

The identification and quantitative determination of all the products of a photochemical reaction is a painstaking process which has been achieved in relatively few instances. By contrast it is a comparatively simple matter to measure the total quantum yield for destruction of the starting material by following its loss of identity. This can be done either by chemical procedures, by paper or column chromatography, or, if the absorption of the photoproduct does not interfere, by absorption spectroscopy. Such procedures have been applied to free amino-acids by Luse and McLaren (1963) with the results reported in Table 3-3. A summary of the literature is given in Table 3-4.

TABLE 3-3. DESTRUCTION OF AMINO-ACIDS AND THE PEPTIDE BOND BY UV OF 2537 Å

Compound	Irradiation conditions	Quantum yield for total destruction	Molecular extinction coefficient,* ε_{2537}
Alanine	in N/10 HCl	~0	—
Cystine	in N/10 HCl, under nitrogen†	0.13	270
Histidine	in water, under nitrogen	<0.03	0.24
Phenylalanine	in N/10 HCl, under nitrogen	0.013	140
Tryptophan	in N/10 HCl, under nitrogen	0.004	2870
	in pH 7 phos. buffer, under nitrogen	~0.01	
	in water, under nitrogen	0.007 ± 0.002	
Tyrosine	in N/10 HCl, under nitrogen	0.0020	320
—CONH—	acetylalanine in water	0.05	0.2

* ε_{CONH} may be taken as 1 in proteins, and Φ_{CONH} as 0.005 (Cf. Chapt II, A, 3).

† Solutions were initially in equilibrium with the atmosphere and so contained dissolved oxygen. Stirring was with a stream of nitrogen during irradiation. Other solutions were nitrogen-saturated before irradiation and similarly stirred.

TABLE. 3-4. PHOTOLYSIS PRODUCTS OF UV IRRADIATED AMINO-ACIDS AND PEPTIDES

Amino-acid	UV source	Irradiation conditions*	Photolysis products and quantum yields	Literature references
1. L-alanine	Hg-l†	in water	L-lactic acid, NH_3($\Phi = 0.027$)	Weizmann *et al.*, 1936, 1938; McLaren, 1949
	Hg-l	in 0.01 N HCl under N_2 or O_2	L-lactic acid, NH_3	Mandl and McLaren, 1951
DL-alanine	Hg-l	in 0.01 N HCl under N_2 or O_2	pyruvic and lactic acids, NH_3, no acetaldehyde	Mandl and McLaren, 1951
			serine, asparagine; aspartic, pyruvic, glyoxalic, and formic acids; NH_3, CO_2	Watanabe, 1960
2. γ-amino adipic acid	Hg-l	in 4 per cent HCl or 4 per cent Na_2CO_3	glycine, β-alanine, acetic and oxalic acids	Morokuma, 1954
3. amino-isobutyric acid	Hg-l	in water	NH_3 and an acetone	(?) Allen *et al.*, 1937
4. L-arginine	Hg-l		β-guanido propionic and δ-guanido butyric acids, guanindine (plus α-keto-δ-guanido valeric acid, ν-guanido-butryic anhydride?)	Hirata, 1958
5. L-aspartic acid	Hg-l	in 0.01 N NaOH	acetaldehyde, NH_3, CO_2	Becker, 1935
	Hg-l	in water	DL-malic acid, NH_3	Weizmann *et al.*, 1936
	Hg-l	in 0.1 N HCl under N_2	L-malic acid, NH_3	Mandl and McLaren, 1951
			oxalacetic, glyoxalic, pyruvic acids, NH_3	Yoshinaga, 1960
6. L-asparagine	Hg-l	in water	acetaldehyde, NH_3	Becker, 1935
7. betaine (trimethyl-glycine	Hg-l	in water, pH 6-8	trimethylamine, NH_3 ($\Phi = 0.06$)	Weizmann *et al.*, 1936; McLaren, 1949

* Irradiation done in air and at room temperature unless otherwise noted.
† l = lamp.

TABLE 3-4—(*contd.*)

Amino-acid	UV source	Irradiation conditions*	Photolysis products and quantum yields	Literature references
8. cysteine	2250 Å	in 0.1 N HCl	cystine, H_2S	Schocken, 1952
	2537 Å	in water, pH 5.5	cystine, sulfates; no H_2S or free S	Deschreider, 1952
cysteine HCl	2537 Å	in water	sulfates; no cystine, H_2S or free S	Deschreider, 1952
9. cystine	Hg-l	in 0.1 N HCl or 0.1 N NaOH	cysteine, H_2S	Szendro *et al.*, 1933
	Hg-l	in 3 per cent HCl	cysteine ($\Phi = 0.02$), NH_3; no H_2S or free S	Lyons, 1948
	2250 Å	in 0.1 N HCl	cysteine, H_2S	Schocken, 1952
	Hg-l	in water under air, N_2 or A	H_2S	Mishchenko, 1954
	2537 Å	in 0.1 N HCl under N_2	cysteine ($\Phi = 0.2$), H_2S ($\Phi = 0.2$), $NH_3(\Phi = 0.04)$	Luse, 1961
10. 3,4-dihydroxyphenyl-alanine	Hg-l		melanins	Schocken, 1951
	Hg-l	in 0.1 N NaOH	aspartic acid	Tominaga, 1958
	Hg-l	in water or 0.1 N HCl	no aspartic acid	Tominaga, 1958
11. glycine	Hg-l		glycollic, acid NH_3, CO	Henri *et al.*, 1934; Ide, 1960
	Hg-l	in water	$NH_3(\Phi = 0.03)$	Weizmann *et al.*, 1938; McLaren, 1949
12. L-histidine	Hg-l	in water	imidazole acetaldehyde	Szendro, 1931
	Hg-l	in water	histamine	Holtz, 1933, 1934
	Hg-l	in water	NH_3, hydroxy-acids (which oxidized to acetaldehydes) (and histamine?)	Weizmann *et al.*, 1939
	Hg-l	in water	histamine	Alechinsky and Reuse, 1946
	Hg-l	in citrate phos. buffer, pH 7	histamine, imidazole, acetaldehyde	Schocken, 1951

* Irradiation done in air and at room temperature unless otherwise noted.

TABLE 3-4—(*contd.*)

Amino-acid	UV source	Irradiation conditions*	Photolysis products and quantum yields	Literature references
	2300–3000 Å	in citrate phos. buffer, pH 7	no destruction noted	Schoken, 1951
	Hg-l	in water, acid and alkali	histamine, imidazole-lactic, imidazole-acetic, and urocanic acids, asparagine, serine	Tomohiro, 1953
	Hg-l	in 4 per cent Na_2CO_3	polymerized 2-hydroxy imidazole	Fujiwara *et al.*, 1954
	2537 Å	in water, with cysteine	histamine	Flemming, 1959
	2537 Å	0.1 N HCl under N_2	histamine, $NH_3 (\Phi < 0.3)$	Luse, 1961
13. homocystine	Hg-l	in 0.1 N HCl	homocysteine	Schocken, 1952
14. hydroxy-proline	Hg-l		destruction	Becker, 1933
15. leucine	Hg-l	in water	NH_3 and unidentified compounds	Allen *et al.*, 1937
16. lysine		in 4 per cent or 4 per cent Na_2CO_3	aspartic, glutamic, ν-amino adipic acids, glycine, alanine, norvaline, norleucine, cadaverine	Morokuma, 1954
17. methionine	2250 Å	in 0.3 N HCl	methylmercaptan (?); no homocysteine	Schocken, 1952
18. DL-ornithine	Hg-l		proline, glutamic, aspartic, α-keto-glutaric, and pyruvic acids, putrescine, NH_3, CO_2	Mukai, 1959
19. L-phenylalanine	Hg-l		NH_3, modified phenyl groups	Rideal and Mitchell, 1937
	Hg-l	in water	NH_3	Weizmann *et al.*, 1938
	Hg-l	in water, acid, and alkali	a phenol and dihydroxyphenylalanine (dropa)	Arnow, 1942
	2700 Å	in citrate phos. buffer, pH 7	tyrosine, mono- or di-hydroxyphenylalanine	Schocken, 1951
	Hg-l	in acid	tyrosine, serine	Matsuda, 1954

* Irradiation done in air and at room temperature unless otherwise noted.

TABLE 3-4—(*contd.*)

Amino-acid	UV source	Irradiation conditions*	Photolysis products and quantum yields	Literature references
	Hg-l	in alkali	tyrosine, serine, alanine	Matsuda, 1954
	2537 Å	in water under O_2 or N_2	NH_3 and two unidentified compounds	Claesson, 1956
	Hg-l		aspartic acid, asparagine, tyrosine; no dopa, serine, or alanine	Tominaga, 1958
	2537 Å	in 0.1 N HCl	tyrosine, dopa $NH_3(\Phi = 0.002)$	Luse, 1961
20. L-proline		in water, acid, alkali	hydroxyproline, pyro-glutamic, α-keto-glutaric, α-ketobutyric, levulic, and acetic acids, glycine, NH_3, CO_2 (least destruction in acid)	Yamada, 1956
21. L-tryptophan	Hg-l	in water	indole-3-acetic acid	Berthelot and Amoreux 1938
	2537 Å	phosphate buffer, pH 7.3	oxidized aromatic nucleus	McLean and Giese, 1950
	Hg-l	in acid under air	serine, aspartic acid, alanine, kynurenine, tryptamine	Matsuda, 1953
	Hg-l	in acid under N_2	as in air, plus NH_3	Matsuda, 1953
	Hg-l	in neutral solution	serine, aspartic acid, kynurenine	Matsuda, 1953
	Hg-l	in alkali	serine, aspartic acid, kynurenine, alanine, 3-hydroxyanthranilic acid	Matsuda, 1953
	2537 Å		formyl kynurenine, which yields upon irradiation alanine and a compound which yields aspartic acid and glycine	Deschreider and Renard, 1955
	2537 Å	in water or acid	tryptamine, $NH_3(\Phi = 0.002)$	Luse, 1961
22. L-tyrosine	Hg-l	in water, 48–50°C	dopa, NH_3 (?)	Arnow, 1937, 1938
	2700 Å	in citrate phos. buffer, pH 7	dopa, which oxidizes to melanins	Schocken, 1951

* Irradiation done in air and at room temperature unless otherwise noted.

TABLE 3-4—(*contd.*)

Peptide	UV source	Irradiation conditions*	Photolysis products and quantum yields	Literature references
	Hg-l	in water	dopa, pigment	Gonnard, 1953; Johnson, 1961
			aspartic acid, asparagine (plus alanine and glycine?); no dopa	Suzatani, 1958
	2537 Å	in 0.1 N HCl under N_2	aspartic acid, $NH_3(\Phi = 0.002)$ no dopa	Luse, 1961
23. L-valine	Hg-l	in water	NH_3 and unidentified compound	Allen *et al.*, 1937
1. Acetyl-DL-alanine	2537 Å	in water	$NH_3(\Phi = 0.05)$, acetaldehyde (0.03), pyruvate (0.02), alanine (*ca.* 0.004)	Esterman *et al.*, 1956
2. Acetylalanylglycine	Hg-l	in water	NH_3 and unidentified compound	Allen *et al.*, 1937
3. Acetylglycyl-DL-alanine	Hg-l	in water	NH_3 and unidentified compound	Allen *et al.*, 1937
4. Acetyl tryptophan	2537 Å	in water	modified indole group; no tryptophan	McLaren *et al.*, 1953
5. Alanylglycine	Hg-l	in water	NH_3 and unidentified compound	Weizmann *et al.*, 1938
6. Alanyltyrosylglycine	2483, 2537 Å	as monolayer	alanine, tyrosine, glycine	Carpenter, 1940, 1941
7. Glycyl-DL-leucine	Hg-l	in water	NH_3 and unidentified compound	Allen *et al.*, 1937
8. Glycyl-tryptophan	2537 Å	in water	Modified indole group; no glycine	McLaren *et al.*, 1953
9. Glycyl-tyrosine	2483–2537 Å	as monolayer	glycine, tyrosine	Carpenter, 1940, 1941
	2537 Å	in water	NH_3, modified phenyl group; no glycine	McLaren *et al.*, 1953
	2537 Å	in water pH 4–8 under air, O_2 or N_2	NH_3, modified phenyl group; no glycine or tyrosine	Claesson, 1956
10. Gramicidin	2537 Å	in ethyl alcohol	destruction ($\Phi = 0.004$)	Setlow and Doyle, 1957
11. Leucyl-glycine	2537 Å	in water	NH_3, traces of leucine and glycine; other unidentified compounds	Claesson, 1956
12. Leucyl-tyrosine	2537 Å	in water	Modified phenyl group; no amino-acids	McLaren *et al.*, 1953
13. Phenylpropionyl alanine	2537 Å	in water	alanine ($\Phi = 0$), NH_3(0.0002)	Estermann *et al.*, 1956
14. Tyrosyl-glycine	2537 Å	as monolayer	tyrosine, glycine	Carpenter, 1940, 1941

* Irradiation done in air and at room temperature unless otherwise noted.

If the product $\varepsilon \times \Phi$ is a measure of the relative probability of the loss in identity of a group R in a protein chain, the two most vulnerable R groups would be those of tryptophanyl and cystyl, followed by those of phenylalanyl and tyrosyl. The behavior of alanyl and its homologous aliphatic R groups is probably without significance. The spectroscopy and photochemisty of each of these groups is different and the important features are as follows.†

Wavelengths of less than 2265 Å are required to induce decomposition of glycine, although absorption begins at about 2400 Å (Weizmann *et al.*, 1936). Glycollic acid, ammonia and carbon dioxide have been identified among the products.

Under nitrogen, pyruvic and lactic acids have been obtained from alanine, but no acetaldehyde. The light source was a polychromatic mercury arc (Mandl and McLaren, 1951).

$$CH_3CH(NH_2)COOH + h\nu + H_2O \rightarrow CH_3CH(OH)COOH + NH_3.$$

The formation of pyruvic acid in the oxygen-free atmosphere cannot be unequivocally accounted for. The optical configuration of L-alanine (and L-aspartic acid) is retained during deamination and must be taken into account in any explanation of the facts. An α-lactone has been postulated as an intermediate.

A simple nucleophilic replacement would yield inverted D-lactic acid (Hammett, 1940) but a two-step mechanism can account for retention of configuration. Both steps cause inversion. First an intermediary α-lactone

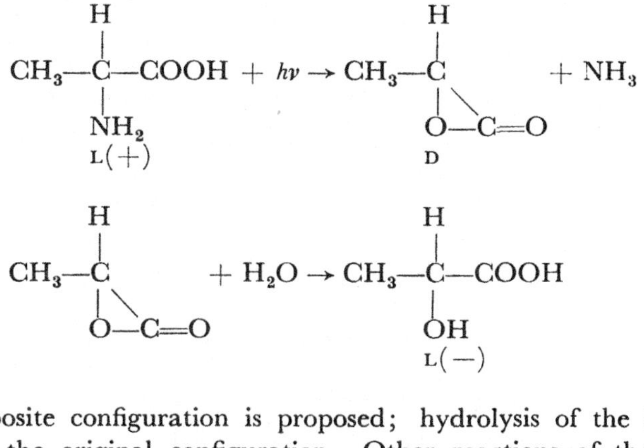

of the opposite configuration is proposed; hydrolysis of the lactone re-establishes the original configuration. Other reactions of the α-lactone could give rise to carbon monoxide, methane, pyruvic acid and acetaldehyde (in the presence of air), four of the known products.

† The photochemistry of amino-acids and peptides prior to 1948 and 1955 has been reviewed elsewhere (McLaren, 1949; Claesson, 1956).

Under nitrogen:

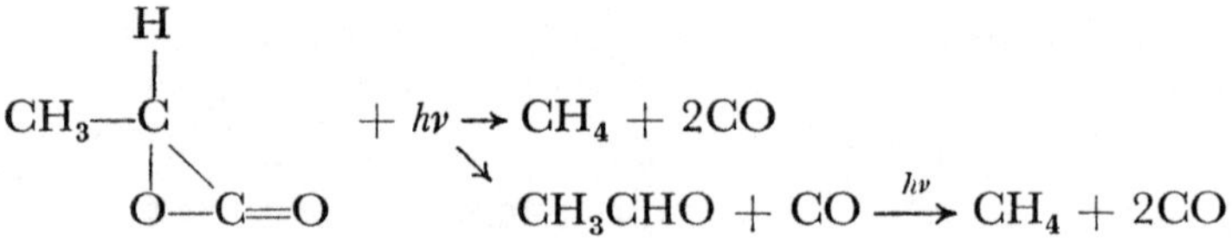

Under air, hydrogen peroxide may become an important reactant:

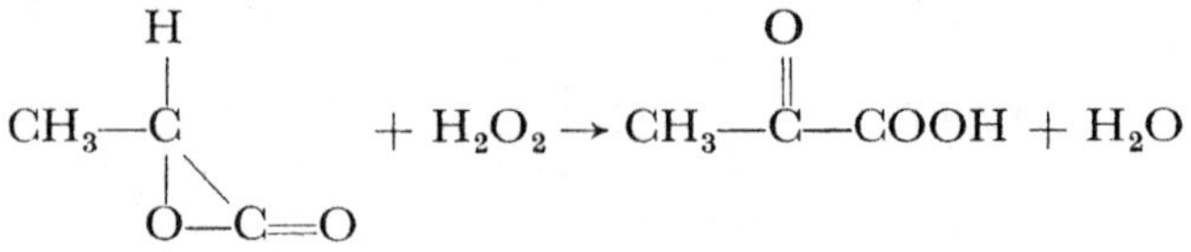

At short wavelengths (e.g. 1849 Å) irradiation of water can lead to H_2O_2. The reaction is $2H_2O \rightarrow H_2O_2 + H_2$. In the presence of oxygen the production of H_2O_2 is increased along with the formation of ozone (Rollefson and Burton, 1949).

In the photolysis of aspartic acid to malic acid, the hydroxy-acid retains the stereochemical configuration of the amino-acid reactant (Mandl and McLaren, 1951). From asparagine ammonia has been found, and from aspartic acid, in addition to malic acid, ammonia and carbon dioxide have been reported. From the latter a new absorption band at 2650 Å has been shown to arise (McLaren, 1949). The photo-decomposition of aspartyl, glutamyl and alanyl residues are probably without significance at 2537 Å owing to the comparatively low values of $\varepsilon\Phi$. The same applies to the amide derivatives of the first two amino-acid residues.

Doubtless all of the aliphatic amino-acids decompose when exposed to the full light of a mercury lamp. Qualitative data for a number of amino-acids and peptides are summarized in Table 3-4.

Histamine, and possibly imidazole aldehyde, has been found in irradiated histidine hydrochloride solution, according to early workers. Cystine and cysteine are "protective" (perhaps by competitive absorption of ultraviolet light) (Schocken, 1951). Under nitrogen the absorption increases by a factor of about 50 in the range of 2600–3400 Å with less than 1 per cent destruction (Luse, 1959).

Cysteine is identifiable by paper chromatography in irradiated cystine solution (Luse, 1961). Based on loss of identity of cystine the quantum yield is 0.12, which is higher than that reported by Lyons, (1948). During irradiation cysteine is likewise destroyed while some of it is also oxidized to cystine. Hydrogen sulfide is one of the products (Mischenko, 1952; Schocken, 1952). Luse has found that the optical density of cystine increases by factors of 1.6 at 2537 Å, 4.8 at 2800 Å and 70 at 3200 Å after 75 per cent decomposition in the presence of air and 0.1 N acid, so as usual not all the products are known.

Cystine, which suffers more degradation during the ultraviolet irradiation of wool than any of the other amino-acids, with the possible exception of tryptophan, is decomposed mainly at the —C—S— bond to give sulfur containing oxy-acids if the irradiation is conducted in the presence of air and in neutral or alkaline solution. In acid solution decomposition of cystine occurs mainly at the —SS— bond and cysteine, lanthionine and alanine are formed (Forbes and Savige, 1962; Eager and Savige, 1963). Free radicals are intermediates (Bogle *et al.*, 1962).

Both oxidized and reduced glutathione give rise to glutamylalanylglycine with a quantum yield of 0.06 in the absence of oxygen. The yield of this product is somewhat smaller when oxygen is present. From oxidized glutathione, reduced glutathione is also formed with a yield of 0.02 (Dose and Rajewsky, 1962). The total yield is therefore 0.08 (cf. Table 3-3 for the yield for cystine, which is of the same order of magnitude).

Irradiated homocystine also yields sulfhydryl (—SH) probably as homocysteine. Methionine is destroyed at 2250 Å (Schocken, 1952). The C—S bond is stronger in methionine than in homocystine (Forbes *et al.*, 1962).

Photo-oxidation has been judged to be less important for both glycyltyrosine and for tyrosine combined in protein than for tyrosine itself (Claesson, 1956). With glycyltyrosine a decrease of the maximum, an increase of the minimum which was shifted towards longer wavelengths and an absorption appearing at λ greater than 3000 Å is observed, Fig. 3-4. No changes in spectra of the irradiated solution were observed on standing. Frontal analysis of the solution revealed that the aromatic character of glycyltyrosine

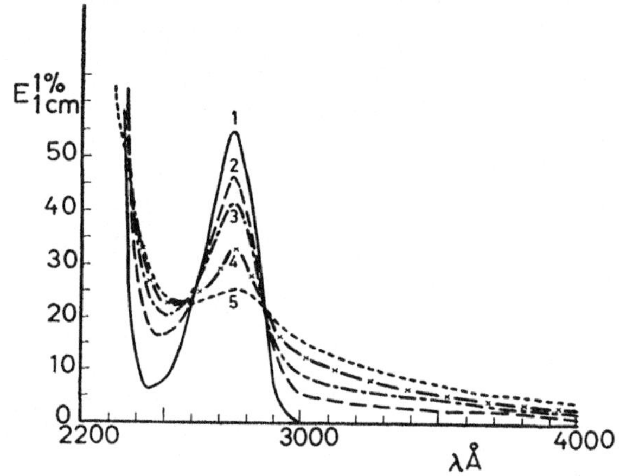

FIG. 3-4. Absorption spectra of glycytyrosine irradiated at pH 6.2 in oxygen saturated solution (Claesson, 1956). Curves numbered in order of increasing dose.

had disappeared completely with a simultaneous formation of ammonia. Similar results were obtained by Claesson with tyrosine, except at 2760 Å where the change is in the opposite direction. The dihydroxyphenylalanine reported among the products from irradiated tyrosine is itself subject to a rapid rate of decomposition (Arnow, 1938).

The difference in behavior of tyrosine and glycyltyrosine is noteworthy and shows that conclusions drawn with model compounds as test substances must be extrapolated with caution to macromolecules. The quantum yield for the destruction of glycyltyrosine based on the decrease in maximum at $\lambda = 2760$ Å is 0.038, which differs in order of magnitude from the yield for the loss of identity of tyrosine during irradiation at 2537 Å. This destruction should be checked by chemical analysis since the difference challenges the conceptual scheme outlined at the beginning of the chapter (cf., however, Chapter V). Arnow reports that tyrosine is very resistant to destruction by ultraviolet light when solutions are irradiated *in vacuo*.

The irradiation of phenylalanine or of propionylphenylalanine produces large increases in absorption of the entire spectra of the parent compounds (Claesson, 1956). A new band around 2100 Å is produced, which disappears slowly if the solution is allowed to stand. From the irradiated solution a substance with a higher ε than phenylalanine can be isolated by frontal analysis. Matsuda *et al.* report the formation of tyrosine and serine from phenylalanine in both acid and alkaline media and of alanine in acid solutions (1959).

Under oxygen or nitrogen the spectrum of tryptophan changes during irradiation to produce an increase at the minimum, a decrease at the maximum and the appearance of an absorption at above 3000 Å. The disappearance of the band at 2800 Å indicates a break-up of the aromatic nucleus. Among the products are kynurenine and indole ethylamine (Masson *et al.*, 1956; Deschreider and Renard, 1955).

Claesson (1956) has concluded that the general changes observed in the spectra are not primarily caused by oxygen, although the presence of oxygen increases the effect. Mischenko (1954) found no changes in the spectra of a number of amino-acids irradiated under air, argon or nitrogen. The amounts of ozone formed during irradiation at sufficiently short wavelengths are too small to be of influence (Giese *et al.*, 1952). A protein, absorbing a dose of ultraviolet light comparable to that required to produce the observed spectral changes in amino-acids, would undergo a spectral increase at the minimum at 2450 Å and above 3000 Å, and a decrease or increase at 2800 Å, depending on which residues are the predominant ones. A flattened-out spectrum results.

In contrast to phenylalanine and tryptophan, the spectral changes with glycyltyrosine seem to be more dependent on pH, under either nitrogen or oxygen. Suzatani (1958a) finds tyrosine to photolyze more rapidly in alkaline than in acid solution; the photolysis products are aspartic acid, asparagine and possibly alanine and glycine, but no dihydroxyphenylalanine.

In connection with these spectral changes it is of interest to note that the change in spectrum of pancreatic trypsin inhibitor protein during irradiation is quite similar to that found with a mixture of leucyltyrosine and propionylphenylalanine (McLaren and Waldt, 1952), Fig. 3-5. The phenomenon

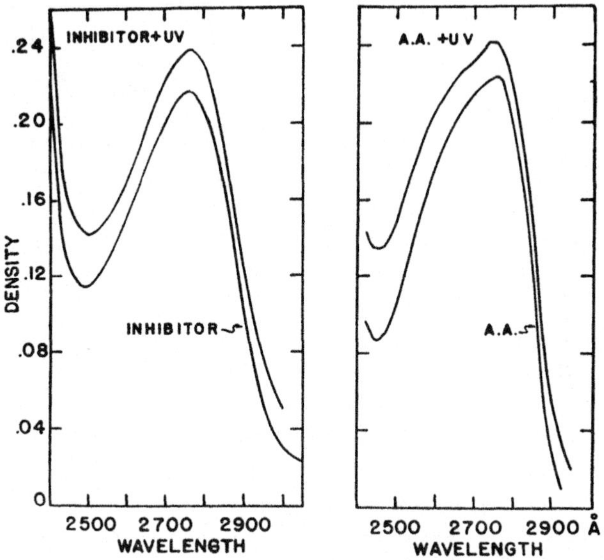

Fig. 3-5. Ultraviolet absorption spectra of (a) irradiated and unirradiated trypsin inhibitor and (b) irradiated and unirradiated solutions of a mixture (A.A) of leucyltyrosine and propionylphenylalanine. Both solutions received the same dose at 2537 Å.

is fairly general (McLean and Giese, 1950). Though it is true that under identical conditions a given number of chromophores absorb nearly the same number of quanta whether present in large or small molecules, the spectral changes may be somewhat different. As Claesson points out, in a solution of irradiated molecules with few chromophores some molecules will have the original extinction coefficient while others will have very different ones. In molecules with a sufficiently large number of chromophores chained together, however, the primarily changed chromophores are for statistical reasons rather evenly distributed in the molecules.

B. PHOTOSENSITIZED DESTRUCTION OF AMINO-ACIDS IN THE PRESENCE OF DYES

Amino-acids can be destroyed by visible light (and ultraviolet) in combination with "photodynamic dyes" such as methylene blue (MB). If these dyes are irradiated in the absence of oxygen there is no evidence of bleaching. When amino-acids are present to serve as hydrogen donors they may be dehydrogenated with a transfer of hydrogen to MB to form a colorless leuco-MB. The scheme is (Blum, 1941):

$$MB + h\nu \rightarrow MB^*$$

$$MB^* + H_nA \rightarrow H_nMB + \text{dehydrogenated amino-acid}$$

If oxygen is now admitted to the system the leuco-dye is oxidized via the reaction

$$H_nMB + \frac{n}{4}O_2 \rightarrow MB + H_2O$$

This is in contrast to photodestruction of the dye by oxygen in the absence of amino-acids or other reducing agents. The scheme outlined is only a rough one and does not include alternative side reactions and the possible formation of hydrogen peroxide.†

The photochemical action of methylene blue dye on amino-acids has been extensively studied by Weil *et al.* (1951). Tryptophan, tyrosine and histidine are particularly sensitive; methionine and cystine are also susceptible; the other amino-acids react sluggishly or not at all. In the first three, the entire oxygen uptake and carbon dioxide evolution are due to the cyclic nucleus and involve rupture of the rings.

The data for methionine are in accord with the following reaction mechanism for conversion to sulfoxide:

$$\text{Methionine} + MB \rightarrow \text{dehydrogenated methionine} + \text{leuco } MB$$

$$\text{Leuco } MB + O_2 \rightarrow MB + H_2O$$

$$\text{Dehydrogenated methionine} + H_2O \rightarrow \text{methionine sulfoxide.}$$

Although H_2O_2 may accumulate during the photo-oxidation of methionine, it is used up during the destruction of the aromatic compounds. Cystine is probably photo-oxidized beyond the cysteic acid stage. Significantly, the peptide bonds in glycylglycine and tetraglycine did not participate in the photochemical action of methylene blue.

These results with the amino-acids have been applied in a systematic manner to the photosensitized inactivation of enzymes by methylene blue,

† The above mechanism applies to the riboflavin sensitized photo-oxidation of indole acetic acid (Galston, 1950), tryptophan and a variety of 3-substituted indole derivatives (Shugar, 1951, 1954). The quantum yields for reactions of this type may, in some instances, be quite high, e.g. about 0.4–0.6 for indole acetic acid with either visible (4300–5500 Å) or near ultraviolet (3600 Å) light (Galston, 1949; Shugar, 1951). For tryptophan Φ is considerably less, about 0.038, whether the amino-acid is in the free state or incorporated in a protein molecule (Shugar, 1951).

as will be illustrated in Chapter V. The behavior of different dyes may differ appreciably, due, presumably, to stereochemical considerations. The fundamentals of photosensitization mechanisms in solution have been reviewed, amongst others, by Bowen (1950), Hochstrasser and Porter (1960), Schenck and Koch (1960), and at a special meeting of the French Soc. Chim. Phys. (*J. chim. phys.* **55**, 869–989, 1958).

CHAPTER IV

Action of Ultraviolet Light on Proteins: General

1. *Site of Action of Ultraviolet Light on Proteins*

If the activity of an enzyme involves a special, relatively small active site or "active center," and ultraviolet light absorbed by other parts of the protein molecule can affect this site, we can consider the energy as being transferred from these parts to the site and the action spectrum of the enzyme should correspond with the absorption spectrum of the whole protein molecule and not with that of the active group only. Experiments by Warburg and others on the splitting off of carbon monoxide from CO-haemin-protein by light absorbed by the haem or by the protein with equal probability (Φ approximately equal to unity at all wavelengths from 2600 to 6600 Å) are elegant in demonstrating the transfer of energy, q, intramolecularly (cf. Chapter I). The action spectrum coincides with the absorption spectrum of the Warburg respiratory enzyme and not with just the haem or protein moieties. Similar results have been obtained by Bucher and Kasper (1947) with CO-myoglobin. Vladimirov and Konev (1957) summarize all of the other relevant data on porphyrin–protein, and porphyrin–peptide systems and it seems that one can safely conclude that most, if not 100 per cent, of the energy absorbed by these systems can result in the cleavage of iron porphyrin-CO to free carbon monoxide. This cleavage requires only about 20–22 kcal, and so the absorbed energy is in 3–4 fold excess of the minimum required. Although all the known ligands able to combine with ferrohemoglobin are qualitatively similar in giving rise to photosensitive compounds, the quantum yields are as low as 0.001 for the others (Gibson and Ainsworth, 1957).

It is difficult to believe that all nine aromatic groups are in contact with the haem and, remarkably, the structure of myoglobin seems to be very "open," Fig. 4-1 (Kendrew *et al.*, 1958). If all proteins are so open, intramolecular photosensitization may not be efficient. Klotz has suggested an electron transfer over a water bridge to explain the results (Klotz *et al.*, 1958). If the absorbed energy produces an —S· free radical by either the reaction —S—S— + $h\nu \rightarrow$ 2—S·, or indirectly from an aromatic group by the intermediate appearance of H· followed by

$$H\cdot + \text{—S—S—} \rightarrow \text{—SH} + \text{—S}\cdot$$

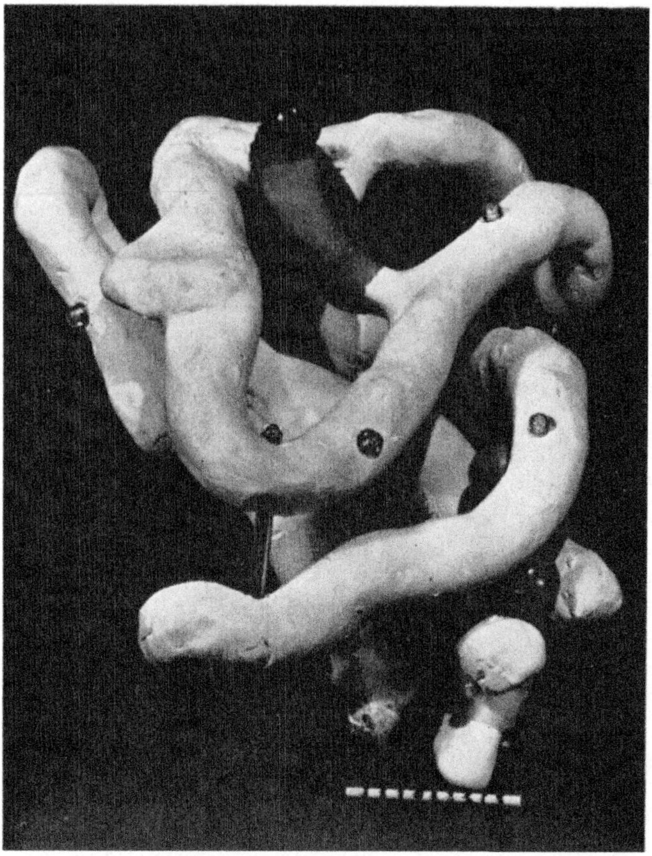

FIG. 4-1. Model structure of the myoglobin molecule. Polypeptide chains are white; the grey disk segment is the haem group. The three spheres show positions at which heavy atoms were attached to the molecule for X-ray diffraction analysis (Kendrew *et al.*, 1958).

the heme iron can be oxidized by the transfer of an electron, by means of H· transport through the water of hydration, to a distant —S· free radical to form the more stable —SH.

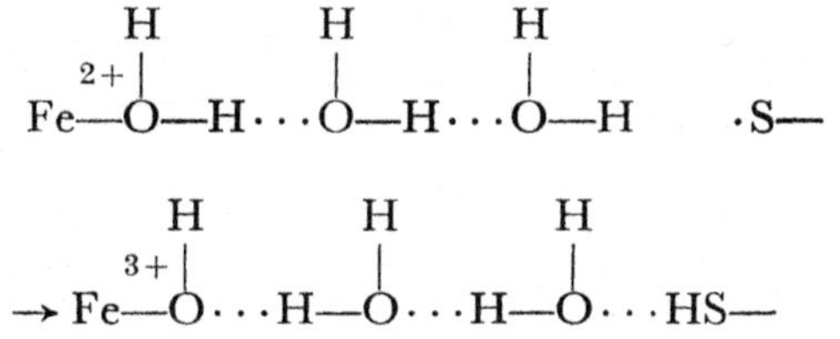

This scheme raises a question about the likely reversibility of the myoglobin-CO dissociation reaction.

Another electron migration hypothesis is that an excited electron from one atom can pass over into an energy level which spreads throughout the whole molecule and may remain there until it encounters a lattice defect or produces a reaction (Vladimirov and Konev, 1957). If the energy levels of the peptide bond do not vary greatly, then it is possible to consider that the activation of a peptide bond may be succeeded by a transfer of the activation energy to an adjacent bond with some or no loss (Goldfarb *et al.*, 1958, 1959). Eventually the absorbed energy could be lost by fluorescence, by chemical reaction or by heat. Now the quantum yields for inactivation of enzymes are nearly the same in solution as in the dry state (Setlow, 1957), whereas the peptide bond is hydrated in solution and therefore not of the same energy level as in the solid state (McLaren and Rowen, 1951) and it is well known that all peptide bonds in a protein cannot be acid hydrolyzed with equal ease, so the postulate does not seem useful or necessary for enzyme inactivation. A serious difficulty is that energy levels proposed for this method of energy transfer are not compatible with those deduced from the absorption spectra of proteins (Vladimirov and Konev, 1957).

A *resonance* transfer of absorbed energy from an excited chromophore remains as the most probable postulate for these results (Shore and Pardee, 1956a, b). The conditions required are (a) the excited chromophore must be fluorescent, (b) the fluorescent spectrum of the energy donor must overlap the absorption spectrum of the acceptor and (c) the donor and acceptor of energy must not be too far apart. All these conditions are fulfilled in protein molecules (cf. Chapter I). Aromatic amino-acids and proteins fluoresce in the near ultraviolet. Some results of Shore and Pardee are summarized in Table 4-1. Somewhat different results were obtained by Teale and Weber (1957), but the order of magnitude is the same. Such fluorescence generally occurs with a quantum efficiency independent of wavelength. Histidine, methionine, cysteine, cystine, and leucylgylcine do not fluoresce nor does ribonucleic acid from tobacco mosaic virus. Thus fluorescence of proteins

TABLE 4-1. FLUORESCENCE EFFICIENCIES BELOW 4000 Å OF SOME AMINO-ACIDS, PEPTIDES, AND PROTEINS IN AQUEOUS SOLUTION (SHORE AND PARDEE, 1956A). EXCITATION AT 2800 Å.

Compound	Efficiency per cent		
	pH 1–1.3	pH 6–7	pH 11
Tryptophan*	2.7	6.0	9.8
Glycyltryptophan	1.7	2.6	3.6
Tyrosine	1.5	4.0	1.2
Leucyltyrosine	1.6	2.0	0.6
Phenylalanine	1.3	1.4	1.3
Ribonuclease	—	1.5	—
Chymotrypsinogen	—	4.6	—
Lysozyme	3.4	2.8	2.6
Tobacco mosaic virus	—	7.0	—
TMV-protein	—	9.3	—

* With tryptophan no oxygen effect was observed.

and nucleoproteins may be due entirely to aromatic amino-acid residues. Haem absorbs fairly strongly in the region in which proteins fluoresce (Steele and Szent-Gyorgyi, 1958), which probably accounts for the nearly 100 per cent energy migration observed by Warburg and Bucher and Kaspers.

Shore and Pardee were able to prepare a conjugate of the dye 1-dimethyl-amino-naphthalene-5-sulfonate with chymotrypsinogen which gave an efficiency of transfer of ultraviolet light energy from protein to dye of from 55 to 94 per cent in the range of excitation by 2537–2750 Å light. Below 2500 Å the transfer efficiency fell rapidly to zero, perhaps because of increased absorption by non-fluorescing cystyl residues. It would be of great interest to know if the addition of such a conjugate dye to an enzyme would reduce the yield of inactivation. No energy transfer from nucleic acid to dye was observed. Dilute sodium hydroxide quenches the fluorescence of the protein and reduces the energy transfer to dye. 6 M urea, a denaturing agent, does not quench the fluorescence of proteins but does reduce the efficiency of transfer to dye by 30–50 per cent. This result seems incompatible with any alternate mechanisms which depend on a highly hydrogen-bonded system in the protein. They also found that light absorbed by the nucleic acid moiety of TMV apparently does not lead to fluorescence of the virus, indicating no transfer of energy from nucleic acid to protein. Conversely, the nucleic acid does not quench the protein fluorescence, indicating no transfer from protein to nucleic acid.

For enzymes which consist only of amino-acid residues (no prosthetic groups), action spectra generally do not correlate well with the absorption spectrum of the entire protein (an exception is aldolase, Setlow, 1957). Setlow has made a detailed analysis of the action spectrum of trypsin. He has concluded that inactivation can occur following the absorption of a quantum by more than one kind of chromophore, and without the necessity of postulating energy migration, but the most sensitive chromophore in trypsin, according to his calculation, is cystine. Taking the chance Φ of inactivation of trypsin by absorption by a cystinyl residue as 0.07, and the chance of inactivation for a quantum absorbed by the rest of the molecule as only 0.002, Setlow could account for the action spectrum of trypsin. This demonstrates a minor role for the aromatic amino-acids in alternate pathways leading to inactivation between 2300 and 2700 Å (cf. McLaren and Luse, 1961). In addition to the fact that quantum yields for inactivation are low compared to fluorescence efficiencies of the dye-protein conjugates and dissociation of heme-protein —CO complexes, they generally vary with wavelength and pH. Energy transfer as an important factor in inactivation would therefore be very difficult to demonstrate. In fairness to any such proposals, however, it should be remembered that only about 1 per cent to 0.1 per cent of absorbed quanta are involved in inactivation, whereas the proposed transfer mechanisms apply to the remaining 99 per cent or so of the absorbed energy.

2. *Changes in the Absorption Spectra of Proteins with Irradiation*

Aside from the obvious precipitation of a protein from solution which occurs when some proteins are irradiated, perhaps the most easily observable change is that in the absorption spectrum. It is well to remember that although a given number of amino-acid chromophores in neutral solution absorb nearly the same number of quanta near 2800 Å whether they are present in the free state or incorporated into a protein molecule, the spectral changes may be different on account of the possibility of energy transfer in macromolecules. It is also possible that the more rigid surrounding of a chromophore in a macromolecule may influence the rate of reaction of excited chromophores with solvent and other molecules: chromophores at the surface of a molecule and those in the interior may react differently (cf. Laskowski, 1961). Thus, although the study of the behavior of chromophores in micromolecules will give valuable information, the information obtained may not be fully applicable to the reaction of the same chromophores in large molecules (Claesson, 1956). In a solution of irradiated micromolecules with few chromophores some molecules will have the original extinction coefficient while others will have different ones. With molecules having a fairly large number of strongly absorbing chromophores, say about 10 per cent as with proteins, the changed chromophores are for statistical reasons probably rather evenly distributed in all molecules.

At a given wavelength, if a reactant goes over to a product with no further change in the product we may write

$$[A_0] - [A] = \frac{1}{\varepsilon_P} (D_t - D_A) \tag{4-1}$$

where ε_P is the molar extinction coefficient of the product, D_t is the optical density of the mixture at time t of irradiation and D_A is the density of the native protein of initial concentration $[A_0]$ in the mixture. The equation is plotted in Fig. 4-2 for chymotrypsin (McLaren and Finkelstein, 1950).

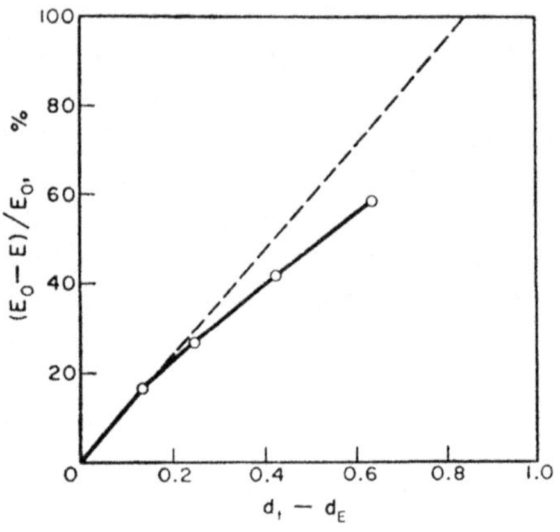

FIG. 4-2. Changes in the absorption of chymotrypsin solution (2500 Å) irradiated in air at 2537 Å at pH 7.5, for an initial enzyme concentration of 2×10^{-8} mole/ml. Here E = A and d = D.

Further photochemical changes in the initially formed product are responsible for an early departure from the simple relationship. In calculating quantum yields one has to take into account these changes. Equation (1-8a) in differential form, namely

$$\frac{-d[A]}{dt} = \Phi I_{abs} \frac{\varepsilon_A[A]}{\varepsilon_P[P] + \varepsilon_A[A]} \tag{4-2}$$

can be rewritten as

$$\frac{-d[A]}{dt} = \Phi I_0[A] \frac{1 - e^{-(\varepsilon_A[A] + \varepsilon_P[P])}}{[A] + ([A_0] - [A])\varepsilon_P/\varepsilon_A} \tag{4-3}$$

where the extinction coefficients are for the wavelength of irradiation. With tobacco mosaic virus $\varepsilon_A = \varepsilon_P$ and the equation simplifies at once to

$$\frac{-d[A]}{dt} = \Phi \cdot I_{abs} \frac{[A]}{[A_0]}$$

the usual first order equation. With chymotrypsin under nitrogen $\varepsilon_P = 2.1\varepsilon_A$ initially and changes with continued irradiation, Table 4-2 (Finkelstein, 1949). Even so, the quantity in brackets changes only from 0.24×10^5 initially to 0.23×10^5 at 23 per cent inactivation, showing the familiar semi-logarithmic plot of inactivation data can sometimes be insensitive to changes in the light-absorbing power of the system (McLaren, 1951). At 23 per cent inactivation, if $\varepsilon_A = \varepsilon_P$, the enzyme in the mixture would absorb 0.77 of I_{abs}. Actually the enzyme receives only

$$\frac{0.77\ \varepsilon_A}{0.77\ \varepsilon_A + 0.23 \times 2.1\ \varepsilon_A} = 0.62$$

of the I_{abs}, yet $K = \ln([A_0]/[A])/t$ varies only four per cent in theory and was found to be experimentally constant. From the fact that ε_P calculates to be 72,000 at 23 per cent inactivation but is found to be only 56,500 at

TABLE 4-2. OPTICAL BEHAVIOR (2537 Å) OF IRRADIATED CHYMOTRYPSIN
$A_0 = 2 \times 10^{-8}$ moles/ml, pH = 3.0. IRRADIATION UNDER NITROGEN

Inactivation, per cent	Apparent ε_{2537}	$\dfrac{1 - e^{-(\varepsilon_A[A] + \varepsilon_P[A])}}{[A] + ([A_0] - [A])\varepsilon_P/\varepsilon_A} \times 10^{-5}$
0	34,200	0.24
23	43,000	0.23
99	56,500	0.34

99 per cent inactivation indicates that the initial products have a higher specific absorption than do the secondary, tertiary, etc., products formed from them during continued irradiation.

With concentrated solutions of equine encephalomyelitis virus the quantity in brackets is rather far from constant (Taylor *et al.*, 1941) but could be solved graphically (Finkelstein, 1949; Englehard and Eikenberg, 1955). Otherwise one can calculate apparent quantum yields and extrapolate these back to zero dose in order to obtain the best value. For aldolase a graphical method has been used; the resultant correction was only about 3 per cent (Labeyrie and Shugar, 1951).

The changes in absorption spectrum during irradiation of protein in solution can result from photochemical reaction and from aggregation which causes light scattering (Pavlovskaya and Pasynskii, 1955). The latter effect will always be toward enhancement. The change in ultraviolet spectrum of chymotrypsin has been studied under both nitrogen and oxygen (Fig. 4-3). After irradiation under nitrogen the spectrum parallels the shape of the

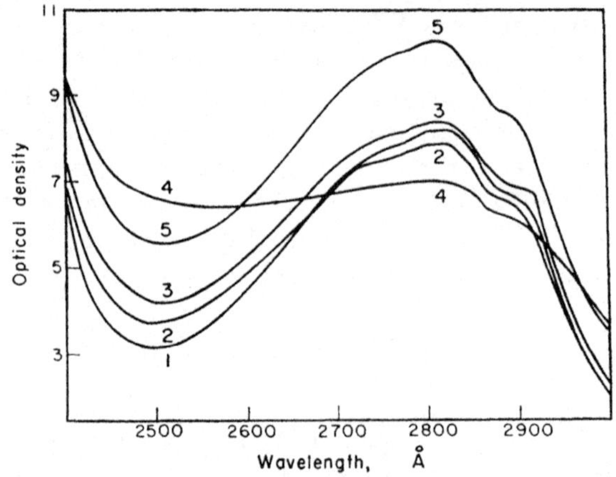

FIG. 4-3. Extinction versus wavelength of irradiated chymotrypsin solution, pH 3.0. 1, control: unirradiated; 2, irradiated under oxygen, 73.5 per cent remaining activity; 3, irradiated under nitrogen, 77 per cent remaining activity; 4, irradiated under oxygen, about 0.3 per cent remaining activity; 5, irradiated under nitrogen, about 0.3 per cent remaining activity (Finkelstein and McLaren, 1950).

unirradiated sample, but is displaced upward, whereas the samples irradiated in oxygen have spectra which dip below the control in the region of the absorption maximum. The quantum yield for inactivation, however, does not depend on the atmosphere during irradiation. Without enzymatic activity as an additional sensitive parameter for observation, one would not be able to distinguish an initial process (defined as loss in activity) from one giving rise to changes in spectra. Earlier work on proteins is similar to the results with chymotrypsin (McLaren, 1949).

By contrast with chymotrypsin, which has a relatively small amount of tyrosine in comparison with the phenylalanine and tryptophan, the absorption of bovine serum albumin is predominantly that of tyrosyl and phenylalanyl residues. The spectrum of irradiated bovine serum albumin shows much larger changes during irradiation (2537 Å) than that of the corresponding aromatic amino-acid mixture. Light scatter was not the cause, so Claesson (1950) suggested that energy transfer to phenylalanine and to histidine (which absorbs very little by itself at 2537 Å) could be partly responsible. An irradiated serum albumin solution was fractionated in an ultracentrifuge and each fraction was found to have the same specific absorption irrespective of molecular weight of the fragment or aggregated material, suggesting that the spectral changes occurred rather uniformly for all molecules.

Giese *et al.* (1950, 1952) and others have attempted to explain some of the changes in protein spectra in terms of changes in the aromatic groups.

The absorption spectra of benzene and phenol before and after irradiation closely resemble those of phenylalanine and tyrosine. Colorimeter tests for phenolic groups with irradiated phenylalanine and ovalbumin showed that these are formed during irradiation. From comparison with the spectra of irradiated quinone and hydroquinone, it was suggested that quinones might be intermediates in the oxidations, which can lead to increases in absorption above 2700 Å . The loss in maximum near 2780 Å during irradiation of tryptophan is a spectroscopic indication that changes of tryptophanyl residues are responsible for the decrease in the spectrum of chymotrypsin at this maximum during irradiation under O_2.

Studies with proteins and amino-acids have usually been made in the presence of air, because of simplicity. Giese *et al.* (1952) observed that absorption spectra of serum and egg albumin and mixtures of aromatic and aliphatic amino-acids are altered by ozone as well as by ultraviolet light. Ozonation and irradiation produce characteristic changes which, however, can be opposite in nature, and it was therefore concluded that the likelihood of an explanation of ultraviolet effects on the basis of ozone produced during irradiation is improbable. It cannot be concluded that the influence of oxygen has been explained quantitatively, however, since published data reveal that the changes in spectra depend on the oxygen content of the atmosphere employed. Hydrogen peroxide is not obtained from water at the wavelengths generally used on proteins and thus cannot be involved.

It has been reported that sodium caprylate has a protective effect during irradiation of serum albumin since the increase in optical density is less in its presence (Duggan *et al.*, 1950). This protection may be really a prevention of aggregation with its concomitant light scatter of solution (Claesson, 1956) . Caprylate does not protect serum albumin against photolysis into protein fragments soluble in 10 per cent trichloroacetic acid (Giese, 1950). It does not prevent at all the changes in spectrum of chymotrypsin accompanying ultraviolet irradiation up to 90 per cent inactivation, so the phenomenon is not general. A practical application of this phenomenon has been suggested (Hinks and Johnson, 1951). The addition of 0.04 M caprylate to serum makes it possible to irradiate the latter so that the icterogenic virus is destroyed, while albumin and other components are unaffected, under the conditions of irradiation employed. Subsequent fractionation of these components results in the reduction of the caprylate concentration to permissible levels. It is claimed that the following plasma components of clinical usefulness can be prepared in this way: albumin, γ-globulin, fibrinogen, and thrombin.

Spectrophotometric measurements of color changes in wool induced by ultraviolet light have been reported (Lennox, 1959). Proline is considered to be a radiation-sensitive amino-acid along with cystine (Hildebrand and Kersten, 1959).

3. *Photolysis and Aggregation of Irradiated Proteins*

On the basis of the action of ultraviolet light on peptides and amino-acids, one may expect that it will produce fragments from proteins either by the photolysis of peptide bonds and/or of —S—S— linkages between polypeptide chains. An absorbed quantum can conceivably instigate any one of a large number of reactions within a protein molecule. Among these possibilities we would like to know the relative probability that fragmentation will occur. Since an enzyme might be inactivated by modification of an active site, or by denaturation of the molecule by a non-hydrolytic reaction (Neurath *et al.*, 1944) or by fragmentation, we would also like to know which mechanism predominates during inactivation, or perhaps takes place during a subsequent dark reaction if one or more inactivation pathways involves both a light and a dark reaction. Several authors have suggested that photolysis of —CONH— linkages is a significant process (cf. McLaren, 1949). Let us examine the evidence.

(*a*) *Photolysis of the Peptide Bond.* Proteins have been irradiated in solution, as monolayers, and in the dry state at 2537 Å. McLean and Giese (1950) tested for hydrolysis of peptide bonds in ovalbumin by ultraviolet irradiation in solution, by formal titration for acidic groups, titration of basic groups and by the method of Van Slyke for amino groups. No increase in basic or amino groups was found during irradiation but acidic groups were produced with a quantum efficiency of 7×10^{-4}. Since it is most unlikely that amino groups would be destroyed as fast as formed (not necessarily the same ones) in the molecule, we are forced to conclude that the simple reaction

$$\text{—CONH—} + H_2O + h\nu \rightarrow \text{—COO}^- + \text{—NH}^+$$

is not taking place. Further, and for the same reason, there is no way to tell what reaction produced the titratable acid groups. They suggest that transient phenolic groups might be responsible, since with the dosage used (250 quanta per chromophore at 2537 Å) the ultraviolet spectrum of the protein increased. Leitner (1950) found no increase in titratable groups during prolonged irradiation of bovine serum albumin and Luse (1961) has obtained the same results with chymotrypsinogen, DFP-chymotrypsin (i.e. chymotrypsin treated stoichiometrically with di-isopropyl fluorophosphate and thereby rendered inactive; Jansen and Balls, 1952), and ribonuclease, with dosages sufficient to inactivate up to 99 per cent of the enzymes. In other words, the quantum yields for production of titratable acidic or basic groups are too low to account for the much higher, by a factor of ten to one hundred, quantum yields for inactivation of the enzymes. Also lysozyme and lactoglobulin solutions were irradiated by a 200 W. Hg lamp from a distance of 15 cm, for 15–20 hr: because the total amounts of acidic and basic groups did not change and because of no change in paper chromatograms of the solutions, it was concluded that peptide bonds were not hydrolysed (Obata and Koshika, 1959).

Kaplan *et al.* (1954) found no increase in acidic groups during the first hour of irradiation of dilute (0.01 per cent) solutions of ovalbumin at an unspecified intensity. Thereafter titratable groups appeared which Kaplan assumed to be —COOH. During the first hour the irradiated protein was rendered coagulable and non-spreading as monolayers at an air-water interface at its isoelectric point (pH 4.9); yet the protein could be spread to an even larger specific area at pH 2.25. After 10 hr of irradiation the protein had lost the capacity to form films, due to a destruction of the molecule with the appearance of small soluble, non-spreading fragments. Also the consumption of base amounted to 2 acidic groups for every 5 residues present in the protein. This is so great that the non-specific oxidation of aromatic groups suggested by McLean and Giese is not adequate to account for the results (Kaplan *et al.*, 1954).

With prolonged irradiation rupture of the primary structure of a protein will take place. The importance of the results with the model compound acetylalanine (Chapter III) are informative here, for from them we predict that simple rupture of —CONH— is probably not taking place. The first action observable on the absorption of ultraviolet light is probably a modification not involving photolysis of —CONH— but rather the rupture of some —SS— and hydrogen bonds which change the *tertiary* and perhaps the *secondary* structure of the protein to an extent similar to that brought about by the usual denaturing treatments (for example, heating in boiling water). Gibson (1959) has described the formation of a quickly reacting form of hemoglobin by flash photolysis. Following the photo-decomposition of carboxyhemoglobin, the reduced hemoglobin formed exhibits, for a brief period, an enhanced rate of reaction with oxygen or carbon monoxide. The product differs only slightly in its spectroscopic characteristics from ordinary reduced hemoglobin. The rapidly reacting product is pictured as a molecule which has lost its ligand molecules within a short time by photochemical action, but has not had time to undergo the structural alterations corresponding to the change from the saturated to the reduced state.

One must be careful in interpreting titration results on an irradiated proteolytic enzyme such as pepsin, which can digest itself with the production of —COOH and —NH_2 groups (Northrop, 1934). Although neither DFP-chymotrypsin nor chymotrypsinogen show an increase in titer following doses equivalent to those causing inactivation of chymotrypsin, irradiation of the latter leads to the production of large amounts of titratable groups. With the acetone acid titration of Linderstrøm–Lang using napthyl red indicator, Leitner (1950) found that during irradiation at pH 7.15 and 4.66, about 23 basic (H^+ accepting) groups appear initially per molecule inactivated (molecular weight taken as 41,000). Since there are about 26 aromatic amino-acid residues per molecule of chymotrypsin, it seems likely that these titratable groups have arisen from the digestion of peptide bonds adjacent to these groups by residual active enzyme. (Peptide linkages involving aromatic residues are specific substrates for chymotrypsin.) Although this explanation is attractive, the increase of basic groups does not occur at a constant rate during inactivation as do acid groups, as determined by formol titration, so some other factor must be operating (cf. Fig. 4-4). The other factor might have been the production of basic groups

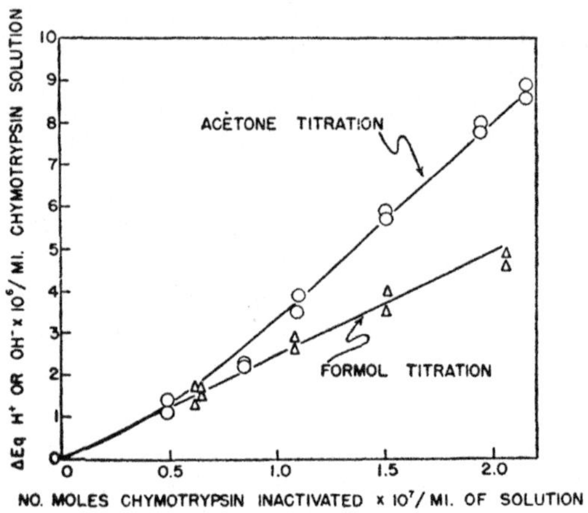

FIG. 4-4. Titration results on a solution of chymotrypsin 16.4 mg/ml irradiated at pH 4.66. Molecular weight taken as 41,000. For titration in acetone, hydrochloric acid and naphthyl red indicator was used; the formal titration was performed with the aid of the glass electrode (Leitner, 1950).

from the destruction of aromatic rings, but the irradiation of a mixture of amino-acids corresponding in composition to the amino-acid content of the enzyme did not result in the appearance of new titratable groups. Native proteins are often digested at a slower rate by trypsin than denatured proteins, and since serum globulin following irradiation is hydrolyzed by trypsin at a higher rate than native serum globulin we have here evidence for the denaturing action of ultraviolet light. Indeed, Haurowitz and Tümer (1949) suggest that the injurious effect of ultraviolet light on living organisms is attributable to the formation of points of attack for intracellular proteolytic enzymes. The action of these enzymes could lead to breakdown of the cellular framework and to the death of the cell. There are many more ideas on this subject however, which will be considered along the way.

An observation that absorption at 2537 Å by peptide bonds probably does not lead to photochemical inactivation is that clupein, unlike lysozyme and ribonuclease, does not lose its ability to degrade apurinic acid when irradiated at this wavelength. Consequently, the lack of inactivation at 2537 Å may be regarded as reasonable proof that the effect of such radiation on enzymes is manifested, not via peptide bond absorption in whole or in part, but rather as a result of aborption by aromatic amino-acids, cystine and/or cysteine (Shugar *et al.*, 1959). Clupein does lose activity against apurinic acid under the influence of radiation below 2400 Å, a region in which CONH begins to absorb appreciably.

(b) *Effect of ultraviolet light on disulfide linkages.* The photolysis of —SS— bonds in proteins, as one of the first reactions that can occur during irradiation, is not precluded by these observations. As already mentioned, the action spectrum of trypsin has been interpreted on this basis, and cystine in wool is lost during irradiation (Haly, 1957). Engelhard and Eikenberg (1955) measured the photolysis of disulfide linkages in bovine serum albumin in water and calculated a quantum yield of the order of 10^{-3} based on the appearance of —SH groups. From each —SS— group one —SH group was formed. They reported a concentration dependence for this reaction and concluded that five simultaneous excitations were involved in each splitting of an —S—S— linkage on the basis of a formal kinetic equation. This conclusion, though tenuous (Setlow, 1957), is difficult to argue with. One might guess that at low intensities some kind of a resynthesis of —S—S— linkages might take place, or side reactions could effect the concentration of —SH groups, although no temperature dependence for the appearance of these groups was found below 45°C. Above 45°C heat denaturation is observable and the efficiency of production of —SH groups with irradiation does decline somewhat. The experience of Englehard and Eikenberg is a noteworthy exception to the general applicability of the reciprocity law to proteins. It may be significant that a yield of 10^{-3} is of the order of magnitude of quantum yields for inactivation of enzymes. The authors also reported (1956) that precipitation of serum albumin at its isoelectric point following irradiation also takes place with a yield of 10^{-3}, showing a close relationship between ultraviolet-induced denaturation and cleavage of —S—S— linkages. These results parallel the reduction of these linkages in chymotrypsin which also results in inactivation and reduction of solubility (McLaren, 1957).

Kofman (1957) and also Pavlovskaya and Passynskii (1956) have observed a photo-oxidation of —SH groups in ovalbumin. The quantum yield is 2×10^{-3}. In the presence of cysteine or ascorbic acid the yield can be held to zero. The photo-oxidation of the protein occurs independently of photo-denaturation (i.e. loss of solubility at the isoelectric point). If an ultraviolet-irradiated solution of egg albumin (pH 8) is first held at 40°C to complete a "dark" reaction state of denaturation and then brought to pH 4.8 a precipitate appears; the denatured protein is removed by filtration and the filtrate contains a photo-oxidized soluble product (SP). Although the soluble product is not precipitated at the isoelectric point, it is less stable than native ovalbumin as regards solubility on standing in ammonium sulfate. Following heat denaturation—precipitation of the soluble product, the solution contains perhaps one per cent of a high molecular weight, soluble product too large to pass through cellophane and lacking absorption by specific aromatic residues. A considerable fraction (34 per cent) of the —SH groups of the (SP) is oxidized, but this fraction is larger (65 per cent) if ovalbumin is heat-denatured prior to irradiation. Photo-oxidation also, as is to be expected, takes place in the denatured fraction. On more prolonged irradiation a further more deep-seated stage of photo-oxidation begins which takes in the remaining —SH groups. Probably this involves the denaturation of the ovalbumin, since the 34 per cent oxidized —SH groups occurs as a plateau of reactivity whereas only 12 per cent denatured protein is present in irradiated ovalbumin. Photo-oxidation of the protein —SH groups shows an induction period, whereas the appearance of denatured protein does not. This induction period parallels the freeing of bound tyrosine groups† in the ovalbumin, which also indicates a loosening of the tertiary structure of the protein under the influence of ultraviolet light. These results are a logical extension of the early work of J. H. Clark (1936), (cf. McLaren, 1949). Since the energy of a quantum of ultraviolet light can be more than that required to break a bond in the primary structure of the protein, there is energy left over. In a simple molecule the "left over" energy may be used to impart translational energy to the products. In a complex molecule the left over energy can be considered to produce alterations in secondary and tertiary structure such as to lead to changes in optical rotation and reduction in loss of solubility (McLaren, 1949; Kauzmann, 1954). These changes are similar to those produced by heat denaturation and perhaps involve disruption of the "ice-berg" structural water of the native protein molecule as well (Klotz, 1958).

(c) *Aggregation phenomena.* That the state of an ultraviolet irradiation denatured molecule is not quite the same as a heat denatured molecule has been shown by a solubility study of denatured soybean trypsin-inhibitor (Katchman, 1955). Although the inhibitor is insoluble in dilute acetate buffer at the isoelectric point, pH 4.5, following heat denaturation, it is still soluble after loss of 30 per cent activity by ultraviolet irradiation. Precipitation at the isoelectric point does occur upon further radiation-inactivation, showing that ultraviolet absorption by inactive protein produces further changes in the initial products. A plot of quantum yield for loss of inhibitor activity versus pH reveals a minimum at about pH 7, and there is no reason, therefore, to assume that the ultraviolet-stability

† Tyrosine groups may also be involved in the supercontraction of wool irradiated with ultraviolet light (Haly *et al.*, 1957).

of this protein is related directly to a condition of net zero charge of the entire molecule. Although at least some of the tyrosyl-OH groups are bound in native inhibitor, the binding is not involved in the stability of the inhibitor toward ultraviolet light; the Φ vs. pH plot shows no inflection in the neighborhood of pH above which the state of binding of tyrosine is changed.

Electrophoretic and sedimentation patterns of some irradiated proteins flatten out and others do not (Obata and Sakamura, 1958; Sakamura and Obata, 1958). Irradiation of bovine serum albumin under nitrogen causes an aggregation of molecules (Rideal and Roberts, 1951). Although there is some aggregation initially, during irradiation under oxygen, this is superseded by a photo-oxidative decomposition of the protein, Fig. 4-5.

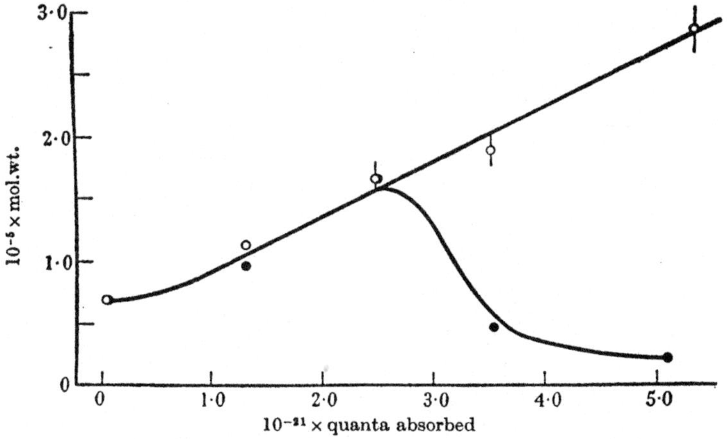

FIG. 4-5. Irradiation of 1 per cent solution of serum albumin under nitrogen, ○, and under oxygen, •.

Since irradiation of the protein under nitrogen and in the presence of acrylonitrile leads to polymerization of acrylonitrile, it is clear that free radicals are produced by ultraviolet light. These free radicals are presumed by Rideal and Roberts to be responsible for the aggregation of the protein molecules. (That two distinct *macro*-molecular free radicals arise from the photolysis of any of the *n*—CONH— linkages present seems most unlikely because of the comparatively small value of the product $n\varepsilon\Phi$ compared to those of other chromophores present (Setlow, 1957).) If protein free radicals react with tyrosyl, carboxyl, amino and thiol groups of other molecules, then the addition of low molecular weight compounds such as hydroquinone and cysteine should reduce the aggregation. In fact acetic acid and tyrosine (also hydroquinone) were effective, but cysteine and methylamine were not.† The suggested reactions of the two free radicals formed, P·, were‡

$$P_1\cdot + HOC_6H_4OH \rightarrow HOC_6H_4OP + H\cdot$$
$$P_2\cdot + RCOOH \rightarrow RCOP + \cdot OH$$

The luminal test for hydroxy radicals was positive in irradiated solution.

The authors carried out a kinetic analysis based on the questionable assumption that each of the two radicals produced was completely selective in reactivity toward a carboxylic acid or a phenolic compound. If one accepts the assumption, the results of the analysis are that the rate of aggregation is proportional to the square root of the product $I_{abs}C_xC_y$ where C_x and C_y are the concentrations of carboxylic and phenolic hydroxyl respectively. Radical recombination seems, on the basis of the kinetic analysis, to be much faster than aggregation.

Claesson (1956) has extended this reasoning by a consideration of the influence of secondary reactions on quantum yields. The most likely secondary reaction which might interfere with a determination of loss of identity of the original reactant is a combination of a denatured or excited molecule or macroradical with the former. Thus, for a native protein A she writes

$$\text{(a)}\quad A + h\nu \rightarrow A^*$$
$$\text{(b)}\quad A^* + A^* \rightarrow A\text{—}A$$
$$\text{(c)}\quad A^* + A \rightarrow A\text{—}A + R\cdot$$
$$\text{(d)}\quad A^* + S \rightarrow A + R\cdot$$

where R· is any possible small radical and S is solvent. If reaction (b) is dominant no problem arises in computing a yield for the primary process. If reaction (d) is dominant, there is no change in molecular weight. If reaction (c) is dominant the situation is complex. Initially when the majority of molecules are unchanged reactant, i.e. monomer, two molecules of monomer disappear for each effective quantum hit. A determination of the quantum yield obtained from the initial slope of a plot of $\ln[A_0]/[A]$ vs. time will then be twice that for the primary reaction (a). Reaction (c) has been tested with chrymotrypsin and, since the quantum yield for inactivation is independent of both initial concentration and extent of inactivation, we may conclude that it is of little or no importance in the primary process of enzyme inactivation. This is reinforced by the observation that quantum yields for inactivation

† It was not stated whether corrections were applied for ultraviolet absorption by the added compounds.

‡ Free radicals do not react with hydroquinone as written, however, but instead produce substituted hydroquinones (Price and Read, 1946).

of enzymes in solution or as solids are not much different (Setlow, 1957). For the reaction in which there is splitting into two large reactive species,

$$\text{(e)}\quad A + h\nu \rightarrow A_1^* + A_2^*$$

with a dominating reverse reaction, the actual concentration of A*, as already pointed out, would be proportional to the square root of the light intensity. This also is not likely with chymotrypsin since the quantum yields for enzyme inactivation are independent of intensity, indicating that if such a splitting reaction occurs the reverse reaction is negligible. Further, Claesson has reconsidered the osmotic pressure measurements of Rideal and Roberts after finding from ultracentrifugations that aggregation is the principal reaction. The equation for the number average molecular weight M_n is

$$M_n = \frac{[A_0]M_1}{[A_0] - \Phi_0\,(\text{dose})}$$

where Φ_0 is the quantum yield obtained from the osmotic pressure data. She points out that if $1/M_n$ is plotted against the dose, straight lines are obtained initially, which also eliminates the notion of an induction period as drawn in Fig. 4-5. The slopes are proportional to Φ and independent of intensity. Thus this straightforward, simple analysis of the data leads to quite a different conclusion from that drawn by Rideal: reaction (e) does not seem likely for serum albumin either.

A systematic study of the sedimentation behavior of irradiated bovine serum albumin gave the following results. Under air or oxygen, only seemingly unchanged albumin (with a molecular weight of *ca.* 69,000) and aggregation products are observed. This renders the fall of M_n in oxygen reported by Rideal and Roberts, Fig. 4-5, difficult to explain. The very small amounts of dialyzable products formed (1–2 per cent of aromatic-group containing compounds) cannot be responsible. From the appearance of her sedimentation diagrams it may be concluded that multiples of the original molecular weight are present in the irradiated solutions with no predominance of any particular component. It is not possible to reconcile the data with a splitting of the monomers into two large radicals which then cause the aggregation. Plots of log $[A]/[A_0]$ vs. dose for the monomer gave straight lines under the various atmospheres, independent of intensities of ultraviolet light. From these the sedimentation quantum yields Φ_s were found to be 2.2×10^{-3} under nitrogen and air and 1.1×10^{-3} under oxygen at pH 6.0–6.2. If the aggregation takes place between an activated monomer and a native monomer, these values will be too high by a factor of 2. If reaction takes place between two activated molecules the yields can be taken at face value. Since Φ under air from the ultracentrifuge data is about twice that obtained from the above equation for M_n from osmotic data ($\Phi_s = 2\Phi_0$), Claesson concludes that an activated molecule aggregates with an intact molecule.

It must be remembered, however, that the sedimentation data and the osmotic pressure data were obtained in two different laboratories with different samples of albumin, and a factor of two in quantum yields under variable conditions is not unusual even with similar assay methods (McLaren, 1949). At present it appears likely that the following reactions are also probable, where P is a product, namely

$$\text{(f)}\ A^* + S \rightarrow P + R$$

$$\text{(g)}\ P + P \rightarrow P\text{—}P\ \text{(aggregation)}$$

and that the quantum yields may be taken as representative of a one-hit process involving one quantum for each molecule lost as measured by sedimentation analysis. With chymotrypsin, for which the quantum yield for inactivation is constant over a wide range in concentration, reaction (f) probably applies as it is concentration independent. Reaction (g) is also applicable if the equilibrium constant heavily favors the product P—P. On the other hand, if reactions (b–d) were prominent, Φ would vary with concentration because of a competition of solvent and protein for the excited reactant. The sedimentation behavior of irradiated serum albumin also seems to be independent of concentration. Furthermore, irradiation of serum albumin at pH 2.2 did not produce aggregation but aggregates appeared if such solutions were dialysed against pH 6.0 buffer, showing the probabilities of reactions (f) and (g) over these involving A* species with A* or A.

Claesson also checked the experiments of Rideal and Roberts involving irradiations of albumin in the presence of tyrosine and hydroquinone. No signs of products with sedimentation constants lower than that of monomeric albumin were observed, hence it is doubtful that the monomer splits into two large free radical fragments. From the polymerization results with acrylonitrile we know that free radicals are formed but, since one does not know how many free radicals are produced per quantum, it is not possible to conclude how important free radical formation is in the denaturation process or whether it is only a minor reaction. A decrease in aggregation products was found, which was expected from the inner optical filter effect of the added phenols. Since the aggregation reactions are not temperature dependent, reaction (g) must have a low heat of reaction. Also the ultracentrifuge diagrams were the same for irradiated material which had stood 3–24 hr (during this time the absorption spectrum changed somewhat) and were not much influenced by moderate changes in ionic strength.

Hemocyanin from *Helix pomatia* undergoes a definite splitting into smaller fragments upon the absorption of ultraviolet light. When the undissociated molecules are irradiated in a system where the stability, i.e. the binding of sub-units, is not too great, a splitting into halves is found with a quantum yield of about 10^{-5}. The halves, upon further irradiation, yield eighths and lower products. Although it is not known which primary bonds are broken, if any, tryptophan groups are destroyed by the light and may be involved in the association of sub-units (Claesson, 1956). An increased tendency toward aggregation also occurs after irradiation.

Many proteins which were once considered to be monomeric in solution have since been found to be dimeric, trimeric, and even tetrameric, depending on pH, ionic strength and composition of buffer. Insulin, chymotrypsin and urease are striking examples. Urease, of course, is a protein with enzyme activity and it is of interest to see if changes in sedimentation behavior accompany or follow inactivation of enzyme activity with commensurate doses of ultraviolet light. Urease is inactivated at 2537 Å with a quantum yield of 9.3×10^{-4} (molecular weight of 480,000). A solution (0.5 per cent) of crystallized urease, pH 7 in 0.02 M phosphate buffer, was sedimented at 60,000 times gravity. It gave a sedimentation constant, S_{20}, of 18.2 Svedberg units, with a minor component with S_{20} equal to 27.5. After a loss in activity of 45 per cent by irradiation, 60 per cent of the main component was detected in the sedimentation diagram together with an increase in amount of material heavier than the main component. Some material lighter than the main component also appeared (McLaren, 1949). (For comparable extents of heat inactivation (at 80°), however, no clearly defined component was visible, the material having become extremely polydisperse and inhomogeneous (McLaren *et al.*, 1948)). The retention of the major component in the ultraviolet inactivated material is consistent with the one-hit kinetics of inactivation. The quantum yield is so low that it is just possible that one or more peptide bonds are broken during inactivation, leading to lower molecular weight split products which are capable of aggregation to give particles of S_{20} higher than 18.2. A more extended study of the sedimentation behavior of irradiated urease is shown in Fig. 4-6 (Sheppard, 1951). Summer *et al.* (1938) originally computed a molecular weight for urease of 483,000 based on the $S = 18$ component but other components were observed in his sedimentation diagrams. The molecular weight of urease from the target theory for ionizing radiation is about 87,000 (Spector, 1956) which corresponds to an $S_{20} = 6$. Some of the $S = 6$ component is generally present in crystalline urease preparations and is enzymatically active. This component also decreases in amount during irradiation. The decline in amounts of all three components is accompanied by aggregation. The quantum yield for inactivation decreases with the degree of inactivation. Since urease is an enzyme whose activity depends on the —SH content, and since —SH groups are vulnerable to ultraviolet light, it is possible the radiation products brought about some inactivation of native urease.

As the activity of chymotrypsin falls from 100 to 32 per cent, the apparent sedimentation constant decreases exponentially from 3.31 to 2.61 Svedberg units (pH 4.5). Since the value of 2.6 is the sedimentation constant for chymotrypsin monomer, the results suggest that chymotrypsin inactivated by ultraviolet light cannot form dimers (Smith *et al.*, 1951). Inactivation may be accompanied by a change of configuration of the molecules since irradiation at this low pH produces a product having reduced solubility in 0.563 saturated ammonium sulphate solution. As we have already seen, if the concentration of chymotrypsin is high during irradiation, inactive

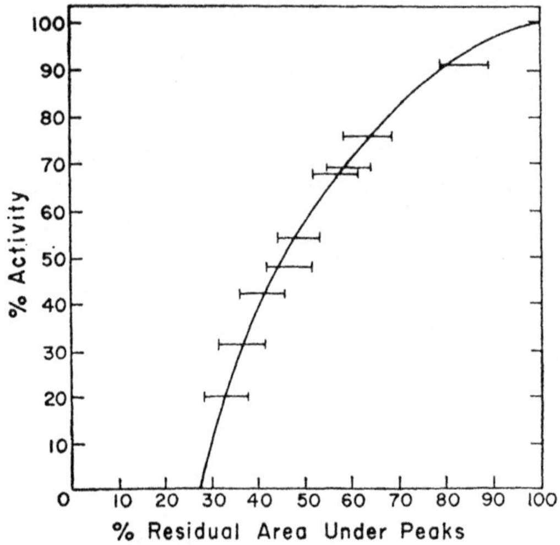

FIG. 4-6. Areas remaining under S = 26 components of urease in ultracentrifuge diagrams as a function of degree of inactivation in phosphate buffer, pH 7 and 0.06 M (Sheppard, 1951). The sedimentation constants for the residual native enzyme in solution did not change from that of the control.

enzyme is digested by native enzyme at a rapid rate. Both low molecular weight products and aggregation products may be seen in sedimentation diagrams. A solubility study with irradiated chymotrypsin of 5 mg/ml, however, showed that it is possible to separate digestion products and any protein photolysis products from native chymotrypsin by precipitation, following the addition of ammonium sulfate (McLaren, 1957). The separation is not exact, since active enzyme is adsorbed to some extent on precipitated, inactive protein. It has also been shown with chymotrypsin that during irradiation with ultraviolet light, an active but thermally labile intermediate is formed. It is probably similar to the soluble nondenatured photo-oxidized ovalbumin of Kofman.

Observations with more complicated systems should also be mentioned. It has been reported that ultraviolet light either impairs or completely destroys the ability of immune serum to lyse, to agglutinate, to combine with toxins, to precipitate, to fix complement, or to confer passive anaphylaxis (Battisto *et al.*, 1953). Ultraviolet irradiation of rabbit serum affects electrophoretic migration velocities of serum components (Larin, 1958). If sufficiently concentrated mixtures of different serum proteins are exposed to ultraviolet light, mixed aggregates composed of molecules of different proteins are formed. Such complexes can also be formed between irradiated, but still active antibodies and other serum proteins (Kleczkowski, 1954).

Structural alterations in horse serum albumin caused by ultraviolet radiation resulted in disappearance of the original determinants and formation of new ones (Kleczkowski and Gold, 1962).

We can summarize the possible changes induced in a typical globular protein (A) by ultraviolet light as follows (assuming that the protein possesses enzyme activity initially). Thermal inactivation is sketched in, and all steps have been found with chymotrypsin (McLaren, 1957). For similar results with trypsin, see Augenstine and Ghiron (1961). The product insoluble in dilute buffer will not be chemically the same by both pathways, of course.

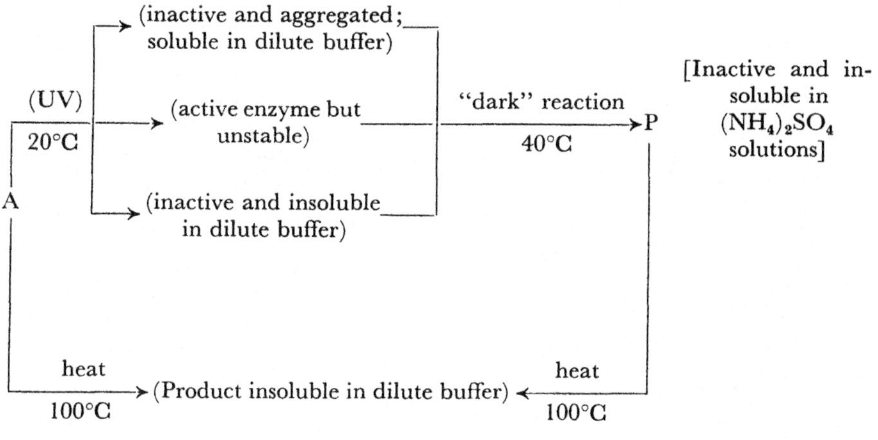

Giese (1950) has summarized these and other often noted changes in proteins following exposure to ultraviolet light in Table 4-3.

4. *Irradiation of Protein Monolayers*

The irradiation of monolayers of proteins has lead to results some of which supplement the information already presented. When a drop of a dilute solution of some of the globular proteins is placed carefully on the surface of a clean salt solution, i.e. with a surface devoid of oil or other surface-active impurity, the protein in the drop sometimes will spread out to cover a large area. The molecules appear to unfold in such instances so as to be only about as thick as single polypeptide chains. Given enough available surface, any enzyme activity which may have been associated with the protein is lost. The tertiary structure and perhaps much of the secondary structure is eliminated by the surface tension (Bull, 1951).

TABLE 4-3. CHANGES IN PROTEIN SOLUTIONS FOLLOWING EXPOSURE TO ULTRAVIOLET LIGHT

Property	Change	Possible meaning
Absorption coefficient	Increase and decrease depending on wavelength and atmosphere	Destruction and formation of chromophores
Acid groups	Increases with low Φ; pH change	Peptide cleavage; formation of phenolic groups
Basic groups	No detectable change unless enzyme is proteolytic	—
Colloid osmotic pressure	Decreases	Aggregation of molecules
Color	Transient color formation in flash photolysis. Browning with very large doses	Free radical formation and photo-oxidation
Conductivity	Increases	Degradation
Digestibility by trypsin	Increases	Denaturation by opening up of molecule
Electrophoretic pattern	Heterogenization	Degradation and re-aggregation
Immunological effects	Destruction of antigenic properties	Aromatic groups destroyed
Non-precipitable N	Increases	Degradation
Optical rotation	Increase in laevorotation	Change in secondary structure
Phenol groups	May increase	Formation *de novo*
Sedimentation	Decrease in amount of original component	Aggregation; depolymerization
Sensitivity to heat	Usually increases	Hydrophobic groups exposed
—SH groups	Increase; oxidized	Change in overall structure
Solubility	Decreases; then increases	Aggregation, then degradation
Surface area	Variable	Photolysis of monolayers; formation of soluble products
Surface potential	Increases	Formation of ionic fragments
Surface tension	Variable	Degradation
Viscosity	Increases	Aggregation

Calculations show that the areas corresponding to a point of minimum compressibility (i.e. corresponding to the smallest area to which a film can be compressed without partial collapse) is about 0.8 m^2/mg. Ovalbumin is a protein which is easily spread in this fashion. Indeed it spreads so easily that a dilute mildly acid solution can be surface denatured visibly simply by shaking in the hand. As fast as protein comes to the air–water interface it spreads to an insoluble film and then aggregates to form a visible precipitate.

A monolayer of protein diminishes the surface tension of water so that a movable surface barrier separating the monolayer from solvent will move tangentially toward the reservoir of solvent. The pressure required to restore a frictionless barrier to its original position before the protein was deposited is the total film pressure (dynes/cm). Kaplan and Fraser (1952) have irradiated ovalbumin monolayers at a constant film pressure of 1 dyne/cm (at 2537 Å). During the initial stages of irradiation a slight increase in area at constant pressure is observed, but soon after a linear decrease occurs on 0.1 M sodium chloride. The scheme proposed by Kaplan and Fraser is as follows:

$$\text{Soluble albumin} \xrightarrow{\text{spreading}} \text{Partly unfolded albumin} \xrightarrow[\text{small dose}]{\text{UV}}$$

$$\text{Further unfolded albumin} \xrightarrow[\text{large dose}]{\text{UV}} \text{Mixture of large and small fragments.}$$

The spreading of soluble protein is complete in about 5 min (Kaplan and Fraser, 1954b). The initial action of ultraviolet light is to more completely unfold the spread protein by breaking a few secondary bonds still remaining, with a small increase in area.† This is consistent with the observation that protein heated or mildly irradiated in solution prior to spreading will spread to a somewhat larger area than native ovalbumin. As we already have noted, in solution some of the bound tyrosyl groups are freed by ultraviolet irradiation prior to denaturation. With large doses large fragments (which cannot be compressed into fibers under high surface pressures) and small fragments (soluble in the substratum liquid) are formed. The small fragments are not surface-active and upon solubilization, lead to the gradual decrease in the area of the irradiated films. The prolonged irradiation of ovalbumin solutions, as we have seen, leads to fragmented protein which cannot form films (Kaplan *et al.*, 1954a), in support of the proposed scheme.

This line of investigation has been extended to trypsin monolayers at the air-water interface (Augenstine and Ray, 1957). Expanded films contain

† It has been found that polycaprolactam (perlon), which has only peptide bonds as chromophores, shows a sharp increase in pressure under irradiation. The effect is reversible; when the ultraviolet lamp is turned off the films immediately recover. Hydrogen bonding between keto and amino groups in adjacent chain segments must be broken by the radiation (Cheesman and Davies, 1954).

only unfolded molecules of trypsin which have permanently lost all activity in the process of unfolding, whereas the film formed and kept under a compression greater than about 0.25 dyne/cm consists of fractions of enzyme in globular and unfolded configurations. At a film pressure of 1–2 dynes/cm, inactivation of trypsin in monolayers seems to be a one-hit event with a quantum yield of 0.01 moles/einstein (M = 20,000) based on loss of ability to hydrolyse casein. This yield is about half that found in solution, but the uncertainties in the calculation render further comparison useless.

Cheesman and Davies (1954) and Danielli and Davies (1951) and McLaren (1949) have discussed the results of numerous investigations of the effects of ultraviolet light upon other physical properties of protein monolayers including surface potentials. The results do not contribute to the development here.

CHAPTER V

Photochemical and Photosensitized Inactivation of Enzymes

1. *Kinetics of Enzyme Inactivation in Solution*

We have been discussing the gross physico-chemical changes observed during and after the irradiation of proteins. Some proteins have enzymatic activities which give us a natural reference point for appraising these changes and we may profitably ask the following questions. Which of the usually observed physical and chemical changes accompany loss in enzyme activity and which are only observable after doses have been absorbed considerably in excess of say, amounts determining 37 per cent remaining activity?†

By way of example α-chymotrypsin is inactivated by ultraviolet light; it also undergoes loss of solubility in 0.56 saturated ammonium sulfate solution and with large doses it tends to become resolubilized. At what rates are secondary products formed which are not precipitated by the addition of ammonium sulfate and how does the formation of precipitable protein proceed with respect to inactivation? The question can be discussed in terms of the following, where A is enzyme.

Combining reactions (a) and (f) of Chapter IV, we have:

$$(1a)\quad A + h\nu \xrightarrow{k_{1a}} P_1 + P_2 + \cdots \text{ (inactivation)}$$

This reaction is probably temperature independent; it has been reported so with amylase (Nikolov and Saev, 1957). Also, it is already known that some irradiated A can become A_1, an unstable intermediate which on long standing loses activity by a "dark" reaction. The dark reaction is temperature dependent with amylase, chymotrypsin and aldolase (cf. Setlow and Doyle, 1957). In other words, if a solution is irradiated at low temperature, say at 3.5° and pH 4.5 and the loss in activity is measured at 0°, the quantum yield is 5.7×10^{-3}. If, before assay, the solution is allowed to warm up to room temperature for a couple of hours and assayed at 35.5° a higher quantum yield is found, 8.9×10^{-3}. We write

$$(1b)\quad A + h\nu \xrightarrow{k_{1b}} A_1 \text{ (instabilization)}$$

† At 37 per cent remaining activity the dose per ml is just equal to $[A_0]/\Phi$, where $[A_0]$ is the initial concentration of the species A.

for the photochemical reaction and

(1c) $A_I \rightarrow P_1 + \cdots$ (thermal inactivation of unstable intermediate)

for the dark reaction (McLaren and Finkelstein, 1950). (Cf. Augenstien, 1959.) Reactions (1a) and (1b) proceed simultaneously whereas reactions (1b) and the thermal reaction are successive.†

On prolonged irradiation degradation is possible, viz.,

$$(2)\ P_1 + h\nu \xrightarrow{k_2} P_a + P_b + \cdots$$

where P_1, P_2, etc., are primary products and P_a, P_b, etc., are various possible degradation products. Finally polymerization reactions are possible, with and without light:

$$(3)\ P_1 + P_2 \rightarrow P_1P_2 \text{ etc.} \qquad \text{(g) of Chapter IV}$$

$$(4)\ A + P_1 \rightarrow AP_1 \text{ etc.}$$

We do not distinguish A from its dimer AA for chymotrypsin since Φ is independent of the concentration from 0.04 mg/ml to 20 mg/ml at pH 4.5. This range of concentration involves monomer at one extreme and dimer at the other; so the loss of A as A or AA is a one-hit process for each molecule of A lost; two enzyme molecules are not inactivated by one quantum in the case of dimerization. Inactivated chymotrypsin apparently does not dimerize (Egin *et al.*, 1957; McLaren, 1957).

The kinetics of reactions (1a), are given by

$$-d[A]/dt = \Phi I_{abs}\left(\frac{\varepsilon_A[A]}{\varepsilon_P[P] + \varepsilon_A[A]}\right) = k_{1a}[A](h\nu) = k[A] \qquad (5\text{-}1)$$

The application of eqns. (5-1) in integrated form (Eqn. 1-8a) to lysozyme is shown in Fig. 5-1. This equation holds under a wide variety of conditions, although Φ, I_{abs} and ε_A are all more or less dependent on pH. As we have seen, ε_P is dependent on the atmosphere present (Finkelstein and McLaren, 1950). Reaction (1b) and the dark reaction have not been studied in detail (Kleczkowski, 1954a; Nikolov and Saev, 1957). At 3.5° reaction (1b) may be about 0.56 times as efficient as reaction (1a) (McLaren, 1957). Reaction (4) is probably not significant for the reason previously discussed, whilst reaction (3) is observable (Chapter IV).

In order to study the relative rates of reactions (1a) plus (1b) and (2) by the solubility method of Bernhart (1939), the products P_1 and P_1P_2 are

† How closely the photochemically unstabilized enzyme compares to the transition state of strictly thermal inactivation of chymotrypsin is not known. Passynsky (1958) gives the following empirical relationship between quantum yields and activation energies ΔH^* for thermal inactivation: $\log \Phi = \log \Phi_0 + k\Delta H^*$. Φ_0 is simply a constant.

considered together as regards loss of solubility in 0.56 saturated ammonium sulfate solution, and P_a, P_b, etc., are soluble. Polymers, P_n, are also insoluble. Native chymotrypsin was irradiated in dilute buffer at room temperature and then brought to 40° to complete reaction (1c). Ammonium sulfate was added and the precipitated protein was removed from solution and estimated. Digestion of denatured protein to give products and the appearance of P_a

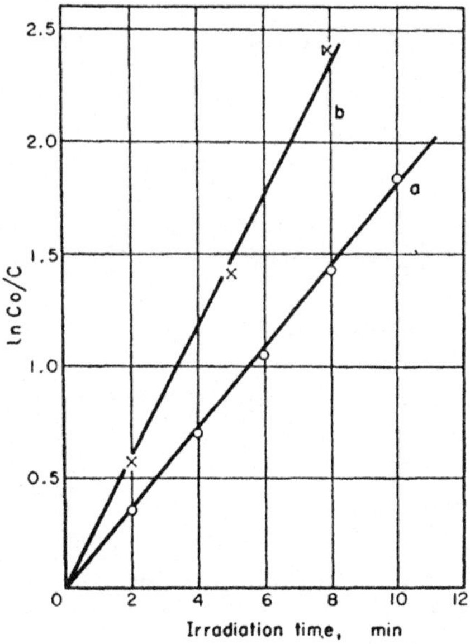

FIG. 5-1. Inactivation of lysozyme by UV at 0.25 mg/ml (pH 7.1, a; pH 12, b).

and P_b, soluble in salt solution, must be slow under the conditions of irradiation (pH 5.5) since none were observable under six hours of irradiation at moderately high intensity (2537 Å). At 6 hr, the residual activity was 7.5 per cent. Since the inactive products P_n can appear by reaction (1) and disappear by reaction (2) we may write, if we neglect self-digestion,

$$[P_n] = \frac{k_1[A_0]}{k_2 - k_1} (C^{-k_1 t} - C^{-k_2 t}) \tag{5-2}$$

Writing initial concentration of A as $[A_0]$ and further values as $[A]$,

$$[A_0] = [A] + [P_1 + P_2 \cdot \cdot \cdot] + [P_a] + [P_b] + \cdot \cdot \cdot,$$

or

$$[A_0] = [A] + [P_n] + [P_x]$$

Allowing for self-digestion, the second term is maximal but still instructive. From the results just stated k_2 must be much smaller than $k_{1a} + k_{1b}$. We can get a rough value of k_2 in terms of $(k_{1a} + k_{1b})$ as follows.

Since reaction (1c) was forced to completion,

$$h\nu(k_{1a} + k_{1b}) = k_1 = \Phi I_{abs}/[A_0]$$

$$\Phi = 2.4 \times 10^{-3} \text{ at pH } 5.4$$

$$I_{abs} \text{ was } = 2 \times 10^{-5} \text{ einsteins/cm}^3 \text{ per hr (a constant)}$$

$$[A_0] \text{ was } 1.22 \times 10^{-7} \text{ moles/cm}^3 \text{ (M} = 41{,}000)$$

so

$$k_1 = 0.394 \text{ min}^{-1}$$

We see from equation (5-2) that for $[P_n]$ to differ much from $[A_0] - [A]$ in less than six hours, k_2 would need to be of the order of $k_1/10$ (McLaren, 1957).

Bernhard found that k_1 was 20–50 times k_2 in air with egg albumin. We may therefore conclude that with ovalbumin and chymotrypsin almost all of the native protein is inactivated or denatured before appreciable degradation is observed. It must be remembered that denaturation is operationally defined here as loss of solubility in salt solution (Neurath *et al.*, 1944). As chymotrypsin undergoes inactivation it also loses its solubility in 0.56 saturated ammonium sulfate. Katchman, however, did not find such a close parallel between inactivation of trypsin inhibitor and loss of solubility, indicating that further changes are possible in P_1, etc., before solubility is lost, which is similar to Kofman's results with egg albumin. Also, the products P_1, P_2, etc., may be inactive as a result of the absorption of a single effective quantum, yet not be identical. As yet we have not distinguished all possibilities experimentally.

It is sometimes found that the thermal stability of an enzyme depends on whether substrate is present or not, and substances have also been reported which "protect" enzymes from inactivation by ultraviolet light. For example, amylase is protected by substrate (Nikolov and Saev, 1957). The addition of the tryptic inhibitor-substrate benzoylarginine (BA) to the extent of 0.01–0.05 mg/ml seems to protect trypsin somewhat from such inactivation, but when equation (5-1) is modified to include a term $\varepsilon_{BA}[BA]$ in the denominator for absorption of ultraviolet light by BA, the quantum yield remains unchanged (McLaren, 1957). Thus the aromatic substrate merely acts as a simple screen or inner filter for the light and cuts down the number of quanta reaching the enzyme. Similar results have been reported with ribonuclease and disulfides (Brighenti and Falaschi, 1962). The quantum yields for the inactivation of pepsin in 4 M urea and in dilute buffer are the same (McLaren and Pearson, 1949).

2. *Mechanism of Inactivation of Enzymes by Ultraviolet Light*

The question of inactivation with respect to the nature of the substrate employed is particularly interesting. For example, chymotrypsin can hydrolyze both amides and esters of aromatic amino-acids (Neurath and Schwert, 1950). Yet it possesses but one active site per molecule (Loewus and Briggs, 1952). The active site in chymotrypsin responsible for the binding of a substrate and cleavage of a —CONH— or —COOR linkage is however not a single amino-acid residue. Histidine, tyrosine, serine, and tryptophan are all implicated (Brouwer, 1957), and Neurath and Schwert have presented evidence of multiple points of attachment between α-chymotrypsin and its substrates. Chymotrypsin oxidized with sodium periodate is transformed into a derivative which possesses about 65 per cent of esterase activity toward tyrosine ethyl ester but only 35 per cent proteinase activity, of the parent enzyme. These results suggest that the oxidized enzyme may have one less point of attachment, thus affecting the hydrolysis of simple esters very little but greatly affecting the hydrolysis of the more complicated peptide structure of hemoglobin. Acetylation of trypsin reduces the proteinase activity but not the esterase activity of this enzyme, as another example. Jansen *et al.* (1951) conclude that proteolytic enzymes contain some groups which are "auxiliary" to the extent that they may be "almost" necessary in some cases. These groups, such as amino, are not absolutely vital as is the hitherto not clearly identified group which reacts with di-isopropyl fluorophosphate. Substitution of this latter group renders the enzyme totally inactive. Reaction of one of two histidine residues in chymotrypsin with 2,4-dinitrofluorobenzene (Whitaker and Jandorf, 1956) results in inactivation. Tryptophan seems to function as an auxiliary rather than a necessary group (Wood and Balls, 1955), tyrosine is essential in chymotrypsin (Sizer, 1945). Off hand, therefore, it would be no surprise to find that chymotrypsin should preferentially lose proteolytic activity more rapidly than esterase activity during exposure to ultraviolet light (McLaren, 1947). By actual test, the quantum yield for inactivation of chymotrypsin was the same with chloroacetyl phenylalanine ethyl ester, phenylalanine ethyl ester, phenylalaninamide, glycylphenylalaninamide and casein (Goldenberg and McLaren, 1951). It seems possible that ultraviolet light can selectively destroy some residue at or about the active site and/or that the entire molecule is denatured by the light, perhaps by cleavage of an —SS— bond. The latter seems to be the case with this enzyme, although the former may proceed simultaneously yet be missed because of the denaturation (McLaren and Finklestein, 1950).

The quantum yield for inactivation of trypsin is also independent of the substrates tried, namely casein, hemoglobin, and benzoylargininamide. The known structural features of trypsin include the following (Neurath and Dixon, 1957; Dixon, Kauffman and Neurath, 1958). Trypsin has 6 intra-molecular disulfide bridges. Two of these are close to the active center, as proposed by Neurath and Dixon, Fig. 5-2.

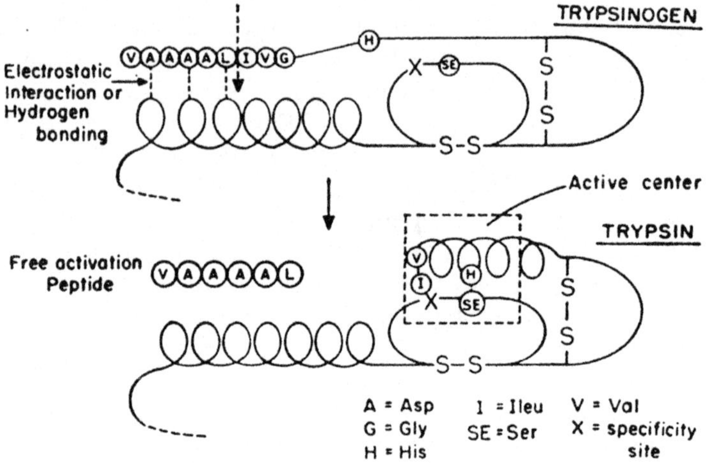

FIG. 5-2. Schematic representation of the structure of trypsin (Neurath and Dixon, 1957). A closely related structure applies to chymotrypsin, with an alternate and as yet undefined specificity site.

Oxidation of the lower loop probably gives the following peptide sequence:

$$\text{—Asp—Ser—}\underset{\substack{|\\ SO_3^-}}{\text{Cys}}\text{—Glu—Gly—Gly—Asp—Ser—Gly—Pro—Val—}\underset{\substack{|\\ SO_3^-}}{\text{Cys}}\text{—Ser—Gly—Lys—}$$

The sequence in chymotrypsin is similar.

According to the scheme, which is based on extensive data, the histidine and serine residues are close together in the active enzymes. Clearly rupture of —SS— by ultraviolet light would play havoc with this structure. Reduction of these bonds leads to loss of activity (Liener, 1957). The roles of tyrosine and tryptophan in enzyme activity are not yet fitted in with this simple structure.

Assuming that a —SS— group *per se* is vital for the enzymatic integrity of a protein molecule, then proteins with high fractions of —SS— groups would expectedly have high quantum yields for inactivation because of the increased probability that absorbed quanta would be absorbed at —SS— linkages. This assumes, of course, that the general proportions of the other amino acids are about the same among the proteins in question. As a matter of fact, Setlow (1955), has tabulated quantum yield data for a number of enzymes and concluded that quantum yields are roughly proportional to the cystine content of enzymes at 2537 Å, i.e.

$$\Phi = \Psi[\text{cys}] \tag{5-3}$$

Let us develop the idea that the breakage of a particular cystine —SS— linkage with a quantum yield of ϕ can lead to inactivation of an enzyme with a quantum yield of Φ. Let

$$\Phi = \frac{\text{bonds broken in active center}}{\text{total quanta absorbed per molecule}} = \frac{\text{bond } \textcircled{H}}{\text{total quanta absorbed}}$$

$$= \frac{\text{bond } \textcircled{H}}{\text{quanta absorbed by } \textcircled{H}} \left[\frac{\text{quanta absorbed by } \textcircled{H}}{\text{total quanta absorbed}}\right]$$

$$= \phi\left(\frac{I_0 - I_{\textcircled{H}}}{I_0 - I}\right) \tag{5-4}$$

where ϕ is the quantum yield for cystine, $I_{\textcircled{H}}$ is the intensity transmitted by the —SS— groups and I is the intensity transmitted by the molecule. For a short path length, at 2537 Å,

$$\Phi = \phi\,\frac{\varepsilon_{\textcircled{H}} C_{\textcircled{H}}}{\varepsilon C} = \phi\,\frac{\varepsilon_{\textcircled{H}}}{\varepsilon} \tag{5-5}$$

where the concentration of bond $\textcircled{H}$, $C_{\textcircled{H}}$, equals the concentration of molecules, C. $\varepsilon_{\textcircled{H}}$ is the molecular extinction coefficient of —SS— and ε that of the whole protein.

From Table 3-3, Chapter III, we have $\phi\ \varepsilon_{\textcircled{H}} = 0.13 \times 2.7 \times 10^2$ at 2537 Å. For a molecular weight of 23,800 for trypsin (Neurath and Dixon, 1957) the molecular extinction coefficient for trypsin at this wavelength is 13,700,† based on the combined data of Goldenberg (1950) and Laskowski and Laskowski (1954).

We thus calculate a quantum yield for inactivation as

$$\Phi = \frac{32}{13{,}700} = 2.3 \times 10^{-3}$$

Comparison with the experimental value, namely 0.015 (Goldenberg and McLaren, 1951) shows the calculated value to be too low. Since there is evidence for the breakage of —SS— bonds during inactivation of this enzyme (Luse and McLaren 1963), we should consider that cleavage of any of the —SS— linkages in trypsin may lead to inactivation. Since there are six such linkages, the calculated value comes to 0.014 in excellent agreement

† The UV absorption spectrum of trypsin has been reported with poor accord by various authors (Uber and McLaren, 1941; Verbrugge, 1943; Kunitz, 1947). Goldenberg has compared the trypsin samples used by these authors, along with twice crystallized trypsin from the Worthington Biochemical Corp., and found the spectra to be identical within 2 per cent from 2400–2900 Å (1950).

with the found value. This leads us to an answer to the question originally posed by Uber and McLaren as to whether some particular residue(s) must absorb a quantum to produce inactivation of the molecule or whether any chromophore may do so, but with the yield found for inactivation. It would seem that the particular chromophores are —SS— linkages. Setlow (1957) has carried out a similar analysis and concluded that inactivation by photochemical changes elsewhere in the molecule is possible but with a probability of only 3 per cent as much. Bearing in mind that tryptophan may be only an auxiliary group in the activity of trypsin and that tyrosine may not be involved (Green and Neurath, 1954), reference to Table 3-3, Chapter III, leads us to conclude that Setlow's reconstruction of the action spectrum of trypsin on this basis is quite reasonable.

Similarly, there are four intramolecular —SS— bonds per molecule of chymotrypsin (Neurath and Dixon, 1957) of molecular weight 25,000. The calculated yield is 0.0056 and the found value near the pH minimum of quantum yield is 0.005, Table 5-1. There are also four disulfide bridges in ribonuclease (but perhaps only two are vital (Anfinsen, 1957), and the calculated yield agrees with the known yield of 0.03 (Shugar and Rzendowska, 1956)).

We can also assess the importance of tryptophan, tyrosine, histidine and other residues, including the peptide bond by means of the data of Table 3-3, Chapter III, and the following equations. Let us assume that destruction of any important chromophores can lead to inactivation. From equation (5-1), for small degrees of inactivation,

$$-\mathrm{d}[\mathrm{A}]/\mathrm{d}t = \Phi I_{\mathrm{abs}} \frac{\varepsilon_{\mathrm{A}}[\mathrm{A}]}{\varepsilon_{\mathrm{A}}[\mathrm{A}_0]} = \phi_{\mathrm{Tyr}} I_{\mathrm{abs}} \frac{\varepsilon_{\mathrm{Tyr}} \cdot n[\mathrm{A}]}{\varepsilon_{\mathrm{A}}[\mathrm{A}_0]} + \phi_{\mathrm{His}} I_{\mathrm{abs}} \varepsilon_{\mathrm{His}} n[\mathrm{A}] + \cdots$$

$$= I_{\mathrm{abs}} \left[\phi_{\mathrm{Tyr}} \frac{n_{\mathrm{Tyr}} \varepsilon_{\mathrm{Tyr}}[\mathrm{A}]}{\varepsilon_{\mathrm{A}}[\mathrm{A}_0]} + \phi_{\mathrm{His}} \frac{n_{\mathrm{His}} \varepsilon_{\mathrm{His}}[\mathrm{A}]}{\varepsilon_{\mathrm{A}}[\mathrm{A}_0]} + \cdots \right]$$

$$= \frac{I_{\mathrm{abs}}}{\varepsilon_{\mathrm{A}}[\mathrm{A}_0]} \sum^{i} n_i \varepsilon_i [\mathrm{A}] \phi_i$$

and

$$\Phi_{\mathrm{enzyme}} = \frac{\sum^{i} n_i \varepsilon_i \phi_i}{\varepsilon_{\mathrm{A}}} \tag{5-6}$$

where [A] is the concentration of the enzyme considered, n_i is the number of chromophores of kind i in the enzyme molecule, having a residue extinction coefficient ε_i and a quantum yield for destruction ϕ_i. Using again data of Table 3-3, Chapter III, McLaren and Luse (1961) obtained the results in Table 5-2. The calculated contribution of tryptophan to the inactivation

Table 5-1. Quantum yields for inactivation of enzymes at 2537 Å

Enzyme	ε_{2537}	Number of —SS— groups		Quantum yield		
		Total, n	Number required	Found	Calculated	
Chymotrypsin	23,400	4	?	0.005	0.0056 with $n = 4$	McLaren and Finkelstein, 1950
Pepsin	16,000	3	2	0.002	0.0040 with $n = 2$	McLaren and Pearson, 1949
Ribonuclease	5000	4	2	0.03	0.026 with $n = 4$	Shugar and Rzendowska, 1956
Trypsin	13,700	6	?	0.015	0.014 with $n = 6$	Goldberg and McLaren, 1951

TABLE 5-2. ESTIMATION OF QUANTUM YIELDS FOR ENZYME INACTIVATION FROM QUANTUM YIELDS FOR MONOMER DESTRUCTION (2537 A; MCLAREN AND LUSE, 1961)

Enzyme: / Amino Acid:	Chymotrypsin $\varepsilon_A = 23{,}000$		Lysozyme $\varepsilon_A = 18{,}000$		Ribonuclease $\varepsilon_A = 4400$		Trypsin $\varepsilon_A = 15{,}500$	
	n	$n\varepsilon\phi$	n	$n\varepsilon\phi$	n	$n\varepsilon\phi$	n	$n\varepsilon\phi$
Cystine	5	175	5	175	4	140	6	210
Histidine	2	<0.015	1	—	1	—	1	—
Phenylalanine	6	11	3	5	3	5	3	5
Tryptophan	7	80	6	69	0	—	4	46
Tyrosine	4	3	3	2	6	4	4	3
—CONH—	~200	~1	~130	~1	~130	~1	~200	~1
$\Sigma n_i \varepsilon_i \phi_i / \varepsilon_A =$								
Φ, calculated	0.01		0.01		0.03		0.02	
Φ, known	0.005		0.024		0.027		0.015	

of chymotrypsin is significant. Obviously, overall energy migration from one chromophore to another, e.g. to —SS— need not be invoked in order to explain Φ for these enzymes (Luse and McLaren, 1963).

If tyrosine were important in inactivation then, because of ionization of tyrosine at high pH, the action spectra should differ as much as the absorption spectra at high and low pH. Ribonuclease shows a large difference in absorption spectra at high and low pH but only a small difference in inactivation spectra, indicating that tyrosine is also relatively unimportant in the inactivation of this enzyme (Setlow, 1957). Since Φ for lysozyme is independent of pH, tyrosyl ionization is not involved (Shugar, 1952). It may be important in the inactivation of pepsin† (Imahori, 1955), but the quantum yield can be accounted for on the basis of two required cystine residues (Bovey and Yanari, 1958). Quantitative data which lead to a direct evaluation of destruction of tryptophan residues as one of the possible ways by which an enzyme might be inactivated have been determined by Luse (1961). He found about 1 tryptophan residue destroyed per molecule of trypsin inactivated and the same for chymotrypsin. The tryptophan content of lysozyme decreases during ultraviolet inactivation (Shugar, 1952).

Luse also found that 2-3 disulfide linkages were photolysed (3.7 SH groups formed/molecule) per molecule of trypsin and disufide linkages were photolysed in chymotrypsin (cf. Augenstine and Ghiron, 1961). The number of —SS— groups photolysed in ribonuclease per molecule inactivated was 1.7 (that is, 2.7 ± 0.4 —SH groups formed). Of the four cystine residues

† Imahori found that dopa was formed during the inactivation of pepsin.

in ribonuclease, perhaps two are preferentially cleaved by ultraviolet, however (McLaren and Luse, 1961; Augenstein and Grist, 1962). No phenylalanine, tyrosine, lysine and/or histidine or arginine residues were changed. With well over 99.9 per cent inactivation, these amino-acid residues were also changed, as is to be expected, however. Those observations followed irradiation in solutions irradiated while being stirred with O_2-free nitrogen

Contrary to photosensitized inactivation of enzymes (Section 4), histidine does not undergo stoichiometric changes during ultraviolet inactivation at this wavelength (cf. Table 3-3, Chapter III).

It is significant that no free amino-acids or peptides are formed during the inactivation of lysozyme (Shugar, 1952) and ribonuclease (Shugar and Rzendowska, 1956). There is also no increase in titratable amino groups, indicating that photolysis of peptide linkages is not important for inactivation (Obata and Koshika, 1959; Luse, 1961).

Continuing from equations (5-3) and (5-5), we see that if cystine residues are so vital in the mechanism of inactivation at 2537 Å, an increase in the ratio $n\,\varepsilon_{Ⓗ}/\varepsilon$ would lead to an increase in the quantum yield for inactivation. This can come about by having a higher percentage of cystine in a protein with the rest of the amino-acid composition remaining essentially the same, at least as regards the light absorption. For a series of enzymes eqn. (5-5) can then be equated roughly with eqn. (5-3) and

$$\Phi = \frac{\phi}{\varepsilon} \cdot n\,\varepsilon_{Ⓗ} \cong \Psi(\text{Cys})$$

which expresses the trend observed by Setlow. On the other hand, taking into account that the optical densities per g per liter among proteins vary by only a factor of two or three, ε is very roughly proportional to molecular weights of proteins, i.e. for unit path length

$$\varepsilon = \frac{D}{C} \cong Q'M \qquad (5\text{-}7)$$

where Q' has the dimensions of l./g protein. Equation (5-5) then becomes

$$\Phi \cong \phi\varepsilon_{Ⓗ} \cdot \frac{1}{Q'M} \cong Q/M \qquad (5\text{-}8)$$

where Q is another constant. Equation (5-8) was found empirically for a number of enzymes with $Q = 450$ (McLaren *et al.*, 1953). For example, with trypsin, $\Phi = 450/23{,}800 = 0.019$ in comparison with the found value of 0.015. The basis for the equation was originally Fig. 5-3. It has been confirmed by quantum yield measurements for inactivation of desoxyribonuclease and catalase (Setlow *et al.*, 1953a) and of antibodies to tobacco mosiac virus and to clover nodule bacteria (Kleczkowski, 1954b). It also states that the number of g enzyme inactivated per einstein is about 450 or, in other words, to inactivate 50 per cent of an enzyme requires about 0.75 J per mg. corrected for light absorbed by inactive protein (McLaren and Takahashi, 1957).

If both empirical equations (5-3) and (5-8) are valid, one would expect to find that molecular weights of the enzymes studied and reciprocals of per cent abundance of cystine tend to be directly related, Fig. 5-4.

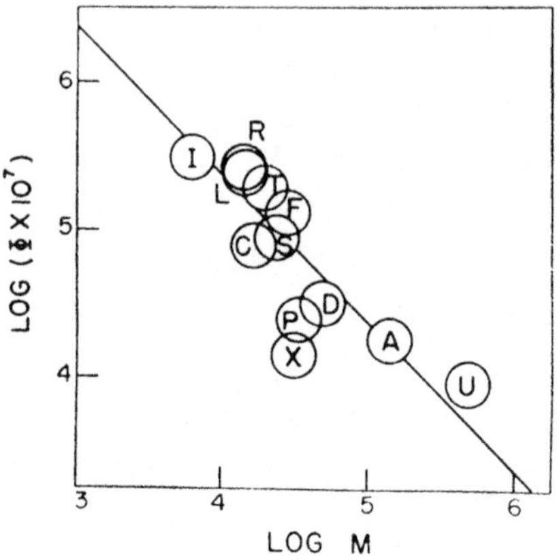

FIG. 5-3. Plot of log ($\Phi \times 10^7$) vs. log M for a number of enzymes and related proteins. (Aldolase, A; Carboxypeptidase, X; Chymotrypsin, C; Triosphosphate dehydrogenase, D; Ficin, F; Pancreatic trypsin inhibitor, I; Lysozyme, L; Pepsin, P; Ribonuclease, R; Soybean trypsin inhibitor, S; Trypsin, T; Urease, U.) (McLaren *et al.*, 1953).

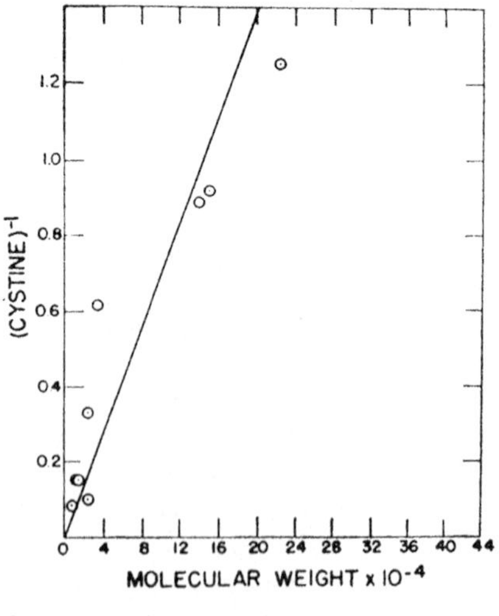

FIG. 5-4. Plots of reciprocals of cystine content of proteins [1/(g cystine in 100 g protein)] versus molecular weights (McLaren, 1957).

The quantum yields for inactivation of enzymes vary more or less with pH, (Fig. 5-5). (For chymotrypsin the minimum is at pH 5.4; for trypsin, at pH nearer 4.0. There is difficulty in interpreting the pH function in terms of the peptide sequence shown above or with the isoeletric points of the enzymes, which are above about pH 9). A variation of quantum yield

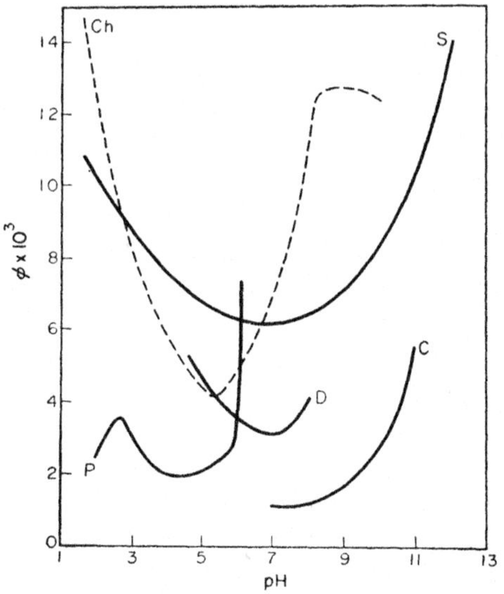

FIG. 5-5. Variation of quantum yield with pH for carboxypeptidase, C; Chymotrypsin, Ch; Triosephosphate dehydrogenase, D; Pepsin, P; Soya-bean trypsin inhibitor, S. Data are collected from original literature and are uncorrected for newer revised molecular weights (McLaren, 1957).

with pH points to an influence of intra-molecular electrostatic repulsion on inactivation, but the pH of maximum stability does not generally coincide with the isoelectric point of the enzyme, indicating that the overall charge on the molecule is not as important as the charge about some local structure. Tyrosine bonding does not seem implicated (Katchman, 1955), as also emerges from the fact that quantum yields for lysozyme and ribonuclease are similar despite the fact that, for the latter, about one-half the phenolic groups are involved in tight bonding and are released only upon denaturation. A partial explanation of this phenomenon may reside in the large intramolecular electrostatic repulsion possible inside proteins (up to 20,000 cal or more) plus the tendency of the three-dimensional protein structure to remain intact if only one bond is broken by ultraviolet light. The repulsion would tend to hold free-radical ends apart until reaction with solvent takes place; the overall structure would tend to hold free-radical ends in a "cage" thereby favoring recombination without a net reaction (McLaren and Lewis, 1950). The cage influence, if real, can be compared with the Franck–Rabinowitch effect. The rate of decomposition of a few amino-acids is also somewhat pH dependent but cystine has not been studied in this way. The variable pH dependence of quantum yields among enzymes indicates either that cystine is not the only important ultraviolet-sensitive residue in enzymes, or that the influence of intramolecular repulsion of ionized chain-segments about broken disulfide linkages is pronounced. Disruption of secondary hydrogen-bonded structures in enzymes alone is frequently not sufficient to inactivate enzymes (Anfinsen *et al.*, 1955). It is also possible that as the pH is raised from low to high values the sensitivity of one vulnerable residue decreases as another increases, giving rise to a minimum in the quantum yield.

Although it cannot be said as yet just what are the detailed factors in the pH effect, the pH independence of quantum yield for lysozyme indicates that the ultraviolet-sensitive region involved in activity is not ionizable (Shugar, 1952). Electrostatic repulsion might expectedly be reduced by the addition of salt, and Rapkine *et al.* (1949) found that the rate of inactivation of triosephosphate dehydrogenase is actually decreased upon the addition of ammonium sulfate. A salt effect was not observed with chymotrypsin and is small, if real, with pepsin. With pepsin one would have expected a large salt effect since it is a polyanion over a wide pH range. The quantum yield of chymotrypsin is somewhat different in acetate and phosphate buffers (McLaren and Finkelstein, 1950), which only complicates the mystery. The pH of maximum thermal stability of chymotrypsin is 3.0–3.5, which is lower than that for the quantum yield (Northrop *et al.*, 1948).

The variation of quantum yield with wavelength for pepsin, ribonuclease and urease in solution is shown in Fig. 5-6. Quantum yields, of course, are determined by means of assay with enzyme substrates. The enzyme may be denatured by ultraviolet light so that it will not combine with its substrate, or it may combine but not be able to act on the substrate. Trypsin thus inactivated can no longer combine with trypsin inhibitor (Soya bean) (Estermann and McLaren, 1962), showing a loss of ability to combine with a potential substrate. How this factor influences quantum yield measurements as a function of pH is not known.

A number of other factors have been examined with regard to the kinetics of inactivation. One of these is light intensity. Using polychromatic light in single bursts of about 0.5 msec duration and 2×10^{-7} einsteins per burst per ml solution, McLaren *et al.* (1952), found that the reciprocity law was obeyed with pepsin. The pepsin solution was about 5×10^{-8} moles/ml, so that the incident quanta reaching the solution per burst were comparable to the number of molecules of pepsin per ml.† This

† Intensity measurement with uranyl-oxalate actinometry also conform to the reciprocity law under these conditions (Christie and Porter, 1952).

intensity is 100,000 times those customarily employed, yet Setlow has calculated that even so only 10^{-5} of the molecules are excited at one time at room temperature. Much higher intensities must be reached before one can expect failure of the reciprocity law because of simultaneous absorption of several quanta. With much longer lifetimes of the excited state, as at low temperature, one might find some dependence of quantum yield on intensity (Setlow, 1957).

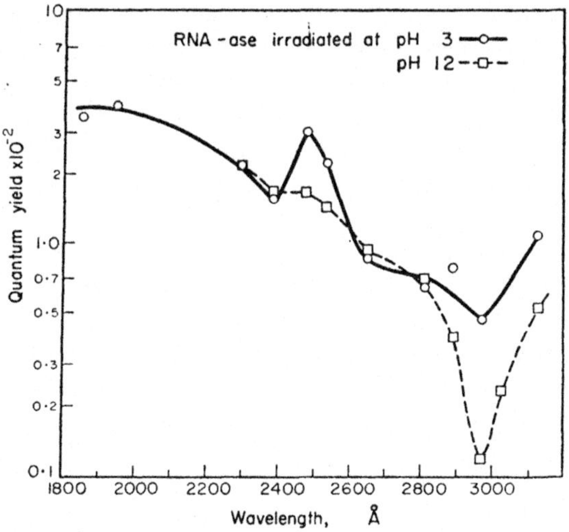

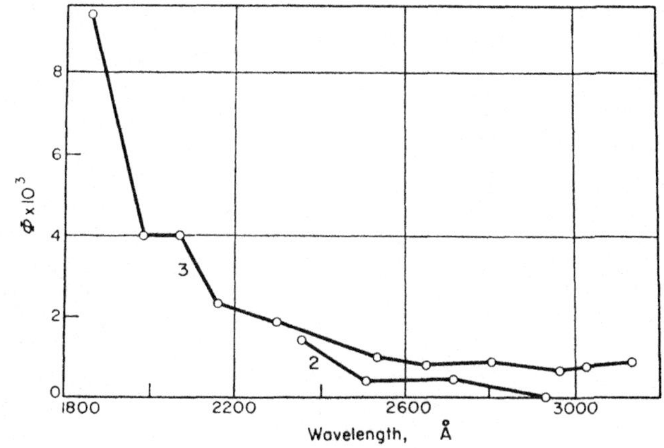

FIG. 5-6(A and B). Quantum yield, Φ, vs. wavelength for the enzymes ribonuclease, 1; pepsin, 2; and urease, 3 (McLaren, 1947) (Setlow and Doyle, 1957).

Although the quantum yield for inactivation of an enzyme may vary with pH, the variation with pH is not necessarily the same for every wavelength. For example, the variation of quantum yield with pH is quite different with polychromatic light as compared with monochromatic light (2537 Å) with triosephosphate dehydrogenase and pepsin (Shugar, 1951). This undoubtedly means that there are possible detailed alternative ways the molecule can be inactivated and different chromophores are involved to different extents at different wavelengths. The quantum yields for inactivation of chymotrypsin and ribonuclease in solution are independent of the presence of oxygen, showing that the primary process of ultraviolet inactivation is not a photo-oxidation. The nature of the products are dependent on the presence or absence of O_2, as already seen. Also the presence of solvent is probably not of paramount importance during irradiation since quantum yields with enzymes are nearly the same irradiated dry or in solution, Table 5-3. The quantum yield for ribonuclease is the same in 95 per cent alcohol as in aqueous solution (Shugar and Rzendowska, 1956); the same is true of desoxyribonuclease (Shugar and Rzendowska, 1959). This also means that the fraction of absorbed energy involved in inactivation, which is of the order of one per cent or less, is not altered by any differences in amounts lost by fluorescence. Any free radicals formed in the solid state may react with water when the protein is dissolved for assay.

TABLE 5-3. QUANTUM YIELDS FOR THE INACTIVATION OF ENZYMES IN AQUEOUS SOLUTION, AS MONOLAYERS OR DRY (2537 Å).

Substance	Quantum yield			Reference
	Dry	Monolayer on water	Solution	
Chymotrypsin (Mol. Wt. 25,000)	0.016		0.005	Setlow, 1955
Ribonuclease (Mol. Wt. 15,000)	0.03		0.03	Shugar, Rzendowska, 1956; Setlow, Doyle, 1957
Trypsin (Mol. Wt. 23,800)	0.018		0.022	Setlow, Doyle, 1957
		0.008		Augenstein, Ray, 1957
			0.015	Goldenberg, McLaren, 1951
Desoxyribonuclease (Mol. Wt. 60,000)	0.0039		0.0034	Shugar and Rzendowska, 1959; Setlow and Doyle, 1953b

That the quantum yield for inactivation of an enzyme varies with wavelength is no longer a surprise, although early results from Svedberg's and Warburg's laboratories were obtained with either the prejudice that the yield should be unity or else constant so that action spectra would parallel absorption spectra (McLaren, 1949). The action spectrum of aldolase has the same shape as the absorption spectrum except above 3000 Å, showing that the quantum yield is fairly independent of wavelength and indicating that all principal chromophores contribute to the inactivation step. The contribution may be that of passing absorbed energy to the weak link rather than photochemical changes in each chromophore, since the latter if true, would not be expected to take place with a constant quantum yield. The yield is about 0.01 at 2482, 2650, and 2804 Å (Setlow and Doyle, 1957).

As we have seen, two important considerations are the variation in extinction coefficients of absorbing residues and the peptide bonds and the variation in quantum efficiencies for rupture of these chromophores with wavelength and pH. Energy transfer, say from an aromatic group to —S—S— linkages is a third; this is a reasonable interpretation of the decreasing fluorescence of a protein toward lower wavelengths (Shore and Pardee, 1956). There is some evidence for a variation in quantum yield with wavelength for tyrosine, and tyrosine is more important in the activation process at longer wavelengths. That is, at 2800 Å cystine rupture represents only one half of the inactivation probability in trypsin, whereas at 2537 Å it represents over 80 per cent. These calculations are supported by the observation that the ultracentrifugal sedimentation diagram of trypsin inactivated 85 per cent at 2537 Å and at 2804 Å are not identical, showing these radiations inactivate the enzyme activity by non-identical mechanisms (Setlow and Doyle, 1957). The destruction of tryptophan residues (absent in ribonuclease) is probably wavelength independent between 2400–3000 Å since the quantum yield for inactivation of gramicidin (containing tryptophan but no tyrosine or cystine) is independent of wavelength in this region. Setlow's result also shows that photolysis of peptide bonds is of negligible importance above 2300 Å (Setlow and Doyle, 1957). At much shorter wavelength, say near 2100 Å, the photolysis of peptide bonds may be of more importance in inactivation because of the large increase in the fraction of absorbed light absorbed by these linkages.

The so-called sulfhydryl enzymes, i.e. those requiring free —SH groups for activity are somewhat more sensitive to ionizing radiation than other enzymes. The enzyme ficin has a quantum yield quite in line with expectation for $\Phi = 450/\text{M}$, however, namely 0.013 (Mandl and McLaren, 1949). Acetylcholinesterase is inactivated by ultraviolet light but cysteine showed no capacity to reactivate the irradiated enzyme, indicating that oxidation of sulfhydryl groups to disulfide groups had not taken place or that —SH groups were not important in enzyme activity. Hargreaves (1955), gives the latter interpretation. His notion, however, that —SH enzymes are more susceptible to UV than non —SH enzymes is not supported by quantitative data. If oxidation of a sulfhydryl group of an —SH enzyme is an essential part of the inactivation step, the product is probably not an —SS— group, for crystalline urease is permanently inactivated by ultraviolet light; the activity cannot be restored by reducing agents as is true of air-oxidized urease (Pillemer *et al.*, 1938). Irradiated urease has not the capacity to form antiurease in rabbits but it will react with serum from animals immunized wich either oxidized or active urease. Probably irradiated urease contains split —S—S— groups; we have seen it is also denatured.

Partial protection of triosephosphate dehydrogenase against ultraviolet light by the presence of cysteine has been found, but the inactivation rate is then no longer first order (Shugar, 1951a). Further, completely oxidized enzyme (the activity of which is evident only in the presence of a reducing agent such as cysteine) inactivates more rapidly than one which is only partially oxidized. In interpreting this information it might be helpful to know whether oxidized enzyme contains inter- or intra-molecular —S—S— linkages. Dimers might have a reduced fluorescence efficiency due to resonance energy transfer and consequently have a higher quantum efficiency of inactivation than oxidized monomers in the same mixture. Mutual oxidation–reduction reactions of irradiated species during the irradiation might also change the kinetics; since the dimer of chymotrypsin does not seem to have a different quantum yield from the monomer, secondary valences are not enough for such energy transfer. As pointed out by Shugar, it is difficult to maintain suitable controls in dilute solution, so the observation that the quantum yield depends on concentration should not be taken as sufficient evidence for the failure of equation (5-1). In view of the fact that about 35 per cent of the absorption of this enzyme at 2537 Å is due to its content of diphospho-pyridine nucleotide (DPN), it is of interest to note that the rate of inactivation is not affected by prior reduction of the enzyme-bound coenzyme DPN, (the DPN is bound 1 : 1 to a protein moiety of $M = 50{,}000$), in agreement with the fact that the absorption of DPNH differs very little from that of DPN at 2537 Å.

The reduction of enzyme-bound DPNH results in the appearance at 3400 Å of the characteristic absorption maximum of DPNH. During irradiation at wavelengths above 3100 Å the activity of TPD-DPNH *increases* and the rate of increase is not dependent on temperature (to as low as 4°). Since the bound DPNH is unaffected by the light, it would appear most likely that some reduction of inactive —SS— groups to the —SH form is involved in photoreactivation. After inactivation at 2537 Å, photoreactivation is not possible at longer wavelengths. Above 3100 Å —SS— groups of TPD-DPNH can be photoreduced by DPNH to give active enzyme. Irradiation at 2537 Å probably does not produce —SS— groups since photoreversal of inactive enzyme by the DPNH present is not found (cf. Shugar, 1951b).

Serum oxidase (or ceruloplasmin of M = 151,000 and with eight atoms of copper per molecule) liberates copper under the influence of ultraviolet light (2537 Å), in proportion to loss in activity (Aprison, 1958). It will be of interest to obtain the quantum efficiency for this reaction.

3. *Kinetics of Inactivation of Dry Enzymes*

The photochemistry of enzymes in solution is, at first sight, complicated by the presence of dissolved gases and ions. Since enzymes *in vivo* are largely surrounded by water, it is precisely the information on enzymes in solution which we want. Irradiation of enzymes in the dry state might be expected to provide additional information on the general mechanisms involved; it possesses also the additional advantage that the range of wavelengths employed may be extended to the quartz-ultraviolet limit without the accompanying free radical production in aqueous medium; and into the vacuum-ultraviolet, where the shortest wavelengths result in ionization as compared to dissociation in the quartz-ultraviolet. This latter factor is of some interest in a comparison of the effects of ultraviolet and ionizing radiations and enzymes, although it must be borne in mind that in the vacuum-ultraviolet the absorption due to the peptidyl linkages of a protein far surpasses that due to other components.

In the quartz-ultraviolet region, the quantum yields for enzyme inactivation in the dry state and in aqueous medium do not differ significantly (Table 5-3), suggesting that the mechanisms involved are not radically different.

Setlow (1955) has carried over the formalism of target theory from ionizing radiation methods to ultraviolet studies of proteins and enzymes. By eqn. (1-4) and Table 1-1 of Chapter I we have

$$I/I_0 = \mathrm{e}^{-n\sigma l} = 10^{-\varepsilon Cl}$$

where σ is the absorption cross-section, and by eqn. (1-8c),

$$[\mathrm{A}]/[\mathrm{A}_0] = \exp - (\Phi\sigma)(\text{dose})$$

where $(\Phi\sigma)$ is the inactivation cross-section in cm^2/quantum. Eqn. (1-8c) is followed well at all wavelengths when, for example, 10 μg of trypsin are dried down on $1/4 \times 7/8$ in. glass cover slips to produce "optically thin" targets for ultraviolet light. It was found that the absorption spectrum of trypsin at liquid air temperatures was the same as at room temperature, so any effects which are observed at low temperatures are not due to a difference in the absorption of radiation but a different mode of energy dissipation. The inactivation cross section is 0.01 Å^2/quantum at 300°K, which of course is much smaller than the size of the molecule both because of the small chance of an incident quantum being absorbed in passing through the molecule, and the small value of the quantum yield (Setlow and Doyle, 1954). The variation in the inactivation cross-section of trypsin as a function of temperature, Fig. 5-7, shows a low value at 90°K, in common with all proteins. It can be assumed that a large fraction of the energy which could go into inactivation (by dissociation of a part?) of the molecule appears in

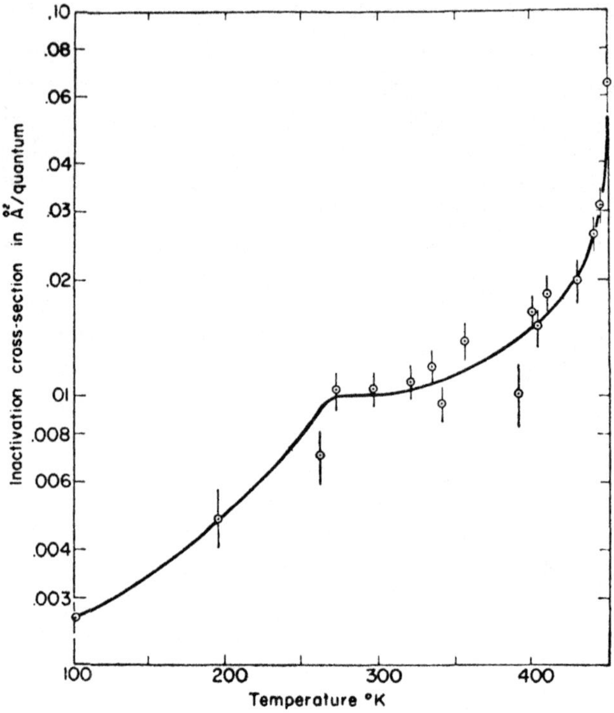

FIG. 5-7. The inactivation cross-section of trypsin with 2537 Å UV as a function of temperature (Setlow and Doyle, 1954).

the form of a long-lifetime fluorescence (Debye and Edwards, 1952). With a rise in temperature, this fluorescence decreases, and the probability of inactivation increases. Near room temperature the inactivation cross-section (and in this case also the quantum yield) is not a marked function of temperature, and the quantum yield is 0.02 for a molecular weight of 20,000 (Setlow and Doyle, 1953). As the temperature rises above 400°K there is a rapid increase in the sensitivity of trypsin to ultraviolet light (and also to heat). The quantum yield rises to a value five times higher than at room temperature. If the inactivation of trypsin is the result of cleavage of predominantly a cystine linkage, the quantum yield of such linkages must also have increased with temperature. A quantum of ultraviolet light becomes very efficient in causing such bonds to pass the energy barrier of inactivation. These ideas are embodied in Fig. 5-8. It is assumed that a transition takes place from the ground state to an upper excited state, the potential energy curve of which may cross those of several other states, one of which is metastable. If the excited molecule passes through the region A it may rearrange or dissociate at some bond. If it crosses over to the

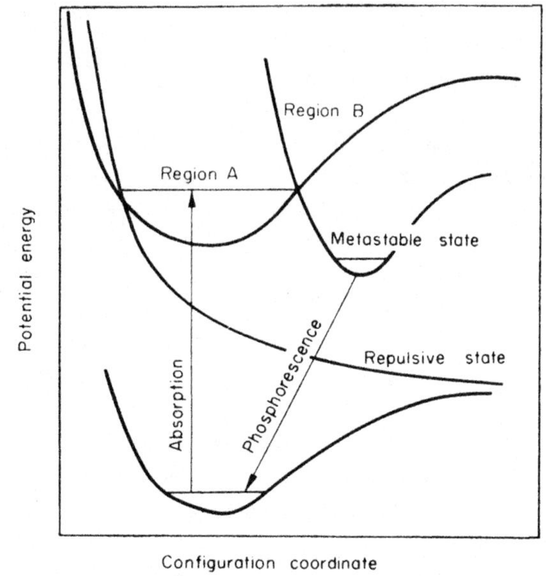

FIG. 5-8. Schematic potential energy curves to explain the decrease in inactivation cross section at low temperature (Setlow and Doyle, 1953).

metastable state it can lose some vibrational energy and stay there for a time which depends on the temperature. At low temperature the time may be of the order of seconds, giving rise to phosphorescence. At high temperatures the phosphorescence is absent and the molecule either radiates, dissociates, or loses energy by collision with other molecules, in very small fractions of a second. Every dissociation that takes place may not end in a net reaction as the free-radical ends may recombine. As the temperature is raised this cage effect would expectedly decrease due to microbrownian motion of chain segments, and the inactivation cross-section increases. This rearrangement or the dissociation can involve more than one kind of bond, and the molecule may not be inactivated as a whole but simply unstabilized until finally inactivated by heat on standing. The unstabilization may consist of disrupted tyrosyl-carboxylate hydrogen bonds, or else cleaved —SS— bonds, which result in subsequent breakdown of secondary and tertiary structure vital to enzyme activity (Anfinsen, 1957).

The action spectrum of dry trypsin between 1900 Å and 3100 Å does not parallel the absorption spectrum at 300°K or at 90°K. This is due to a pronounced variation of quantum yield with wavelength. The yield has a maximum at both 2500 Å and at 3100 Å. Setlow analyses the situation as follows. Light absorbed by an individual unit of a protein molecule, e.g. the ith unit, will have a probability ϕ_i causing inactivation with an inactivation cross-section of $\phi_i \sigma_i$. The inactivation cross-section (ICS) for the molecule as a whole is, therefore,†

$$(\text{ICS}) = \Sigma \phi_i \sigma_i$$

Because the action spectrum of trypsin looks more like the absorption spectrum of cystine than of trypsin, Setlow assumed that the effect of ultraviolet light on trypsin could be resolved into two components, one representing photons absorbed by cystine residues, the other representing photons absorbed by the remainder of the molecule, i.e.

$$(\text{ICS}) = n\phi_{\text{cystine}} \times \sigma_{\text{cystine}} + \phi_m \sigma_m$$

where the subscript m represents the molecule minus the corresponding values for cystine and n represents the number of cystine residues present, i.e., 6 for trypsin (Setlow used the number 4 which is now known to be in error). In Fig. 5-9 the curve marked (trypsin–cystine) is a plot of the second term of this equation and the curve marked cystine represents the first term only. The solid curve is the sum, with assumed values for ϕ_i which would have to be revised because of the discrepancy in the preliminary analytical data for cystine, but the conclusion that ϕ_{cystine} is considerably higher than ϕ_m, which represents aromatic groups and peptide bonds, is consistent with the data for the amino-acids, Chapter II. The fact that the action spectrum can be resolved into these components indicates that the aromatic residues (which absorb strongly at longer wavelengths) and the peptide bonds (which absorb strongly below 2200 Å) make relatively more important contributions to inactivation above and below 2537 Å.

The quantum yields for the inactivation of trypsin and chymotrypsin at 1215 Å are ten times higher than those at 1600 Å. This is thought to be due to the fact that 1215 Å radiation ionizes the molecules whereas at 1600 Å only excitation takes place (Setlow, Watts, and Douglas, 1959). By comparison, trypsin in solution has a somewhat lower ratio of $\phi_{\text{cystine}}/\phi_m$, indicating a somewhat larger increase in the contributions of aromatic residues to the inactivation mechanism. As already seen, however, at 2537 Å

† For the symbol σ see Table 1-1 of Chapter I. It stands for a different quantity than that used by Setlow and Doyle (1957).

the quantum yield for inactivation of trypsin in solution can be accounted for quite well by assigning the major role to cystine, in keeping with the approach of Setlow and Doyle.

Based on results with aldolase, gramicidin, ribonuclease, and trypsin, Setlow has concluded that enzymes with a high cystine concentration have action spectra which are not parallel to absorption spectra. The cystine component shows up most where its absorption is relatively high compared

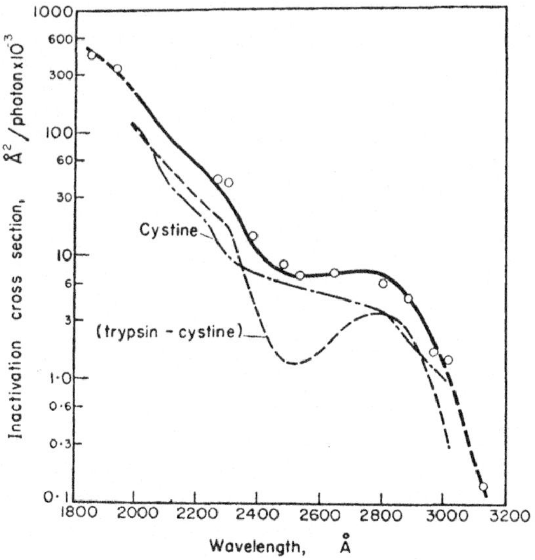

FIG. 5-9. Action spectrum of dry trypsin. The solid curve is the sum of the two component curves and the points are experimental (Setlow and Doyle, 1957).

to the aromatic amino-acids. Since —SS— groups are used to hold polypeptide chains together by intramolecular cross links, it is no wonder that photolysis of these links leads to changes in a number of properties such as solubility and digestibility by proteolytic enzymes, all of which come under the heading of denaturation.

On the whole it must be admitted that our knowledge of the photochemical reactions resulting from the ultraviolet irradiation of proteins and enzymes leaves much to be desired. During the past 15 years, general interest in photobiology has switched from proteins to nucleic acids. It is, however, to be hoped that this situation will be at least partially rectified in the near future, since our understanding of photochemical effects *in vivo* requires a more adequate interpretation of the chemical modifications resulting from irradiation of proteins.

It would be desirable, in this connection, to make use of the many polypeptides which can be prepared by synthetic methods, of which a great variety has now been developed. Some of these should be especially useful because of their biological activity (e.g. Katchalski *et al.*, 1955; Merrifiell and Woolley, 1958; cf. Shugar *et al.*, 1959).

4. *Photosensitized Inactivation of Enzymes*

As early as 1879 it was found that sunlight, particularly the blue-violet region of the spectrum, can result in the aerobic photo-inactivation of crude enzyme preparations (Downes and Blunt, 1879). Subsequently it was observed that the addition of fluorescent dyes, such as eosin, greatly accelerates such inactivation (Schomer, 1936) and it is now fairly clear that light-sensitized inactivation of crude enzyme preparations is due to the presence of pigment impurities, as is probably also the case for a number of crude virus preparations (Section VIII E). The biological and medical significance of reactions of this type, known as photodynamic action, have formed the subject of a separate monograph (Blum, 1941). The detailed mechanism of the way in which the light-activated dye acts as a hydrogen carrier between substrate and oxygen is still in doubt, although there appears to be reasonable evidence that free radicals are involved (Weiss, 1949; Bowen, 1950), probably reduced dye radicals.

On the other hand a good deal of information has now been accumulated with regard to the photo-oxidation of various amino-acids and some of their analogues, as well as for the oxidation of proteins and inactivation of enzymes, both of which apparently involve the sensitized photo-oxidation of several specific amino-acid residues, in particular the aromatic amino-acids and histidine, and possibly S-containing amino-acids (Vodrazka and Pristonpilova, 1957).

Of the amino-acids, probably tryptophan has been the most extensively investigated, along with a variety of indole derivatives (Shugar, 1954) and, of these latter, particularly indoleacetic acid (Galston, 1950; Ferri, 1951; Denffer and Fisher, 1952; Brauner, 1952), to a considerable extent because of its important role as a plant hormone (Galston, 1950). The reaction is usually first-order with respect to substrate and the quantum yield for indoleacetic acid in the presence of riboflavin as sensitizer, in the wavelength region 3950–5200 Å, is about 0.44 (Galston, 1949; cf. Shugar, 1952b), while at 3666 Å it is 0.43 (Shugar, 1952b). This independence of quantum yield with wavelength is typical of most photosensitized reactions.

Galston and Baker (1949) have shown that a number of enzymes, including α-amylase, tyrosinase, and urease, are photo-inactivated in the presence of riboflavin, the reaction being first order in all cases. The riboflavin-sensitized photo-inactivation of lysozyme is also first order, but ribonuclease is apparently unaffected under these conditions, over a wide range of wavelengths (Shugar,

1952b). This was ascribed to the presence of tryptophan in lysozyme, since this amino-acid is absent in ribonuclease; and, in fact tryptophan is quantitatively destroyed during lysozyme inactivation. Independent confirmatory evidence for this suggestion has been forthcoming from the studies of Orlans (1953) on the photosensitized inactivation of antibodies, where changes in ultraviolet absorption also led to the conclusion that inactivation involved modifications in the tryptophan residues. As will be shown below, however, the nature of the sensitizer used may markedly influence the nature of the amino-acid residue affected.

The quantum yield for the riboflavin sensitized photo-inactivation of lysozyme is 0.006 at 3666 Å and is practically unaltered at 2537 Å, if a correction is made for the light absorbed by the enzyme itself at this latter wavelength. For tryptophan the quantum yield at 3666 Å is 0.038. Bearing in mind the fact that the lysozyme molecule contains six tryptophan residues, the quantum yield for destruction of tryptophan in the lysozyme molecule comes to 0.036, in fairly close agreement with the value obtained for tryptophan directly.

Although the quantum yield for photosensitized oxidation of lysozyme is only 0.006 as compared to 0.024 for ultraviolet inactivation, it has been pointed out (Shugar, 1952b) on the basis of the qualitative observations of Galston and Baker (1949) for a number of enzymes, that for larger enzyme molecules such as urease the quantum yield for photosensitized oxidation may be as much as one order of magnitude greater than that for direct ultraviolet inactivation, a result of some significance in the interpretation of photochemcial effects *in vivo*.

The riboflavin-sensitized photo-inactivation of snake venoms has been extensively investigated by Guidolin and Ferri (1950) and Ferri and Guidolin (1951), who found that a number of other sensitizers such as quinine, eosin, fluorescein, and chlorophyll will likewise destroy the toxicity. A similar observations has been made by Weil *et al.* (1957) who found that the photo-oxidation of crystalline *Clostridium botulinus* type A toxin, sensitized by methylene blue, results in the destruction of toxic, but not antigenic, activity. The potential practical applications of these observations do not appear to have been exploited as yet (Ferri and Guidolin, 1951).

For chymotrypsin, methylene blue-sensitized oxidation leads to the simultaneous destruction of both proteolytic and esterase activities (Fig. 5-10), a result in agreement with other observations indicating that the reactive groups responsible for both these activities are the same. Photo-inactivation is accompanied by the elimination of the reactive site of attachment of di-isopropyl phosphate, half of the histidine residues and almost one-half the tryptophan. No hydrolysis of peptide bonds was observed, nor were there any modifications in enzyme solubility or viscosity, again testifying to the localized character of the reaction (see above). Loss of ability to dimerize after photo-oxidation is also associated with the destruction of one of the two histidine residues in chymotrypsin (Egan *et al.*, 1957).

In addition to chymotrypsin, extensive studies have been made of the photosensitized oxidation of lysozyme and ribonuclease in the presence of methylene blue (Weil *et al.*, 1952, 1953, 1955). In the case of ribonuclease, total inactivation occurred following uptake of 3 moles of oxygen per mole of enzyme, the only observable chemical change being the oxidation of 3 of the 4 histidine residues. Up to 50 per cent inactivation, the decrease in

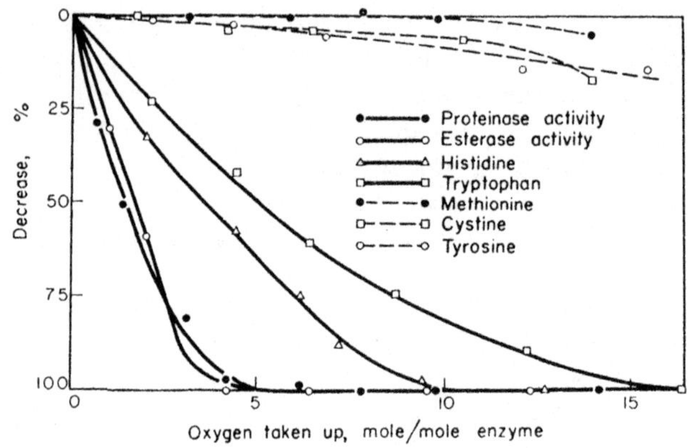

FIG. 5-10. Decrease of various amino-acid residues and enzymatic activities produced by the photo-oxidation of chymotrypsin (Weil *et al.*, 1953).

histidine content was proportional to the decrease in enzyme activity. Since no changes in physical properties could be observed (viscosity or aggregation), while irradiation under a nitrogen atmosphere under identical conditions was without effect, the importance of histidine as at least part of the catalytic centre of the ribonuclease molecule was suggested. In the case of lysozyme, however, the findings are not so clear-cut, a 70 per cent reduction in enzyme activity being accompanied by the destruction of the 1 mole of histidine present, but also of 1 of the 8 moles of tryptophan; only considerable additional radiation resulted in any effect on the tyrosine residues. A large fraction of photo-inactivated lysozyme is denatured (J. Ulrich, unpublished observation).

An elegant application of photosensitized inactivation to the study of the need of histidine for phosphoglucomutase activity has been described by Koshland *et al.* (1958). The enzyme contains 11 histidine groups, but the photosensitized inactivation in the presence of methylene blue is much more rapid than loss of histidine groups. However a semi-log plot of loss of histidine vs. time of irradiation (top curve in Fig. 5-11) can be divided into two portions, as shown in the figure, one of which corresponds to 5 histidines which are inactivated "fast" and the other to 6 which are inactivated "slowly." It was found that loss of enzymatic activity corresponded exactly to the "fast" curve in Fig. 5-11, and the data were further analyzed to show that only one of the "fast" histidines is involved in enzymatic inactivation (the type of analysis used in this case should be applicable to other enzymes containing several histidine or other residues involved in

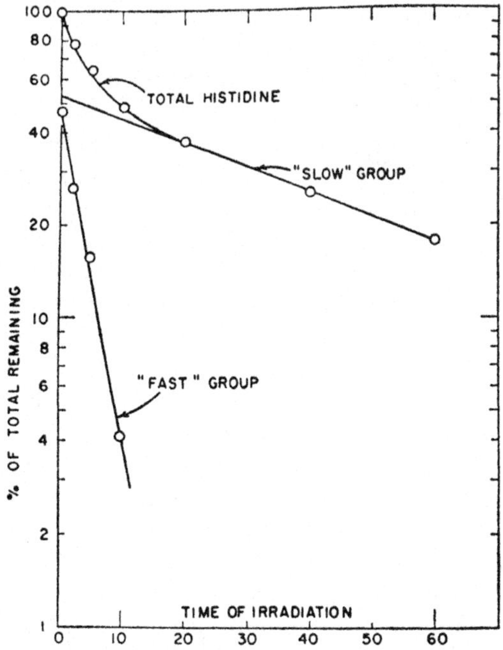

FIG. 5-11. Effect of irradiation time on histidine content of phosphoglucomutase (Koshland, 1958).

enzymatic activity). It was further found that the rate constant for destruction of the "fast" histidine corresponded to the rate constant for photooxidation of free histidine, and this was interpreted as supporting evidence for the conception that the "fast" histidines are on the surface of the molecule, while the slow ones are in the interior. The requirement of only one histidine residue for enzymatic activity, in conjunction with analogous results for chymotrypsin (Weil *et al.*, 1953) was also regarded as supporting evidence for the existence of a common catalytic unit for the two enzymes. Some exception may be taken with regard to this conclusion, however, since no information is given as to whether other amino-acid residues are affected during photosensitization of phosphoglucomutase. The usefulness of the above type of approach is, nonetheless, beyond question.

It has been applied by Nakataru (1961) to show that two histidine residues are essential to the enzymatic activity of catalase; and by Belitser and Kotkova (1960) to demonstrate that 8 of the 56 histidyl residues of fibrinogen are linked with its ability to undergo conversion to fibrin. The latter authors have also shown how to evaluate the effects on other amino-acids. Using methylene blue as photosensitizer, fibrinogen "activity" is lost following oxidation of 8 histidyls and 7 tryptophanyls at slightly alkaline pH. If, however, the fibrinogen was irradiated at pH 5.9, only tryptophan residues were oxidized; under these conditions the fibrinogen retained its ability to be converted to fibrin. Consequently it is the oxidation of several histidine residues which lead to loss of fibrinogen "activity."

The results of Weil *et al.*, for ribonuclease are of some interest in relation to those of Shugar (1951) who, as mentioned above, found no influence of riboflavin on this enzyme. This question of dye specificity is a subject of general interest in photosensitization mechanisms in solution (Weiss, 1949; as well as appended discussion). It has been investigated semi-quantitatively for enzymes in a study of the photo-oxidation of desoxyribonuclease by a wide variety of dyes (Shugar and Rzendowska, 1959). The rate of inactivation in the presence of methylene blue, for example, was so rapid, even in ordinary daylight, that it was difficult to measure; on the other hand such dyes as eosin, rosaniline, auramine, toludine blue (the latter of which bears a close resemblance to methylene blue) were quite inactive at 3660 Å and longer wavelengths. Under these conditions riboflavin and fluorescein were almost equally effective, with quantum yields for desoxyribonuclease inactivation of the order of 10^{-3} and 10^{-4}.

These findings have their counterpart *in vivo*. An interesting example is the study of Santamaria and Angelotti (1957) on the efficiency of various dyes in the photosensitized killing of *E. coli*. Of 24 dyes examined, methylene blue was most efficient; rose bengal was only one-fourth as active, while a dye such as benzopyrene was completely inactive. The foregoing results are not fully quantitative, since the wavelength of the incident light, and the absorption spectra of the dyes, were not taken into consideration.

The efficiency of a number of different dyes in sensitizing the photo-oxidation of a simple substrate, indoleacetic acid, has been investigated by Brauner (1952) who found a very wide range of activities, the most efficient being riboflavin followed, in order of decreasing efficiency, by thionine, methylene blue, eosin, fluorescein and Rhodamine B. On the other hand, the course of destruction of histidine and tyrosine in ovalbumin is very similar with methylene blue, eosin, riboflavin, and photoporphyrin. Apparently no one has yet attempted to make a comparative study of the effect of different dyes on a protein together with a simultaneous study on those amino-acids in the protein which are susceptible.

The photo-oxidation of serum proteins has been given a good deal of attention by Czech workers (Prauss, 1955; Vodrazka and Sponar, 1957; Vodrazka and Pristonpilova, 1957). The methylene blue-sensitized photo-oxidation of human globin requires the uptake of 95 moles of oxygen per mole enzyme; at 80 moles oxygen uptake, all the histidine and tryptophan residues are eliminated, but only 50 per cent of the tyrosyls. Of considerable interest is the observation that the reaction apparently involves binding between globin and methylene blue, but such binding is not a necessary prerequisite since human albumin and γ-globulin are also photo-oxidized but do not significantly bind the dye. A general review of the subject of protein photosensitized oxidation has been prepared by Vodrazka (1959).

CHAPTER VI

Photochemistry of Purine and Pyrimidine Derivatives

In any attempts to interpret the effects of ultraviolet irradiation on nucleic acids, it would appear to be obvious that a knowledge of the photochemical behavior of the individual components is an essential prerequisite. On the other hand, it must not be expected *a priori* that the photochemical transformation of purine and pyrimidine bases (or their nucleosides and nucleotides) will be unaffected when they are incorporated in polymer chains where their environmental conditions are profoundly modified compared to those of the free derivatives in solution. Studies on individual derivatives should therefore be expected to serve more as a guide to their possible behavior in polymer chains.

It must also be emphasized that, as in the case of proteins and enzymes, the photochemistry of nucleic acid derivatives has been frequently investigated in the past from two rather different points of view, one of which was concerned principally with the physical chemistry of the reactions and employed such doses of radiation as were necessary to produce some appreciable physico-chemical modifications readily subject to measurement; and the other in which observers were primarily interested in seeking clues to effects resulting from the use of radiation doses in the biological range. The distinction between these two approaches will be frequently stressed in this and succeeding chapters, since it is largely the initial effects which are of importance in interpreting the biological effects of radiation.

From what is to be described in the succeeding sections of this chapter, it will be seen that our understanding of the photochemical transformations undergone by nucleic acid derivatives has been progressing at a rapid pace during the past few years; and interest in this field remains not only unabated, but is rather on the increase. Attention should therefore be drawn to two important gaps which have hitherto remained unfilled. First, most investigations have been carried out with mercury-resonance lamps at only one wavelength, 2537 Å. It is most desirable to investigate the possibility of wavelength effects in the entire quartz-ultraviolet. Secondly, we have now reached the point where an adequate interpretation of the *mechanisms* of the photochemical reactions undergone by pyrimidine derivatives requires the initiation of investigations on the nature of the photochemically excited states of these compounds, either by flash photolysis methods (Section II.A.8) or by fluorescence measurements on trapped radicals, such as has already been done in the case of amino-acids and proteins.

Apart from the foregoing, it must also be borne in mind that living cells contain a considerable pool of free nucleotides, nucleotide analogues, and nucleotide coenzymes, the latter of which are frequently found in combination with proteins or enzymes. It is consequently conceivable that some biological radiation effects will manifest themselves via these compounds. One such example has already been referred to, viz. the role of DPN in the photochemical behavior of triosephosphate dehydrogenase (Chapter V). Another illustration is the correspondence which has been reported between the destruction of intracellular ATP and the mobility of fertilizing capacity of frog spermatozoa under the influence of radiation (Kanazir and Errera, 1955). Yet another example is that of uridine-diphosphate glucose (UDPG) which has been shown to be involved in sucrose synthesis in plants; Zill (1957) and Zill and Tolbert (1958) suggest that it is more than a coincidence that the rate of photochemical transformation of UDPG is comparable to the rate of radiation-induced inhibition of sucrose synthesis in plants.

1. *Carbohydrates*

In view of what has previously been pointed out with regard to the transparency of carbohydrates in the quartz-ultraviolet (Section II.B.2), it is pertinent to inquire to what extent they are affected by irradiation, particularly as they occur in nucleic acids or their derivatives. No systematic studies appear to have been made on this subject and, as will be seen, a good deal of the existing confusion is due largely to the use of ill-defined sources.

There have been a number of reports on the destructive effects of ultraviolet on mucopolysaccharides (e.g. Mogilevskii and Laufer, 1951; Howe *et al.*, 1958; Balazs *et al.*, 1959; Hvidberg *et al.*, 1959), but no serious effort appears to have been made to pin down the wavelengths responsible. For example, Hvidberg *et al.* (1958) found that the viscosity of hyaluronate solutions could be eliminated by irradiation and that the actinic wavelengths were below 3000 Å. However, the source used was a Phillips 125 W HP lamp with the glass filtering envelope removed, and there is consequently little doubt but that the effective wavelengths were well below 2300 Å.

Rice (1952) irradiated neutral solutions of ribose, glucose, cytidylic and adenylic acids, and yeast RNA and DNA. Glucose and ribose were rendered incapable of reducing alkaline copper solutions or of giving a Molisch test, while adenylic, cytidylic and ribonucleic acids lost the ability to form furfural and DNA no longer reacted to the Stumpf test. However, the additional observation that a protective effect prevailed in the presence of a number of substances known to be protective against ionizing radiations, raises some doubts as to the nature of the primary reaction involved. When to this is added the fact that irradiation was by means of an "unscreened quartz lamp," it becomes almost certain that the effects were due to absorption of wavelengths below 2300 Å as well as to free radicals formed from the solvent and dissolved oxygen (cf. Barszcz and Shugar, 1961).

Holtz (1936) had found that ultraviolet irradiated solutions of glucose possessed an oxidation-reduction system resembling that of ascorbic acid, while the irradiated solution absorbed ultraviolet light with a maximum at 2900 Å. The additional observation that X-irradiation induced the same results (Holtz and Becker, 1936) again implies the use of an ultraviolet source emitting at short wavelengths. This conclusion is supported by Laurent and Wertheim (1952), who found no changes in ultraviolet absorption of irradiated neutral solutions of glucose.

On the other hand, irradiation of alkaline solutions of hexoses and pentoses even at wavelengths above 2537 Å results in the formation of photoproducts with absorption maxima in the neighborhood of 2600 Å and with extinctions dependent on the sugar concentration, pH, temperature and the nature of the atmosphere under which the solutions are irradiated (with higher yields of photoproducts under nitrogen than under oxygen). The photolysis products of glucose exhibit a maximum at 2450 Å in acid solution and a p*K* of 4.5 (Laurent and Wertheim, 1952; Laurent, 1956). The same changes may be observed in solutions of monosaccharides under the influence of alkali alone, at a lower rate, but equally dependent on pH and temperature (Berl and Feazel, 1954; Peruely and Meizner, 1953). Analogous effects are likewise observed when neutral solutions are subjected to γ-irradiation (Khenokh, 1955) or upon exposure of neutral or alkaline solutions to X-rays (Bothner-By and Balazs, 1957). The latter authors demonstrated that the reaction products are the same as those produced by ultraviolet light on alkaline solutions of glucose. The effects of ionizing radiations on carbohydrates have recently become the object of intensive study (e.g. Phillips *et al.*, 1958; Phillips and Moody, 1958, 1960a, 1960b; Wolfrom *et al.*, 1959). The results of these investigations, in which a number of the decomposition products have been identified, should be useful in attempts to identify the products of alkaline decomposition of carbohydrates as well as the enhanced yields of such products by ultraviolet irradiation due, most likely, to photochemically excited hydroxyl ions.

Prolonged irradiation at 2537 Å has been shown to lead to the formation of uracil from uridine (Rapport *et al.*, 1955) but it was not reported whether the rupture of the glycosidic linkage was accompanied by destruction of the ribose. Since the primary reaction in the photolysis of uridine leads to saturation of the 5,6 double bond (Sinsheimer, 1954) with the resultant labilization of the glycosidic linkage as in dihydropyrimidine glycosides, it is conceivable that the ribose was liberated intact. On the other hand, the ultraviolet inactivation of the co-enzyme activities of DPN and TPN has been reported to be accompanied in both cases by a partial destruction of ribose, but involving only that proportion linked to the pyridine moiety (Seraydarian *et al.*, 1954; Seraydarian, 1955).

On the whole, it appears reasonable to assume that, in neutral aqueous media, wavelengths longer than 2200 Å are without direct effect on sugars. The same may be said to apply in alkaline solutions insofar as nucleosides and nucleotides are concerned, since involvement of the carbohydrate in a glycosidic linkage renders it stable to alkali. In support of this it has been shown that irradiation of adenosine at pH 12, under conditions where wavelengths below 2400 Å are eliminated, and at an intensity of 10^{17} quanta/cm^2 per min for a period of one hour, is without effect on the ribose content as measured by the orcinol reaction (Shugar and Wierzchowski, 1957). Additional evidence for the absence of any direct effect on the sugar components of nucleotides and nucleosides is forthcoming from the high degree of reversibility prevailing for the photoproducts of pyrimidine glycosides (Sections 5, 10).

This does not necessarily exclude the possibility of destruction of carbohydrate components in polynucleotides, where the sugar phosphate links two absorbing constitutents between which energy transfer may occur with resultant "strain" on the internucleotide linkages. Such energy transfer has actually been demonstrated in a dinucleotide analogue, DPN (Weber, 1958) and, as mentioned above, irradiation of DPN apparently does result in destruction of some of the ribose; on the other hand, it is only the ribose which is linked to the nicotinamide moiety which apparently undergoes destruction. The experiments of Rice (1952) on nucleic acids, referred to above, throw no light on this point because of the ill-defined nature of the irradiation source used. Extensive experiments with model oligo- and polynucleotides, to be discussed in detail in Chapter VII, however, remove all doubts as to the lack of any direct effect on the carbohydrate moieties, at least for radiation doses in the biological range and several-fold higher.

2. *Purine Derivatives*

Irradiation of a solution of DNA to the point where 30 per cent of the pyrimidine bases are destroyed leads to a photoproduct from which the entire guanine and more than 90 per cent of the adenine may be recovered (Errera, 1952). This relative resistance of purines to irradiation as compared to pyrimidines has been repeatedly observed by a number of observers (Sinsheimer and Hastings, 1949; Carter, 1950; Christensen and Giese, 1954). Runnstrom *et al.* (see Carter, 1950) had early observed that although irradiation of red cells leads to a complete loss of co-enzyme activity, cophosphorylase activity was hardly affected, indicating that the adenylic acid moiety remained intact, a result which is, however, only qualitatively in agreement with those of Carter (1950), Seraydarian *et al.* (1954) and Seraydarian (1955).

There is a paucity of quantitative data on the photochemistry of purine derivatives which renders a comparison of the results of different observers rather difficult, and this difficulty is further accentuated by the frequent use of ill-defined ultraviolet sources. Qualitative observations on the photochemical destruction of naturally occurring purine derivatives have been made by a number of observers, a commonly applied criterion being destruction of the characteristic absorption spectrum between 2000 and 3000 Å (Heyroth and Loufbourow, 1931; Rapport and Canzanelli, 1950; Canzanelli *et al.*, 1951). Irradiation for 12 hr with a GE germicidal lamp of 5×10^{-5} M solutions of adenine, adenosine, xanthine, hypoxanthine, uric acid, guanine and caffeine, suffices in all cases to disrupt completely the absorption spectrum (Rapport and Canzanelli, 1950). Isoguanine also loses its characteristic absorption under the influence of irradiation (Stimson, 1942). On the other hand, Christensen and Giese (1954) report that prolonged irradiation of adenine, adenosine and adenylic acid leads in all cases to an increase in absorption over the range 2300–2900 Å, a result rather at variance with those of most other observers. Canzanelli *et al.* (1951, Table 6-1, p. 366) show a seven per cent increase in adenine absorption following irradiation at wavelengths greater than 2500 Å, but apparently attach no significance to this.

For purposes of comparison with pyrimidine derivatives in studies on the photochemistry of nucleic acids, Shugar and Wierzchowski (1958) made some rough measurements of the quantum yields at 2537 Å for adenine and guanine, the values obtained being 0.6×10^{-4} and 2×10^{-4} respectively. In the light of the studies of Kland and Johnson (1957), to be discussed below, these values should be considered as only approximate, although they are in accord with the qualitative result of Canzanelli *et al.* (1951). From the data presented by these latter authors it may be concluded that the quantum yields for xanthine and hypoxanthine are somewhat higher, while uric acid is probably the most susceptible of all the naturally occurring purines (Kland and Johnson, 1957). Using a series of cut-off filters, Canzanelli *et al.* (1951) found that, at wavelengths longer than 2300 Å, adenine and yeast adenylic acid were little affected as compared to other purine derivatives. It would obviously be desirable to examine quantitatively the effect of monochromatic light of varying wavelengths, but the high intensities required to achieve a reasonable degree of degradation render such studies technically difficult.

Loofbourow and Stimson (1940) concluded, on the basis of changes in absorption spectra, that the lability of purines and pyrimidines to irradiation appeared to parallel the number of carbonyl groups in the ring. This is, however, too broad a generalization; although it does appear to apply roughly for purines alone since the stability of some of these is, in decreasing order, adenine, hypoxanthine and guanine, xanthine, uric acid. Of more importance is the presence of a carbonyl group in the 2-position (Canzanelli *et al.*, 1951).

The first and, to our knowledge, only attempt to study the kinetics of ultraviolet decomposition of purines is that of Kland and Johnson (1957). They used a bank of 15 W GE germicidal lamps (unfiltered) and irradiations were carried out under oxygen and nitrogen. Some typical results are illustrated in Fig. 6-1. With the exception of adenine under nitrogen there is an initial

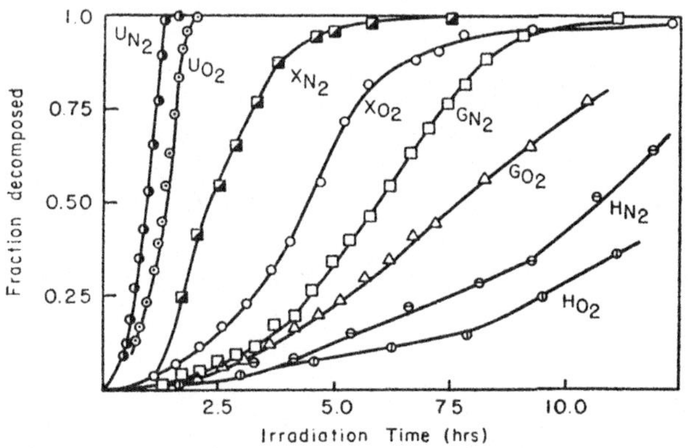

FIG. 6-1. Relative rates of decomposition, as measured by loss of absorption at maxima, of uric acid (U), xanthine (X), guanine (G), and hypoxanthine (H). The subscripts N_2 and O_2 indicate that irradiation was under an atmosphere of nitrogen or oxygen (Kland and Johnson, 1957).

"induction" period, following which the rate increases and then levels off. At high irradiation intensities, decomposition is proportional to the square of the exposure time, with the exception of adenine. The data were used to divide the five purines studied into two classes, i.e. those substituted in the 6-position (adenine and hypoxanthine) which decompose more slowly under nitrogen, and disubstituted purines (guanine, xanthine) and uric acid for which the rates are slower under oxygen.

It should, however, be noted that all the above compounds were irradiated in unbuffered solution; the initial pH values of which varied from 5.35 (uric acid) to 5.94 (adenine). Both these compounds have pK values at 5.4 and 4.15 (Bendich, 1955), so that a mixture of ionic species with different absorption spectra (Beaven *et al.*, 1955) was being studied. Furthermore, the liberation of ammonia taking place during decomposition of purines (Canzanelli *et al.*, 1951) could have resulted in pH changes during the reaction. The importance of this point may best be assessed from the fact that in the experiments of Canzanelli *et al.* (1951), where concentrations irradiated were of the same order of magnitude (10^{-4}–10^{-5} M), the pH of phosphate buffered solutions (pH 7.2, molarity not given) fell by 1 pH unit during the 12 hr irradiation.

Since under aerobic conditions adenine is completely destroyed after 24 hours irradiation, whereas under nitrogen only 10 per cent disappears, one of the reaction mechanisms suggested is oxidative breakdown at the amino group

$$R \cdot NH_2 \xrightarrow[{[O]}]{h\nu} [R \cdot NH{:}OH] \xrightarrow[H_2O]{h\nu} R \cdot OH$$

and this is supported by the isolation of hypoxanthine from the reaction mixture. Hypoxanthine was also isolated, however, from adenine irradiated under nitrogen. Furthermore, the quantities of hypoxanthine isolated in both cases were very small, although this could have been due to the greater sensitivity to irradiation of this compound.

The photochemical decomposition products of purine derivatives have received very little attention. The liberation of ammonia by a number of irradiated purines and pyrimidines was qualitatively investigated by Lieben and Getreuer (1933) in connection with a study of the photolysis of amino-acids. Canzanelli *et al.* (1951) subsequently examined in greater detail the production of urea and ammonia from a variety of irradiated purine and pyrimidine derivatives, some of their findings being illustrated by Table 6-1. It was found that urea liberated is derived almost exclusively from the pyrimidine ureide group and the amount of urea produced was greatest for those derivatives with a keto (or hydroxyl) group in the 2-position, as might be expected, since the pyrimidine ureide then most nearly resembles urea itself (note from Table 6-1, for example, that uric acid produces no more urea than xanthine, although it contains an additional ureide group in the imidazole ring). Ammonia liberated was also found to be derived mainly from the pyrimidine ureide group. Despite the higher yield of ammonia from adenine (Table 6-1), cytosine does not give a significantly greater yield than uracil (cf. Table 6-6).

One general conclusion of some importance is that the greatest formation of urea is associated with the least production of ammonia, so that the production of each of these substances probably involves a different decomposition path. It should, however, be noted that urea and ammonia determinations were made on all the above compounds only following complete alteration of the absorption spectrum, so that little can be concluded from this with regard to the intermediate products of photodecomposition.

TABLE 6-1. AMMONIA AND UREA PRODUCTION FROM PURINE DERIVATIVES IRRADIATED AT A CONCENTRATION OF 5×10^{-5} M WITH AN UNFILTERED GERMICIDAL LAMP TO THE POINT OF COMPLETE DESTRUCTION OF ABSORPTION SPECTRUM (CANZANELLI, GUILD AND RAPPORT, 1951)

Compound 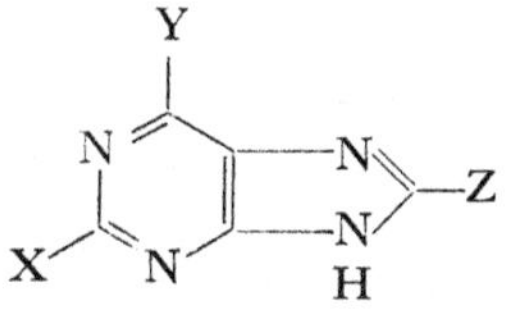		NH_3 Moles per mole compound irradiated	Urea Moles per mole compound irradiated
X = Z = H, Y = NH_2	Adenine	1.01 ± 0.07	0.17 ± 0.04
	Adenosine	1.02 ± 0.08	0.15 ± 0.02
	Adenosine-2′(3′)-phosphate	1.10 ± 0.07	0.17 ± 0.04
X = NH_2, Y = OH, Z = H	Guanine	0.50 ± 0.04	0.17 ± 0.01
	Guanosine	0.35 ± 0.05	0.20 ± 0.20
	Guanosine-2′(3′)-phosphate	0.46 ± 0.08	0.15 ± 0.03
X = Y = OH, Z = H	Xanthine	0.19 ± 0.08	0.39 ± 0.01
	Xanthosine	0.39 ± 0.10	0.61 ± 0.02
X = Z = H, Y = OH	Hypoxanthine	0.79 ± 0.09	0.14 ± 0.02
	Inosine	1.10 ± 0.06	0.16 ± 0.01
X = Y = Z = OH	Uric acid	0.23 ± 0.11	0.40 ± 0.02

Adenine-1-*N*-oxide has been found to be much more photosensitive than the parent adenine. Its rate of photochemical transformation (presumably at 2537 Å) is first-order for the first 50 per cent of the reaction, with a quantum yield of 0.1, which is considerably higher than for most pyrimidines. The products resulting from photolysis included isoguanine, adenine and an unidentified product in the ratio 4:5:1 (Levin and Brown, 1962).

Extensive irradiation of uric acid leads to the formation of a variety of photoproducts in addition to urea and ammonia. One of these was isolated by Fellig (1954), following irradiation of 2 l. of a neutral solution of uric acid, 0.1 mg/ml, with a bank of 7 germicidal lamps. Concentration of the irradiated solution led to the isolation of crystalline triuret in about 10 per cent yield with respect to the original substance. It was suggested that the reaction proceeds via formation of allantoin and allantoic acid, followed by decarboxylation and oxidation:

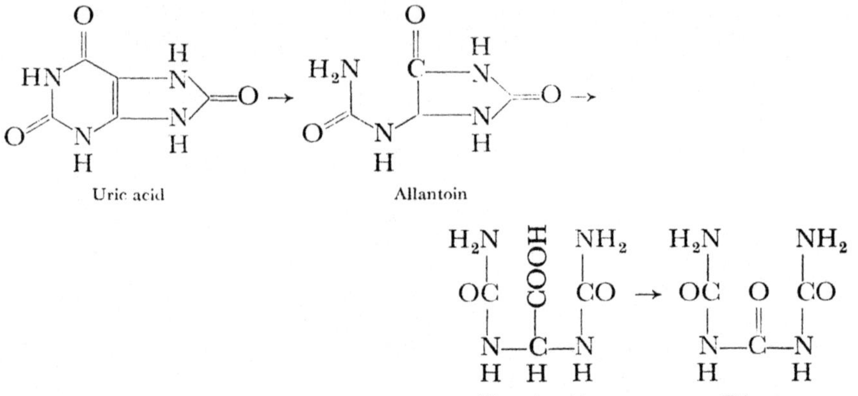

Uric acid Allantoin

Allantoic acid Triuret

In the absence of any evidence as to the existence of allantoin or allantoic acid during irradiation, the foregoing scheme cannot be considered as established, although it is reasonable. Undoubtedly the simplest procedure for testing this would be by irradiation of the intermediates themselves. This, or some analogous procedure, is, in fact, necessary and desirable since it would materially assist in establishing the decomposition paths of other purines. Furthermore, since only 10 per cent of the initial product was isolated as triuret amongst the photoproducts, other decomposition paths are most likely involved as well. Cyanuric acid

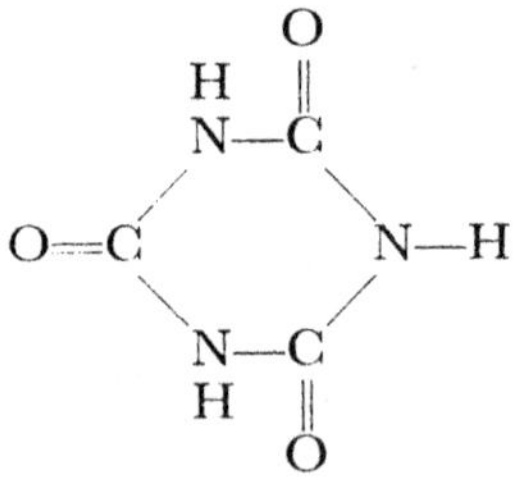

was also isolated from among the photoproducts and it was presumed that this arose from triuret; the latter, upon heating in air or in 1 N NaOH, is transformed to cyanuric acid and ammonia.

3. *Purine Nucleotides*

Nucleosides and nucleotides of adenine and guanine, like the parent purines, are relatively resistant to radiation (Canzanelli *et al.*, 1951; Christensen and Giese, 1954). At 2537 Å, the radiation induced destruction of yeast adenylic acid, as measured by decrease in optical density at the absorption maximum, proceeds with an initial quantum yield of 0.5×10^{-4}, or about the same as for free adenine; for guanylic acid the quantum yield is 1.5×10^{-4} as compared to 2×10^{-4} for free guanine (Wierzchowski and Shugar, 1960b).

Under conditions where adenosine-2′(3′)-phosphate, AMP and adenosine were unaffected by irradiation, ATP was found to undergo considerable degradation to adenine; in the case of DPN the photolytic products included adenine and small amounts of inorganic P. It was concluded from this that the glycosidic linkage in ADP and ATP is labilized by the pyrophosphate group.

Several workers have reported some type of "reactivation" of irradiated ATP. All of these are of doubtful validity, but, in view of the considerable interest attaching to biological photoreactivation, a few words about them is undoubtedly warranted. Kita *et al.* (1954) report that the decrease in absorption of ATP resulting from irradiation (by means of a quartz lamp which is otherwise unidentified) may subsequently be regained under the influence of visible light. Garay and Guba (1954), using a high-pressure Hanau lamp without filter, claim that irradiation of ATP leads to an increase in absorption, and conclude that this is due to some change in configuration resulting from interaction with free radicals formed from the irradiated aqueous medium; with time the photoproduct is said to revert to the original ATP. A similar effect was reported to occur under the influence of ultrasonic irradiation. These findings are contrary to those of other observers and, in fact, an examination of the above publications has failed to reveal any data substantiating the conclusions drawn. Wells and Johnson (1953) had earlier attempted to demonstrate a reversal of photolysis of ATP by exposure to visible light, with negative results. It would, indeed, have been surprising if positive results had been obtained, since the photolysis products of ATP do not absorb in the visible region of the spectrum.

4. *Nucleotide Co-enzymes*

An extensive investigation of the decomposition products of irradiated DPN has been made by Seraydarian *et al.* (1954), with paper and column chromatography techniques. Following 100 per cent inactivation of co-enzyme activity, the four main degradation products were found to be ADP, AMP, adenine and nicotinamide; about 50 per cent of the total ribose, almost all of it from the nicotinamide moiety, no longer responded to the orcinol reaction. In agreement with an earlier finding of Carter (1950), no adenosine was found in the irradiation products despite the presence of AMP and adenine. It therefore follows that the ribosidic linkage is more vulnerable to irradiation in the nucleoside than in the (mono) nucleotide form. The 7.5 per cent decrease in optical density following

100 per cent inactivation was ascribed entirely to disruption of the nicotinamide ring, again in agreement with the earlier results of Carter (1950).

The behavior of irradiated TPN is analogous to that for DPN, with the exception that only about 30 per cent of the ribose (again mainly that from the nicotinamide moiety) is destroyed following complete loss of co-enzyme activity. In this case, however, the splitting of the ribose phosphate ester bond of the nicotinamide moiety liberates, in place of ADP, a compound similar to the latter, but with an additional phosphate group at the 2′ (or 3′) hydroxyl of adenosine, as follows:

```
Nicotinamide ribose—¦—P—P—ribose—adenine
                                 |
                                 P
```

The identification of this product on the basis of enzymatic tests was regarded only as tentative (Seraydarian, 1955). It is, however, fully in agreement with the structure of TPN proposed by a number of observers since 1950 and may, in fact, be regarded as additional supporting evidence for such a structure (Shugar, 1960b).

Shigemoto (1958) has reported results which are not entirely in agreement with the above. The products of photochemical degradation of DPN were reported to include nicotinamide-ribose-5-phosphate, AMP and adenosine; for ATP the corresponding photoproducts were AMP and adenosine. However, in the absence of the original publication, it is not possible to comment on this discrepancy. The same author also examined the effects of irradiation at 2600, 3700, and 4500 Å (using a mercury lamp with appropriate filters for isolation of these lines) on flavin mononucleotide and flavin adenine dinucleotide. These were reported to be decomposed at all three wavelengths, but most effectively at 2600 Å, the decomposition products including riboflavin, lumiflavin, and lumichrome. The visible light photosensitivity of riboflavin had been observed as long ago as 1932 by Warburg and his collaborators, and the photochemistry of various flavin derivatives has been subjected to extensive investigations, particularly by Karrer and his co-workers in Switzerland (see Halwer, 1951).

In view of what has been said above with regard to carbohydrates, the destruction of nicotinamide-bound ribose is rather unusual, but may be due to energy transfer with resultant strain rupture of the carbohydrate ring. It would clearly be desirable to investigate the photochemical behavior of isolated nicotinamide-ribose as well as its phosphate ester. The resistance of the adenine-bound ribose is, on the other hand, in accord with the fact that very high radiation doses are necessary for destruction of ribose in nucleic acids.

The above work was carried out with the use of germicidal lamps as sources. Seraydarian (1956) subsequently showed that complete inactivation of co-enzyme activities of DPN and TPN could be achieved with the use of wavelengths including only the range above 2100 Å. Actually the influence of wavelength on the inactivation of co-enzyme activity of cozymase had been much earlier investigated quantitatively by Warburg and Christian (1935) over the range 1830–2830 Å. With decreasing wavelength the "apparent" quantum yield was found to increase considerably, so that at 1830 Å it was almost 120 times that at 2830 Å (Table 6-2). Assuming that

TABLE 6-2. APPARENT QUANTUM YIELD FOR INACTIVATION OF CO-ENZYME ACTIVITY OF DPN (WARBURG AND CHRISTIAN, 1935)

λ, Å	Φ
1860	0.65
1960	0.16
2080	0.016
2270	0.016
2537	0.005
2650	0.005
2830	0.011

the quantum yield was independent of wavelength, and that no direct inactivation resulted from irradiated solvent water molecules, they concluded that only a portion of the co-enzyme molecule was affected by irradiation, an inference surprisingly close to the truth (see above). The assumptions on which this conclusion was based are nonetheless questionable. There is really no basis for the supposition that the quantum yield is independent of wavelength over the range 1860–2830 Å; while at wavelengths below 2000 Å not only is there appreciable absorption by the carbohydrate moiety, but the solvent (Golding, 1960; Barrett and Mansell, 1960) and buffer molecules themselves absorb appreciably and, in the presence of dissolved oxygen, give rise to a number of free radicals including peroxides (Barszcz and Shugar, 1961).

It has been reported by Wells (1956) that the biological activity of irradiated DPN may be partially restored by subsequent irradiation with light of longer wavelength, an observation which, if substantiated, would be of considerable importance in relation to the phenomenon of biological photoreactivation. The following considerations suggest, however, the need for independent verification of this result. The ultraviolet irradiated (inactivated) DPN solution was exposed to the illumination from a GE CH-4 spotlamp through a Corning 3060 filter. Such a filter, according to catalogue data, transmits only 0.5 per cent at 3650 Å and zero below this wavelength; but neither DPN nor its photoproducts absorb at 3650 Å. It is consequently most likely that whatever effect was observed was in some way related to the crude tissue extracts used to test for DPN activity. Ekert and Monier (1958) attempted to duplicate Wells' finding, using an established enzymatic assay procedure for DPN activity, with negative results.

Reduced DPN, or DPNH, exhibits an absorption band with a broad maximum at 3400 Å, but irradiation in this wavelength region is without effect on its co-enzyme activity (Shugar, 1951a, 1951b).

5. *Pyrimidine Derivatives*

The remainder of this chapter will deal with the photochemistry of pyrimidine derivatives, including glycosides. The rapid developments which have been taking place in this field during the past few years, and which are continually attracting new workers to the field, render it impossible to organize the accumulated data in chronological order, and we shall therefore subdivide it into appropriate sections, as required.

Up to 1954 it was generally assumed that the loss of the absorption spectrum of irradiated pyrimidines was the result of irreversible degradation of the pyrimidine ring. As a result of this, we now know that some of the earlier studies are only of historical importance. It is indeed curious that the excitement aroused by Kelner's report of photoreactivation in micro-organisms in 1949 somehow obscured the discovery in the same year, by Sinsheimer and Hastings (1949), that the "photodecomposition" of uracil, uridine and cytidylic acid could be subsequently reversed in the "dark" by heating or acidification, the criterion for reversal being the reappearance of the original absorption spectrum. It was only after Sinsheimer himself (1954) again took up this question five years later that its significance began to be appreciated in connection with the photochemistry of the nucleic acids and its bearing on the fascinating problem of biological photoreactivation. We now know that the first question to be resolved in an examination of the photochemistry of a pyrimidine derivative is whether the reaction leads to the reversible or irreversible formation of a new derivative, or to irreversible degradation. Furthermore, it is somewhat risky to regard the loss of absorption spectrum as the sole criterion for whatever photochemical transformation is taking place.

The photochemical behavior of uracil was first investigated by Heyroth and Loofbourow (1931), who followed the modification in absorption spectrum resulting from exposure of this compound to a watercooled Victor mercury arc and found that the principal absorption maximum gradually decreased to zero with a simultaneous increase in absorption at 2200–2300 Å and above 2900 Å. Subsequently, Heyroth (1932), seeking an explanation for the appearance in irradiated blood of substances capable of reducing arsenophosphotungstic acid, observed that the irradiation of uracil resulted in the formation of a substance (or substances) also capable of reducing the above-mentioned acid, particularly during the initial stages of irradiation. Discussing the possible mechanisms of photolysis of uracil, Heyroth and Loofbourow (1941) advanced the hypothesis that the initial stage of this reaction could be due to the formation of photoproducts such as 2,5-dioxy-pyrimidine (isouracil) since, according to Johnson and Johns (1914), only pyrimidines possessing an amino or hydroxyl group in position 5 of the ring are able to reduce alkaline solutions of phosphotungstic acid. It is curious that this observation of Heyroth's has gone unnoticed, since it may have some relation to the primary photoproduct of uracil which may revert again to uracil in the "dark" under suitable conditions. It should be noted in this connection that the presence of a hydroxyl or amino group in position 5 of the pyrimidine ring is not a sufficient condition for the ability of the latter to reduce alkaline phosphotungstate (or molybdate). An additional hydroxyl or amino group must also be present at positions 4 or 6 (Bendich and Clements, 1953).

The first attempt to quantitatively follow the course of photolysis of a pyrimidine is that of Uber and Verbrugge (1940) on the pyrimidine component of thiamine, viz. 2-methyl-4-amino-5-ethoxymethylpyrimidine, at pH 4.8, using a Hanovia resonance lamp. The course of the reaction was followed spectrographically and the quantum yield was calculated from the loss in biological activity, the spectral measurements being used to correct for the decrease in absorption of the solution during the course of irradiation. The reaction was found to proceed with a quantum yield of 1.84×10^{-2} and was accompanied by a decrease of the entire absorption spectrum without any change in shape. From ammonia determinations the authors concluded that 1/3 of the amino nitrogen was released as a result of photolysis. Since at any given stage during irradiation the optical density of the solution was always higher than what it should have been according to the biologically determined quantum yield, the suggestion was advanced that the reaction mechanism involved the removal of the amino and, perhaps, other side groups either simultaneously with or preceding the rupture of the pyrimidine ring. However, removal of side groups as the initial reaction resulting from irradiation does not appear very plausible, since such a process should be accompanied by a shift in the absorption spectrum; no such shift is observable in the spectra presented by the authors following various periods of irradiation. Furthermore, the appearance of ammonia in the photolyzed solution is by itself no conclusive evidence for removal of amino groups since, as we shall see below (Table 6-10), ammonia frequently appears as a photolysis product of pyrimidine rings, even of those not containing amino groups, and usually only after extensive irradiation. Particular attention should be directed to the fact that, since the pK of the compound is about 6.4 (this may be deduced from the variation of the spectrum with pH presented by the foregoing authors), the amino group was almost fully protonated at the pH used for irradiation, 4.8. Indeed, if irradiation is carried out at pH 8.0, where the compound is entirely in the amino form, its photochemical behavior is radically altered; under these conditions the photoproduct spontaneously reverts to the original compound in the "dark" as may be verified both physico-chemically and biologically (Wierzchowski and Shugar, 1960c). We shall come back to this in more detail below, in connection with the photochemistry of 4-aminopyrimidines.

Stimson and Loofbourow (1941) observed that 2-chloro-4-aminopyrimidine is as resistant to radiation as adenine, from which they concluded that susceptibility to photolysis parallels the number of carbonyl groups in the ring. This assumption was apparently based on a qualitative comparison of the behavior of this compound with that of uracil (Heyroth and Loofbourow, 1931) and barbituric acid (Loofbourow and Stimson, 1940) under similar experimental conditions. It is, however, not strictly applicable. As in the case of purines, it is rather the presence of a carbonyl group in the 2-position of the ring which influences the susceptibility to irradiation. Even this generalization is limited in applicability if we compare the quantum yield for 2-methyl-5-ethoxymethyl-4-aminopyrimidine, above, with those for some uracil analogues (Table 6-3).

Before going on to a discussion of the final photoproducts resulting from degradation of the pyrimidine ring, it is desirable, for the sake of clarity, to first consider the initial effects of radiation. These are also the most important from a biological point of view.

6. *Reversible Photoproducts of Uracil Derivatives*

As already mentioned, in 1949 Sinsheimer and Hastings reported that the photolysis of uracil, uridine and cytidylic acid could be reversed by subsequent acidification to pH 1, the criterion for reversal being the reappearance of the original absorption spectrum of the irradiated substance. They pointed out that the course of the reaction was markedly dependent on the nature of the irradiation source and, in particular, on the presence or absence of wavelengths below 2300 Å (such as the Hg lines at 1849, 1942 and 2242 Å). If these lines were not filtered out from the source, the quantum yield for photochemical transformation was considerably higher than at 2537 Å, while dark reversibility was either considerably reduced or entirely eliminated.

The photochemical transformation of uracil (Sinsheimer and Hastings, 1949; Sinsheimer, 1954) in neutral, aqueous, medium proceeds with a quantum yield of 0.052, calculated from the rate of disappearance of the principal absorption maximum at 2600 Å. About one-half the uracil is converted to a product which reverts to uracil on acidification to pH 1 or upon heating at neutral pH. The remainder of the uracil is transformed to a number of other products (see below). A qualitatively similar behavior is exhibited by 1-methyluracil and 1,3-dimethyluracil (Moore and Thompson, 1955, 1956, 1957; Wang *et al.*, 1956; Wierzchowski and Shugar, 1956, 1957), with the marked exception that reversibility for these derivatives following irradiation may be as high as 100 per cent.

The course of the reaction for 1,3-dimethyluracil as well as the pH dependence of the subsequent reverse acid–base catalysed reaction for uracil and dimethyluracil is shown in Figs. 6-2 and 6-3. Moore and Thompson's (1955) choice of dimethyluracil for attempted elucidation of the

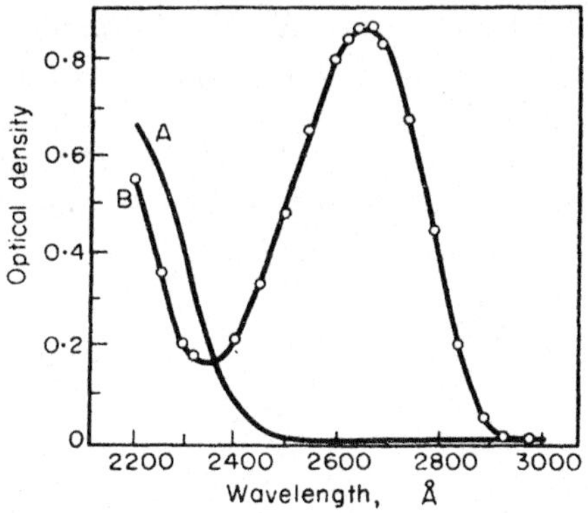

FIG. 6-2. Photolysis, and subsequent regeneration, of 1,3-dimethyluracil (10^{-4} M in H_2O). Curve B is that for dimethyluracil prior to irradiation, curve A the product of irradiation, and the points on curve B are obtained after heating the irradiation product for 5 mins at 100°C at pH 1 (Moore and Thomson, 1955).

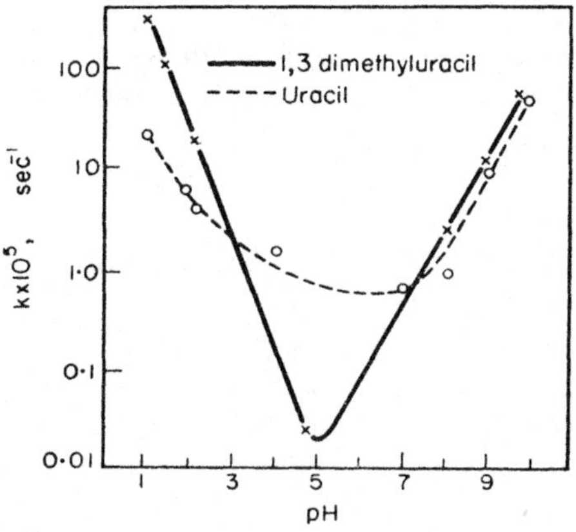

FIG. 6-3. Influence of pH on the first-order rate constant k (at 20°C) for regeneration of uracil (-------) and dimethyluracil (———) from their respective photoproducts (Moore and Thomson, 1956).

nature of the photoproduct was a fortunate one in that the latter is completely reversible to dimethyuracil, while the physico-chemical properties of the photoproduct made it relatively easy to isolate by fractionation and chromatography. Its relative stability at room temperature was also a useful contributing factor. Elementary analysis of the photoproduct, together with its molecular weight of 158.3 as compared to 140 for dimethyluracil, indicated uptake of a water molecule by the pyrimidine ring during irradiation. That the point of attachment was the 5,6 double bond was shown not only by the disappearance of the characteristic absorption maximum (see Fig. 6-2) but also by the absence of the characteristic reaction of unsaturated pyrimidine rings with bromine water. The infrared absorption spectrum of the photoproduct in alcohol-free chloroform exhibited also an intense band at 2.98μ, corresponding to —OH frequencies, and absent in dimethyluracil. On the basis of these facts Moore and Thomson (1955) proposed that the photoproduct is either 5- or 6-hydroxy-1,3-dimethylhydrouracil:

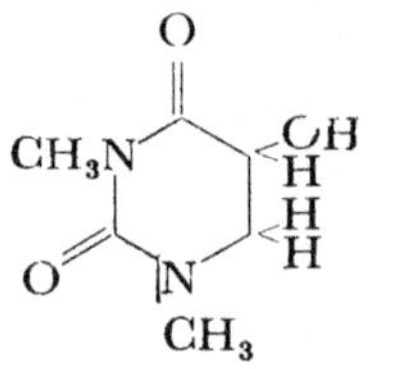

1,3-dimethyl 5-hydroxy-6-hydrouracil

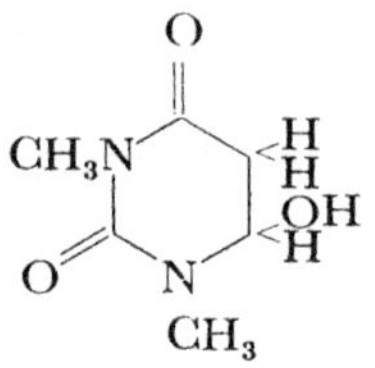

1,3-dimethyl-6-hydroxy-5-hydrouracil

Since the properties of the known compound, 1,3-dimethyl-5-hydroxy-6-hydrouracil, did not correspond to those of the photoproduct, it was concluded that the latter must be 1,3-dimethyl-6-hydroxy-5-hydrouracil.

This suggestion was subsequently confirmed by Wang *et al.* (1956) and independently by Moore and Thomson (1957), both of whom succeeded in synthesizing authentic 1,3-dimethyl-6-hydroxy-5-hydrouracil and showing that its properties corresponded completely with those of the photoproduct of 1,3-dimethyluracil.

By analogy one would expect the reversible photoproduct of uracil to be 6-hydroxy-hydrouracil. Moore (1958) synthesized this compound and demonstrated that it did, in fact, correspond to the photoproduct of uracil which reverts to uracil in the dark.

Studies on uridine and uridylic acid (Sinsheimer, 1954) (see Fig. 6-4), 2′:3′-isopropylidene-uridine (Wierzchowski and Shugar, 1959), uridine-2′:3′-phosphate (Shugar and Wierzchowski, 1958b), uridine-5′-pyro and triphosphates as well as uridine-diphosphate-glucose (Zill, 1957) all show that substitution of the N_1 position in the uracil ring by carbohydrates, in place of a methyl group, does not affect the nature of the reaction qualitatively. By contrast, esterification of the sugar moiety, as well as the position of esterification, may markedly influence the stability of the photoproduct, i.e. the rate at which it reverts to the parent substance. The rate of the reverse reaction is greater for uridine than for uridylic acid; for uridine-3′-phosphate it is greater than for uridine-2′-phosphate (Sinsheimer, 1954), and the rates for these two latter are exceeded by that for uridine-2′:3′-phosphate (Shugar and Wierzchowski, 1958b). The pH of maximum stability of the photoproduct of uridine-3′-phosphate, i.e. the pH at which the rate of the reverse reaction is minimal, is 5.2 (Sinsheimer, 1954) or roughly the same as for the photoproduct of dimethyluracil (Fig. 6-4).

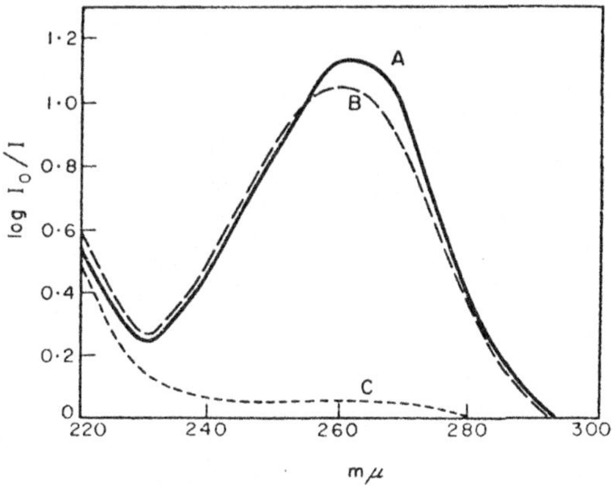

FIG. 6-4. Effect of irradiation and subsequent acidification on absorption spectrum of uridine-3′-phosphate. Curve A, before irradiation; B, irradiated solution acidified to pH 0.8 and left 42 hours at 20°C (Sinsheimer, 1954).

Supplementary evidence regarding reversion of the photoproducts to the parent compounds was furnished by Mitchell (1955) and Rapport *et al.* (1955), both of whom demonstrated that the growth requirements of a pyrimidine-requiring mutant of *Neurospora crassa* could not be satisfied by irradiated uridine, but could be if the photoproduct were first acidified, heated at neutral pH, or made alkaline.

A rather unusual observation has been made by Prusoff (1963), who found that the hydrated photoproducts of uracil and deoxyuridine, while stable at room temperature or at −70°C, rapidly reverted to the parent substances at −20°C. The rate of reversal at −20°C was much the same as if reversal were achieved in liquid medium *via* an increase in hydrogen ion concentration of 100–1000 fold, and it is suggested that the rapid reversal rate at −20° is due to the free migration of hydrogen ions in the imperfections of the crystal structure of ice at this temperature. While this phenomenon is of undoubted interest in relation to the crystal structure of ice and the interdispersion of solute in the latter, it does not add anything new as regards the reverse process itself.

TABLE 6-3. QUANTUM YIELDS AT 2537 Å FOR SOME PYRIMIDINE DERIVATIVES IRRADIATED IN AN ATMOSPHERE OF AIR; IN NEUTRAL OR AQUEOUS MEDIUM AT ROOM TEMPERATURE, AT CONCENTRATIONS OF 10^{-3} TO 10^{-4} M; AND NATURE OF REACTION (WHETHER REVERSIBLE OR NOT BY HEATING AT NEUTRAL pH OR BY ACIDIFICATION OR, IN SOME INSTANCES, BY ALKALIZATION). CALCULATED QUANTUM YIELDS ARE FOR INITIAL STAGE OF REACTION AND ARE BASED ON CHANGES IN ABSORPTION SPECTRUM.*

Compound	$\Phi \times 10^3$ (moles/einstein)	Reversibility
Uracil	5.2(*a*), 6.0(*b*), 2.5(*c′*)	~50%
1-methyluracil	12.5(*b*)	High
1,3-dimethyluracil	10.4(*c*)	100%
Uridine	21.6(*a*, *b*), 16.0(*c′*)	90–100%
2′:3′-isopropylidene-uridine	~21(*b′*)	90–100%
Uridine-2′-phosphate	21.6(*a*)	90–100%
Uridine-3′-phosphate	21.6(*a*), 19.0(*c*)	90–100%
Uridine-2′:3′-phosphate	21.6(*b′*)	90–100%
UDP, UTP, UDGP	—	High
2-ethoxyuracil	8.0(*b′*)	Nil
4-ethoxyuracil	very low(*b′*)	Nil
2,4-diethoxypyrimidine	very low	Nil
Thymine	0.4(*b*), 0.03(*c′*)	Nil†
Thymidine	0.65(*b′*), 0.025(*c′*)	Nil†
Thymidylic acid	1.0(*b′*), 0.3(*c′*)	Nil†
5-bromouracil	1.7(*c′*)	Nil
5-nitrouracil	1.0(*c′*)	Nil
5-bromo-1,3-dimethyluracil	10.4(*c′*)	Nil
6-methyluracil	0.83(*b′*), 0.23(*c′*)	Nil(*b′*), slight(*c*)
1,3,6-trimethyluracil	low(*b*)	>60%
Orotic acid	1.2(*c′*)	Nil†
Uracil-5-acetic acid	1.2(*c′*)	Nil
2-thiouracil	1.7(*b′*)	Nil
6-propylthiouracil	0.6(*c′*)	Nil
Barbituric acid (pH 5.2)	1.7(*c′*)	Nil
1,3-dimethylbarbituric acid	4.6(*b′*)	Nil
1,3-dimethyl-6-methoxybarbituric acid	4.1(*b′*)	Nil

* For influence of concentration on quantum yield see Section 9.

† Under certain conditions photolysis is reversible (Section 8).

(*a*) Data from Sinsheimer (1954).

(*b*) Data from Wierzchowski and Shugar (1957, 1960a, 1960b) and Shugar and Wierzchowski (1958a) and (b′) unpublished.

(*c*) Data from Moore and Thomson (1956).

(*c′*) Calculated from 1st-order rate constants of Moore and Thomson (1956), using their value of Φ for 10^{-4} M uridylic acid as standard.

(*d*) Data from Zill (1957).

Not all uracil analogues behave photochemically like the foregoing compounds. Table 6-3 lists a number of derivatives which have been examined with respect to reversibility, as well as the quantum yields for the photochemical reaction as measured by the decrease in absorption spectrum resulting from irradiation.

In making use of this table, it is extremely important to bear in mind the experimental conditions given therein for, as will be shown in the following sections, modifications of the conditions of irradiation may alter not only the nature of the photoproducts but also the conditions for reversibility. It is nonetheless clear from the data presented in Table 6-3 that the substitution of the 5 or 6 position of the pyrimidine ring appreciably affects both the quantum yield and potential reversibility. In addition, a number of other derivatives are irreversibly photolyzed under these conditions, e.g. 2-thiouracil, 2-thiocytosine, 2,4-diethoxypyrimidine. For these latter it is clear that this cannot be due to steric factors and it is therefore the electron distribution density in the pyrimidine ring which is of importance; reversible photolysis is the result of excitation of the pyrimidine ring leading to an

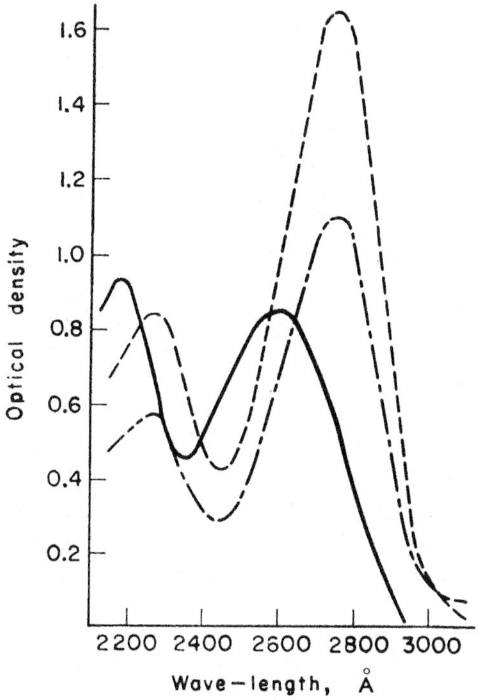

FIG. 6-5. 2-ethoxyuracil in 0.2 M phosphate buffer pH 7.2: ——— prior to irradiation; ------- after 150 min irradiation at 2537 Å ; —·—· following heating of irradiated solution for 140 min at 85°C (Shugar and Wierzchowski, 1958).

altered distribution of electron density facilitating the uptake of a water molecule (Shugar and Wierzchowski, 1957a, 1958).

The behavior of 2-ethoxyuracil requires special comment. An examination of the photochemical transformation of this compound in light and heavy water at neutral pH suggests that the reaction involves uptake of a water molecule at the 5,6 double bond, but the photoproduct will not revert to the original compound either by acidification or heating (see Table 6-3 and Shugar and Wierzchowski, 1958). In addition, photochemical transformation does not result in destruction of the absorption spectrum; instead the photoproduct exhibits a new maximum at 2740 Å, with an extinction twice that of the original compound and an additional maximum at 2270 Å (Fig. 6-5). Heating or acidification results only in destruction of this photoproduct. Attempts to purify it in larger quantities for identification have hitherto been unsuccessful because of its relative instability.

Even for those derivatives for which the reaction is reversible, prolonged irradiation may substantially reduce the degree of reversibility, due, of course, to the eventual destruction of the initial photoproduct. Furthermore, for some compounds the initial reaction may be accompanied by side reactions leading to other products; e.g. for uracil only about 50 per cent is reversibly photolysed. We shall consider these cases in more detail below. On the other hand, for compounds such as uridine and 1,3-dimethyluracil, if irradiation is terminated at the point where the selective ultraviolet absorption maximum just disappears, subsequent reversibility to the original compound may attain close to 100 per cent (see Fig. 6-4).

7. *Photochemically Induced Additions to Double Bonds*

Reference should be made to the monograph of Ingold (1953) for a presentation of the theoretical aspects of chemical additions to conjugated mono-olefinic carbonyl compounds. The subject of additions to double bonds of π-electron deficient *N*-heteroaromatic rings has been discussed by Albert (1958; see also discussion following Albert's article). In general the reagents which may add are not those such as Br_2 or HCl, which readily undergo addition to ethylenic bonds, but rather those that add to an ethylenic double bond conjugated to an electron attracting group(s). These include HCN, sodium bisulfite, H_2O, alcohols, etc., the general character of which is nucleophilic. A number of examples of these additions are cited by Albert (1958), including hydration which occur without irradiation.

Photochemically induced nucleophilic additions to double bonds are by no means rare, and a number of experimentally observed examples will be found in the reviews of Masson *et al.* (1956), Schonberg (1958), and Schenck (1960). In the case of hydration (addition of a water molecule), it has been suggested by Schenck (1960), from energetic considerations, that this must proceed as a single and not a two-stage process (cf., however, Ingold, 1953).

Indirect evidence has been obtained that ethanol, like water, will add reversibly to the 5,6 double bond of ultraviolet-irradiated uracil and dimethyluracil (Moore and Thomson, 1956; Shugar and Wierzchowski, 1958) as well as to uridine and uridylic acid (Shugar, unpublished). Like the hydrated derivatives, the alkoxy ones are also relatively stable at room temperature and should therefore be readily susceptible to isolation following irradiation of the appropriate derivative in ethanol. In all likelihood similar photoproducts would result from irradiation in methanol.

More direct evidence has been produced by Moore (1959) for the photochemically induced nucleophilic addition of HCN to uracil, 1,3-dimethyluracil and uridine. In fact, for all the three foregoing compounds, the photochemically excited pyrimidine ring reacts preferentially with HCN, as compared to H_2O, in aqueous medium. This is particularly marked for uracil where calculation showed that the rate constant for HCN addition to the 5,6 double bond is of the order of 10^3 times that for addition of a water molecule. The HCN addition products differed significantly from the hydrated ones in that they would not revert in the dark to the parent substances under those conditions where the hydrated ones would by elimination of a water molecule. That HCN addition was involved was demonstrated directly by the use of $H^{14}CN$ and uracil-2-^{14}C, while involvement of the 5,6 double bond of the pyrimidine rings was shown by the disappearance of the principal absorption maximum. Chromatography, however, showed that uracil and uridine yielded not one but three principal photoproducts accounting for 90 per cent of the total in each case. The precise nature of these products was not elucidated. It therefore appears that the primary photochemical reaction of uracil analogues in solution is nucleophilic addition to the 5,6 double bond. In line with the general scheme of Mulliken (1953) and Braude *et al.* (1954), according to which the first electronically excited state of compounds such as aromatic ketones may be represented as a zwitterion with the negative charge on the oxygen and the positive charge on the carbon at the opposite end of the conjugated chain, Moore (1959) has proposed that irradiation results in the formation of such an intermediate excited state in the case of the pyrimidine ring, as follows:

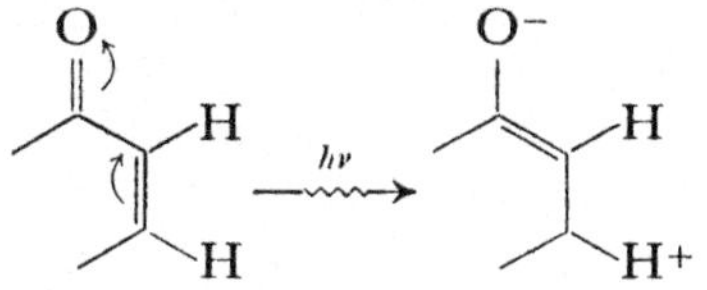

which then undergoes the nucleophilic addition reaction.

The fact that thymine *apparently* does not undergo photochemically induced hydration *in solution* (Wierzchowski and Shugar, 1957) has been proposed by Moore (1959) as supporting evidence for such an intermediate since the hyperconjugating effect of a methyl group at position 5 of the ring would be expected to oppose the development of a positive charge on carbon 6. Wang (1959), on the other hand, suggested that the failure of ultraviolet irradiated thymine to undergo hydration is due to the relative ease with which it undergoes dehydration.

Wacker *et al.* (1960c) claim that there is no uptake of a water molecule by a number of irradiated uracil analogues. This conclusion was based on the failure of the photoproducts to exhibit any radioactivity following irradiation in T_2O. No experimental details have as yet been published, and it is therefore difficult to comment on this finding, which is indeed at variance with results for uracil and dimethyluracil where the principal photoproducts have actually been isolated and chemically identified. Attention should, however, be drawn to the potential importance of the hydrogen isotope effect in hydration with T_2O, which would materially reduce the uptake of T_2O as compared to H_2O, and to which we shall refer in the next section.

8. *Dimer Photoproducts*

An additional possibility which must be considered is the addition of one double bond to another with the formation of 4, 5, or 6-membered rings. The theoretical aspects of such chemical additions have been considered by Ingold (1953). In the case of pyrimidines, mutual addition of the 5,6 bonds would lead to the formation of a cyclobutane ring system, as follows:

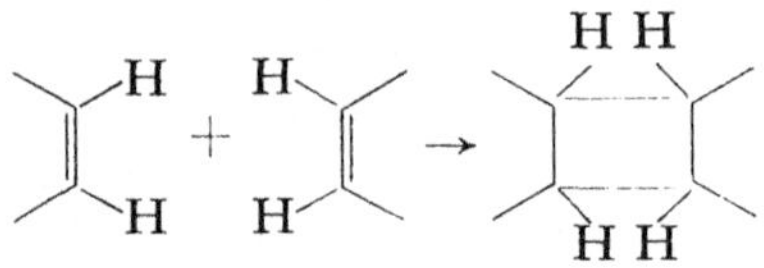

Radiation induced reactions of this type are fairly common (Schonberg, 1958; Schenck, 1960) and have been postulated as occurring between adjacent pyrimidine rings in irradiated polynucleotide chains in aqueous solution (Shugar and Wierzchowski, 1960; Shugar, 1960b) as well as being the source of the cross-linking resulting from irradiation of nucleic acids in the form of air-dried films (Baranowska and Shugar, 1960). This type of reaction is apparently not reversible by acidification, heating, or alkalization, as for hydrated pyrimidine rings. But, under certain conditions, it can be reversed by other means and we shall now consider this in more detail. It is pertinent to emphasize, in this connection, that dimer formation, which was first demonstrated experimentally by Beukers and Berends (1960b, 1961), has now been shown to be one of the major factors responsible for many biological effects resulting from the irradiation of DNA, both *in vitro* and *in vivo*.

If a solution of thymine in water is frozen and then subjected to irradiation at 2537 Å in the *frozen* state, an extremely rapid photochemical transformation takes place, as may be verified by thawing of the solution and measuring the optical density of the principal absorption maximum. If now the thawed solution is again irradiated with the same source, the optical density increases until it approaches the initial value (Fig. 6-6A).

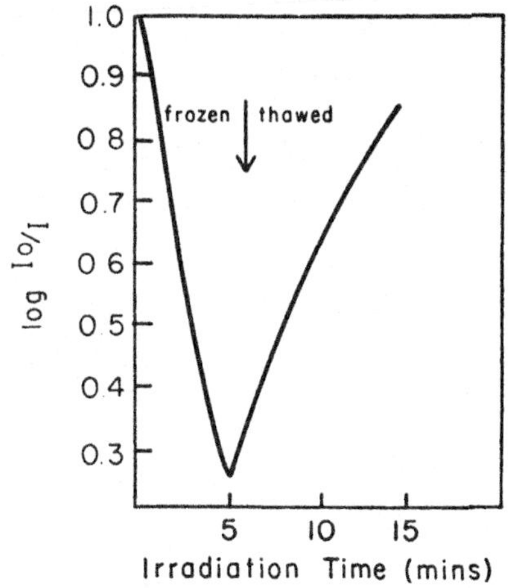

FIG. 6-6A. Influence of irradiation on thymine in frozen solution and on photoproduct in liquid medium. Solution frozen and irradiated 6 mins, resulting in drop of optical density at maximum to 0.25; at point indicated by arrow, solution thawed and again irradiated (Beukers *et al.*, 1959c).

The photoproduct exhibits only end absorption starting at about 2650 Å, while the reactivated product possesses the same absorption spectrum and the same R_f value chromatographically as thymine (Beukers *et al.*, 1959, 1959c). The photoproduct itself is stable in acid and alkali even at elevated temperatures. It was subsequently isolated by irradiation of large volumes of frozen solution, followed by extraction of the dried residue with absolute ethanol, in which the photoproduct is insoluble. Following crystallization from water the photoproduct was found to have the same chemical composition as thymine, indicating that it was either an isomer or polymer of thymine. Rough molecular weight and crystallographic measurements suggested a dimer. Since the characteristic ultraviolet absorption of thymine was absent, saturation of the 5,6 double bonds was also implicated. It was therefore proposed that the photoproduct is

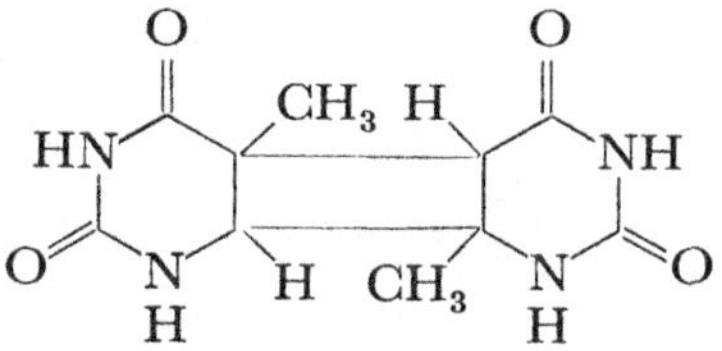

in which the 5,6 double bonds of the two thymine rings are saturated by formation of the cyclobutane ring system (Beukers and Berends, 1960b, 1961).

On the basis of nuclear magnetic resonance (NMR) measurements, formation of the isomer with methyl groups at adjacent C atoms was excluded. This argument was, however, shown to be invalid (Wulff and Fraenkel, 1961), since the hydrogen atoms are in equivalent magnetic environments irrespective of isomer, so that no spin-spin splitting is to be expected. Actually there are four possible isomers, as pointed out by Wulff

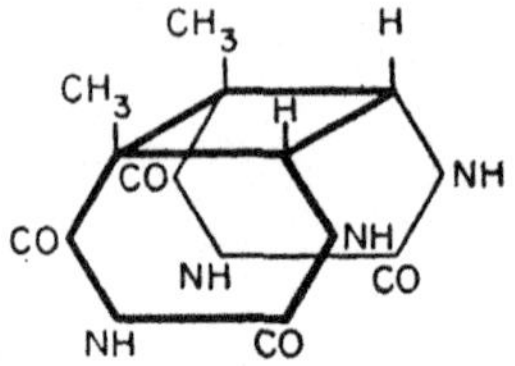

(meso)
I

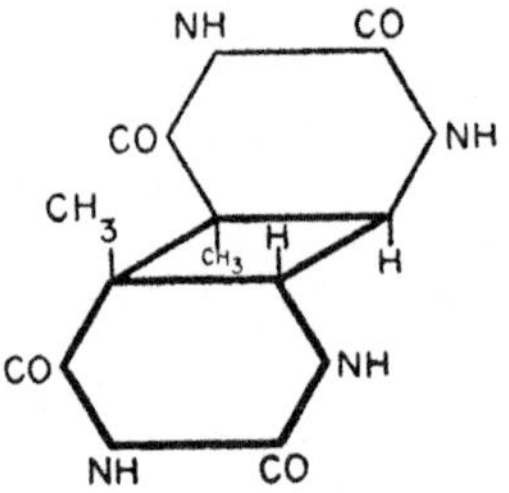

(d,l)
II

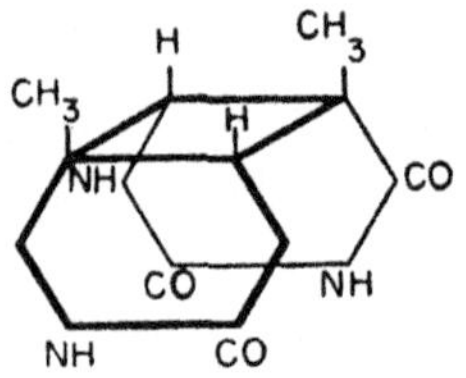

(d,l)
III

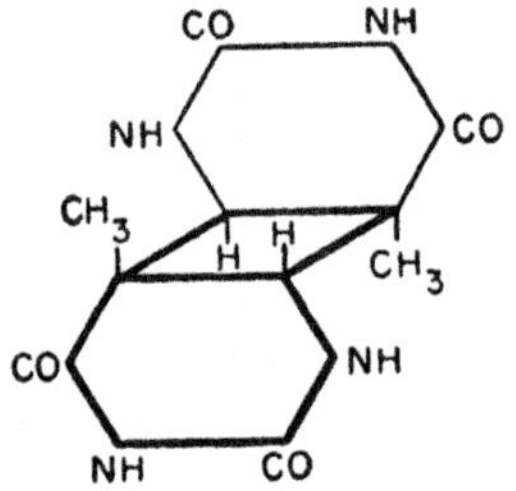

(meso)
IV

and Fraenkel (1961). The authors were able to isolate two such isomeric dimer photoproducts from irradiated *N,N'*-dimethylthymine and showed, furthermore, that one of them was equivalent to the product of *N*-methylation of the dimer photoproduct of thymine. On the other hand, NMR spectra alone were insufficient to distinguish between the different isomers.

By the use of ^{14}C-labelled pyrimidines it has been demonstrated indirectly that mixed dimers may be formed; e.g. one of the photoproducts formed by irradiation of a frozen mixture of uracil and thymine is a dimer of uracil and thymine (Beukers and Berends, 1960b; Wacker *et al.*, 1961a).

The above dimer photoproducts were formed mainly in aqueous, *frozen*, solutions. Wacker *et al.* (1961a) report that irradiation of 10^{-4} M solutions of thymine, thymidine or thymidylic acid in the *liquid* state at room temperature with an incident intensity of 5×10^5 ergs/mm^2 gives a 3 per cent yield of dimer photoproducts. The same authors (Wacker *et al.*, 1961b) irradiated a variety of uracil analogues at 10^{-3} M concentration under similar conditions and claimed that dimer formation was the *only* product in all instances. This result has already been referred to, and discussed in part, above; it remains to add that the presumed dimer of uridine, formed in 30 per cent yield under the above conditions, could be reconverted quantitatively to uridine upon heating. It is consequently of importance to emphasize that, although unlike the photoproducts of uracil and dimethyluracil, that of uridine in *liquid* medium has not been directly isolated and chemically characterized, the evidence is overwhelmingly in accord with its structure as a hydrated derivative. The observation of Wacker *et al.* (1961b) that their presumed uridine dimer reverted to uridine on simple heating in aqueous medium is the most convincing argument against its being a dimer. To this must be added the fact that, although the major photoproduct of uracil in liquid medium has been isolated and identified chemically as the hydrated derivative (Moore, 1958), the product of irradiation of uracil in T_2O obtained by Wacker *et al.* exhibited no radioactivity. Hence their mistaken conclusion with regard to the dimerization of uridine must be due to some error in technique, the origin of which is undoubtedly the hydrogen isotope effect involved in the use of tritium. It should be noted that hydration in the presence of 100 per cent D_2O proceeds at only one-half the rate in H_2O. In T_2O the rate of hydration will be much lower, probably of the order of 10 per cent of that in H_2O. In addition, the tritiated water used in the experiments could hardly have contained more than a few per cent T_2O, say 5 per cent at most. The resultant uptake of tritiated water by uridine photoproduct would consequently be of the order of $0.1 \times 0.5 = 0.005$, or 0.5 per cent, and it is therefore not surprising that none was observed.

Actually, under certain conditions photodimerization might be expected in aqueous medium (Smietanowska and Shugar, 1961), and such dimer formation has been observed in high yield on irradiation of very concentrated solutions of thymidine (Wulff, personal communication). Accumulation of the dimer in this instance was undoubtedly due to internal shielding by the high concentration of unreacted molecules. On the other hand, irradiation of uridine solutions at concentrations as low as 10^{-3} M has been found to result in formation of up to 5–6 per cent dimer, which may be resolved from hydrated and unreacted uridine by paper chromatography (Sztumpf and Shugar, unpublished). The formation of even low yields of dimer photoproducts in relatively dilute solutions suggests that we may be dealing with long-lived excited states, since the molecules must be suitably orientated to react with each other instead of with the more readily available water molecules. A quantitative study of dimer formation in solution at various concentrations should furnish some useful information as to the relative rates of dimerization and hydration. Wacker (personal communication) claims to have demonstrated that the initial reaction in aqueous solutions of uracil is dimerization, followed in turn by hydration. If this were indeed so, it would imply that hydration proceeds at least in part via excited molecules resulting from dimer dissociation (cf. Smietanowska and Shugar, 1961).

It might therefore be asked, why does photodimerization of some pyrimidine derivatives proceed so readily in frozen, but not in liquid, medium? Before attempting to clarify this apparent anomaly, a few additional facts will be presented.

Using a procedure analogous to that of Beukers and Berends (1960b), the photodimer of uracil has been isolated in crystalline form in fairly good yield from an irradiated, frozen, solution of uracil (Smietanowska and Shugar, 1961). The same authors concluded, from spectral observations, that the uracil and thymine dimers obtained by irradiation in frozen medium are the *cis* isomers. The quantum yields for dimer formation in ice were estimated to be of the order of 0.1–0.2; but these values are only approximate, since the amount of light lost by scattering from the ice crystals is considerable.

Both uracil and thymine dimers are resistant to concentrated $HClO_4$ at 100°. They are stable in alkali at room temperature, but at 100° in 0.1 N KOH they are completely degraded in 1 hour.

Since dimer formation is accompanied by saturation of the 5,6 double bonds of the pyrimidine rings, the dimers should exhibit the spectral characteristics of 2,4-diketo-dihydropyrimidines, i.e. end absorption in the quartz-ultraviolet in neutral medium and an absorption maximum at about 2300 Å in alkaline medium. This is, in fact, what is observed, as will be seen from Fig. 6-6B for thymine dimer (completely analogous curves obtain for uracil dimer, see Smietanowska and Shugar, 1961). The variation in extinction of the maximum at 2350 Å with pH is due to dissociation involving the number 4 carbonyls (Janion and Shugar, 1960). Irradiation of the dimer at 2537 Å at pH 12 results in quantitative conversion to thymine, as shown by the curve on the right hand side of the figure.

The quantum yields for dissociation of uracil and thymine dimers at alkaline pH are, respectively, 0.52 and 0.73. Since these values are relatively independent of pH in the alkaline range, they probably prevail also at neutral pH. Using a crude preparation of thymine dimer, Setlow (1961) measured the quantum yield for thymine dimer photodissociation at neutral pH and at several wavelengths to the violet of 2600 Å; the yield was approximately independent of wavelength, with an average value of the order of 0.6.

It was also observed by Smietanowska and Shugar (1961) that a variety of alkylated 2,4-diketopyrimidines undergo photodimerization in frozen medium, the yields varying somewhat for the different derivatives. The criterion used as a test for dimerization was subsequent dissociation of the suspected dimer by irradiation in liquid medium, with regeneration of the original substance. The importance of some such criterion must be emphasized, since the mere demonstration of a reaction in frozen medium is by no means evidence that this is either in whole, or in part, dimerization.

We shall now come back to the question posed above, viz., why does dimerization proceed so readily in frozen medium? It should be clear by

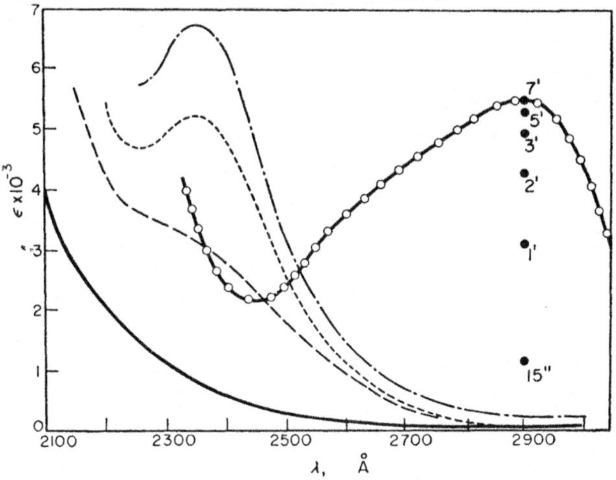

FIG. 6-6B. Absorption spectrum of thymine dimer at neutral and alkaline pH values, and photochemical conversion of dimer to thymine at pH 12:

——— dimer at pH 7
— — — dimer in 0.01 N NaOH
------ dimer in 0.1 N NaOH
—·—·— dimer in 1 N NaOH

o-o-o absorption spectrum resulting from irradiation of dimer at pH 12 for 7 min at incident intensity of 0.9×10^{17} quanta/cm^2/min at 2537 Å. Spectrum corresponds accurately to that of thymine at pH 12 and figures under maximum of curve show amount of thymine regenerated at various times of irradiation.

now why dimerization does not normally take place in aqueous medium. The quantum yield for dimer dissociation is so much greater than for dimer formation that any dimer formed would tend to dissociate. In frozen medium, on the other hand, dimer dissociation is rendered difficult since the dimers, once formed, are "trapped." By the same criterion, dimer formation must occur readily in frozen medium because the trapped monomer molecules are suitably oriented with respect to each other. This is indeed what one would be led to expect from the finding of Gerdil (1961), according to which thymine monohydrate forms crystals the structure of which are such as to permit formation of isomer I of Wulff and Fraenkel, i.e. 1,2-dimethyl-cyclobutane-*cis*-dithymine; in other words the dimer which would be formed by adjacent pairs of pyrimidine bases in a single-stranded polynucleotide chain. It remains, of course, to establish whether such crystals are actually formed in frozen solution.

Wang (1961) has formulated some broad generalizations with regard to a presumed special role of frozen media in photochemistry, based in part on some similarity in dimer photoproduct formation in ice and in dry films. In the light of the foregoing, however, these are perhaps redundant. It is a fairly general observation that a given photoproduct, or a specific isomer of a dimer photoproduct, may be formed preferentially, depending on the conditions or irradiation. The dimer photoproducts of anthracene provide one of a number of good illustrations of this; if irradiated in solution, only a *cis* isomer is formed; but irradiation in the crystalline form leads to a mixture of the *cis* and *trans* isomeric dimers (Greene *et al.*, 1955).

Insofar as dimer dissociation is concerned, the possibility of this occurring under the influence of factors other than irradiation must be considered. The chemical retrogression of addition products to double bonds has been discussed in detail by Ingold (1953), who presents representative examples of such reactions. Examples of the retrogression of various dimer photoproducts by acid and heat are by no means uncommon (Mustafa, 1955; Schonberg, 1958). In fact thymine and uracil photodimers are rather an exception to the general rule with regard to their stability to acid, alkali, and heat.

Reference was made above to the finding of Setlow (1961) that photodissociation of thymine dimer is essentially wavelength independent. The dimerization reaction itself, on the other hand, proceeds with a quantum yield which is dependent on wavelength (Johns *et al.*, 1962). Furthermore since the dimers exhibit decreasing extinction coefficients with increasing wavelength, then at constant light intensity the degree of dimer dissociation will decrease with increasing wavelength. Hence at any given wavelength there will be an equilibrium set up between dimer formation and dissociation; e.g. at wavelengths to the red of 2900 Å, where the dimers exhibit negligible absorption at neutral pH, dissociation will be small and dimerization should proceed in high yield. These considerations have been applied to the study of dimerization in model thymine oligonucleotides (Johns *et al.*, 1962; Sztumpf and Shugar, 1962; Deering and Setlow, 1963). The high extinction of uracil and thymine dimers in alkaline medium has also been made use of to follow dimer photodissociation (Sztumpf and Shugar, 1962).

At wavelengths to the violet of 2250 Å, reactions other than hydration and/or dimerization are encountered. In particular, for thymine irradiated at 1849 Å in the presence of oxygen or air, Daniels and Grimison (1963) have found that the principal product of irradiation, formed with the high quantum yield of about 0.2, is a hydroperoxide, believed to be 6-hydroperoxy-5-hydroxythymine. The evidence for the location of the hydroperoxy group was not given; it is of considerable interest in that the product of X-irradiation of thymine in the presence of oxygen is known to be 5-hydroperoxy-6-hydroxythymine (Ekert and Monier, 1959). In both instances, of course, the reaction proceeds via free radical attack.

Two details of interest to the experimentalist merit mention. The alkaline absorption spectra of uracil and thymine dimers can be demonstrated in irradiated oligo- and polynucleotides of these bases and therefore provide an additional tool for studying dimer formation in irradiated polymers. Furthermore dimer photoproducts of pyrimidines may be readily revealed on paper chromatograms by irradiation of the latter with an ordinary germicidal lamp. Several minutes irradiation of a wet chromatogram suffice to bring out as little as 2.5 μg of uracil or thymine dimer. Dry chromatograms require longer irradiation times (Smietanowska and Shugar, 1961).

9. *Kinetics and Mechanism of Photochemical Transformation of Some Uracil Analogues*

Some data have been accumulated on the kinetics of photochemical transformation of several uracil analogues (Sinsheimer, 1954; Wierzchowski and Shugar, 1959), largely with a view to their application to the kinetics of photolysis of uracil residues in model oligo- and polynucleotides (Wierzchowski and Shugar, 1959, 1960a, 1960b). It was assumed in these studies that the basic reaction involved is the nucleophilic addition of a water molecule at the 5,6 double bond, as substantiated for uracil and dimethyluracil by isolation and identification of the photoproducts. Insofar as the kinetics for other analogues are in agreement with such an assumption, no major difficulties arise. If, however, appreciable deviations from the theoretically predicted course of a reaction are encountered, it is now clear that any conclusions drawn therefrom cannot be regarded as unequivocal if spectral methods alone are employed as criteria for identification of the photoproducts. This follows from the fact that disappearance of the principal absorption maximum at 2600 Å of a uracil derivative, resulting from saturation of the 5,6 double bond, may be due not only to nucleophilic addition of a water molecule but also to dimerization or, in some instances, to ring rupture. For uracil derivatives there is usually the possibility of isolation and identification of photoproducts, a fact of considerable importance in kinetic studies. We shall see later that the situation is considerably more complicated for cytosine and its glycosides where the instability of the initial photoproducts has hitherto prevented their isolation and, consequently, their identification by chemical analysis. Under these circumstances more reliance must be placed on kinetic data.

The following treatment is that used by Sinsheimer (1954) for uridylic acid. If we irradiate a solution of optical density D at a given wavelength with light of the same wavelength, the energy absorbed per unit time is

$$I_{abs} = I_0(1 - 10^{-D}) \tag{1-4a}$$

If the photoproduct formed does not absorb at this wavelength, the rate of formation of photoproduct is given by equation (4-2)

$$\frac{dC}{dt} = -k_1 I_{abs} = k_1 I_0 (1 - 10^{-D})$$

or, since $D = k_2 C$,

$$\frac{1}{k_2}\frac{dD}{dt} = -k_1 I_0(1 - 10^{-D})$$

whence

$$\frac{dD}{dt} = -kI_0(1 - 10^{-D}) \tag{6-1}$$

so that a plot of dD/dt vs. $(1 - 10^{-D})$ should be a straight line.

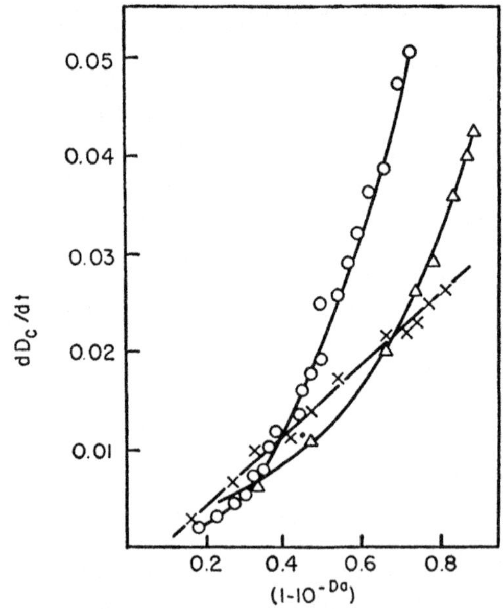

FIG. 6-7. Course of photolysis of uridine-2′(3′)-phosphate in 0.02 M phosphate buffer pH 7.2 for three different initial concentrations, C_0, and plotted as a 1st order reaction according to eqn. (1):

× × × × × × $C_0 \sim 10^{-4}$ M
△ △ △ △ △ △ $C_0 \sim 10^{-3}$ M
○ ○ ○ ○ ○ ○ $C_0 \sim 10^{-2}$ M

(Wierzchowski and Shugar, 1959).

For optically thick solutions, i.e. $D \gg 2$, $10^{-D} \ll 1$ and

$$\frac{dD}{dt} = \text{const} \tag{6-2}$$

if no concentration effects intervene. When the reaction order is not a simple one, and not in accord with eqn. (6-1), the van't Hoff difference method may be used to evaluate the order (Wierzchowski and Shugar, 1959).

The kinetics of photochemical transformation of dimethyluracil, uridine, isopropylidene-uridine and uridylic acid have been examined in aqueous neutral medium by the above procedures over the concentration range 10^{-5} to 10^{-2} M. Some results for uridylic acid, plotted according to eqn. (6-1), are shown in Fig. 6-7, from which it is clear that the reaction order is unity only at concentrations $\lesssim 10^{-4}$ M. At higher concentrations marked deviations are observed, which are accompanied by corresponding increases in initial quantum yields, which decrease during the course of the reaction. Moore and Thomson (1956) had earlier drawn attention to the fact that the quantum yield for uridylic acid was doubled when the concentration was increased from 10^{-3} to 10^{-2} M. The following table, taken from Wierzchowski and Shugar (1959), illustrates how the calculated quantum yield for uridylic acid varies during the course of the reaction for two different initial concentrations.

TABLE 6-4. VARIATION OF QUANTUM YIELD DURING THE COURSE OF PHOTOLYSIS OF URIDYLIC ACID AT NEUTRAL pH FOR TWO DIFFERENT STARTING CONCENTRATIONS

Experiment	C_0	Per cent photolysis	$\Phi \times 10^3$
1	0.89×10^{-4} M	6.0	19.5
		20.0	19.5
		31.0	18.8
		47.0	18.4
		64.5	18.2
		77.3	16.7
		83.5	14.4
		90.0	10.6
2	1.06×10^{-3} M	2.0	29.0
		10.0	28.2
		28.5	22.8
		38.5	21.5
		49.0	18.3
		61.0	16.0
		72.5	14.1
		82.5	11.5

It will be seen that, at an initial concentration of 10^{-4} M, the quantum yield is nearly constant for three quarters of the course of the reaction. At 10^{-3} M, on the other hand, the variation is appreciable after about 10 per cent photolyisis. Uridine and isopropylidene-uridine behave similarly.

For dimethyluracil the reaction follows the course indicated by eqn. (6-1) up to a concentration of 10^{-3} M, with departures from linearity at higher concentrations. A fifty-fold variation in the intensity of the incident light was without effect on the reaction order.

For all the above compounds, irrespective of the initial concentration over the range 10^{-5} to 10^{-2} M, the photochemical reaction was largely reversible in the dark either by acidification or heating. This was regarded as evidence that the ultimate reaction involved was uptake of a water molecule at the 5,6 double bond. The kinetic data and, in particular, the increase in quantum yield with concentration, were therefore interpreted in terms of the transition of excited singlet state molecules to a long-lived state as a result of collisions of the second kind with molecules in the ground state. There is as yet no *direct* experimental evidence, one way or the other, for the existence of metastable states in photochemically excited pyrimidines (see below). A similar dependence of quantum yield on concentration has been reported by Millich and Oster (1959) for the photoreduction of 3,6-diaminoacridine and the concentration dependence on quantum yield interpreted in the same way as for the above compounds; the appreciable fluorescence of diaminoacridine made it possible to verify this interpretation quantitatively by measurement of the self-quenching of fluorescence with increase in concentration.

An additional possibility is the formation, at higher concentrations, of dimer photoproducts. By analogy with thymine dimer, such photoproducts might be expected to undergo dissociation; but the resulting excited monomers might then readily take up a water molecule, the net result being a more rapid rate of hydration of the molecules in the irradiated solution (Smietanowska and Shugar, 1961). The reported isolation from uracil and uridine solutions of dimer photoproducts, already referred to above, provides support for this latter postulate. It should be noted, in this connection that, at the time the foregoing kinetic studies were carried out, pyrimidine photodimerization had not yet been demonstrated experimentally, and was ignored. A re-examination of this question is certainly warranted.

The kinetics of photochemical transformation of uracil itself have not been investigated, but it is certainly more complicated than for the above derivatives. Rorsch *et al.* (1958) have shown that the initial stage of the reaction involves the transformation of about 6 per cent of the uracil to a product which cannot revert to uracil by heating or acidification. On the other hand, following irradiation to the point where the absorption spectrum has completely disappeared, only about 50 per cent of the photoproducts revert to uracil in the dark (Moore, 1958). Furthermore, irradiation in an atmosphere of oxygen was found to decrease the radiation sensitivity of uracil, while irradiation in the presence of nitrogen or hydrogen resulted in a more rapid photochemical transformation under otherwise identical conditions; this effect was even more marked for orotic acid (Beukers *et al.*, 1959b). Particularly striking in the case of orotic acid is the effect of consecutive irradiation of the same solution in nitrogen, and then in oxygen, saturated solution (Beukers *et al.*, 1959c), illustrated in Fig. 6-8. There

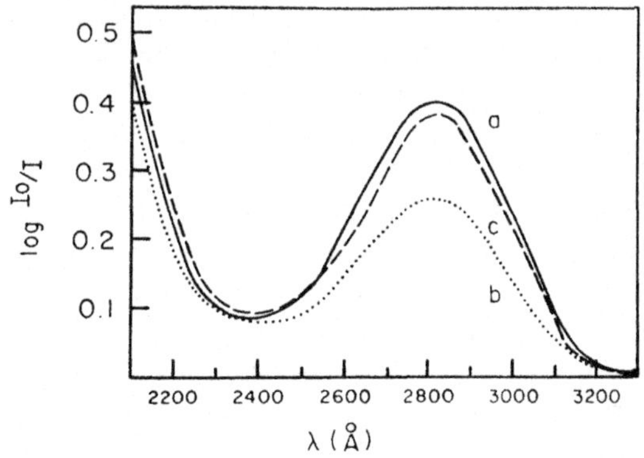

FIG. 6-8. Spectrum of orotic acid at pH 1: (a) prior to irradiation; (b) after 15 mins irradiation in an atmosphere of N_2; (c) irradiated solution flushed, and saturated with, O_2 and again irradiated for 15 min (Beukers *et al.*, 1959c).

appears to be very little doubt from the figure but that the photoproduct resulting from irradiation under nitrogen has been regenerated to the original substance under oxygen. It remains to establish the nature of the photoproduct, e.g. whether it is hydrated orotic acid or a dimer. Isolation of the photoproduct in quantity may, in this instance, be somewhat complicated. Kinetic studies may assist in identifying the nature of the photoproduct, if low concentrations are used. It was suggested by Beukers *et al.* (1959b) that the above effects are due to a modification of the probability of singlet-triplet transitions. Some evidence does, indeed, exist for interaction between excited molecules and oxygen, but the nature of this interaction remains to be clarified (Noyes, 1959). In those cases where flash photolysis has been used to investigate the influence of oxygen on metastable, presumably triplet, states of molecules, no such states have been found in the presence of oxygen. This may involve not only quenching but also chemical interaction. Some evidence has recently been found for triplet state quenching by paramagnetic species (Moon *et al.*, 1960). On the other hand, for those cases hitherto examined, no effect of oxygen on singlet states has been noted (Noyes, 1959; cf. Cherkasov and Vember, 1959).

The foregoing interpretation as to the effect of oxygen on orotic acid has derived additional support from the observation that a variety of paramagnetic, but not diamagnetic, ions can replace oxygen for regeneration of the photoproduct of orotic acid to its parent substance as described in Fig. 6-8 (Beukers and Berends, 1960b). It should not, however, be regarded as applying indiscriminately to all pyrimidines; irradiation of dimethyluracil in an atmosphere of nitrogen is without effect on the kinetics (Wierzchowski and Shugar, 1959), nor is there any oxygen or nitrogen effect observable in the photochemical transformation of uracil and cytosine containing oligo- and polynucleotides (Wierzchowski and Shugar, 1960b). Finally, ultraviolet induced biological inactivation of micro-organisms or viruses, or of transforming DNA, likewise does not exhibit an oxygen effect.

It remains, nonetheless, clear that investigations on the excited states of pyrimidine derivatives are urgently required.

Kinetics of Decomposition of Photoproducts. The kinetics of the reverse reaction for uracil derivatives, in those instances where it has been investigated, has been found to be 1st order and is an acid–base catalyzed reaction, the nature of which we shall return to below in connection with the analogous reverse reaction for the photoproducts of cytosine derivatives. For dimethyluracil the activation energy for the reverse reaction is 22.6 kcal/mole (Moore and Thomson, 1956). No measurements have been reported for the conversion of the photoproducts of uracil glycosides to the parent compounds, but Wierzchowski and Shugar (1960b) were able to calculate it for the regeneration of the photoproduct of the Up component of the dinucleotide ApUp. The value obtained was 26.2 kcal/mole and is close enough to that for dimethyluracil to permit the assumption that similar values prevail for free nucleotides of uracil.

The rates of the reverse reaction for uracil glycosides have not been extensively investigated as for cytosine glycosides. For the photoproduct of uridylic acid, the rate of conversion to uridylic acid is independent of concentration, but it is more rapid for the 3′ than for the 2′ isomer. For uridine-2′:3′-phosphate and isopropylideneuridine the rates of the reverse reaction are even more rapid, the half-time for completion of the reaction being 5 min at 75° as compared to 10–15 min at 80° for uridylic acid.

10. *Reversible Photoproducts of Cytosine Derivatives*

We turn now to an examination of the photochemistry of cytosine and some of its derivatives and, in particular, of its glycosides. The latter and, to a lesser extent, cytosine and 1-methylcytosine, form photoproducts which, like the corresponding uracil analogues, revert to the parent substances in the dark under the influence of acid, base or heating at neutral pH. These photoproducts are, on the other hand, less stable than the corresponding uracil derivatives so that the reverse reaction often proceeds spontaneously at room temperature, even at neutral pH. Consequently, isolation of the photoproducts has not been achieved as yet and their identification is, for the moment, based on indirect methods. Furthermore, the instability of the photoproducts is frequently such as to complicate the kinetics of the photochemical reaction since, in fact, an equilibrium is established between the forward (photochemical) and reverse (dark, thermal) reactions.

If C_0 is the initial concentration of the cytosine glycoside, C_e the concentration at equilibrium, then $(C_0 - C_e)$ is the concentration of the photoproduct at equilibrium. If k_1 and k_r are the rates of the forward and reverse reactions, assumed to be first-order, then at equilibrium

$$k_1 I_0 \alpha C_e = k_r(C_0 - C_e)$$

where α is a proportionality constant to allow for the amount of light absorbed by C_e. It follows that

$$C_e = \frac{k_r C_0}{k_r I_0 \alpha - k_r}$$

assuming, of course, that the photoproduct itself is radiation resistant. Since k_r is the rate constant for a temperature-dependent reaction, the equilibrium concentration C_e, at a given pH, will be a function of the incident light intensity and the temperature of the solution. With a sufficiently high light intensity the value of C_e may be reduced to the point where photoproduct formation is almost complete. In practice it has been found more convenient to reduce the temperature of the solution during irradiation so that the rate of the reverse reaction is reduced.

The photochemical behavior of cytosine and 1-methylcytosine differs in some respects from that for cytosine glycosides and, for purposes of simplification, it will be more convenient to discuss first the behavior of a typical glycoside such as cytidine-3′-phosphate, or Cp,† illustrated in Fig. 6-9A, where curve *a* is that for Cp and curve *b* is that for Cp plus its photoproduct at equilibrium, irradiation being at 7°C to reduce the rate of the reverse reaction. Removal of the irradiation source and warming of the solution results in the reappearance of the original absorption spectrum, curve *c*. Curve *d* is the calculated absorption spectrum of the photoproduct obtained from curve *b*, the absorption of which in the neighborhood of 2700 Å indicates that six per cent of the Cp had not reacted at equilibrium. From the difference between curves *a* and *c* it follows that 13 per cent of the Cp has reacted irreversibly in this experiment; the overall percentage reversibility is nonetheless quite high, and, in fact, is usually well over 90 per cent if irradiation is not unduly prolonged. The reverse reaction is quantitative

† The following nomenclature will be used: C stands for cytidine; Cp is cytidine-3′-phosphate; pC is cytidine-5′-phosphate; a prefix *d*- refers to the corresponding deoxy compound; Cp! is cytidine-2′:3′-phosphate.

so that curve *d* corresponds to curve *c* decreased by six per cent. The extinction of the photoproduct may therefore be calculated from that for Cp and the absorption spectra of both of these are shown in Fig. 6-9B.

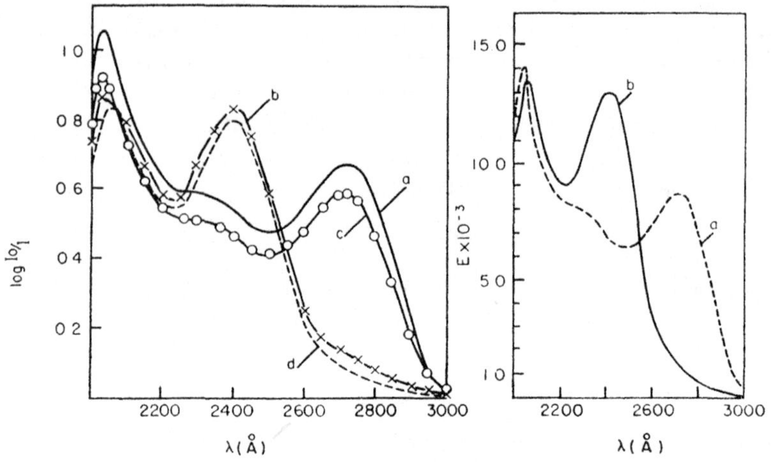

FIG. 6-9A. Photochemical transformation of cytidylic acid in 0.02 M phosphate buffer pH7, temperative 7°C, incident intensity 2.6×10^{-7} einsteins/cm^2 per min: (a) prior to irradiation; (b) following attainment of equilibrium between forward and reverse reactions; (c) following removal of irradiation source and heating to cause reverse reaction to go to completion; (d) absorption spectrum of photoproduct alone.

FIG. 6.9B. Absorption spectra of (a) Cp and (b) its photoproduct (Wierzchowski and Shugar, 1961a).

The optical constants for other isomeric nucleotide photoproducts have been obtained in the same way (Wierzchowski and Shugar, 1961a), and differ from that for Cp only in the value of the extinction coefficient at 2400 Å.

(a) *Nature of Photoproducts of Cytosine Glycosides.* It will be seen from Fig. 6-9A that, as for uracil derivatives, irradiation of cytosine glycosides results in a disappearance of the principal absorption maximum in the ultraviolet, at 2700 Å. This is accompanied, however, by the simultaneous appearance of a new maximum at about 2400 Å. But this is precisely what would be expected from saturation of the 5,6 double bond of the pyrimidine ring as may be seen from an examination of Fig. 2-15(b) for dihydrocytosine derivatives. By analogy with the situation for the corresponding uracil derivatives, it might therefore be expected that the photoproduct is also the result of the nucleophilic addition of a water molecule across the 5,6 double bond. Dimer formation, on the other hand, also leads to saturation of the 5,6 double bond and would equally well account for the spectra of the photoproducts; but two facts argue against this latter possibility, at least for irradiation at concentrations of the order of 10^{-4} to 10^{-3} M solutions. The first of these is the kinetics of formation of the photoproduct. Nucleophilic addition of a water molecule should be a first-order reaction, dimer formation a reaction of the second order. Fig. 6-10 shows the course of the

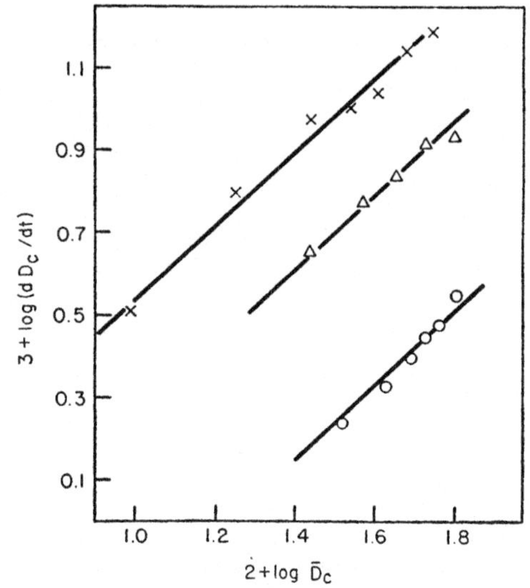

FIG. 6-10. Kinetics of photochemical transformation of C2′p (× × × × ×), pC (△ △ △ △ △) and d-pC (○ ○ ○ ○ ○), plotted according to van't Hoff difference method. Concentration ~ 10^{-4} M in 0.02 M phosphate buffer pH 7 and at temperature of 7°C to slow down reverse reaction; incident light intensity 2.6×10^{-7} einsteins/cm² per min (Wierzchowski and Shugar, 1961a).

photochemical reaction for three cytosine nucleotides, plotted by the difference method of van't Hoff (Wierzchowski and Shugar, 1959, 1961a). In all instances the slope is unity, indicating a first order reaction with constant quantum yield.

The hydrogen isotope effect for photoproduct formation provides additional supporting evidence for nucleophilic addition of a water molecule (Shugar and Wierzchowski, 1957; Wierzchowski and Shugar, 1957), the reaction being as follows:

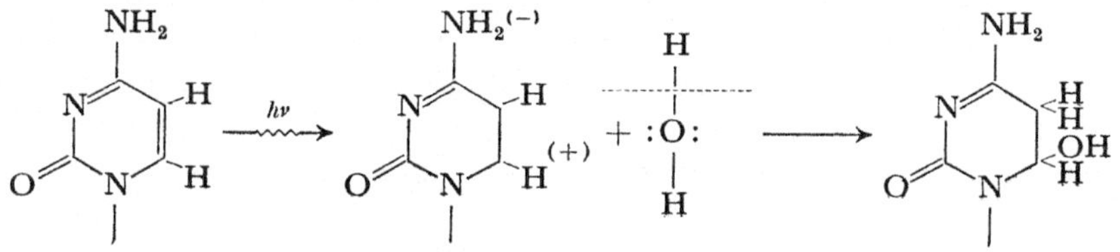

Opening of the D—O bond requires a higher activation energy than does the corresponding H—O bond, so that quantum yields should be lower in D_2O than in H_2O. The quantum yields for several uracil and cytosine analogues in light and heavy water have been measured and are as follows (Table 6-5):

TABLE 6-5. QUANTUM YIELDS FOR PHOTOLYSIS OF SOME PYRIMIDINE DERIVATIVES IN LIGHT AND HEAVY WATER (SHUGAR AND WIERZCHOWSKI, 1957a, 1958a)

Compound	$\Phi \times 10^3$	
	in D_2O	in H_2O
1-Methyluracil	5.7	12.5
2-Ethoxyuracil	4.0*	8.0*
Cytosine	0.7	1.7
Cytidine	4.2	9.0
2-Methoxycytosine	5.0*	10.0*
Thymine	0.9†	0.4†

* pH 7.2 phosphate buffer 0.02 M.
† Reaction not reversible in the dark.

It will be seen that the isotope effect for 1-methyluracil, where photoproduct formation is known from chemical evidence to involve addition of a water molecule, is $\Phi_{H_2O}/\Phi_{D_2O} \sim 2.2$. A similar value prevails for cytosine and cytidine, but not for thymine where it is known that photolysis in aqueous medium does not lead to uptake of a water molecule. As regards the other two compounds in the table, for which the isotope effect is also about 2, the nature of the photoproducts is as yet unknown (see Fig. 6-5). The results in Table 6-5, although not conclusive, are nonetheless in agreement with the formation of hydrated products as a result of irradiation.

The hydrogen isotope effect for the reverse reaction is more complex than originally suggested (Shugar and Wierzchowski, 1957) and will be taken up below in connection with the kinetics of the reverse, thermal reaction.

(b) *Cytosine and 1-methylcytosine.* It might be expected that cytosine and, in particular, 1-mothylcytosine would behave photochemically like cytosine glycosides, but this is only partially so (Wierzchowski and Shugar, 1957, 1961a). It should be recalled that the isotope effect for photoproduct formation is the same for cytosine (Table 6-5), implying nucleophilic addition of a water molecule; spectral data are only partially in agreement with this.

Irradiation of cytosine in neutral, unbuffered medium (Fig. 6-11) results in a decrease in optical density of the principal absorption maximum; there is also a simultaneous increase in absorption at 2400 Å, as for glycosides, but without formation of a clearly defined maximum. Upon removal of the irradiation source, the reverse reaction is placed in evidence by a decrease in absorption at 2400 Å and a concomitant increase at 2700 Å until, finally, the

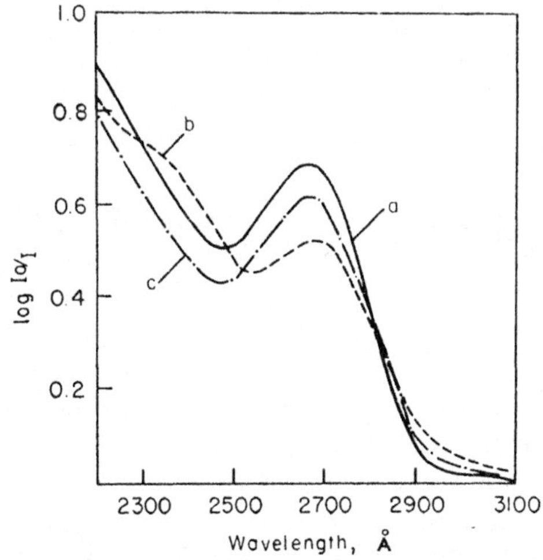

FIG. 6-11. Photochemical transformation of cytosine in unbuffered, aqueous medium, pH 7, room temperature, incident light intensity 3×10^{-7} einsteins/cm^2 per min: (a) prior to irradiation; (b) following 90 min irradiation; (c) irradiated solution heated until reverse reaction completed (Shugar and Wierzchowski, 1957; Wierzchowski and Shugar, 1961a).

ratio of the extinction at 2700 Å to that at 2400 Å is identical with that for the parent compound. The degree of reversibility is dependent on the extent of photoproduct formation; following 25 per cent photolysis, reversibility is about 60 per cent and is even higher for smaller doses, but entirely absent after about 80 per cent photolysis. The ionic strength of the medium is quite important; e.g. in the presence of 0.1 M NaCl no reversibility whatever can be observed.

The behavior of 1-methylcytosine is identical to that for cytosine, even to quantum yield which is quite low, 2×10^{-3}. Although both compounds form reversible photoproducts, the available evidence shows that more than one reaction takes place on irradiation (Wierzchowski and Shugar, 1961a).

Because of the unusual behavior of cytosine as compared to its glycosides (Wierzchowski and Shugar, 1957, 1961a), it was postulated by Wang (1959) that the photoproducts of cytosine and its glycosides are not the 5-hydro-6-hydroxy derivatives but tautomers in which the proton on the N_1 nitrogen has been shifted to position 5 with the accompanying displacement of the 5,6 double bond to N_1, C_6 as follows:

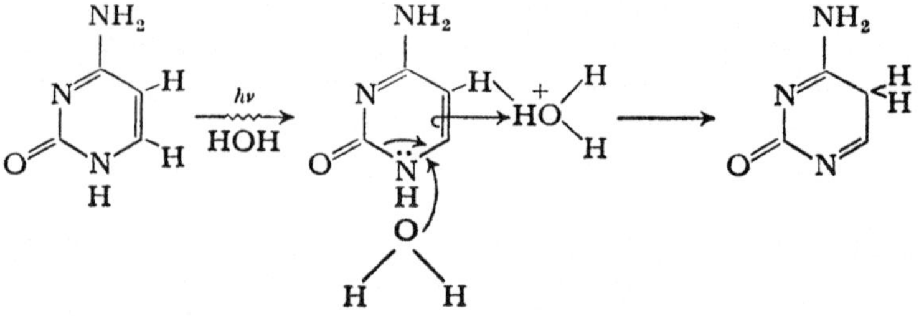

The assumptions on which this proposal are based are, however, of doubtful validity. For example, the foregoing reaction sequence is obviously inapplicable when the number 1 nitrogen is substituted, as in glycosides; it is also contrary to the known fact that cytosine and 1-methylcytosine exhibit identical photochemical behavior. According to Wang the photoproducts of cytosine glycosides cannot be the 5-hydro-6-hydroxy derivatives since they exhibit an absorption maximum at 2400 Å; this claim is, however, based on an unwarranted generalization of the observation of Batt *et al.* (1954) that the dihydroderivatives of uracil analogues exhibit only end absorption in the quartz ultraviolet, and certainly does not apply to dihydrocytosine derivatives which should, and do, exhibit an absorption maximum at about 2400 Å. It must nonetheless be admitted that the nature of the photochemical transformations undergone by cytosine and 1-methylcytosine is not as clearly established, above, as for cytosine glycosides.

This difference in behavior between cytosine and 1-methylcytosine on the one hand, and cytosine glycosides on the other, was subsequently considerably clarified by an examination of the photochemical behavior of alkylamino cytosine derivatives such as the following:

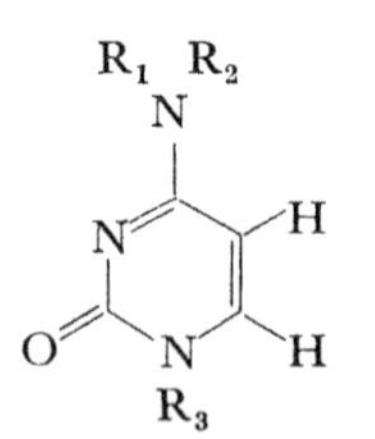

R_1 = H or alkyl
R_2 = H or alkyl
R_3 = alkyl or ribose or ribosephosphate.

Mono- or di-alkylation of the amino group in the foregoing series of analogues renders them much more radiation resistant; but the resulting photoproducts are completely analogous to those formed from the unsubstituted amino derivatives, in addition to which they are also much more stable;

i.e. the reverse thermal reaction is very slow. This was found to be particularly marked for 1-methyl-monomethylaminocytosine ($R_1 = H$, $R_2 = R_3 = CH_3$), the photoproduct of which was shown spectrally to be almost identical with that for cytidine or cytidylic acid (Fig. 6-9A), as it should be. It may therefore be concluded that the "anomalous" behavior of 1-methylcytosine (and cytosine) is due to the inherent instability of their photoproducts, i.e. the ease with which these photoproducts undergo dehydration (Fikus *et al.*, 1962). It remains to add that the photoproducts of some of the above alkylamino cytosine analogues proved to be sufficiently stable as to envisage the possibility of their isolation for more detailed investigations by NMR, etc. In particular they are ideal model compounds for detailed investigations of the dehydration reaction under various conditions.

(c) *Quantum yields for cytosine derivatives.* The quantum yields for cytosine, 1-methylcytosine and the various cytosine glycosides, over a wide range of pH values, are presented in Table 6-6. Most of these values were originally measured at a time when the velocities of the reverse reaction had been observed for only a few of the compounds at neutral pH (Shugar and Wierzchowski, 1957; Sinsheimer, 1957; Wierzchowski and Shugar, 1956). Subsequently it was found that, for a compound such as deoxycytidine-3′,5′-diphosphate, the rate of the reverse reaction is so high in comparison to the quantum yield that irradiation on a hot summer day is almost without effect! All the values, as well as some additional ones, have therefore been measured once more using two different irradiation procedures: (a) irradiation at a temperature of 22°C and an incident light intensity of 6.5×10^{-7} einsteins/cm^2 per min which is sufficiently high to minimize the effect of the reverse reaction; (b) irradiation at a temperature of 7°C, to reduce the velocity of the reverse reaction, and at an incident intensity of 2.7×10^{-7} einsteins/cm^2 per min. The results obtained in this way for several compounds differed from the mean by only 2–3 per cent, which is within the limits of experimental error (Wierzchowski and Shugar, 1961a; see also Fikus *et al.*, 1962).

It must nonetheless be emphasized that the velocity of the reverse reaction is so high at pH values above 11 that not too much confidence should be placed in the measured quantum yields. They should be regarded rather as "apparent" quantum yields and most likely represent the yields, not for addition of a water molecule, but rather for destruction of the initial photoproduct.

It will be noted that the quantum yields for cytosine and 1-methylcytosine are almost identical, except at pH 13. It is consequently quite striking to find that, in the neutral pH range, the yields for most of the glycosides are considerably higher than for 1-methylcytosine. At acid pH, on the other hand, quantum yields are almost identical for all compounds.

TABLE 6-6. QUANTUM YIELDS (IN MOLES/EINSTEIN × 10^3) AT 2537 Å FOR PHOTOCHEMICAL TRANSFORMATION OF CYTOSINE DERIVATIVES AT A CONCENTRATION OF 10^{-4} M AT VARIOUS pH VALUES[a]

Compound	pH								
	1	2	5.6[b]	7.1	8.9	10.5	11	13	14
Cytosine	—	1.3	1.7[c]	1.3[b,c]	—	—	—	6.3	8.0
1-Methylcytosine	—	1.4	2.1[c]	1.3[b,c]	—	—	—	2.1	—
Cytidine	—	1.6[c]	9.0[c]	10.0[c]	—	8.9[c]	6.8[c]	3.6	2.9
Deoxycytidine	—	1.6[c]	9.5[c]	8.6[c,f]	—	8.3[c]	6.0[c]	0.5	—
Pyranosylcytosines	1.6[c]	—	16.0[c]	13.0[c]	15.0[c]	11.0[c]	7.9[c]	1–2[d]	0.25–2.5[d]
C2′p	—	—	—	14.1[c,e]	—	—	—	—	—
Cp	—	—	—	9.5[c,e]	—	—	—	—	—
C2′(3′)p	1.1	—	—	12.5[c]	—	9.4[c]	6.3[c]	1.4	—
Cytidine-2′:3′-phosphate	—	—	—	11.5[c]	—	—	—	—	—
pC	—	—	—	5.2[c]	—	—	—	—	—
d-pC	—	—	—	3.0[c]	—	—	—	—	—
d-Cp	—	—	—	10.6	—	—	—	—	—
d-pCp	—	—	—	3.4	—	—	—	—	—

[a] Wierzchowski and Shugar (1957, 1961a), Shugar and Wierzchowski (1958a, 1958b) and Shugar (1960b). Values above pH 11 are uncorrected for reverse reaction.

[b] Unbuffered solution; for influence of buffer see Wierzchowski and Shugar (1957, 1961a).

[c] Reaction reversible in the dark; for cytosine and 1-methylcytosine only partial reversibility.

[d] Quantum yield increases appreciably during course of reaction.

[e] Sinsheimer (1957) reports a quantum yield of 17 × 10^{-3} for the 2′ and 3′ isomers of cytidine.

[f] Moore and Thomson (1957) report a quantum yield of 7 × 10^{-3} at a concentration of 10^{-2} M.

These observations were regarded as evidence for some type of interaction between the pyrimidine and carbohydrate rings and, in particular, between the 2-carbonyl of the pyrimidine rings and one of the carbohydrate hydroxyls (Wierzchowski and Shugar, 1957). Insofar as this suggestion was based in part on the reduced quantum yields at pH 13, however, it requires modification: as pointed out above, the reduced quantum yields at pH 13 are only apparent, and due to the extremely rapid reverse reaction at high alkaline pH. A more detailed study of the kinetics of the reverse reaction for cytosine glycosides has now revealed that such interaction extends likewise to nucleotides and involves the phosphate hydroxyls as well.

(d) *Kinetics of reverse reaction.* For all cytosine derivatives hitherto examined, and which undergo reversible photochemical transformation, the reverse reaction has been found to be strictly first order with a rate-constant dependent on pH, temperature, and to some extent the ionic strength of the medium. The pH dependence of the velocity of the reverse reaction is illustrated by that for cytidine in Fig. 6-12. The complex nature of the curve on the acid side of neutrality is probably due to the fact that, in addition to

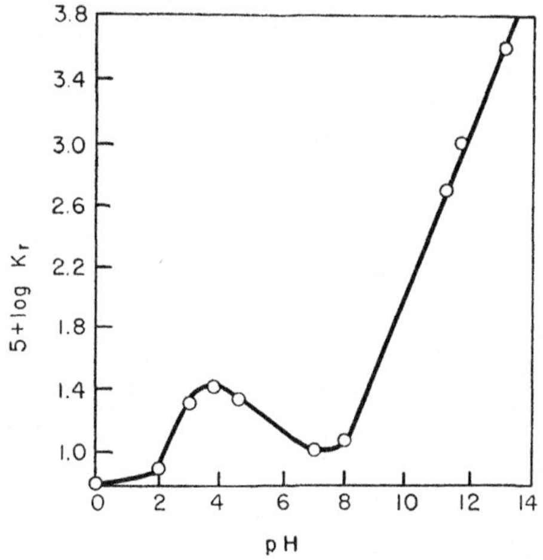

FIG. 6-12. Rate of transformation of photoproduct of cytidine to cytidine at 25° as a function of pH; k_r is the 1st order rate constant for the reverse reaction (Wierzchowski and Shugar, 1961a).

the catalytic effect of hydrogen ions on the reaction, we are dealing also with a mixture of ionic species resulting from the dissociation of the amino group. The curve on the alkaline side likewise exhibits some complexity up to pH 8, due in the case of nucleotides to dissociation of the secondary phosphate hydroxyls (see Table 6-7); this has now been examined in some detail (Wierzchowski and Shugar, 1961a; Fikus *et al.*, 1962) and will be referred to in the following section.

The rate of the reverse reaction for Cp at neutral pH has been measured over the temperature range 25°–50°C and adheres closely to an Arrhenius plot with a Q_{10} of 2.6 and an activation energy of 15.5 kcal/mole (Wierzchowski and Shugar, 1961a). Using the data of Sinsheimer (1957), a value of 15 kcal/mole is obtained. None of the other isomers have been examined but it is most likely they would all exhibit similar activation energies. The application of such measurements to the behavior of cytosine residues in model oligonucleotides has proven extremely useful (Wierzchowski and Shugar, 1960b).

The rate constants for the reverse reaction of almost all cytosine nucleosides and nucleotides have been measured at acid, neutral, and slightly

TABLE 6-7. FIRST-ORDER RATE CONSTANTS AT 25°C FOR REVERSAL OF PHOTO-PRODUCTS OF CYTOSINE GLYCOSIDES TO PARENT COMPOUNDS

Photoproduct	k_{-1} (sec^{-1} × 10^4)		
	pH 2	pH 7	pH 8.1
Cytidine	0.77	1.10(1.4)*	1.25
Glucopyranosyl-cytosine	—	1.50	—
C2′p	1.76	4.00	0.64
C3′p	0.74	1.00	0.64
pC	1.27	6.50	0.83
pCp, pC2′p	1.90	4.60	1.40
Cp!	—	6.30	0.85
2′,3′-isopropylidene-cytidine	—	1.20	0.96
Deoxycytidine	0.94	1.70	1.48
d-Cp	0.94	1.34	0.83
d-pC	1.60	5.50	1.06
d-pCp	1.63	5.00	2.30

* Figure in brackets is for solution in 0.1 M NaCl.

alkaline pH and the data are presented in Table 6–7. It should be noted that the values for the nucleosides exhibit only minor differences, whereas appreciable variations prevail between the nucleotides at a given pH as well as for a given nucleotide at different pH values. The influence of ionic strength, as illustrated by the 30 per cent increase in rate for cytidine in 0.1 M NaCl, is apparently not very great, but is deserving of further study, particularly for the nucleotides where the interaction between the charged amino and phosphate groups might be expected to be significant. For cytosine at neutral pH (not shown in Table) the rate constant of 1.6×10^{-4} sec^{-1} at zero ionic strength is increased to 10^{-3} in 0.02 M buffer and is so rapid as not to be measurable in 0.1M salt solution. This effect of ionic strength on the reverse reaction may be of some significance biologically, in view of the reported effect of salt concentration on photoreactivation of micro-organisms (e.g. Kaplan and Gunkel, 1960).

The influence of the phosphate groups on the reverse reaction for nucleotides, i.e. on the stability of the photoproducts, is obvious on an examination of the rate constants at pH 7 and 8.1 in Table 6-7. In this pH range the only modification resulting from a change in hydrogen ion concentration is the dissociation of the secondary phosphate hydroxyls. Bearing in mind the influence of the location of the phosphate group on the quantum yield for photoproduct formation, it follows that the phosphate hydroxyls influence both the excited state of the cytosine ring as well as the stability of the resulting photoproduct. In part, this interaction is most likely of the dipole–dipole type between the two functional, polar phosphate and amino groups, and it has been pointed out that the most unstable photoproducts are those in which these two functional groups are closest, i.e. in the 5′ nucleotides. The increase in the rate of dehydration of the photoproducts, accompanying the appearance of a proton on the phosphoester group, may also conceivably provoke an intramolecular proton transfer from the phosphoester group to the hydroxyl on C_6 of the photoproduct, thus catalyzing the elimination of the water molecule on the 5,6 bond. And, in fact, etherification of the two phosphate hydroxyls in cytidine-5′-phosphate, to give cytidine-5′-dimethylphosphate, eliminates the pH dependence of the rate of dehydration of the photoproduct. For a more extensive discussion of the quantum yields and photoproduct stability of cytosine nucleosides and nucleotides in relation to their stereochemistry, reference should be made to two publications (Wierzchowski and Shugar, 1961a; Fikus *et al.*, 1962).

Hydrogen isotope effect of reverse reaction. The hydrogen isotope effect for

TABLE 6-8. HYDROGEN ISOTOPE EFFECT FOR ELIMINATION OF WATER MOLECULE FROM PHOTOPRODUCTS OF URIDINE AND CYTIDINE

Photoproduct	$k_{70°} \times 10^4$ (sec^{-1})	
	in H_2O pH 1.6	in D_2O pD 1.6
$U \cdot H_2O$	7.4	13.4
$U \cdot D_2O$	—	7.1
	$k_{40°} \times 10^4$ (sec^{-1})	
	in H_2O pH 1.6	in D_2O pD 1.6
$C \cdot H_2O$	5.1	1.5
$C \cdot D_2O$	3.4	1.1
	$k_{25°} \times 10^4$ (sec^{-1})	
	in H_2O pH 11.6	in D_2O pD 11.6
$C \cdot H_2O$	84.5	7.8

the reverse reaction is more complex than originally assumed and has now been investigated in greater detail (Wierzchowski and Shugar, 1961a), bearing in mind that it involves two effects: (a) dissociation of the C—H (or C—D) bond at positions 5 of the photoproduct, and (b) the effect of the acid–base properties of the light or heavy water, in which the elimination reaction is carried out, on the rate of a reaction catalyzed by hydrogen or hydroxyl ions. A separation of these two effects has been achieved for uridine and cytidine as follows: the photoproduct of each substance was obtained in light and heavy water and lyophilized; each photoproduct was then divided into two portions, one of whch was dissolved in light, the other in heavy, water for measurements of the rates of the reverse reaction. The results obtained, although not as extensive as might have been desired, are shown in Table 6-8.

It is clear from the table that the elimination of D_2O from the photoproducts is slower than for elimination of H_2O, as is to be expected for a reaction involving dissociation of a C—D, as compared to a C—H bond. If, however, the photoproduct in heavy water is as follows:

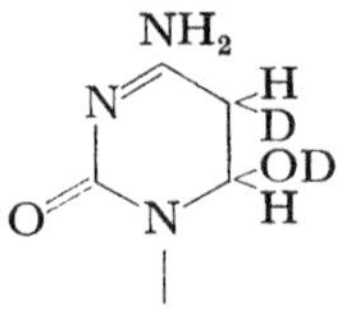

the measured isotope effect should be lower than that expected theoretically for dissociation of an isolated C—D bond, since some molecules will eliminate an H atom instead of a D from position 5. For cytidine, on the other hand, where the rate of the reverse reaction is appreciable at room temperature, the photochemical equilibrium established in heavy water during irradiation should lead to the exchange of both H atoms for D at position 5. The isotope effect for cytidine photoproducts should consequently be greater than for the photoproducts of uridine. Actually the reverse is true, the values being 1.9 for uridine and 1.5 for cytidine.

If we examine the effect of the medium, it will be seen from Table 6-8 that the photoproduct of uridine undergoes the elimination reaction faster in heavy water so that k_{D_2O}/k_{H_2O} is 1.8 whereas for the photoproduct of cytidine it is 0.3 under the same conditions and as low as 0.1 in alkali. The catalysis mechanism must therefore be different for each photoproduct. For reactions catalysed specifically by H_3O^+ or OH^-, the isotope effect is usually $k_{H_2O}/k_{D_2O} < 1$ whereas for general acid–base catalysis this ratio is greater than unity. Consequently the elimination reaction for the photoproduct of uridine (and most likely other uracil derivatives) is catalysed specifically whereas the reaction for the cytosine photoproducts belongs to the class of more general acid and base catalyzed reactions. By analogy with the scheme presented by Ingold (1953) for the monomolecular elimination of water from an aliphatic double bond (mechanism E_1), the reverse reaction would be as follows:

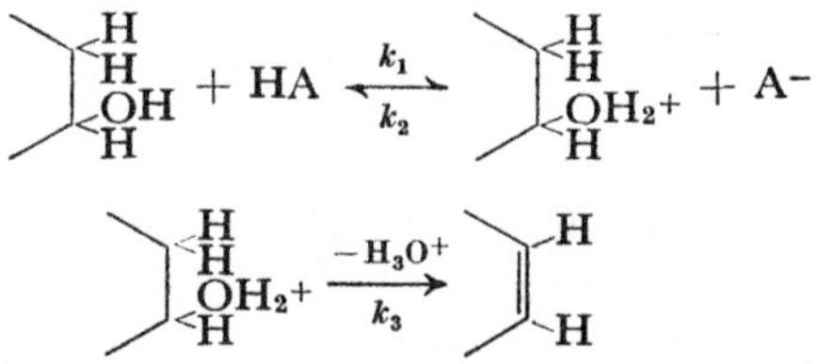

so that for cytosine derivatives $k_a > k_2$, since only then will $k_{H_2O}/k_{D_2O} > 1$ (Brodskij, 1952). For uracil derivatives, on the other hand, where the value of k_a is about 2 orders of magnitude lower than for cytosine analogues, $k_3 < k_2$ and the elimination reaction, which is the slowest, determines the kinetics of the overall process. The concentration of the conjugate acid

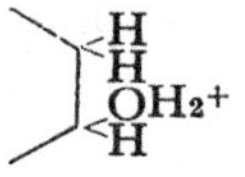

is higher in D_2O than in H_2O since heavy water possesses less basic properties than ordinary water (Brodskij, 1952; Wiberg, 1955). The photoproduct therefore competes more effectively for the deuteron in D_2O than for the proton in H_2O so that the reaction is more rapid in heavy, than in ordinary water.

Finally, attention should be drawn to the fact that the hydrogen iostope effect for the photochemical reaction has been utilized to study the nature of the photochemical transformation of uracil residues in model oligonucleotide chains (Wierzchowski and Shugar, 1959). The reverse reaction might equally profitably be applied for this purpose.

11. *Primary Photoproducts of Substituted Pyrimidine Derivatives*

Additions to the 5,6 double bond are by no means the only primary reactions resulting from the irradiation of pyrimidine derivatives. One apparent exception has already been cited, viz. 2-ethoxyuracil, illustrated in Fig. 6–5; the absorption spectrum of the primary photoproduct in this case is difficult to reconcile with simple saturation of the 5,6 bond. A number of additional examples may be found in publications by Wierzchowski and Shugar (1960a, 1960c). Spectral methods alone have proven inadequate for identification of these products. Even in those instances where irradiation results in destruction of the absorption spectrum, it would be imprudent to assume from this that the reaction involves either the formation of adducts or rupture of the pyrimidine ring.

Because of the various substituted pyrimidine derivatives participating in the metabolism of, or incorporated as minor components in, nucleic acids, one is naturally led to explore the photochemical behavior of a variety of substituted pyrimidine derivatives, and, in particular, the influence of specific substituents in the pyrimidine ring on the transformations the latter may undergo. An excellent illustration of this was the fortunate choice of 1,3-dimethyluracil for elucidation of the photochemical transformation of uracil and its glycosides. Another illustration is that of 1-methylcytosine which, while behaving photochemically like cytosine, differs so markedly from the corresponding glycosides as to emphasize the importance of the interaction between the aromatic and carbohydrate rings in the reaction.

We shall now present a brief description of some of the photochemical reactions undergone by a variety of substituted pyrimidine derivatives. Some of these are of potential importance in an interpretation of the photochemical behavior of nucleic acids. Furthermore, in a number of instances it has been possible to draw some conclusions regarding structure from photochemical behavior (Wierzchowski and Shugar, 1957; 1960a, 1960c). In some instances it has also been possible to separate stable photoproducts on a moderate scale, so that this is a matter of some interest to the organic chemist as well as to the biochemist interested in potential antimetabolites.

One simple example is that of 6-methyluracil, the photoproduct of which was reported to exhibit "slight" reversibility (Moore and Thomson, 1956); this could not be confirmed but led to a trial with 1,3,6-trimethyluracil with results illustrated in Fig. 6-13. It is clear that more than one photoproduct is formed but that the major one is capable of reverting to the parent substance in the dark. Although the nature of this photoproduct has not been definitely established, the quantum yield and time of heating at neutral pH for accelerating the reverse reaction suggest that it involves water addition to the 5,6 bond.

Another illustration is provided by 6-aminouracil and 6-aminothymine. In acid medium each of these compounds is photolyzed with a quantum yield of the order of 10^{-3}. At pH 8.5, where the amino group is not protonated, 6-aminothymine is very rapidly transformed on irradiation, with a quantum yield of the order of 0.15, the highest value yet reported for photochemical transformation of any pyrimidine derivative. It is most likely that we are dealing here with an example of dimerization in *liquid* medium, as suggested by changes in absorption spectrum of the photoproduct, but confirmatory evidence is yet to be supplied.

Widespread interest centres around the identification of the photoproducts of 5-halogeno substituted uracils. This is due to the fact that such compounds may be biologically incorporated into the DNA of microorganisms, viruses and other cells in place of thymine, the net result of such incorporation being usually a marked increase in sensitivity to ultraviolet radiation. This subject is extensively reviewed by Smith (1963) and Wacker (1963). Very little is, however, known about the photoproducts of the halogenated uracils or their glycosides, apart from the fact that some dehalogenation takes place in the case of 5-bromouracil. One of the principal difficulties involved in identification of halogeno-uracil photoproducts is that the quantum yields for photochemical transformation are so low. In all

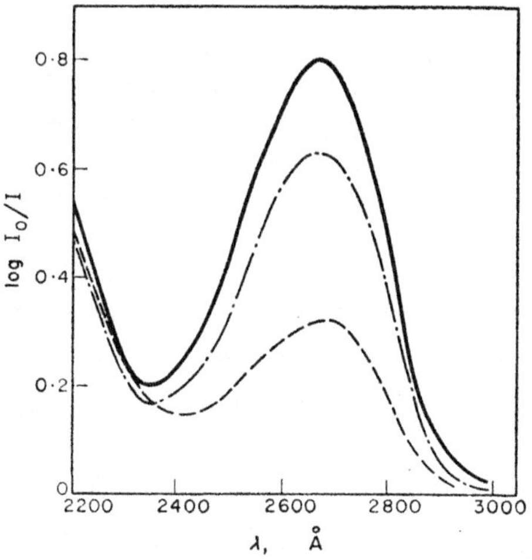

FIG. 6-13. Photochemical transformation and thermal reversibility of 1,3,6-trimethyluracil in 0.02 M phosphate buffer pH 7.2:

——— before irradiation
------- after irradiation for 3 hrs at incident intensity of 2.6×10^{-7} einsteins/cm² per min
—·—·— irradiated solution heated 75 min at 80°C (Wierzchowski and Shugar, 1960a).

probability it would be more promising to examine the photochemical behavior of synthetic polynucleotides of the halogeno-uracils, and to compare these with synthetic polynucleotides of uracil and thymine. Even such model systems may not be fully satisfactory since biological incorporation of halogeno-uracils is never equal to 100 per cent so that the resulting halogeno-DNA's must also contain adjacent residues of thymine and 5-bromouracil. This is certainly a fertile field for further investigation.

Spectral evidence indicates strongly that the primary photoproduct of isocytidine results from the addition of a water molecule at the 5,6 double bond, with a high quantum yield (0.054), but the reaction in this instance is apparently not reversible.

A preliminary investigation has been made of the primary photoproducts of 5-substituted cytosine nucleosides and nucleotides. This is of some significance in relation to biological photoreactivation in the T-even bacteriophages which contain only 5-substituted cytosines such as 5-hydroxymethylcytosine and the glucosylated analogue. While none of these derivatives was found to give rise to photoproducts which could revert to the parent substance as a result of heating or change in pH, they all formed photoproducts relatively stable at room temperature and neutral pH (Wierzchowski and Shugar, 1960a).

Since 5-substituted cytosine derivatives are relatively expensive compounds and not readily available, while it is also difficult to prepare the photoproducts in quantity, attempts have been made to elucidate the nature of the photoproducts of some other 4-aminopyrimidines, with the hope that this would materially assist in identifying the photoproducts of the 5-substituted analogues (Wierzchowski and Shugar, 1960c). Although this study has not yet been completed, the results are of considerable interest. It has been found that a number of 4-aminopyrimidines give rise to photoproducts which, from spectral evidence, resemble each other in structure and which, in some instances, will partially or wholly revert to the parent compound in the dark. None of these photoproducts result from simple saturation of the 5,6 double bond, as may be seen from the absorption spectrum of the photoproduct of 2-methoxycytosine, Fig. 6-14, which is

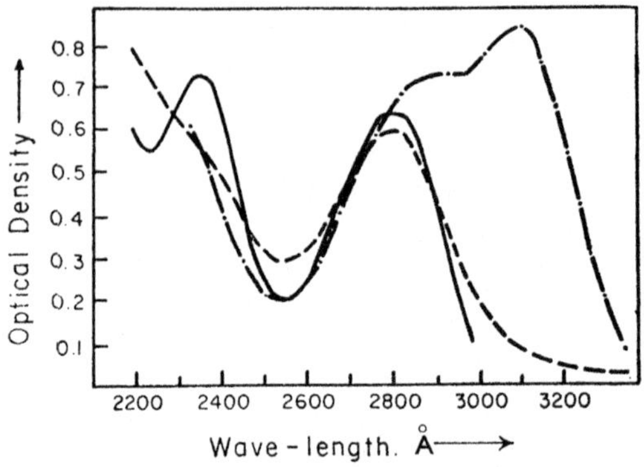

FIG. 6-14. 2-methoxycytosine in 0.02 M phosphate buffer pH 7.2:

——— prior to irradiation
-·-·-· following 150 min irradiation
------- following heating of irradiated solution 40 min at 85° (Shugar and Wierzchowski, 1958).

also typical of that for such compounds as 2,6-dimethyl-4-aminopyrimidine and the pyrimidine component of vitamin B_1 (2-methyl-4-amino-5-ethoxymethylpyrimidine). The latter compound reverts almost quantitatively to the parent substance in the dark at room temperature, but only at slightly alkaline pH; at acid pH, where the amino group is protonated, no reversible photoproduct is formed (cf. Section 5).

The importance of the above photoproducts in relation to those of cytosine and 5-substituted cytosines is demonstrated by the fact that cytosine and 1-methylcytosine (but not cytosine glycosides) form completely analogous photoproducts in anhydrous medium (ethanol), the characteristic feature of all of these being that they exhibit a new absorption maximum at longer wavelengths than the parent substance (see Fig. 6-14). This type of reaction is exhibited by some derivatives only in aqueous medium, by others only in anhydrous medium. There is some evidence that the solvent itself does not participate in the reaction, e.g. 2,6-dimethyl-4-amino-pyrimidine forms the same photoproduct in water, ethanol or hexane. Particularly interesting is the fact that the photoproduct formed by 5-methylcytosine in ethanol, when transferred to aqueous medium, is spontaneously transformed to the photoproduct formed by irradiation in aqueous medium. Some of the characteristic features of these reactions undergone by 4-amino-pyrimidines are summarized in Table 6-9. It is clear that identification of any one of these photoproducts would materially assist in resolving the structures of the others. It is consequently of some significance that the photoproduct of 2,6-dimethyl-4-aminopyrimidine has been isolated in crystalline form (Wierzchowski and Shugar, 1960c).

The structure of the primary photoproduct (II) of 2,6-dimethyl-4-aminopyrimidine (I) was subsequently fully elucidated. This was achieved by the successive deamination of II→III and III→IV. The latter compound turned out to be identical to the cyanoacetylacetone (3-cyanopent-2-en-2-ol-4-one) synthesized many years earlier by Traube (1898), and confirmed by chemical analysis, infrared and ultraviolet spectroscopy, and NMR (Wierzchowski *et al.*, 1963). Once the structure of IV had been determined, it proved relatively straightforward to establish the structure of II (2-amino-3-cyanopent-2-en-4-amine), the transformation of which via III (2-amino-3-cyanopent-2-en-4-one) gives IV, as follows:

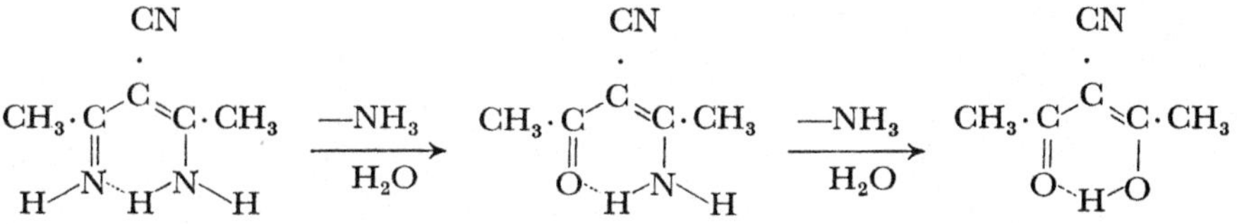

II III IV

All three compounds are intramolecularly chelated, as indicated in the foregoing structural formulae, and demonstrated by the infrared evidence for the existence of intramolecularly hydrogen bonded N—H groups in II and III, and an OH group in IV. Such intramolecular chelation accounts for the very high extinction coefficients of these compounds in the ultraviolet since it markedly enhances the degree of resonance conjugation of the system of unsaturated bonds resulting, in effect, in formation of cyclic benzenoid-like systems (Wierzchowski and Shugar, 1963).

It remains to elucidate the mechanism of the photochemical transformation of I → II. This involves the opening of two bonds, C_1—N_2 and N_3—C_4, and the formation of a new bond, C_2—C_5. This is rather difficult to interpret in terms of the usually accepted mechanism for the photochemical transformation of six-membered rings, resulting in the initial opening of one bond, and followed by some dark reaction(s) frequently involving the irradiation medium. Such a mechanism is also difficult to reconcile with the fact that the photochemical transformation I → II is (a) essentially quantitative; (b) first-order; (c) concentration independent; (d) independent of the irradiation medium; (e) exhibits no oxygen effect.

The simplest mechanism which takes into account all of the above effects would appear to be carbon bridging, i.e. the initial formation of the C_2—C_5 linkage, a type of transformation independent of factors which normally influence dark reactions. It was consequently suggested that the transformation of I → II proceeds as follows:

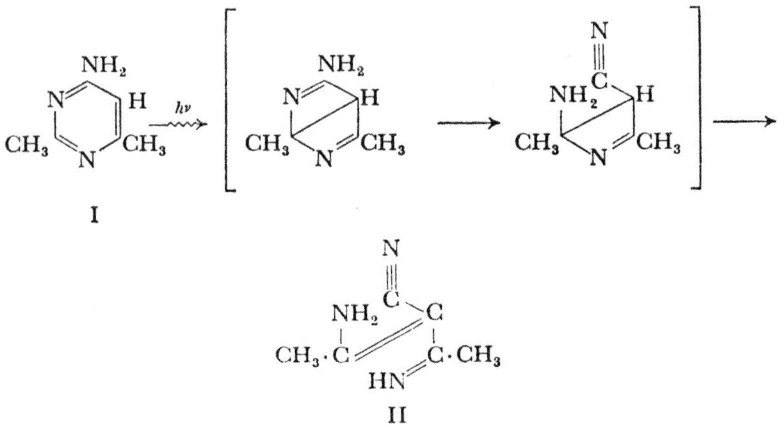

The proposed Dewar-like intermediates are bracketed since there is no direct evidence for their existence. It should be noted that various attempts have been made to detect or isolate such types of ring systems. A case in point is that of Taylor *et al.* (1961a), who claimed to have isolated a stable valence-bond tautomer of pyridone formed by ultraviolet irradiation, but this was subsequently shown to be incorrect (Taylor *et al.*, 1961b; for full elucidation of pyridone photoproducts, see Taylor and Kon, 1962). A more recent illustration is that of van Tamelen and Pappas (1962), who have isolated from irradiated 1,2,4-tri-t-butylbenzene a photoproduct which,

Table 6-9. Photochemical reactions of some 4-aminopyrimidines in aqueous and anhydrous solvents (all wavelengths are in Å units)

Compound	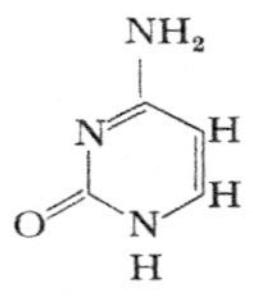	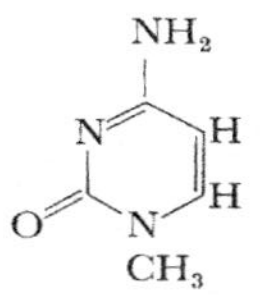	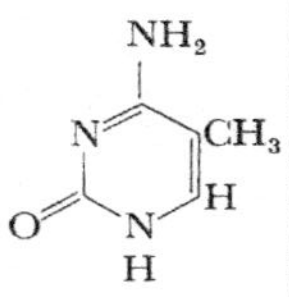	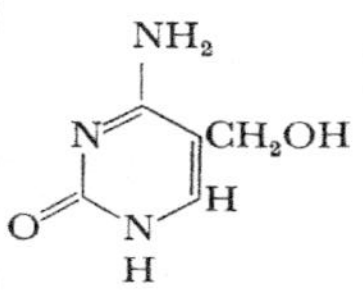	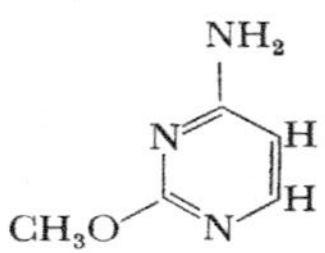	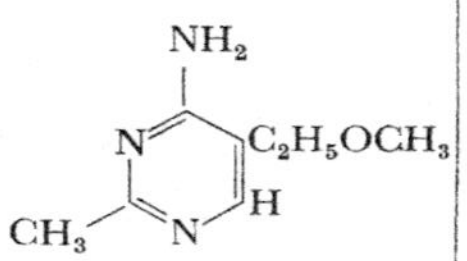	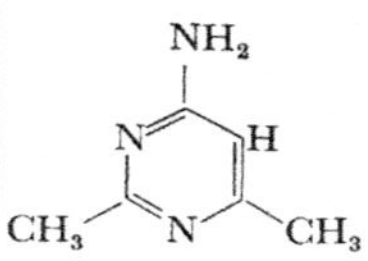
Principal long-wave absorption maximum in water at pH 7	2670	2740	2735	2700	2700	2720	2660
Principal absorption maximum of photoproduct in water	2400	2400	2850	2820	3020	3050	2950
Reversibility	50%	50%	nil	nil	partial	100%	nil
Principal absorption maximum of photoproduct in ethanol	3000	3100	3100	—	Only slow drop in absorption	Only slow drop in in absorption	2950
Reversibility	partial	partial	nil	—	—	—	nil
Remarks	At pH 9 in water there is slight increase in absorption to red of 2700 Å but no formation of maximum	—	Reaction optimal at pH 9 in H_2O. Photoproduct formed in alcohol, on transfer to aqueous medium, spontaneously transforms to photoproduct formed in water	Reaction optimal at pH 9 in water	—	Reversibility gives parent compound with identical physicochemical and biological properties. In acid medium reaction is irreversible	Photoproducts identical in water, ethanol or hexane. Heating of photoproduct in water results in rearrangement to a similar compound (see pp. 213, 219)

from the available evidence, appears to be a stable Dewar-like isomer of the parent compound. This bears only indirectly on the above proposed mechanism for the transformation of I → II, since the postulated Dewar-like intermediates probably exhibit only a transitory existence. A detailed investigation of the excited state of I would be required to provide evidence for or against it.

It remains to add that the photochemical transformation of I is essentially unaltered by protonation of one of the ring nitrogens, so that excitation of one of the paired electrons on the nitrogens is not a prerequisite for the reaction. It is known that an electronic transition of *n*-electrons in diazines, which are generally of low energy, increases the probability of bipartition between the lowest excited singlet and triplet states of the molecule (Kasha, 1961). Hence the lack of any influence of protonation on the course of the reaction, plus the lack of an oxygen effect, imply that we are dealing with an excited singlet state of the basic aromatic system, which may involve a predissociation transition from the excited singlet state potential surface of I to the mutually intersecting potential surface of II.

Attention was drawn above to the similarity in photochemical behavior of various 4-aminopyrimidines. It is, on the other hand, also necessary to emphasize some of the differences, such as the effect of the irradiation medium (see Table 6-9). For a closely related analogue of I, viz. 2-methyl-5-ethoxymethyl-4-aminopyrimidine, the photochemical behavior of the monoprotonated form (Uber and Verbrugge, 1941) is appreciably different from that for the neutral form (Wierzchowski and Shugar, 1961). Bearing in mind the discussion, above, on the nature of the excited state involved in the photochemical transformation of I, it is perhaps not unreasonable to expect that a comparison of the absorption spectra of various 4-aminopyrimidines in different solvent media with their photochemical behavior in the same media may provide supplementary data for the assignment of electronic transitions in these molecules as well as the nature of their excited states.

An additional effect due to the irradiation medium has been briefly reported by Moore (1962). Following up an earlier observation on the influence of the nature of the buffer component on the photochemistry of 2-methoxycytosine in aqueous medium (Shugar and Wierzchowski, 1958), it was found that the course of the reaction is dependant on the anionic components present in the irradiation medium. Three main reaction patterns were distinguished: (a) in water or tris buffer; (b) in phosphate or arsenate; (c) in pyrophosphate, cacodylate or citrate. The photoproduct formed in phosphate buffer was isolated, and did not contain phosphate, but its structure remains to be elucidated.

The differences in behavior between various aminopyrimidines are further illustrated by the fact that, whereas the photoproduct of 2-methyl-5-ethoxymethyl-4-aminopyrimidine apparently is capable of reverting to the parent substance, the photochemical transformation of I → II does not exhibit reversibility and, indeed, would hardly be expected to do so under normal conditions. It will consequently be of interest to determine the specific behavior of some of the 5-substituted cytosine analogues.

12. *Photodegradation Products of Pyrimidine Derivatives*

For many pyrimidine derivatives, including nucleosides and nucleotides, irradiation to the point where the absorption spectrum just disappears leads to the formation of a photoproduct from which the original substance may be regenerated in almost quantitative yield. For some derivatives, such as uracil, cytosine, 1-methylcytosine, the formation of a reversible photoproduct is accompanied by the appearance of degradation products; it is, of course, possible that the former are intermediate in such reactions and this could be most readily examined in specific cases by isolation of the reversible photoproducts and studying their subsequent behavior (see below).

On the other hand, if irradiation is sufficiently prolonged, a number of degradation products always result. For example, intense irradiation of 1,3-dimethyluracil results in the formation of a series of colored degradation products which may be separated on an Al_2O_3 column; one of these is found also in irradiated uracil and uridylic acid (Moore and Thomson, 1956). Extensive irradiation of uridine is reported to lead to the appearance of several degradation products, including uracil (Rapport *et al.*, 1955), although this latter finding probably requires confirmation.

The ultimate degradation products of irradiated pyrimidine derivatives, like those of the purines, include appreciable amounts of urea and ammonia. Of only historical interest in this connection is the early work of Bass (1924) who found that irradiation of thymine in the presence of ferrous salts and oxygen led to ring rupture and the formation of urea and pyruvic acid; what we are dealing with here is the action of free radicals resulting from the action of light on the ferrous ions.

Quantitative measurements of the ammonia and urea liberated from pyrimidine derivatives, following complete destruction of the absorption spectra, have been made by Canzanelli *et al.* (1951) and are presented in Table 6-10. As in the case of purines, it was concluded that most of the urea and ammonia originate from the pyrimidine ureide group and this is supported by a comparison of the results for uracil and cytosine. A rather striking observation is the influence of a ribosidic linkage on the formation of ammonia from cytosine.

Conrad (1954) has isolated several of the decomposition products of uracil by irradiating 10^{-3} M solutions with a GE 15 W germicidal lamp, followed by evaporation of the solution to dryness under vacuum. From C and N determinations of the photoproducts thus obtained it was deduced that 25

TABLE 6-10. AMMONIA AND UREA PRODUCTION FROM PYRIMIDINE DERIVATIVES IRRADIATED AT A CONCENTRATION OF 5×10^{-5} M WITH AN UNFILTERED GERMICIDAL LAMP TO POINT OF COMPLETE DESTRUCTION OF SELECTIVE ABSORPTION (CANZANELLI *et al.*, 1951)

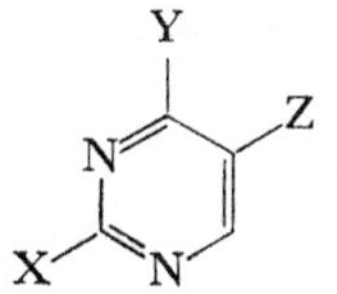	Compound	NH_3 (Moles/mole compound irradiated)	Urea (Moles/mole compound irradiated)
X = Y = OH, Z = H	Uracil	0.14 ± 0.03	0.35 ± 0.08
	Uridine	0.12 ± 0.02	0.41 ± 0.05
X = OH, Y = NH_2, Z = H	Cytosine	0.17 ± 0.03	0.43 ± 0.05
	Cytidine	0.65 ± 0.03	0.30 ± 0.01
	Cytidine-2′(3′)-phosphate	0.59 ± 0.04	0.18 ± 0.04
X = Y = OH, Z = CH_3	Thymine	0.11 ± 0.01	0.30 ± 0.01

per cent of the total N and 75 per cent of the total C were lost in the form of volatile products either during irradiation or during evaporation. From the remaining material 6 products were isolated in relatively pure form, of which two were crystallized and identified as oxamide and parabanic acid:

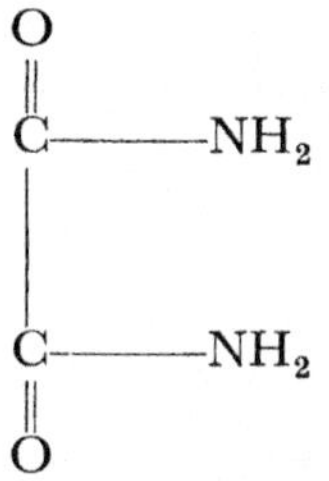

Oxamide

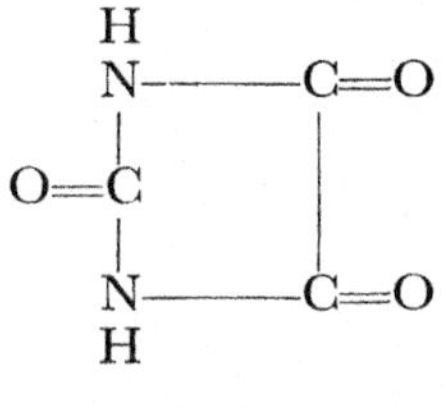

Parabanic acid

At first sight it is difficult to see how either of these products could be formed directly from irradiated uracil as a result of simple ring rupture, and one is tempted to invoke secondary reactions of the excited molecules or free radicals formed in the initial stages of the reaction. Considerable light is, however, thrown on the mode of formation of the above products by the studies of Wang (1958) on the irradiation of 1,3-dimethyluracil (I) which we shall first summarize. The initial step, (see above) as first elucidated by Moore and Thomson (1955), is the formation of 6-hydroxy-1,3-dimethyl-hydrouracil (II) by a 1,4 addition of a water molecule:

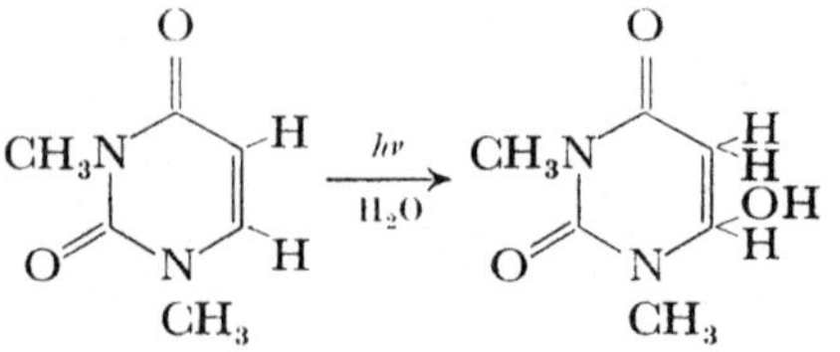

Both Moore and Thomson (1955) and Wang (1958) synthesized 5-hydroxy-1,3-dimethyl-hydrouracil and eliminated it as a possible intermediate. However, further irradiation of 6-hydroxy-1,3-dimethyl-hydrouracil results in the formation of *N,N'*-dimethylmalonamide (IV), a product which was also formed by irradiation of 1,3-dimethyl barbituric acid (III). The reaction sequence is therefore as follows:

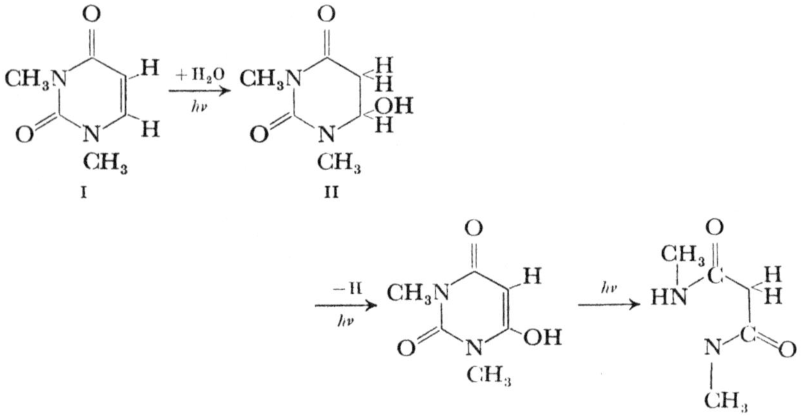

I II III IV

We have written III here in its enol form, because dimethylbarbituric acid does not absorb above 2400 Å in its keto form (Fox and Shugar, 1952b). Moreover, the extinction coefficient of the enol form of III is quite high. Hence if the reaction is conducted in neutral medium, step III-IV must be a rapid one since the absorption of the solution, as observed, is low.

Since 1,3-dimethylbarbituric acid is an intermediate in the above reaction, it becomes clear how parabanic acid may arise during the photolysis of uracil, the reaction probably being as follows:

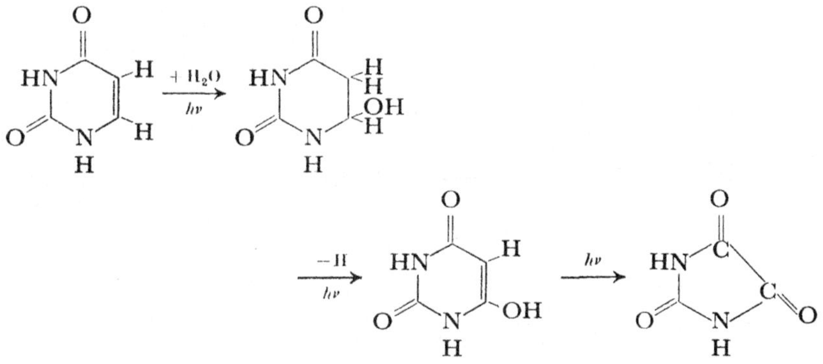

It is therefore likely that oxamide is a product of degradation, as follows:

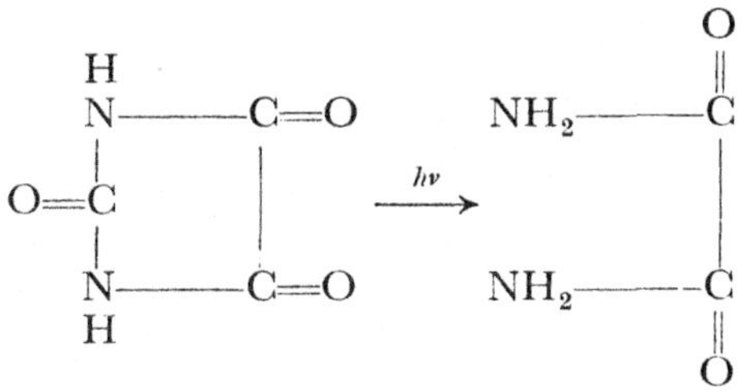

Coming back now to the question of urea production during photolysis, it is conceivable that this product could arise from the simultaneous rupture of two —C—N— bonds in the pyrimidine ring (or in the imidazole portion of the purine ring in the case of uric acid) as a result of dissociation or predissociation of the excited molecule. This is energetically possible since the —C—N— bond energy is about 55 kcal/mole as compared to 110 kcal for 2537 Å radiation; where shorter wavelengths are not filtered out, the probability of such a primary reaction may be even greater. Bearing in mind the symmetry of the molecule and the influence of the symmetry of a given excited state on its dissociation into several different fragments, one can equally well envisage other mechanisms. For those compounds which give rise to reversible photoproducts, it is also conceivable that urea arises from parabanic acid, dimethylmalonamide, or other secondary products formed. The answer to this problem can be obtained only by a study of the kinetics of formation of the various photoproducts or by the use of suitably labelled derivatives, or both.

A rather interesting observation has been reported by Wang (1959), who found that one of the final degradation products of 1,3-dimethylthymine was *N,N′*-dimethylmethylmalonamide. By analogy with the situation for dimethyluracil outlined above, the most logical explanation for the appearance of such a product is the initial uptake of a water molecule at the 5,6 double bond. The inability to observe this intermediate is due to the greater ease with which it undergoes dehydration as compared to the corresponding uracil derivatives. It might of course be argued that the unobserved intermediate is the dimer of thymine, but this is most unlikely because of the high quantum yield, 0.5 (Setlow, 1961; Smietanowska and Shugar, 1961), with which the dimer is dissociated to thymine.

CHAPTER VII

Nucleic Acids and Oligo- and Polynucleotides

A. GENERAL

As pointed out in the previous chapter, a distinction must be made between studies on the photochemical degradation of polynucleotide chains, which are of interest primarily to the physical chemist from the point of view of the structure of nucleic acids, and those investigations which attempt to correlate the effects of radiation with the biological properties of macromolecules and living organisms. Practically all the earlier studies on the photochemical behavior of nucleic acids were based largely upon observations of chemical and physico-chemical modifications resulting from the use of radiation doses now known to be from one to four orders of magnitude greater than those required to induce important biological effects. These studies were considerably hampered by the lack of well-defined model compounds, such as di- and tri-nucleotides, an examination of which might be expected to facilitate the interpretation of the behavior of longer chains. This situation has now been radically improved as a result of the development of chemical and enzymatic procedures (a) for the isolation of well-defined oligonucleotide fragments from natural nucleic acids and (b) for the synthesis, by both chemical and enzymatic techniques, of a wide variety of oligo- and polynucleotides ranging from dinucleotides to chains with molecular weights of 10^6 or more and with base compositions and sequences which can be varied to a considerable extent. When to this we add the remarkable advances currently being made on the biological properties of nucleic acids, it appears only logical to expect that increasing attention will be devoted to the search for some correlation between photochemical effects *in vitro* and *in vivo*.

Notwithstanding the complexities involved in the extrapolation of *in vitro* findings to the photochemical behavior of the living cell, some success in this direction has been attained during the past two or three years in connection with the phenomenon of photoreactivation (Dulbecco, 1955; Jagger, 1958), the physico-chemical behavior of nucleic acids and model oligo- and polynucleotides (Shugar and Wierzchowski, 1958a, 1960; Wierzchowski and Shugar, 1959, 1960a) and the biological behavior of transforming DNA (Rupert, Goodgal and Herriot, 1958). Current progress in these and related fields is such that one is justified in expressing a considerable degree of optimism with regard to the ability to interpret complex *in vivo* photochemical effects in terms of chemically induced modifications of nucleic acids and proteins.

Particular attention will be devoted in this chapter to the possibilities of reversal of the effects of ultraviolet damage to nucleic acids (a) since this provides an additional technique for pinpointing the sites affected, and (b) because of the significance of this phenomenon in the interpretation of genetical effects. For more extensive details of some aspects of nucleic acid photochemistry, reference should be made to the reviews of Loofbourow (1948), Giese (1950, 1953), Errera (1952c, 1953), Butler (1955), Kleczkowski (1957), McLaren (1957), Shugar and Wierzchowski (1958a), Setlow (1958), Jagger (1958), Shugar (1960a and 1960b), Smith (1963), and Wacker (1963).

B. PHOTOCHEMICAL DEGRADATION OF NUCLEIC ACIDS

1. *Deoxyribonucleic Acids*

It was first reported by Greenstein and Jenrette (1941) and Hollaender *et al.* (1941) that irradiation of DNA led to a decrease in viscosity and streaming birefringence, which was interpreted as being due to depolymerization of the molecule. A decrease in viscosity alone, however, is an insufficient criterion of depolymerization of DNA, since the same effect results from mild acidification or from heating at neutral pH. The latter treatments result in a reduction in viscosity without any appreciable change in molecular weight. Although the source used was a resonance lamp, no filter was employed and irradiation times were from 22 to 83 hours so that the doses used must have been enormous.

The effect of irradiation at neutral pH on the viscosity and ultraviolet absorption, at different wavelengths, of DNA was re-investigated by Errera (1952a, 1952c), the results were essentially the same with both an unfiltered resonance lamp and a high-pressure mercury arc (Fig. 7-1). It will be seen from the figure that the viscosity drops almost to zero before any changes in absorption spectrum are apparent (cf. Hollaender *et al.*, 1941). It should, however, be recalled that approximately 50 per cent of the hyperchromicity of DNA is due to hydrogen bonding between the twin strands, so that if one of the bases in a given base pair is affected by the radiation, the resultant *expected* decrease in absorption would be partially compensated for by the elimination of hyperchromicity of the base pair; the resultant decrease in absorption would consequently be masked and this has actually been shown to be so by examining the behavior of poly-A-poly-U.

The quantum yield for loss of viscosity, based on a calibration curve of viscosity vs. DNA concentration, was found to be about 10^{-6}; from the data of Hollaender *et al.* (1941), analogous calculations gave a Φ of about 10^{-5}. These values are considerably lower than those for the individual nucleotides as well as for loss of biological activity, in agreement with the findings that much higher doses are required to affect the viscosity of either RNA or DNA than their biological properties.

The influence of irradiation on the viscosity and molecular weight (measured by light scattering) of DNA has been investigated in greater

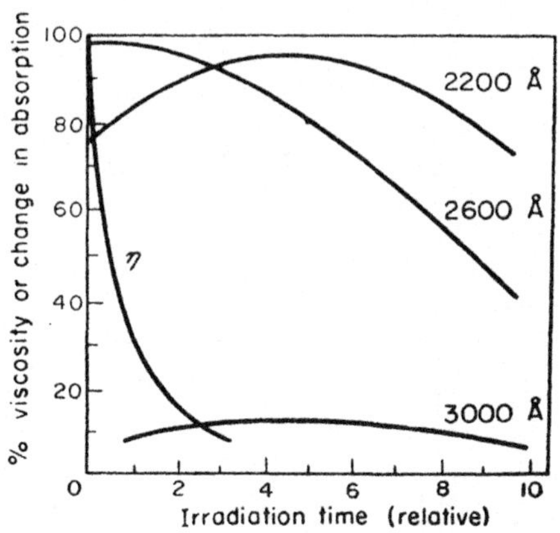

FIG. 7-1. Effect of ultraviolet irradiation of DNA (concentrations up to 0.6 mg/ml) on the absorption at various wavelengths and on relative viscosity (Errera, 1952a).

detail by Alexander and Moroson (1960, 1961). Under nitrogen the initial effect involves a decrease in viscosity with no change in molecular weight; as the dose is increased the molecular weight begins to fall, but lags considerably behind the decrease in viscosity (Fig. 7-2). This is most likely due to an increase in flexibility of the molecule as a result of the rupture of twin strand hydrogen bonds. Under oxygen the molecular weight decreases rapidly, but less so than would be expected from the accompanying decrease in viscosity, from which it is concluded that main-chain scission occurs along with disruption of the twin stranded structure. It is suggested the main-chain scission under oxygen is due to direct interaction between oxygen molecules and singlet excited states to form peroxides which are further decomposed by irradiation; and this is considered to be supported by the fact that dimethylaniline, glutathione, glucose, and catalase are protective against the oxygen effect. However, catalase also protected under nitrogen, whereas the other substances did not. Furthermore, the doses used were very large (see abscissae of Fig. 7-2) compared to those necessary to produce important biological effects. Finally, as the authors themselves point out, biological effects of ultraviolet light are not influenced by oxygen;

and they therefore suggest that cell constituents act as protective agents. While this may be so, it is nonetheless a fact that biological inactivation of transforming DNA, TMV, and bacteriophages (all of which are extracellular) likewise exhibit no oxygen effect. It would undoubtedly have been extremely useful if the authors had conducted simultaneous measurements under an atmosphere of air. Furthermore, an entirely independent investigation of the effects of irradiation on pneumococcal transforming DNA indicated

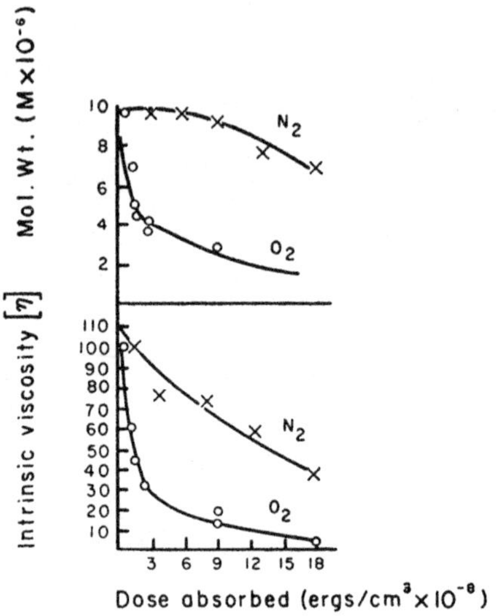

Fig. 7-2. Effect of irradiation at 253.7 mμ on change in weight-average molecular weight (upper diagram) and intrinsic viscosity (lower diagram) of DNA (100 μg/ml in 0.2 M NaCl with citrate buffer) under oxygen or nitrogen (Alexander and Moroson, 1960).

the complete absence of any oxygen effect on the viscosity (Marmur *et al.*, 1961).

It remains to add that Scholes *et al.* (1956) had previously reported the formation of organic peroxides from nucleic acids and their derivatives when irradiated with ultraviolet light in an oxygen atmosphere, but no experimental details were given. In the work of Alexander and Moroson (1960) cited above no direct evidence is given for formation of such peroxides and their suggestion that they are further decomposed by irradiation at least requires independent substantiation for the following reasons: the purines do not form peroxides, while those formed by the pyrimidines are, in the case of thymine, the *cis* and *trans* isomers of 5-hydroperoxy-6-hydroxythymine (Scholes and Weiss, 1959; Ekert and Monier, 1959); i.e. the 5,6 double bond is saturated as in dihydrothymine and these peroxides should consequently exhibit little or no absorption at 2537 Å. They should therefore be relatively resistant to radiation. An examination has actually been made of the possibility of hydroperoxide formation from various nucleic acid derivatives; at wavelengths longer than 2300 Å, formation of peroxides was negligible (Barszcz and Shugar, 1961).

Under the influence of the high radiation doses required to destroy the viscosity of DNA, one may frequently encounter secondary dark reactions which may involve long-lived intermediates, as can be readily observed following removal of the radiation source (Fig. 7-3). Such an effect is also

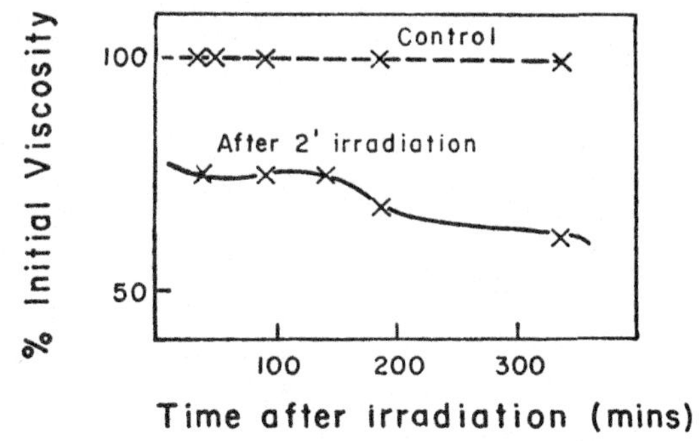

FIG. 7-3. Decrease in viscosity of 0.05 per cent DNA solution at 40°C following 2 min irradiation in quartz tube under nitrogen (Butler, 1955).

occasionally noted with the lower doses required for biological inactivation, but the nature of these intermediate metastable states remains to be clarified.

Using loss of aborption at 2600 Å as a criterion, the sensitivity of DNA has been reported to be greater in acid than in neutral medium (Christensen and Giese, 1954), notwithstanding that quantum yields for cytosine nucleosides and nucleotides are appreciably lower at acid than at neutral pH (Wierzchowski and Shugar, 1957). This is, however, readily interpretable when it is recalled that the hyperchromicity of DNA is partially abolished by acidification; hence, photolysis of the bases will show up more readily by spectral methods.

With prolonged irradiation to the point where only 70 per cent of the pyrimidines could be recovered from a hydrolyzate of irradiated DNA, the recoveries of adenine and guanine were, respectively, over 90 and 100 per cent (Errera, 1952b). This is to be expected in the light of the lower quantum yields for purines as compared to pyrimidines (Section VI-2). In view of the differences in quantum yields between thymidylic and cytidylic acids, it would have been of interest to determine the proportions of thymine and cytosine in the irradiated acid in addition to total pyrimidines. Under the influence of high doses of ultraviolet radiation, a small percentage of dialysable products makes its appearance, while about 10–15 per cent of the irradiated DNA can no longer be precipitated by acetone. The electrophoretic mobility of the remaining 85–90 per cent of the DNA is, however, unchanged. Under these conditions no inorganic phosphate could be detected.

Somewhat different results were obtained by Seraydarian *et al.* (1953), who irradiated 10 mm thick layers of both RNA and DNA in shallow pans at a concentration of 100 μg/ml. Following irradiation, the solutions were concentrated 20-fold and titrated. After 45 min irradiation there was a 12 per cent decrease in absorption and no noticeable effect on the alkali uptake or on formation of inorganic phosphate. Following a 15–30 per cent decrease in absorption, increased alkali uptake demonstrated rupture of internucleotide linkages, while after 18 hr irradiation about one-third of the total P had been liberated as inorganic P. No attempt seems to have been made in these studies to isolate some of the products of degradation following the liberation of appreciable quantities of inorganic P. But it may be recalled that irradiation of either DPN or TPN, both of which contain internucleotide linkages, leads to formation of inorganic phosphate as well as of nucleosides and nucleotides of adenine.

Analogous extreme conditions of irradiation were employed by deCourcy *et al.* (1953) to investigate the extent to which DNA may be degraded under conditions resembling those *in vivo*. Homogenates of rat thymocytes were irradiated and the DNA was then extracted. The yield of polymerized DNA from the irradiated thymocytes was only 50 per cent of that obtained from a non-irradiated control homogenate, but the properties of both preparations were practically identical as regards relative and structural viscosities and sedimentation behavior. Evidence for fragmentation of DNA in the ultraviolet treated thymocytes was based on the qualitative determination of large amounts of thymine recovered from the supernatant fluids during the extraction procedure. These findings suggest a rather high sensitivity of DNA to irradiation *in vivo* as compared to *in vitro*, a fact which, as will be shown below, derives support from other observations.

As against the above, Loofbourow *et al.* (1947) and Loofbourow and Joyce (1941) claim, on the basis of microspectrophotometric observations of irradiated yeast cells, that those cells which have undergone injury exhibit an increased production of nucleic acid-like material (Loofbourow, 1948). Conclusions based purely on spectral observations of fixed cells are, however, somewhat tenuous (Beaven and Holiday, 1952). On the other hand, Loofbourow *et al.* (1941) demonstrated that an apparently higher yield of crude RNA could be extracted directly from ultraviolet irradiated yeast cells which had been subjected to a dose sufficient to kill only 50 per cent of the organisms. As compared to non-irradiated controls, the increase in yield amounted to as much as 100 per cent. It should be noted, however, that the extraction procedure was not a very efficient one, the maximum yields of crude RNA from the irradiated organisms being only about one per cent. It is consequently conceivable that the increase yield from irradiated cells was due to the greater ease of extraction of smaller oligonucleotides which could have resulted from RNA depolymerization as a result of irradiation. Such an explanation derives some support from the observation by the same authors that the extinction of the RNA from irradiated yeast was somewhat higher than that from non-irradiated organisms, as would be expected if the former were partially depolymerized. In a recent investigation Suzuki and Ono (1960) have actually shown that, following irradiation of *E. coli*, RNA synthesis is radically modified and includes at least two components, one of which is "unstable."

Devreux *et al.* (1951) demonstrated that the staining affinity of DNA for methyl green is decreased as a result of exposure to ultraviolet light. Irradiation of a 0.3 per cent solution to the point where its viscosity is diminished by 75 per cent led to a decrease in staining affinity of 30 per cent. While this does not necessarily indicate a degradation of polynucleotide chains, it does suggest a separation of the twin strands of the two-stranded Watson–Crick structure (Rosenkranz and Bendich, 1958); the latter authors have shown that the binding of methyl green by DNA is independent of chain length if the double helical structure is intact. Photolysis in acid medium led to a more rapid decrease in staining affinity (cf. Christensen and Giese, 1954, above).

A similar loss in staining affinity for methyl green is reported for cell nuclei as a result of irradiation, the decrease being approximately logarithmic with respect to dose (Errera, 1952c); the Feulgen reaction was apparently unaltered under these conditions. A rough attempt was made to compare the amount of energy required to give comparable reductions in staining ability *in vivo* and *in vitro*, from which the rather surprising result emerged that about 50 times less energy was required *in situ*. Although the calculations of the amount of light energy absorbed by the cell nuclei cannot be considered as very accurate, it is most unlikely that the error involved is sufficient to account for such a large difference.

2. *Nucleic Acid Films*

The foregoing result may conceivably be linked to the state of DNA in living cells as compared to that in solution. Of undoubted interest in this connection are some experiments of Setlow and Doyle (1953, 1954) on the irradiation of dry films of DNA, as a result of which the irradiated nucleic acid forms a gel insoluble in aqueous or salt solutions. The efficiency of gel formation was found to be practically independent of wavelength over the spectral range 1850–3000 Å, with a quantum yield of the order of 10^{-2}, which is 3–4 orders of magnitude greater than that for loss in viscosity (see above) and about the order of magnitude normally prevailing for ultraviolet inactivation of enzyme molecules. The technique for measuring the absorption spectra of dried films on cover glasses, as well as estimation of the possible error in optical density due to non-uniformity of the films, has been described by Appleyard (1952). It should be noted, in this connection, that the absorption spectrum of a dried film of DNA does resemble that of DNA in solution; the same applies to cytosine, but not to adenine (Preiss and Setlow, 1956).

In the experiments of Setlow and Doyle, cited above, the DNA films were irradiated under dry nitrogen or in a vacuum. The phenomenon has been further examined by Kaplan (1955b) in connection with his observation on the dependence of the ultraviolet radiation sensitivity of micro-organisms on humidity (Kaplan, 1955a). The rate of gelation of films of DNA was found to be markedly dependent on their moisture content, the rate increasing by a factor of 22 as the humidity is decreased from 97.5 to 33 per cent. A quantitatively similar relationship was found to prevail for the S-mutation and killing rates of *Serratis marcescens*, from which it was postulated that a common mechanism exists for all three processes, and involving hydrogen bond rupture in the DNA molecule. This is, however, rather an over-simplification and provides no clue as to the nature of the reaction involved. The entire subject was subsequently re-examined by Shugar and Baranowksa (1960) and Baranowksa and Shugar (1960), whose initial objective was to investigate the staining behavior of irradiated DNA films with the aid of new staining techniques (Shugar and Baranowksa, 1957; Shugar, 1958) with labeled dyes (Bitny-Szlachto and Shugar, 1958, 1959). It was found that the affinity of such films for methyl green increased considerably as a result of irradiation (Fig. 7-5); similar results were obtained with highly polymerized DNA from three different sources, calf thymus, bacteria, and bacteriophage. Furthermore, exposure of irradiated films to water resulted in the formation of insoluble threads which were also completely resistant to the action of deoxyribonuclease. It was found that the degree of initial polymerization of the DNA was of no importance; e.g. apurinic acid (APA) is obtained by mild acid hydrolysis of DNA which results in the complete removal of purines as well as chain scission; the resulting apurinic acid has a molecular weight of about 15,000 or less than one per cent of that of the parent DNA from which it was derived. Nonetheless, irradiated APA films, which initially show practically no affinity for methyl green, take up as much methyl green as irradiated DNA (Fig. 7-4), and form fibers equal in magnitude to those obtained from DNA (Fig. 7-5). The quantum yield for fiber formation was about the same as the value of Setlow and Doyle (1954) for the gelling reaction (see above). Since APA contains no purines, experiments were performed with model oligo- and polynucleotides. Poly-A and poly-G were found to be inert, but irradiated poly-U exhibited increased dye uptake (Fig. 7-4) and fiber formation; poly-C did so as well but to a small extent, as did also RNA (Fig. 7-4). The results with model oligonucleotides, implicating specifically pyrimidine rings, led to the formation of the following reaction mechanism, involving the formation of cross-links between the 5,6

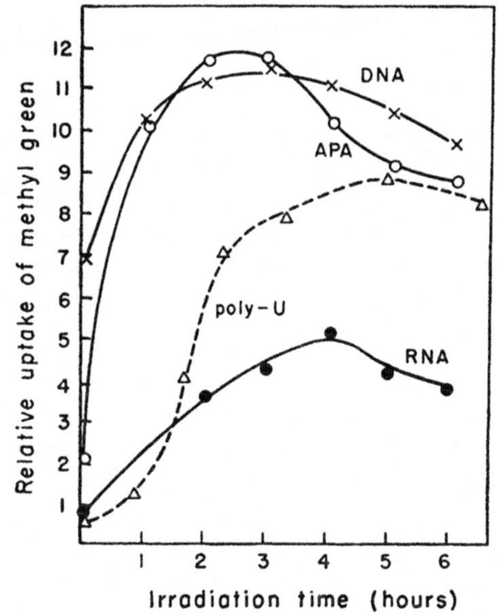

FIG. 7-4. Effect of ultraviolet irradiation (253.7 mμ) on affinity for methyl green of dried films of DNA, RNA, APA (apurinic acid) and poly-U; dose rate 100 ergs/mm^2 per sec (Baranowksa and Shugar, 1960).

double bonds of pyrimidine rings in adjacent chains. Since this involves

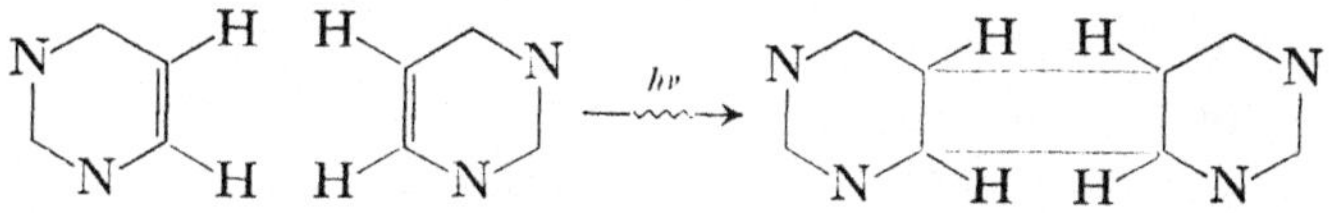

saturation of the 5,6 double bonds, one would expect the principal absorption maximum of poly-U, for example, to decrease as for dihydrouracil; such was indeed found to be the case. Extensive cross-linking between chains would also account for the insolubility of the photoproducts. Furthermore considerable experimental evidence has now been accumulated which shows that a similar reaction takes place between the adjacent pyrimidine residues in either single or twin-stranded polynucleotide chains in aqueous medium, as well as between the residues of the twin strands in DNA so that the strands are, in effect, cross-linked. The biological importance of these reactions has now been well established (Marmur and Grossman, 1961; Wulff and Rupert, 1962; Bollum and Setlow, 1963).

A somewhat analogous mechanism has been postulated for the ultraviolet induced cross-linking of linear polyesters (Voigt, 1958, see also Schonberg, 1958). The fact that polymers such as nucleic acids may, under the influence of irradiation, undergo cross-linking, is of considerable theoretical interest inasmuch as it has been shown (Alexander and Stacey, 1959) that ionizing radiations may induce cross-linking of DNA in fish sperm heads, i.e. under conditions resembling those *in vivo*, and where the DNA is combined with protein. Indirect evidence, based on staining reactions, indicated that ultraviolet induced cross-linking may also take place *in situ*, although with a rather low efficiency (Baranowksa and Shugar, 1960).

The action of ultraviolet light on films of DNA and proteins is deserving of wider study and it should be emphasized that the application of quantitative staining techniques would undoubtedly considerably facilitate such investigations (Baranowksa and Shugar, 1960; Shugar, 1961).†

3. *Ribonucleic Acids*

Although some reference has already been made to the photochemistry of RNA in the preceding two sections, it must be admitted that considerably less attention has been devoted to the photochemical degradation of RNA. This is due in part to the greater interest attaching to the genetic functions of DNA and to some extent because of the difficulties of obtaining highly polymerized and uniform preparations of RNA.

Oster and McLaren (1950) studied the degradation of RNA from tobacco mosaic virus (TMV), obtained from the virus by heat denaturation of the latter. The RNA so obtained has a molecular weight in the neighborhood of 15,000 and an intrinsic viscosity of about 13, which corresponds to an axial ratio of about 10 for a rod. Prolonged irradiation of a 0.3 per cent solution leads to a drop in intrinsic viscosity to practically zero, suggesting breakdown to smaller molecules. The optical density of the ultraviolet absorption maximum simultaneously increased by about 10 per cent; a similar increase in optical density resulted from irradiation of a more dilute solution of the same RNA preparation. This increase in absorption could conceivably be due to disruption of the secondary structure of the molecule, followed by a trend toward random coil formation, which would also explain the decrease in intrinsic viscosity in place of the breakdown to smaller fragments suggested by the authors. However, the foregoing is about the only reported instance of an increase in ultraviolet absorption of a nucleic acid preparation as a result of irradiation, and its validity has since been questioned by one of the authors as a result of subsequent investigations (McLaren and Takahashi, 1957) on infectious RNA from TMV. Prolonged

† A general review on the subject of photochemically-(including photosensitized) induced polymerization reactions has appeared (Delzenne, 1959).

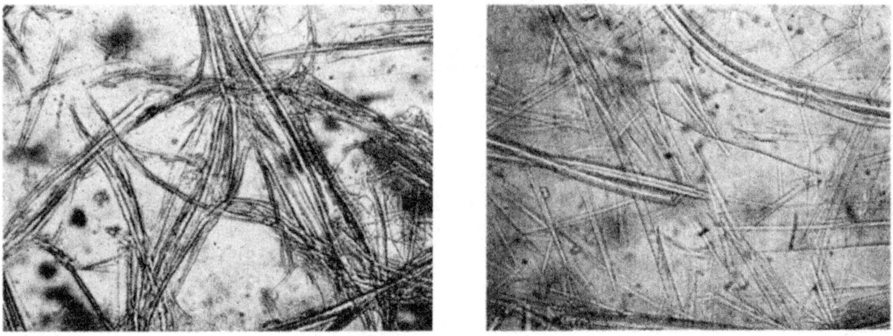

(a) (b)

Fig. 7-5. Appearance under the microscope of fibers formed by irradiation of dried films of (a) DNA and (b) APA. Following irradiation, a drop of water was placed on each film prior to photographing (110x) (Baranowksa and Shugar, 1960).

irradiation of dilute solutions of RNA was, however, found to lead to a loss in ability to stain metrachromatically with safranin, an observation which suggests that fragmentation most likely did occur.

C. NUCLEOPROTEINS

In view of the fact that nucleic acids exist in living cells in combination with proteins, one might have expected some attention to be devoted to the effect of irradiation of isolated nucleoprotein preparations. Efforts in this direction have, however, been limited with both ultraviolet and ionizing radiations.

Some early observations of Errera (1952a) on deoxyribonucleohistone in salt solutions of different concentrations suggested little difference in behavior from that for free DNA. Giese (1947) examined the behavior of nucleoprotein from sea-urchin sperm, which forms relatively long fibres when transferred from 2 M saline solution to distilled water. Under the influence of irradiation the length of the threads formed upon transference to water gradually decreased until finally only a turbid solution was formed. Since, however, the nucleoprotein was irradiated in strong salt solution, where it is largely dissociated, it is difficult to decide whether the major effect was on the free DND or the nucleoprotein, or even to some extent on the free histone; unpublished experiments (Kochanska and Shugar, 1956) have shown that ultraviolet irradiation of either nucleic acids or proteins has an appreciable influence on the nature of the artificial complexes subsequently formed by mixing the two aqueous solutions. On the other hand, if only biologically effective doses are used, a nucleoprotein such as TMV may be partially reformed from its protein component and its irradiated RNA component (Chapter VIII). The measurements of McLaren and Takahashi (1958) on TMV and infectious RNA from TMV appear to indicate that the nucleic acid moiety is somewhat more stable to irradiation when combined with protein. That this result cannot be generalized is shown by the fact that infectious RNA from polio virus is less radiation-sensitive than when incorporated in the virus (Norman. 1960).

As has been described above, DNA is, at least in some instances, apparently more readily degraded *in situ* than *in vitro*. Further evidence has been provided by Beiser and Ellison (1958) and Ellison and Beiser (1960) from studies on the radiation sensitivity of transforming DNA in intact cells as compared to that of the active DNA isolated from the same organisms. While the comparison may not be fully quantitative, the results lead to the conclusion that the biological activity of the intracellular DNA may be considerably more susceptible to irradiation than the isolated DNA. It is, of course, difficult to conclude from this experiment whether the difference in sensitivity is due solely to the existence of the DNA in the cells as nucleoprotein, although the available evidence suggests that more than nucleic acid-protein linkages are involved. It is even conceivable that the inactivation mechanism is somewhat different in the two cases, e.g. some cross-linking *in vivo*. Clearly it would be desirable to compare the radiation sensitivity of transforming DNA within the cells not only with that extracted from the cells, but also with nucleoprotein from the same cells, due precautions being taken to avoid dissociation of the DNA–protein complex during the extraction and irradiation procedures. It would obviously be useful to know more about the photochemical behavior of the nucleic acid moieties of cellular nucleoproteins. But this will be no simple task in the case of the nucleoprotamines and nucleohistones because of the ease with which they dissociate following their extraction from the cell. Some of the more recently isolated and stable nucleoproteins may be more suitable models for such studies. On the other hand, viruses and bacteriophage provide admirable models for investigation since they are, in effect, nucleoproteins possessing a variety of biological functions susceptible to quantitative measurements.

D. BIOLOGICAL INACTIVATION

From the point of view of gaining a better understanding of the effects of radiation on living organisms, there is probably no more fruitful line of investigation than that involving an examination of the effects of ultraviolet light on the biological activity of isolated nucleic acids along with the accompanying modifications in chemical and physicochemical properties. Two of the biological properties of nucleic acids have been widely utilized for photochemical studies: the transforming activity of bacterial DNA (T-DNA); and the infectivity of isolated viral RNA. Preliminary trials have been made with the yeast RNA core (the dialysis residue remaining following exhaustive enzymatic hydrolysis of RNA with ribonuclease), which is active in promoting streptolysin-S formation in hemolytic streptococci. Several additional biological activities of DNA have been reported (e.g. Blix *et al.*, 1954; Phillips *et al.*, 1957; Zahn *et al.*, 1958), but some of these require confirmation; in addition they are not readily susceptible to quantitative measurement.

1. *Infectious RNA*

The development of techniques for isolation in biologically active form of RNA from plant and animal viruses has paved the way for a renewed attack on the photochemical behavior of ribonucleic acids. The value of such RNA samples is still further enhanced by the fact that they may be once more combined with the protein moieties to give a reconstituted virus exhibiting biological activity. A detailed discussion of infectious RNA's will be reserved for Chapter VIII, but it may be noted here that the inactivation of RNA from several strains of TMV has been shown to follow a first order reaction (Siegel *et al.*, 1956; McLaren and Takahashi, 1957; Siegel and Norman, 1958).

An apparent dependence of quantum yield on concentration has been described, as has also been reported for transforming DNA. It is of some interest that the quantum yield for inactivation of TMV-RNA was found to be independent of wavelength over the range 2300–2800 Å, a result of considerable value for action spectra investigations. The degree of photoreactivation was also found to be independent of the wavelength used for inactivation in the same range. Bearing in mind the fact that the quantum yields for adenylic and guanylic acids are much lower than those for uridylic and cytidylic acids, McLaren and Takahashi (1957) interpret the 0.7 per cent decrease in absorption of inactivated TMV-RNA as due to the reaction of uridylic and cytidylic acid residues. Particularly instructive is the observation that total inactivation of TMV-RNA is not accompanied by any decrease in viscosity or intrinsic viscosity; furthermore, the radiation dose required for total inactivation of RNA is about one-hundredth that required to provoke depolymerization of the molecule. It consequently follows that damage due to biological inactivation must be highly localized in character, a conclusion we shall repeatedly return to in subsequent sections. Further support for such a conception is provided by Fraenkel-Conrat and Singer (1959) and Fraenkel-Conrat *et al.* (1959), who demonstrated that TMV-RNA which had undergone more than 99 per cent ultraviolet inactivation was still capable of combining to an appreciable extent with TMV-protein to give a reconstituted virus with viral antigenic activity of high specificity; whereas even slight depolymerization induced by traces of ribonuclease abolished the ability of the viral RNA to recombine with the protein moiety.

We shall see subsequently that photochemical studies on model oligonucleotides enable us to form a clearer idea as to the real meaning of quantum yields in terms of the nature and type of base residues in a nucleic acid chain affected by radiation.

2. *Transforming Deoxyribonucleic Acids* (*T-DNA*)

These are purified deoxyribonucleic acids, isolated from a bacterial strain referred to as the "donor" and which, upon addition to "competent" cultures of another strain, (not too distantly related) impart to the latter, or recipient, strain one or more of the phenotypic properties of the donor strain. The property of transformation is undoubtedly the most useful, as well as the most promising, characteristic of nucleic acids for experimental investigation of the nature of the chemical and physico-chemical modification accompanying the biological effects of radiation.

By contrast with the situation prevailing for viruses, all observers are in agreement that ultraviolet inactivation of T-DNA, whatever its source, is of a complex, usually multi-component type. For example, the inactivation at 2537 Å of T-DNA for capsule *b* formation in *Haemophilus influenza* has been found to follow a "multi-hit" type of process (Zamenhof *et al.*, 1956), as illustrated by curve a in Fig. 7-6. But this is apparently not a general phenomenon. The form of survival curve usually encountered is that shown for inactivation of T-DNA for streptomycin resistance in type *d influenza*

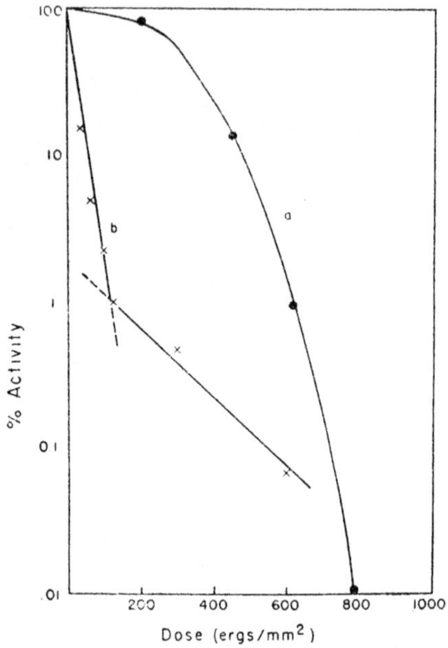

FIG. 7-6. Inactivation by ultraviolet light of *Haemophilus influenza* transforming DNA (a) for type *b* capsule formation (Zamenhof *et al.*, 1956) and (b) for streptomycin resistance in type *d* (Pakula *et al.*, 1960a).

cells (curve (b) of Fig. 7-6); i.e. with increasing dose the slope of the survival curve decreases. This is what one would expect if the irradiated solution is heterogenous as regards radiation sensitivity, each class of molecules undergoing exponential inactivation. At high doses, where the surviving molecules are the most resistant, the plot becomes a straight line. As shown in curve (b) of Fig. 7-6, and in subsequent figures below, the tendency has been to divide these types of survival curves into two or more straight-line portions, although this is not strictly justified. Quantum yields are also usually calculated for the initial and final slopes.

Figure 7-7 exhibits the course of ultraviolet inactivation of group H hemolytic streptococci T-DNA for homologous and heterologous transformation of streptomycin resistance. It will be observed that the differences in radiation sensitivity for the two types of transformations are appreciable;

they were much less so for pneumococcal T-DNA, where the heterologous receptor was the *sbe* type I/II strain of group H hemolytic streptococci (Pakula *et al.*, 1959).

Similar types of inactivation curves have been reported for a variety of genetic markers by a number of observers, e.g. Beiser and Ellison (1958), Lerman and Tolmach (1959), Marmur *et al.* (1961). Figure 7-8 illustrates

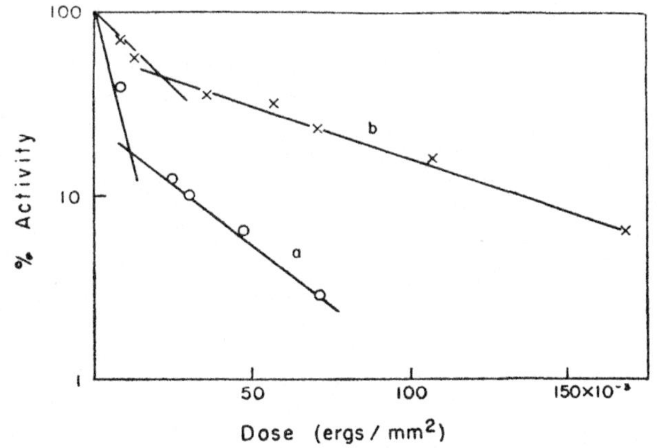

FIG. 7-7. Inactivation by ultraviolet light of transforming DNA from hemolytic streptococci group H, strain Challis: (a) for intraspecific transformation (of streptomycin resistance) and (b) for interspecific transformation, using receptor strain *sbe* type I/II (Pakula *et al.*, 1960a).

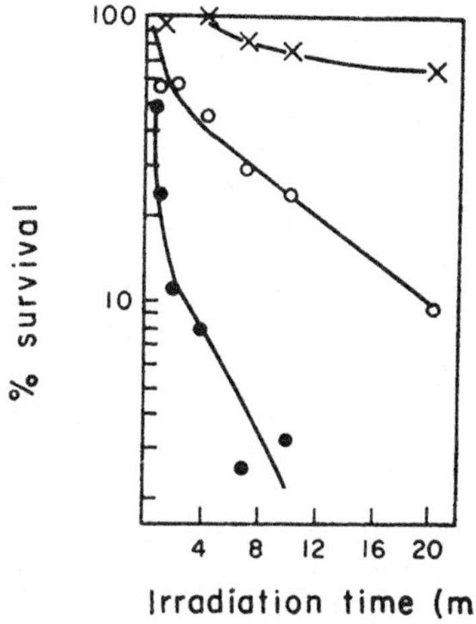

FIG. 7-8. Ultraviolet inactivation at room temperature, of pneumococcal transforming DNA activities to streptomycin resistance (• • • •), optochin resistance (o o o o) and of ability of P^{32}-DNA to be incorporated into cells (× × × ×); DNA concentration 90 μg/ml in thin layer, dose rate 47 ergs/mm^2 per sec (Lerman and Tolmach, 1959).

the behavior of pneumococci T-DNA activities for streptomycin resistance, optochin resistance and the ability of the irradiated nucliec acid to be incorporated into the recipient cells. Note, in particular, the pronounced radiation resistance of the ability of the DNA to undergo incorporation.

A concentration dependence for inactivation rate constants has been reported by two groups of observers (Zamenhof *et al.*, 1956; Lerman and Tolmach, 1959). These results should, however, be treated with reserve, since they are based on single measurements at low residual activities, hence only on a single point of the final portions of the inactivation curves. To establish the validity of any concentration effect, it would be necessary to run a complete survival plot for each concentration. In the heading to Table I of the paper by Lerman and Tolmach (1959) it is suggested that the "small increase in survival in the more concentrated solution may be attributed to attenuation of the radiation by absorption in the upper layers of the sample," but this is hardly valid since it is emphasized in the text that the DNA solutions were "exposed in an evaporating dish, mechanically agitated to provide mixing." Consequently, as in the case of TMV-RNA, it remains to be established whether a concentration effect does, indeed, exist; and, if it does, to what extent it affects the various components of the inactivation curve.

In view of the considerable interest and attention currently being devoted to the nature and origin of the multi-component inactivation curves (not only with ultraviolet, but also with ionizing, radiations), particularly valuable are the findings of Beiser and Ellison (1958) and Ellison and Beiser (1960), who compared the survival plot for the stretomycin resistance marker of pneumococcal T-DNA with those for various fractions of the same DNA isolated on an ECTEOLA column. The results are shown in Fig. 7-9, from which it will be observed that one of the fractions exhibited only the second, slower, component of the inactivation curve for the unfractionated

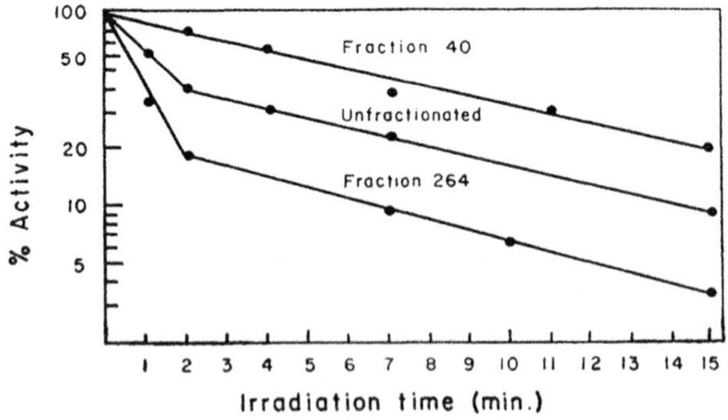

FIG. 7-9. Ultraviolet inactivation curves for streptomycin resistance transforming marker of pneumococcal DNA and two of the fractions isolated from it on an ECTEOLA column (Ellison and Beiser, 1960).

DNA. The remaining fractions exhibited two-component plots; but no mention is made as to whether these, like the representative one shown in Fig. 7-9, *all* differed from that for the unfractionated DNA. The significance of this point is rather obvious since, if such is the case, it implies that the survival plot for the T-DNA in question is the average of that for a number of different fractions, a fact of considerable importance in any attempt to interpret the kinetics of inactivation.

The foregoing findings may be explained by invoking a heterogeneity in molecular weight of the irradiated T-DNA. Marmur *et al.* (1961) have employed sonic vibration, which results in double chain scission of DNA molecules, to prepare a homologous series of pneumococcal T-DNA samples of varying molecular weights from a single preparation of initial molecular weight of 8.5×10^6. It was found that a given genetic marker (the results were the same for three different markers), residing on molecules of differing molecular weights, becomes more resistant to radiation as the molecular weight decreases. Qualitatively similar results have been reported by Setlow (1961). On the other hand, multi-component inactivation plots have been reported for T-DNA isolated from irradiated cells (Stuy, 1961), a finding which argues against molecular weight heterogeneity as the principal source of the multi-component nature of the inactivation curves. A repetition of the experiments of Ellison and Beiser (1960), accompanied by molecular weight determination of some of the fractions, might be expected to clarify this situation.

Pronounced differences in radiation sensitivities likewise prevail between different genetic markers. A direct comparison between pneumococcal and influenza T-DNA is provided by Fig. 7-7 and curve b of Fig. 7-6, for both of which irradiation conditions were identical (Pakula *et al.*, 1959). Latarjet and Cherrier (1957) have emphasized the remarkable radiation resistance of pneumococcal T-DNA, for which they calculate a quantum yield of 10^{-5} (mol. wt. of DNA not given). For the optochin marker of pneumococcal T-DNA (Fig. 7-8), assuming that the number of nucleotides involved in genetic activity is 1.6×10^3, Lerman and Tolmach (1959) calculate a quantum yield of 3.8×10^{-3} for the initial rate of inactivation. For the initial rate of the streptomycin marker the yield is about 1.2×10^{-3}.

Attempts to demonstrate differences in stability of different genetic markers in a single DNA preparation towards ionizing radiations have been unsuccessful (Marmur and Fluke, 1955; Lerman and Tolmach, 1959). The same applies with regard to ultraviolet radiation if loss of viscosity is used as the criterion of sensitivity. An entirely different situation prevails with respect to biological activity, as already illustrated in Fig. 7-8 for pneumococcal T-DNA. Indeed, of a variety of agents possessing the ability to produce differential inactivation of several markers, ultraviolet was until very recently by far the most effective (Zamenhoff *et al.*, 1957).

It has now been established that, under carefully controlled conditions, thermal inactivation is a much more efficient method for the selective destruction of different genetic markers, and the results thus obtained have done much to clarify our conceptions of the mechanisms involved. The thermal inactivation of T-DNA involves two main processes: (a) heating at temperatures below the melting point, T_m, provokes a slow loss of purines leading to an irreversible loss of activity of all markers; (b) if heating is conducted for short periods at temperatures above T_m, it is found that each marker is inactivated at a given characteristic temperature. This latter process is reversible and there is very little doubt but that it is due to strand separation; and that the higher the temperature required for inactivation, the higher must be the G-C content of the marker involved. For full details, reference should be made to the papers of Ginoza and Guild (1961), Ginoza and Zimm (1961) and, especially, Roger and Hotchkiss (1961), who succeeded in "separating" five different markers in a single DNA preparation by heating at defined temperatures above T_m.

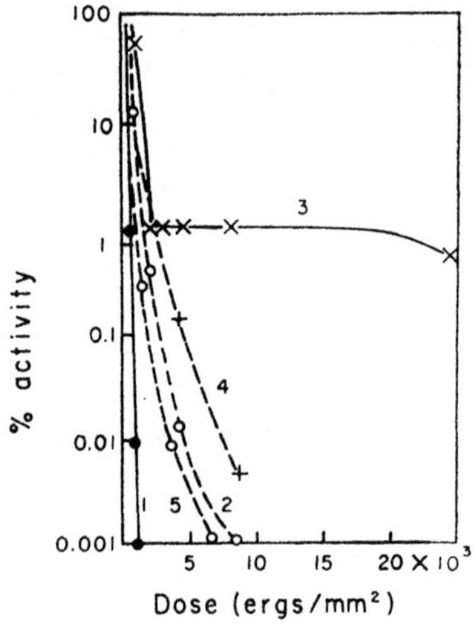

FIG. 7-10. Ultraviolet inactivation of various markers in transforming DNA from *Haemophilus influenza*:

1. Capsule *b* or *c* formation
2. Capsule *d* formation
3. Streptomycin resistance in type *b*
4. Streptomycin resistance in type *d*
5. Streptomycin resistance in type *a*

(Zamenhof *et al.*, 1957).

In Fig. 7-10 are pictured the differential stabilities to ultraviolet radiation of individual heredity determinants in *H. influenza* T-DNA. The experiments involved the preparation of purified T-DNA, the donors being the streptomycin-resistant encapsulated strains of pneumococci types *b*, *c* and *d*, each DNA preparation thus carrying two markers. Non-encapsulated streptomycin strains R_b or R_d served as receptors. From the results shown in the figure it will be seen that there are pronounced differences in stability between various markers; e.g. a 60-fold larger dose is required to produce the same percentage decrease in activity for stretomycin resistance (from type *b*) as for *b* or *d* capsule formation.

It is clearly of the utmost importance to attempt to pin down the origin of this heterogeneity in radiation resistance. Control experiments showed that the foregoing results were independent of the choice of receptor used for measuring activity. It might be argued that the high resistance of the one per cent residual activity of marker 3 in Fig. 7-10 implies the existence of two types of molecules, one much more stable than the other and involving different heredity determinants; the fractionation experiments of Ellison and Beiser (1960), referred to above, supporting such a possibility. This was, however, excluded by preparing new T-DNA from recipient cells transformed with residual activity only and showing that the resulting new T-DNA exhibited a precisely similar survival plot. It should be noted that the existence of such highly-resistant residual activities may also be observed with some viruses as well as with the X-ray inactivation of pneumococcal T-DNA (Ephrussi-Taylor and Latarjet, 1955). The differential ultraviolet stabilities of three markers of pneumococcal T-DNA have also been established and examined by Litman and Ephrussi-Taylor (1959).

Objection has been taken to the use of the term "differential stability of heredity determinants" coined by Zamenhof *et al.*, (1957) on the grounds that it presupposes the mechanism of the effect (Ravin, 1958). Ravin suggests that there may be no difference in stability of the various markers but only a variation in the ability of different markers in a given sample of irradiated T-DNA to undergo incorporation in the host genome. Some validity may attach to this argument, but it should be recalled that the ability to undergo cellular incorporation is highly resistant to radiation (see Fig. 7-8), so that ultraviolet inactivated determinants do at least enter the bacterial cell (Tolmach and Lerman, 1957; Lerman and Tolmach, 1959). Furthermore, as will be shown subsequently, differences in stability of various markers may readily be interpreted in terms of differences in base sequences. Marmur *et al.* (1961) have pointed out that no correlation has been found between the ultraviolet sensitivity of genetic markers and their susceptibility to thermal inactivation, and suggest that no such correlation would be expected since thermal denaturation is a function of the average base composition of the DNA molecule and the base compositions of DNA molecules of bacterial origin are fairly homogeneous. The results of Roger and Hotchkiss (1961), referred to above, indicate that a re-examination of this point might be fruitful.

The behavior of linked markers is particularly interesting in this connection. Marmur *et al.* (1961) examined the relative radiation sensitivity of the erythromycin-resistance linked markers ery_2 and ery_3 of pneumococcal T-DNA and found that the former was more resistant, while each marker was more resistant than the linked ery_2-ery_3 marker. In this instance one cannot invoke differences in size of the DNA molecule, since both genetic markers are on the same molecule. For the same reason Marmur *et al.* (1961) exclude differences in base composition as a possible source of the difference in radiation sensitivity; this argument is perhaps debatable since, as will appear from studies on model polymers, a slight difference in composition can provoke a marked difference in behavior. The location of the marker in the molecule, as well as the size of the marker, may also be factors determining its behavior. Indirect evidence does, in fact, lead to the conclusion that the ery_2 marker occupies a smaller region than the ery_3 (Roger and Hotchkiss, 1961), hence in accord with the higher radiation resistance of the former; but the assumption of the authors (Marmur *et al.*, 1961) that the smaller marker offers a smaller sensitive target to ultraviolet light is fallacious, inasmuch as we are dealing with a photochemical reaction involving specific absorption of light · · · the "smaller" the target, the less light absorbed.

We shall return to this problem of differential stability below in connection with inactivation kinetics. It must, however, be emphasized that the use of linked markers in such investigations is much to be desired.

If we now turn our attention for a moment to the T-DNA molecule as a whole, attention should be drawn to the fact that, as in the case of TMV-RNA, the order of magnitude of the doses required to inactivate biologically are considerably less than those which provoke observable physico-chemical modifications. For *H. influenza* DNA, Zamenhof *et al.* (1956) found it necessary to increase the biological dose by a factor of 500 to produce an observable (3 per cent) decrease in viscosity. Lerman and Tolmach (1959), by contrast, report that a drop in viscosity accompanies the ultraviolet biological inactivation of pneumococcal T-DNA; this may conceivably be related to the much higher radiation resistance of this DNA (see above). However, the authors mention that the concentration of the irradiated solution in this experiment was "exceedingly" high and the rate of biological inactivation slower than in the usual experiments; hence this observation must be accepted with some reserve, the more so in that it is assumed by the authors to reflect a decline in the weight average degree of polymerization, i.e. involving rupture of internucleotide linkages, a result contrary to expectations for biological inactivation doses.

The low values of the quantum yields for T-DNA inactivation are themselves suggestive of a small amount of damage, which must therefore be highly localized. A similar conclusion applies to ultraviolet inactivated bacteriophage which is, in this case, further substantiated by studies of "cross-reactivation" (Stent and Fuerst, 1960). But by far the most significant observation yet made on the degree of localization of ultraviolet radiation injury with doses resulting only in biological inactivation is that of Marmur and Doty (1959), who found that the denaturation temperature T_m of 90 per cent ultraviolet inactivated pneumococcal T-DNA is reduced by about

1.5°C, indicative of the breakage of only a few twin-strand hydrogen bonds and most certainly excluding rupture of internucleotide linkages. More extensive data on the decrease in T_m with ultraviolet dose for different types of T-DNA will be found in the paper by Marmur *et al.* (1961). This behavior can be duplicated with synthetic twin-stranded polynucleotides (Wierzchowski and Shugar, 1960), thus providing independent evidence that *one* of the initial effects of irradiation of DNA is the dissociation of twin-stranded hydrogen bonds.

At least two other reactions result from the ultraviolet inactivation of T-DNA: hydration of cytosine residues and dimerization of thymine residues. For the sake of clarity, we shall defer our discussion of these until we have dealt with the photochemistry of model oligo- and polynucleotides.

According to Ellison and Beiser (1960), intracellular T-DNA is considerably more radiation-sensitive than extracted and purified samples. When pneumococcal organisms were submitted to a radiation dose sufficient to reduce survival to 0.005 per cent, the T-DNA subsequently extracted from these organisms exhibited no activity; but isolated T-DNA was hardly affected by such a dose. This provides an additional example of the apparently greater sensitivity of nucleic acids *in situ*, but it would be premature to draw any generalizations from this. Stuy (1961) did not find T-DNA of *H. infl.* to be especially more sensitive in the intracellular state.

Lerman and Tolmach (1959) observed that the quantum yields for two markers of pneumococcal T-DNA are appreciably reduced by a decrease in temperature from 275° to 77°K during irradiation (Fig. 7-11). The

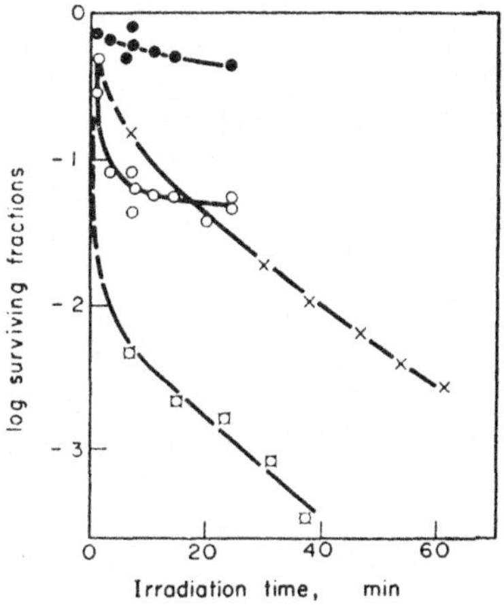

FIG. 7-11. Ultraviolet inactivation of pneumococcal transforming DNA activity to streptomycin resistance at 77°K (● ● ● ●) and 275°K (o o o o) and to optochin resistance 77°K (× × × ×) and 275°K (□ □ □ □); dose rate 47 ergs/mm² per sec (Lerman and Tolmach, 1959).

significance attaching to this observation is not immediately apparent; the more so in that the observed effect is in the opposite direction from that to be expected on the basis of the behavior of cytosine-containing model oligonucleotides. On the other hand, irradiation at the lower temperature was in 80 per cent glycerol, and a change of state was certainly involved; furthermore, at 77°K excited states resulting from irradiation are effectively "trapped." The nature of the reactions resulting on subsequent thawing may be quite different. In addition, the nature of the excited states, and hence also the accompanying photochemical reactions, in 80 per cent glycerol at 77°K most likely differ from those prevailing in aqueous medium at room temperature. For TMV-RNA over the temperature range 6–13°C no difference in quantum yields could be detected (Rushizky *et al.*, 1960).

3. *RNA Core*

If yeast RNA is exhaustively digested with ribonuclease and submitted to dialysis against water, a dialysis residue of about 15 per cent of the material will not diffuse; this residue has been shown to consist of oligonucleotides with a chain length of 5 to 6 on the average, each of which consists of purine nucleotides with one terminal pyrimidine nucleotide (Markham and Smith, 1952). This residue is referred to as the RNA core and has been shown to stimulate streptolysin S formation in hemolytic streptococci when added to the culture medium. In unpublished experiments (Shugar, Wierzchowski and Tomerska, 1959), it has been shown that this oligonucleotide is inactivated only very slowly by ultraviolet light with a quantum yield considerably below 10^{-4}. Following 50 per cent loss in activity, there is only a 10 per cent loss in metachromasia against basic dyes, indicating relatively little chain degradation. Furthermore, since we are dealing with a purine oligonucleotide containing a single pyrimidine residue, the latter of which is much more susceptible to irradiation, it may be concluded that it is not necessary for the biological activity of the RNA core, a conclusion reached independently by direct removal of the terminal pyrimidine nucleotide (Tanaka, 1958). Finally, the above findings confirm biologically the physico-chemical evidence for the high radiation resistance of purine nucleotides in polynucleotide chains.

E. KINETICS OF BIOLOGICAL INACTIVATION OF NUCLEIC ACIDS

In view of the complex nature of the inactivation curves for T-DNA, as well as for some of the biological properties of viruses and other organisms, it is pertinent to examine this question from a theoretical point of view.

The photochemical inactivation or destruction of a sytem of molecules or other organisms exhibiting biological activity frequently follows a first order course with respect to time (or, what is the same, irradiation dose), eqn. (1-8c). There are a number of instances (see above) where the survival curves are of the "multiple-hit" type. The curve is initially flat until some of the molecules have accumulated a sufficient number of "hits" to cause inactivation, at which point the curve starts its downward course. The general formula for such a process which requires m inactivating hits for each of N necessary parts of the molecule, with a probability a per unit dose applied is (Timofeeff-Ressovsky and Zimmer, 1947)

$$\frac{[A]}{[A_0]} = 1 - \left\{1 - e^{-aI} \sum_{0}^{m-1} \frac{aI_0}{m-1}\right\}^N \qquad (7\text{-}1)$$

The constant a depends on the energy absorbed and the quantum yield. If only one essential part of the molecule requires only one hit for inactivation, the formula reduces to equation (1-8c), namely

$$\frac{[A]}{[A_0]} = e^{-aI_0}$$

Inactivation curves for antigenic properties of viruses are usually of the multiple-hit type (see, e.g. Fig. 8-6).

Extrapolation of the exponential portion of a multihit survival curve back to the ordinate gives the so-called "hit-number" for the inactivation process. Alper *et al.* (1960) object to the use of this term on the grounds that its quantitative value is affected by the conditions under which an irradiated sample is treated following irradiation, and propose to replace it with the term "extrapolation number." We should be inclined to accept such a suggestion with reserve. Bawden and Kleczkowski (1960) drew attention to the fact that ultraviolet radiation affects tobacco leaves to a variable extent and that neglect of this has occasionally led to erroneous conclusions with respect to the direct effects of ultraviolet on virus particles *in vivo*; no need was, however, envisaged for a supplementary terminology. It was suggested merely that account be taken of the effects of radiation on leaves.

In the case of ionizing radiations considerable supplementary evidence has been adduced in support of the one-hit type of process for those instances where the survival curve is semi-logarithmic. But the same does not hold rigorously for irradiation with ultraviolet light (Lea, 1955). A one-hit process is generally assumed because it provides the simplest and most logical formulation; but supplementary evidence is still to be desired for ultraviolet inactivation processes. It is worth noting that, for the irradiation-induced formation of mutants, an equally good fit to a logarithmic dependence of rate on dose may be obtained theoretically if each inactivation is assumed to be the result of a number of random events (Opatowski, 1950; Opatowski and Christiansen, 1950).

Insofar as quantum yields are concerned, it is clear that the existence of a sigmoidal survival curve renders it difficult to either calculate or ascribe some definite meaning to Φ. Atwood and Norman (1949) have suggested an interpretation of such survival curves on the basis of the exponential inactivation of the individual units of a multi-curve system, Φ being calculated for each of the individual units.

For T-DNA where, as pointed out above with numerous examples, the survival curve is continuous and convex to the origin, the course of inactivation may be described by more than one rate constant. The usual practise has been to divide such curves into two, or occasionally three, straight-line components, and to calculate a quantum yield for each component. We have already discussed above the interpretation of such curves on the basis of the existence of a molecular population with varying radiation resistance.

Another interpretation has been proposed by Lerman and Tolmach (1959b). Only a single hit within the genetic region embraced by a marker is required to produce inactivation. In addition, the energy of an absorbed photon may be transmitted along the DNA molecule to produce a photochemical lesion at some point separated from the marker. The energy of a second photon cannot be transmitted past this point, hence a second hit must occur between the initially affected locus and the active centre in order to produce a damaged genetic marker. Such a mechanism is conceivable; but one argument against it is that quantum yield measurements in single (Wierzchowski and Shugar, 1959) and twin-stranded model polynucleotides (Shugar and Wierzchowski, 1960) indicate that, whatever energy transfer does take place in such chains, it appears not to provoke photochemical reactions (see Marmur *et al.*, 1961, for additional hypotheses).

On the other hand, there is also evidence that, whatever the nature of the inactivation curves, some photons are absorbed by the structure in question to produce damage which is not decisive. For instance, partially photolysed enzymes and proteins are much more sensitive to heat than non-irradiated controls (Chapter V), suggesting partial disorganization of secondary structure, without inactivation, for some of the molecules. A similar situation applies to TMV (Oster and McLaren, 1950) and other viruses and bacteriophage (Kleczkowski, 1954; Kleczkowski and Kleczkowski, 1953), a rather valuable finding, which has been taken advantage of in the preparation of vaccines (Chapter VIII). It applies also to the physico-chemical properties of DNA (Butler and Conway, 1953; Butler, 1954), nucleoproteins (Giese, 1947), as well as to *H. infl.* T-DNA (Zamenhof *et al.*, 1956) but not to the streptomycin resistance factor of type R_d *H. infl.* T-DNA (Rupert *et al.*, 1958). Kleczkowski (1954) has shown that this "weakening" of the activity of an enzyme or virus, without complete inactivation, is not due to the effect of inactivated molecules on active molecules. Lerman and Tolman (1959a) have also shown that multi-component inactivation curves of T-DNA are not due to any effect of photo-chemically inactivated molecules on the surviving active ones. One practical conclusion to be drawn from the foregoing is that assay conditions for biological activity, following irradiation, must be clearly specified; e.g. the use of a higher temperature for measurement of the activity of an irradiated enzyme, virus, or DNA would result in inactivation of some "weakened" molecules and give an apparent higher quantum yield.

Two rather interesting empirical relationships have been reported by Rupert and Goodgal (1960) to describe the inactivation curves of T-DNA. In one of these log (activity) is plotted vs. log (dose) to give a straight line for the entire course of inactivation and suggesting a power-law dependence of activity on dose. In the second relationship the reciprocal of the square root of the activity gives a straight line over the entire range when plotted vs. dose, so that

$$\frac{I}{I_0} = \frac{1}{(1 + CD)^2}$$

where I_0 is the initial activity, I is the activity remaining after a dose D and C is a constant for a given marker. No theoretical explanation is forth-coming for this relationship but its practical advantages are stressed from the point of view that one may use it to express the relative sensitivities of different markers; e.g. the relative sensitivities of the streptomycin and cathomycin resistance markers of *H. infl.* T-DNA are found in this way to be 10 : 1, a fact difficult of presentation from the usual complex inactivation curves.

Finally it should be noted again that inactivation of T-DNA by ionizing radiations gives survival curves entirely analogous to the ultraviolet inactivation curves. It is most unlikely that this is coincidental. Latarjet *et al.* (1959) present extensive experimental evidence to show that this two-stage inactivation curve is not due to artifacts involved in the measurement of transforming activity, and suggest the following mechanism for inactivation: the first slope of the inactivation curve corresponds to a hit outside the "active" site and the second component to a direct hit on the marker. Whatever damage is caused outside the marker, the latter still has a 5–20 per cent chance of being incorporated into the genome of the recipient cell and it is this additional hypothesis which accounts for the break in the inactivation curve. However, at the time this proposal was made, the authors were unaware of the work of Lerman and Tolmach (1959) referred to above (since it was submitted for publication at about the same time) and in which it was demonstrated that the ability to undergo cellular incorporation is much more resistant to ultraviolet light than inactivation (see Fig. 7-8). While incorporation into the cell does not necessarily imply ability to be incorporated into the genome, the fact remains that if the latter were so, then, according to the above mechanism, the inactivation curve should exhibit first a slow component succeeded by a rapid one. Actually the converse is true.

F. REVERSIBILITY OF NUCLEIC ACID PHOTOLYSIS

In the previous sections we have been dealing with the overall photochemical reactions of nucleic acids. The phenomenon of biological photoreactivation will be taken up in a subsequent chapter, but it must be emphasized that it has played a central role, and continues to do so, in planning experiments designed to elucidate the primary effects of ultraviolet light. It therefore seems logical at this point to examine the possibilities of reversal of the effects of irradiation on polynucleotide chains *in vitro* in the light of what has already been presented regarding reversible photolysis of pyrimidine derivatives. Since the photochemistry of model polynucleotide chains has been investigated primarily from the point of view of reversibility, we shall, for the sake of clarity, discuss these jointly in the following sections.

1. *Physico-chemical Evidence for Reversibility*

Recalling the relatively high radiation resistance of purines as compared to pyrimidines, as well as the fact that for a number of the latter (including their nucleosides and nucleotides) the photo-chemical reactions are reversible, one may readily envisage the possibility that irradiation of nucleic acids would affect those pyrimidine nucleotides the photochemical transformation of which could subsequently be reversed. It is, naturally, assumed *a priori* that the doses used are not sufficiently intense to lead to damage to the polynucleotide chain backbones. Excluding 5-substituted cytosine derivatives, the hitherto identified reversible photoproducts of pyrimidine residues are the result of (a) uptake of a water molecule at the 5,6 double bond, a reaction reversible by acid, alkali or heat, and (b) cross-linking of the 5,6 double bonds of two pyrimidines residues, a reaction which is readily reversed by irradiation. Neither the number 5 nor the number 6 positions in pyrimidine rings are involved in internucleotide linkages or in secondary hydrogen bonds in polynucleotide chains, so that formation of the foregoing photoproducts in nucleic acids might reasonably be expected. This is not to be construed as excluding additional reactions. In fact, a new reaction, involving thymine residues in oligonucleotide chains, the nature of which has as yet not been clarified, has been recently reported (Sztumpf and Shugar, 1962). But the almost 100 per cent recovery of purines from ultraviolet degraded DNA (Errera, 1952b) substantiates the contention that it is primarily the pyrimidine rings which are first affected by radiation. Theoretically, therefore, we should expect to encounter some reversal of photochemical transformation in nucleotide chains.

(a) *Natural nucleic acid preparations.* Initial experiments designed to test the possibility of reversal of photolysis effects in polynucleotide chains were confined to those obtainable from natural sources. The behavior of a commercial sample of yeast RNA submitted to a radiation intensity of about

5 × 10 quanta/cm² per min, is shown in Fig. 7-12. Following about 3 hr irradiation, there is an almost 50 per cent drop in extinction at 2600 Å, while the characteristic absorption maximum has been "flattened." At this stage the solution is quite stable since, on standing for 18 hr at room temperature, its absorption is only slightly altered. But heating for 5 min at 85°C results in a restoration of more than 50 per cent of the absorption loss

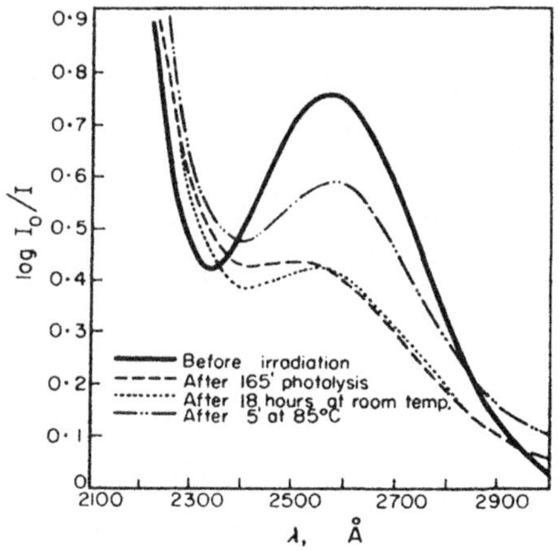

FIG. 7-12. Illustrating the possibility of reversal by heat of the effect of ultraviolet irradiation on yeast RNA in 0.02 M phosphate buffer pH 7. Irradiation was with mercury resonance lamp and, in this case, without filter so that traces of radiation below 2400 Å were present (Shugar and Wierzchowski, 1958b).

provoked by irradiation. The non-reversibility of the increase in absorption at shorter wavelengths than 2400 Å is most likely due to irreversible photolysis of some of the residues in the chain. If the RNA is more extensively photolysed the degree of reversibility is correspondingly reduced and, following irradiation to the point where the extinction of the absorption maximum is reduced to 25 per cent of its initial value, no reversibility can then be detected.

In the preceding example we are dealing with a nucleotide chain containing two components which may be expected to undergo reversibility, viz. uridylic and cytidylic acids. Although apurinic acid (APA) can hardly be considered a "natural" nucleic acid chain, it does offer the advantage of providing a polynucleotide of molecular weight of about 15,000 and containing only thymine and cytosine residues, of which the latter is known to form a hydrated photoproduct which is thermally reversible in the dark. Figure 7-13 illustrates the behavior of a typical APA preparation. It will be observed that the degree of reversibility obtainable by heating is about

50 per cent or somewhat less than for RNA. Particularly interesting is the fact that it was necessary to use a higher temperature for reversal (or longer heating time at the same temperature) than for RNA, in agreement with the behavior of free deoxycytidylic acid as compared to cytidylic and uridylic acids. Assuming only the deoxycytidylic acid component to undergo photolysis in APA, the quantum yield works out to about 3 times that

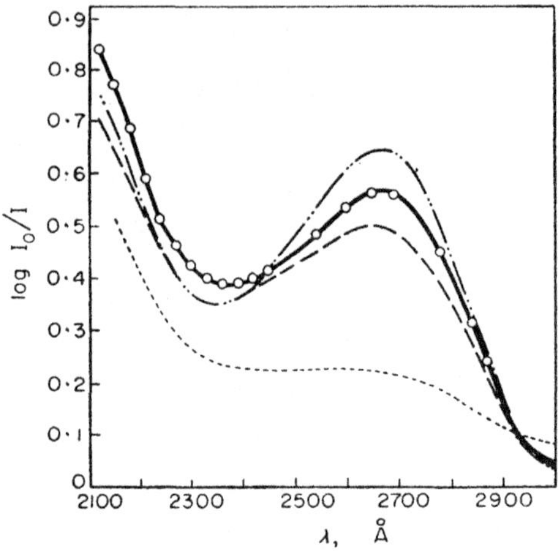

FIG. 7-13. Heat reversal of effects of irradiation on apurinic acid in 0.02 M phosphate buffer pH 7: (a) before irradiation; (b) following 30 min irradiation; (c) after 10 min heating of irradiated solution at 92°C; (d) following prolonged irradiation with germicidal lamp, following which no reversibility with heating can be observed (Shugar and Wierzchowski, 1958b).

which it exhibits in the free state, an apparently not inconsistent result in view of the possible modifications due to incorporation in the nucleotide chain. It is, nonetheless, erroneous since it was subsequently shown by Beukers *et al.* (1959) that thymidylic acid in APA is exceptionally radiation sensitive, and by Wierzchowski and Shugar (1960b) that adjacent thymine residues in model oligonucleotide chains are more readily photolysed than free thymidylic acid. The foregoing consequently points to the necessity of examining the behavior of each nucleic acid derivative when it is incorporated in a polynucleotide, which can be most readily achieved with model nucleotide chains.

Additional natural polynucleotides for reversibility studies may be obtained by deamination of RNA and APA. Although initial experiments with these failed to reveal any thermal reversibility (Shugar and Wierzchowski,

1958b), subsequent examination demonstrated that, with the exercise of suitable precautions, such reversibility obtains for these polymers as well.

Some exception may be taken to the above results from a quantitative point of view, since no account was taken of hyperchromicity. They do establish, however, that hydration of uracil and cytosine residues proceeds to an appreciable extent in irradiated nucleic acid, and that these reactions are thermally reversible. We shall see below that a quantitative interpretation is more readily forthcoming from investigations on model polymers.

(b) *Pyrimidine dimer formation in irradiated DNA.* It is, on the other hand, obvious that photochemically induced hydration is not the only reaction undergone by pyrimidine residues in nucleic acids, since otherwise we would expect a considerably higher degree of thermal reversibility than that actually observed.

The first concrete evidence for the existence of some other type of photoproduct in irradiated nucleic acids was obtained by Beukers *et al.* (1960), who demonstrated the formation of thymine dimers. The procedure employed was to irradiate DNA, following which it was hydrolyzed in concentrated perchloric acid and the hydrolyzate submitted to paper chromatography. The chromatogram was then cut into strips, each strip eluted with water, and the eluates irradiated at 2537 Å until one was found which exhibited the characteristic absorption maximum of thymine on irradiation. It will be recalled that the formation of such dimers had been invoked to explain the cross-linking of nucleic acid films (Baranowska and Shugar, 1960) and the rapid photochemical transformation, not reversible by heat, of some pyrimidine residues in model oligonucleotides (Shugar, 1960; Wierzchowski and Shugar, 1960). The success achieved by Beukers *et al.* (1960) in actually isolating these photoproducts from irradiated DNA was due to the remarkable thermal stability of the dimers, which enable them to withstand the hydrolysis conditions involved (see Section VI-8).

These results were soon confirmed by Wacker *et al.* (1960, 1961a), using chromatographic methods with tritium and ^{14}C-labelled DNA. The same authors also isolated thymine dimers from various micro-organisms which had been inactivated with doses of the order of 10^4 ergs/mm^2; up to 9 per cent of the total thymine of the DNA of *E. coli* was isolated as the dimer, whereas for enterococci the percentage was as high as 18.

Dimer formation may be envisaged as proceeding via several different pathways: (a) between adjacent pyrimidine residues in a single chain; (b) between individual chains, this being the predominant reaction in nucleic acid films, and (c) between the complementary strands of a DNA chain. Dimerization of the latter type was first demonstrated by indirect methods by Marmur and Grossman (1961) and Marmur *et al.* (1961); but it is obvious, from steric considerations, that it must be preceded by localized denaturation of the twin-stranded structure in order to enable the bases to become properly orientated for dimerization. The techniques used involved the demonstration of the formation of cross links between the complementary strands of transforming DNA, such that the molecules became resistant to strand separation under conditions where the latter normally can be induced, e.g. by formamide denaturation. The cross-linked strands resulting from irradiation were also found to exhibit a lower buoyant density in the CsCl gradient; and, following submission to denaturing agents, to resist digestion by a specific phosphodiesterase which attacks only denatured or single-stranded DNA. Furthermore, the higher the thymine content of a given DNA sample, the higher the proportion of cross-linked material obtained as a result of irradiation. Cross-linking between complementary strands was also observed in DNA extracted from irradiated cells.

The presumption, in the foregoing work, that the cross-linking observed is due to pyrimidine dimer formation, and specifically between thymine residues, can by no means be considered as established. Even if it were, it is by no means certain that it is one of the initial effects of radiation, since appreciable collapse of the twin-stranded structure must occur to permit orientation of the bases such as to make possible dimerization. It was, in fact, noted that the degree of cross-linking resulting from irradiation would not adequately account for the accompanying biological inactivation. The significance of these observations stems rather from the finding that cross-linking does accompany inactivation of transforming DNA or bacterial cells. However, the further conclusion that these cross-links are abolished by photoreactivation, while plausible, cannot be considered as established since it was not demonstrated directly. Further evidence for UV-induced cross-linking between complementary DNA strands has been provided by Klouwen *et al.* (1962) and Glisin and Doty (1962), the latter of whom showed that local collapse of the helix is a prerequisite for the formation of such links. If it is assumed that this is so, then the degree of cross-linking should be increased by the partial strand separation such as results from heating in the neighbourhood of the melting temperature T_m. It was, in fact, found that maximum cross-linking occurred when at about 20 per cent heat denaturation. On heating to higher temperatures, so that strand separation increased, the degree of cross-linking decreased, in accordance with expectation.

A phenomenon of equal, if not greater, importance is the finding that UV irradiation provokes the formation of cross-links between DNA and protein both *in vivo* and *in vitro*. The nature of these cross-links is as yet unknown but the fact that they occur *in vivo* with a much higher quantum yield than thymine dimer formation suggests that they may be of considerable biological importance. This subject has been reviewed in detail by Smith (1963).

The evidence for formation of dimers between adjacent thymine residues in the individual strands is much more extensive and convincing (Wulff and Rupert, 1962; Setlow and Setlow, 1963; Bollum and Setlow, 1963; Setlow and Carrier, 1963), and also indicates that this reaction is responsible for up to 50 per cent of the biological inactivation resulting from UV irradiation. It is supported also by the finding that the ultraviolet sensitivity of bacterial cells increases approximately linearly with increasing A-T base pair content of the bacterial DNA (Kaplan and Zavarine, 1962; Haynes, 1963), although it would be desirable to check this result with transforming DNA of varying base compositions.†

(c) *Natural and synthetic oligo- and polynucleotides.* The extent to which the effects of radiation on RNA and APA may be reversed by subsequent heating, as described above, is approximately that prevailing for biological photoreactivation. There is little doubt that we are dealing here with the reversible hydration of uracil and cytosine residues. If, however, the behavior of these residues were the same as for the free nucleotides in solution, we should expect almost 100 per cent heat reversibility instead of about 50 per cent. It is therefore clear that the behavior of individual residues in a polynucleotide chain is not necessarily the same as for the free residues, a result which may equally well be inferred from the behavior of the thymine residues in APA. That such is the case is by no means unexpected. We have already seen that the photochemical behavior of a pyrimidine base is profoundly modified not only when the number 1 nitrogen is involved in a glycosidic linkage (Section VI-6) but is even markedly dependent on the position of esterification in a nucleotide. In a nucleotide chain each residue is esterified at two position, as illustrated in Fig. 7-15 for a trinucleotide. In addition, the twin stranding in the Watson–Crick structure for DNA involves hydrogen bonding of base pairs which, in the case of cytosine-guanine, may involve three bonds in place of the two normally expected (Pauling and Corey, 1956), as shown in Fig. 7-14. Furthermore, we have seen that, as deduced from the hyperchromic effect, even a dinucleotide possesses a more or less defined configuration determined in part by an interaction between the π-orbitals of adjacent aromatic rings so that, in a longer chain, the interaction between neighboring rings will vary from a minimum at either end to a saturation value about 7 or 8 residues from either end, and will also be dependent on the base sequence. Finally in very long chains the configuration is still further modified to produce hydrogen-bonded secondary structures.

Again we must take into account the more and more frequent discovery in natural nucleic acids of bases other than the four normally encountered; in some instances the content of these "abnormal" bases may be considerable, e.g. 5-methylcytosine in wheat-germ DNA, 5-ribosyluracil in the amino-acid activating S-RNA; in others a normal base may be completely replaced

† It should be noted that with ionizing radiations the sensitivity of bacterial cells decreases linearly with increasing A-T content of their DNA (Kaplan and Zavarine, 1962).

by another as in the T-even bacteriophages which contain only 5-hydroxymethylcytosine in place of cytosine. Since, as shown by studies on biologically active nucleic acids, the effects of radiation are highly localized, the presence of even small proportions of foreign bases may influence considerably both

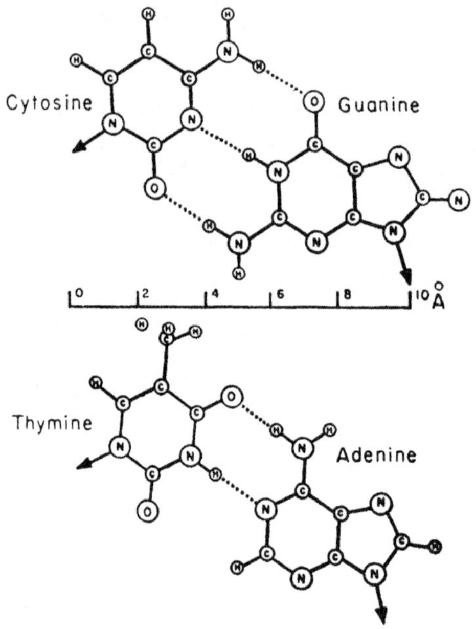

FIG. 7-14. Hydrogen bond formation between base pairs (guanine–cytosine and adenine–thymine) in DNA. The arrows show the points of linkage to the chain.

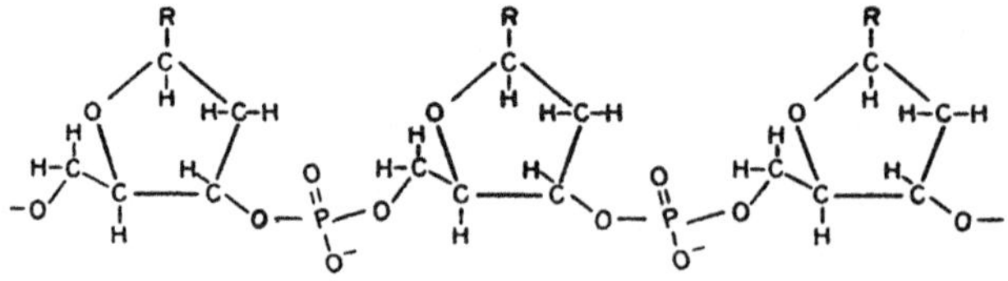

FIG. 7-15. Showing three residues in a polynucleotide chain; R represents purine or pyrimidine bases.

sensitivity as well as the ability to revert to the original compound following irradiation. We shall return to this latter point in some detail in connection with the photochemical behavior of the T-even bacteriophages (Chapter VIII). Consequently, in order to obtain even a rough understanding of the effects of radiation (and this includes ionizing as well as ultraviolet on nucleic acids), it is almost essential to examine the behavior of model chains with known base sequences and various chain lengths.

(i) *Purine oligonucleotides*

Repeated reference has been made to the relatively high radiation resistance of purines and purine nucleotides as compared to the corresponding pyrimidine derivatives, both in the free form as well as in DNA. Nonetheless, it was considered desirable to establish this quantitatively. Interestingly enough, it was found that the purine nucleotide residues in both poly-A and poly-G were even more radiation resistant than the free nucleotides themselves (Wierzchowski and Shugar, 1960b). Thus, whereas Φ for Ap was of the order of 0.5×10^{-4}, or about the same as for adenine, it is at least one order of magnitude lower for Ap in ApA or poly-A, at neutral pH. Furthermore, at acid pH where poly-A is known to form a double helix, its radiation resistance was equally high. For Gp the value of Φ at neutral pH is 1.5×10^{-4} as compared to 2×10^{-4} for guanine. For poly-G, however, the quantum yield is even less, about 0.5×10^{-4}.

It must be noted that the foregoing results are based on modifications in absorption spectra induced by radiation. If the primary effect of radiation were on the imidazole ring, one might expect only minor changes in absorption spectra upon irradiation. As will appear below, the behavior of purine-containing hetero-oligonucleotides provides even more convincing evidence for the relative inertness of purine residues to radiation.

(ii) *Oligonucleotides of Uridylic acid*

At first sight it might appear that the simplest model polymers for photochemical investigations are homopolymers of uridylic acid, since these exhibit relatively low hypochromicity. Modifications in absorption spectra resulting from irradiation consequently reflect reasonably accurately the extent of the reaction. If we recall, however, that uracil and its glycosides undergo not only photochemically induced hydration but also dimerization, that the latter reaction is subject to reversal by irradiation, and that this reverse reaction may also lead to at least partial hydration, it becomes obvious that a complex series of reactions may be involved. It should, therefore, be emphasized that the kinetics of photolysis of polymers of Up, now to be described, were carried out at a time when the nature of the dimerization reaction had not yet been established. Because of this dimerization reaction, we also know now that it is not sufficient to examine the kinetics at one wavelength but both at short wavelengths, where dimer photodissociation predominates, and at longer wavelengths where dimerization is favoured. We shall revert to these points at the end of this section.

Irradiation of a dinucleotide such as UpUp leads to a more rapid decrease in the absorption spectrum than in the case of Up; if, following almost complete disappearance of the absorption spectrum, the radiation source is removed and the solution heated, about 80–85 per cent of the absorption is regenerated (Fig. 7-16) as compared to 100 per cent for Up. If we examine the kinetics of the photochemical reaction, assuming that all the radiation absorbed is photochemically effective, the reaction is found to

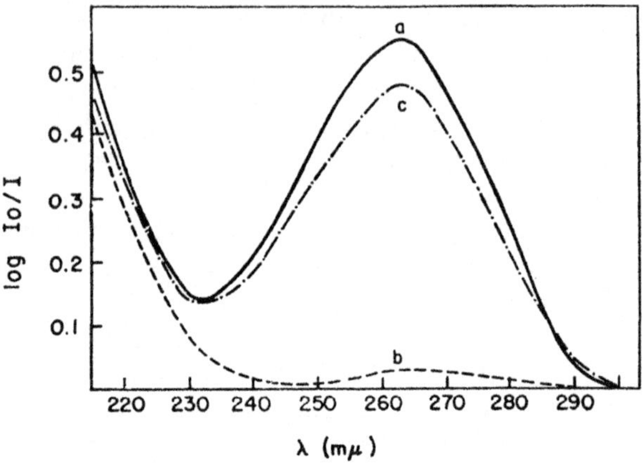

FIG. 7-16. Absorption spectrum of UpU in 0.02 M phosphate buffer pH 7:

(a) prior to irradiation
(b) following 30 min irradiation at an intensity of 6.5×10^{-7} einsteins/cm^2 per min
(c) following 4 hours heating of the photoproduct at 80° (Wierzchowski and Shugar, 1959).

proceed in two stages, of which the initial one is the more rapid (Fig. 7-17). If the data from Fig. 7-17 are transformed to a semi-log plot, the reaction is found to correspond to two first order reactions with a break, as seen from Fig. 7-17, after the transformation of about 50 per cent of the uracil rings. The calculated quantum yield, which initially is 40×10^{-3}, decreases during the course of the reaction until, towards the end, it drops to 5×10^{-3}, a value considerably less than that for free Up. The quantum yield is also found to be independent of concentration over a range of $10^{-5} - 10^{-3}$M. The high degree of thermal reversibility (Fig. 16) implies that the principal photochemical transformation involves addition of a water molecule at the 5,6 double bond of the uracil rings. This is further substantiated by two additional observations: (a) the isotope effect $\Phi_{H_2O}/\Phi_{D_2O} = 2$ as with free Up; (b) the regenerated UpU is chromatographically the same as for the starting product.

It was also found that the photochemical kinetics, and the degree of reversibility, were exactly the same for UpU, UpUp and UpUp! as well as for the same compounds when the internucleotide linkages were 2′,5′ instead of 3′,5′. It is therefore clear that the differences in behavior from free Up are due primarily to the presence of neighboring pairs of uracil rings, which in some way influence each other. This is further illustrated by the kinetics of the reverse reaction, which are first order up to about

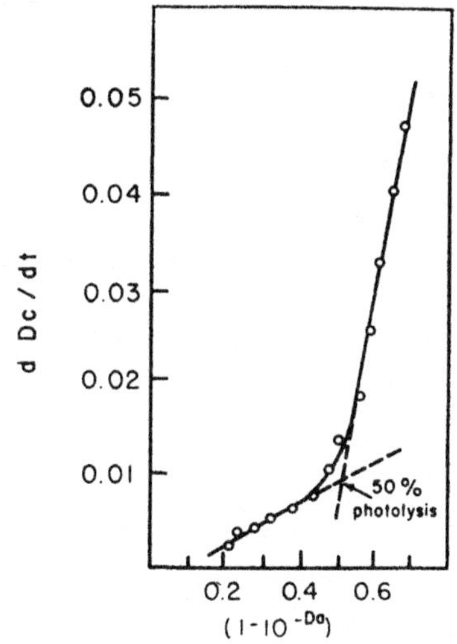

FIG. 7-17. Rate of photochemical transformation, dD_c/dt, of UpU as a function of absorption, $(1 - 10^{-Da})$, of the solution; concentration of solution 10^{-4} M with respect to mononucleotides; 0.02 M phosphate buffer pH 7; irradiation intensity 2.6×10^{-7} einsteins/cm^2 per min. Results are same at a concentration of 10^{-3} M, if the initial optical density is the same. D_c is the optical density at the absorption maximum at time t and D_a is the optical density at the irradiation wavelength, 253.7 mμ, at time t, both of these being therefore proportional to the concentration (Wierzchowski and Shugar, 1959).

50 per cent reversibility, following which there is a break in the curve and a slowing down in the rate.

Particularly interesting is the per cent thermal reversibility as a function of the degree of photolysis. If photolysis is not carried beyond 50 per cent, then only slightly more than one half the drop in absorption can be reversed by heating. However, beyond 50 per cent photolysis the reaction is completely reversible. This indicates that the failure to observe 100 per cent reversibility is due to some subsidiary, thermally non-reversible, reaction which occurs only during the first half of the photochemical transformation of UpU (see below).

For a *trinucleotide*, UpUpU or UpUpUp, the behavior is similar to that for UpU with the exception that the break in the reaction course (cf. Fig. 17) occurs following 60 per cent photolysis. Furthermore, for small ultraviolet doses, the per cent thermal reversibility (TR) is somewhat higher.

The behavior of longer chains is qualitatively similar, but the "break" in the kinetic curve is more gradual (Fig. 7-18). The reality of this break

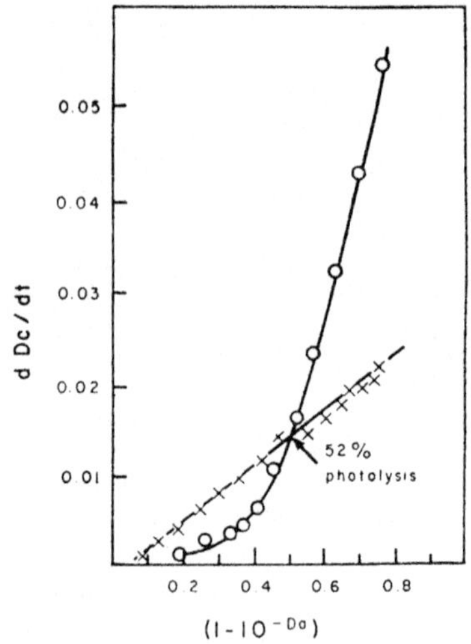

FIG. 7-18. Rate of photochemical transformation of:

(a) ○ ○ ○ ○ highly polymerized poly-U
(b) × × × × same solution first hydrolyzed quantitatively to mono-nucleotides, Up, with ribonuclease.

Concentration in both instances 10^{-3} to 10^{-4} M in Up. Other conditions and symbols as in Fig. 7-17 (Wierzchowski and Shugar, 1959).

in the kinetic curve is testified to by the fact that, if the poly-U is quantitatively hydrolyzed to mononucleotides, the complex kinetic behavior completely disappears, as can be seen from Fig. 7-18. The behavior of poly-U shown in this figure is identical for chain lengths varying from 6 residues up to very highly polymerized chains. There is also no evidence for rupture of nucleotide linkages.

The most significant departure in behavior of these longer chains from di- and tri-nucleotides is in the degree of TR following photolysis, which does not exceed 55 per cent, thus suggesting a more extensive participation of the thermally non-reversible reaction referred to above for UpUp. That this "non-reversible" reaction occurs mainly during the initial stages of photolysis is again supported by an examination of the kinetics in light and heavy water; the isotope effect is found to be as low as 1.3 at the start and gradually increases to the value of 2 expected for Up. An additional supporting argument is the fact that, following extensive photolysis and thermal reversibility, subsequent photolysis is completely thermally reversible; i.e. the latter stage of the reaction is completely reversible thermally.

It follows from the above that one of the dominant reactions in irradiated oligonucleotides of Up is the thermally reversible uptake of a water molecule at the 5,6 double bond as for Up. However, this reaction is in all instances accompanied by a second one which likewise leads to a decrease in absorption in the ultraviolet, but which is thermally non-reversible; the relative extent to which this "non-reversible" reaction occurs increases with chain length. When to this we add the fact that the initial quantum yields are much higher than for free Up, and the marked difference in kinetics as compared to comparable concentrations of free Up, it is clear that the photochemical transformation of the uracil rings in oligonucleotides must be due in some way to their proximity to each other, entailing some type of mutual interaction. That some type of interaction exists between adjacent purine and pyrimidine rings in oligonucleotides is obvious from the hyperchromic effect. While it is true that the hyperchromicity of poly-U is relatively low by comparison with other oligo- and polynucleotides, this does not exclude the importance of interaction between adjacent rings.

Accepting, therefore, the existence of such ring interaction, the initial higher quantum yield of UpU (as compared to Up) is due to some influence of one of the uracil rings on the excited state of the other, leading to a more efficient uptake of a water molecule (cf. Section VI-9 on the influence of concentration of Up on its quantum yield). The precise mechanism by which this occurs remains to be established; one possibility is the resonance transfer of energy from one ring to the other during the absorption of the incident radiation. That some such mechanism is involved is, however, amply demonstrated by the marked change in reaction rate after 50 per cent photolysis; for if, at this point, one ring in each dinucleotide molecule has reacted, then the rate of reaction of the second ring should be markedly decreased since it no longer is in proximity to an *aromatic* ring; such is, indeed, the case. The reaction would consequently be represented as follows:

$$\text{UpUp} \xrightarrow{h\nu} \text{Up}\cdot\text{H}_2\text{O}\cdot\text{Up} \xrightarrow{h\nu} \text{Up}\cdot\text{H}_2\text{O}\cdot\text{Up}\cdot\text{H}_2\text{O}$$

and, in fact, the kinetics of the reaction can be formally represented very well by two successive first-order reactions (Wierzchowski, Ph.D. thesis, 1960). A direct proof of this reaction sequence would require photolysis of UpUp to the point where 50 per cent of the rings have reacted and then demonstrating that the product is UpUp·H_2O; this might be done chromatographically since only one spot should then be visible on the chromatogram; attempts to date have failed because of the inability to find a solvent which would separate the photoproduct from UpUp. In addition, we have for the moment ignored the fact that the initial half of the reaction also involves a "non-reversible" stage; we can, nonetheless, assume that the rate of this non-reversible reaction is at least approximately the same as the rate of the initial portion of the reversible reaction.

If we accept the above reaction sequence for photolysis of UpU (or UpUp), we must re-examine the quantum yields calculated above on the assumption that all the absorbed radiation is photochemically effective during the entire course of the reaction. Actually, only about one half the energy absorbed during the first stage is photochemically effective, the quantum yield for the first ring being then $\Phi_1 = 70 \times 10^{-3}$, which is almost three and one half times as high as for Up alone at the same concentration. For the second ring $\Phi_2 = 8 \times 10^{-3}$. The effect of interaction between rings is also placed in evidence by the kinetics of the reverse reaction which, for free Up, is strictly first order. For UpU the reverse reaction is first order only to the point where 50 per cent of the initial absorption has been restored; beyond this point the reaction order is complex and the rate much slower, indicating that the photoproduct of one ring is more stable when it is in proximity to a second *aromatic* ring.

The above treatment may be extended qualitatively to higher oligonucleotides. For example in UpUpUp, if the middle ring reacts first, the other two should then react more like free Up. If, however, one of the outside rings reacts first, then the other two will behave like UpUp. If each possibility is equally likely, then 2/3 of the rings should react at a higher rate. The actual break in the kinetic curve occurs after 60 per cent photolysis instead of 67 per cent. For higher oligonucleotides the situation will not change markedly, since a longer chain consists merely of tracts of shorter oligonucleotides; in highly polymerized poly-U about 75 per cent of the rings react with a higher quantum yield. If it is assumed that only 75 per cent of the absorbed radiation is photochemically effective, it can be shown that the kinetics for photochemical transformation of poly-U is first order and the quantum yield for each uracil ring about 55×10^{-3}.

"Non-reversible" Photolysis in Oligonucleotides of Up: It remains to examine the nature of this reaction. To begin with, it might be assumed that the non-reversible part of the reaction is only apparent and that full reversibility is inhibited by interaction between rings. This was excluded by submitting irradiated poly-U to hydrolysis by ribonuclease and by 1N HCl. In either case we would expect hydrolysis to mononucleotides, if the photochemical reaction had resulted only in the formation of hydrated pyrimidine rings; subsequent heating should give 100 per cent reversibility as for Up. Experimentally the per cent reversibility was found to be unchanged by either of the above treatments (Wierzchowski and Shugar, 1959).

As regards the mechanism of this "non-reversible" reaction, now that its reality has been established, two or three possibilities may be considered:

(a) It will be recalled that the initial phase of photolysis of uracil and orotic acid involves a non-reversible reaction which, under certain conditions, is partially reversible, depending on whether the initial irradiation was under oxygen or nitrogen. However, an examination of the kinetics of photolysis of oligonucleotides of Up showed no effect whatever of either oxygen or nitrogen.

(b) The photochemical addition of water to Up results in the formation of a 5-hydro-6-hydroxy derivative. In all oligonucleotide chains, on the other hand, there is no *a priori* reason for the exclusion of partial formation of 5-hydroxy-6-hydro photoproduct. The latter is known to be stable and not subject to heat reversibility. This possibility has been examined in irradiated poly-U and excluded (Shugar and Wierzchowski, 1960).

(c) Since the photochemical transformation of poly-U is unaltered in buffered medium and is independent of concentration, it appears most likely that the "non-reversible" reaction is intramolecular. Since it is also accompanied by the disappearance of the characteristic absorption maximum, it must involve saturation of the 5,6 double bonds (if we exclude ring rupture). A reasonable mechanism for such a reaction is the formation of dimers by the cross-linking of the 5,5 and 6,6 positions of adjacent rings, as was postulated for *inter-molecular* cross-linking of pyrimidine rings in dry polynucleotide films (Baranowska and Shugar, 1960). Such a reaction is stereochemically possible in oligonucleotides and is in fairly good agreement with all the experimental data. It is also supported by the reported isolation of thymine dimers from irradiated DNA, discussed above.

A similar equilibrium may be expected in a dinucleotide such as UpU or in poly-U. However, as pointed out above, the photochemical dissociation of dimerized UpU may very well lead to the hydration of one or both of the uracil residues. It will consequently be simpler to defer the discussion of this problem to the next section, following a description of the behavior of TpT and oligo- and poly-T.

It remains to resolve the question as to whether dimer formation in poly-U may be reversed by physico-chemical methods. It will be recalled that, in aqueous liquid medium, both uracil and thymine dimers may be photochemically dissociated in high quantum yield to regenerate uracil and thymine. In a polynucleotide chain, such dimer dissociation may conceivably be hindered by the internucleotide linkages, which would prevent the two pyrimidine rings from separating; or, if they did separate, would combine once more under the influence of the incident radiation. Experiments with TpT, to be described in the next section, demonstrate that dimer dissociation in polymer chains does occur quite readily; and that, under the influence of irradiation, an equilibrium is set up between dimer formation and dissociation.

(iii) *Oligonucleotides of thymidylic acid*

Since thymine and its glycosides do not normally undergo photochemical hydration across the 5,6 double bond, it might be expected that polymers of thymidine would be more suitable for investigating the dimerization reaction between adjacent residues in a single-stranded chain.

It has already been pointed out in Section F-1a that apurinic acid (APA) is more susceptible to ultraviolet light than would be expected from the radiation sensitivity of its deoxycytidylic acid component alone and that this was due in part to the increased sensitivity of the thymidylic acid residues in this oligonucleotide. If APA is heated at 100°C at neutral pH, it is degraded to oligonucleotides of thymidylic acid of various chain lengths (Adamiec and Shugar, 1959), which may be fractionated on an ion exchange or ECTEOLA column. The longest chain obtained in this way contains at least 10 residues of Tp and one of Cp so that, essentially, it may be regarded as an oligonucleotide of Tp or, in other words, as oligo-T.

The course of photolysis of oligo-T, based on decrease in ultraviolet absorption with time of irradiation, is shown in Fig. 7-19. It will be observed

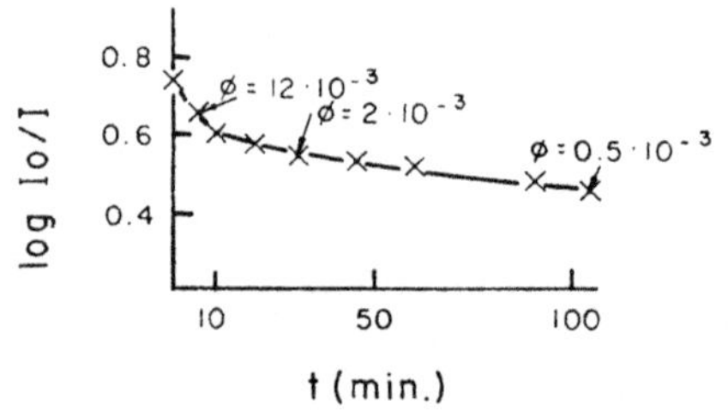

FIG. 7-19. Course of photodecomposition of oligo-T in 0.02 M phosphate buffer pH 7, followed by decrease in optical density of principal absorption maximum at 267 mμ with time of irradiation, at intensity of 2.6×10^{-7} einsteins/cm^2 per min. The calculated quantum yield varies during the course of the reaction, as indicated by the arrows, and decreases to a limiting value of 0.5×10^{-3} equal to that for free Tp after about 35 per cent photolysis (Wierzchowski and Shugar, 1960b).

that the initial quantum yield is about 25 times that for free Tp; following about 35 per cent photolysis, the quantum yield levels out to approximately the value for free Tp. Since the initial reaction is mainly dimerization of adjacent residues, the quantum yield is actually twice that indicated in the figure. The photolysis of oligo-T is not heat-reversible.

The foregoing data were obtained with rather ill-defined preparations of oligo-T and at only one wavelength, 2537 Å. Subsequently the photochemical behaviour of TpT was examined in detail at different wavelengths by Johns *et al.* (1962), Sztumpf and Shugar (1962a, 1962b) and Deering and Setlow (1963). The quantum yield for dimerization was found to be approximately 0.01, varying somewhat as a function of wavelength. On the other hand, the quantum yield for dissociation of TpT-dimer is of the order of 0.5 and relatively wavelength-independent in the range 2400–2900 Å. At neutral pH, and a given wavelength, an equilibrium is set up between dimer formation and dissociation, the equilibrium point being a function of the light intensity, the quantum yields for the two reactions, and the relative absorptions of TpT and TpT-dimer. When these quantities are known, the kinetics of the reaction and the equilibrium point may be calculated quantitatively at any given wavelength (Johns *et al.*, 1962). Or, the TpT dimers may be isolated by paper chromatography and estimated by elution and spectrophotometry.

At 2537 Å, the wavelength used in most photochemical work on nucleic acids, the equilibrium point for TpT is at about 25–30 per cent dimer formation. At wavelengths above 2850 Å, dimerization may attain a value of 90 per cent or more. A more careful examination shows that, to a smaller extent, some reaction other than dimerization also intervenes, but the nature of this reaction is as yet unknown (Sztumpf and Shugar, 1962a; cf. Smith, 1963).

Particularly significant is the fact that, if TpT is first irradiated at wavelengths above 2800 Å, so that dimer formation is high, subsequent irradiation at a shorter wavelength establishes a new equilibrium by photodissociation of some of the dimers formed at the longer wavelength. This led to the suggestion that, if dimer formation were important in the biological inactivation of DNA, it should be possible to inactivate at, say, 2900 Å and then restore an appreciable portion of the biological activity by subsequent irradiation at shorter wavelengths (Shugar and Sztumpf, 1962a). This was, in fact, subsequently shown to be the case for transforming DNA (Setlow and Setlow, 1962) and for the primer activity of DNA with calf thymus polymerase (Bollum and Setlow, 1963) and forms one of the most convincing pieces of evidence for the important role of thymine dimerization in the biological transformation of irradiated DNA, as well as in the phenomenon of photoreactivation.

It is, of course, to be expected that in natural DNA preparations the degree of dimer formation will be dependant on the distribution of thymine residues in the individual chains; this fact has even been applied to examine the distribution of thymine residues in natural DNA chains (Bollum and Setlow, 1963; Smith, 1963; Wacker, 1963).

Dimerization in oligo- and polynucleotides of uridylic acid: It should now be clear from the foregoing why the photochemistry of polymers of uridylic acid can be expected to exhibit considerable complexity. There will be two initial reactions, hydration and dimerization. The latter reaction will be accompanied by photodissociation of dimers. Finally one or both of the excited uracil residues, resulting from photodissociation of a dimer, may then undergo hydration, thus preventing further dimerization at this point.

Irradiation of UpU should therefore lead to the formation of Up·H_2O·U, Up·H_2O·U·H_2O and UpU·dimer, and these three photoproducts have in fact been isolated from irradiated UpU (Johns, personal communication, 1963). The degree of dimerization is, however, much less than for TpT (Sztumpf and Shugar, unpublished) and even in poly-U is quite low, even at longer wavelengths. Swenson and Setlow (1963) have reported in preliminary form that the predominant reaction at 2650 Å is dimerization, and at 2380 Å it is hydration. More quantitative data from the foregoing experiments will be awaited with interest.

Particularly interesting in relation to the above are some findings on the effects of irradiation on the biological activity of uracil-containing polymers, of which two illustrations are available.

It was mentioned above that photo-inactivation of biologically active DNA could be reversed at shorter wavelengths by dissociation of dimers formed at longer wavelengths. A similar behaviour might theoretically be expected with biologically active ribonucleic acids. Such an experiment has been performed by Kleczkowski (1963). Infectious RNA from TMV was irradiated at 2850 Å to different degrees of inactivation; at this wavelength dimer formation should be high. The inactivated RNA was then irradiated at various shorter wavelengths to as low as 2350 Å and reactivation sought for. None was, in fact, found. The most logical explanation for this negative result is that dimer photodissociation at shorter wavelengths was accompanied by hydration of the photodissociated uracil residues. Similar experiments on infectious RNA from other plant viruses are under way and may further clarify the subject.

In a cell-free system for protein synthesis, poly-U is known to code predominantly for phenylalanine (Nirenberg and Matthei, 1961). It was first shown by Grossman (1962) that irradiation of poly-U at 2537 Å results in a loss of the ability to code for phenylalanine, as might be anticipated. The surprising result was, however, obtained that the irradiated polymer exhibited the ability to code for serine. The significance of this finding can hardly be overestimated, particularly in relation to the potential biological effects of irradiation on the transfer RNA in micro-organisms. Since UUU codes for phenylalanine and UUC for serine (Crick, 1963), hydration would appear to be the most logical reaction which would account for the ability of irradiated poly-U to code for serine, since hydrated uracil is structurally closer to cytosine than is uracil dimer. This conclusion derives support from the subsequent finding that irradiation at 3000 Å results in loss of ability to code for phenylalanine without initiation of activity for serine; whereas irradiation at 2500 Å resulted in a change largely in coding ability from phenylalanine to serine (Grossman and Delbruck, 1963). The foregoing should provide some incentive for an examination of the photochemical behaviour of various transfer RNA's (cf. Smith, 1963).

(iv) *Oligonucleotides of cytidylic acid*

For oligonucleotides of Cp, as well as for all hetero-oligonucleotides, the hyperchromicity is appreciable and must be corrected for when using spectral methods to follow the course of the reaction. The procedure adopted has been to assume, as a first approximation, that the hyperchromicity decreases linearly with the course of the reaction, and has been shown to be reasonably well justified; the one exception is for twin-stranded chains, such as poly-A-poly-U, where other methods must be used (see below). In addition, as for cytosine mononucleotides, the rate of the reverse, thermal reaction for all oligonucleotides containing cytosine residues is fairly high at room temperature; hence the photochemical reaction must be conducted in cuvettes thermostated at low temperatures, to inhibit the reverse reaction as much as possible.

Qualitatively the spectral changes accompanying the irradiation of CpCp are completely analogous to those for Cp; the absorption maximum at 2710 Å disappears and is accompanied by the appearance of a new maximum at about 2400 Å (cf. Fig. 6-9B) with an extinction coefficient of 11.7×10^3, or 5 per cent lower than that for the cytidine-2′-phosphate photoproduct. The kinetic data, based on the rate of decrease of the absorption maximum at 2710 Å, are found to give a reasonably good fit to a first order reaction on the assumption that each of the rings is equally capable of reacting. The initial 20 per cent of the curve deviates somewhat from that to be expected and this is reflected in a somewhat increased quantum yield of 11×10^{-3} which, following 20 per cent photolysis, reaches a value of 7.5×10^{-3}, and this remains sensibly constant for the remainder of the reaction.

The almost completely photochemically transformed CpCp will revert in the dark to the parent dinucleotide with an efficiency of about 80–90 per cent. Of the 10–15 per cent which is transformed irreversibly, about one third undergoes this transformation during the initial 20 per cent photolysis, and this may be the source of the initial higher quantum yield.

The kinetics of the reverse thermal reaction are somewhat more complicated than for UpUp and this may be related to the marked influence of the location of the phosphate group in isomeric mononucleotides of Cp on the rate of the reverse reaction. As a result of this the kinetics of the reverse reaction would be expected to be dependent on whether the internucleotide linkages are 2′,5′ or 3′,5′ or a mixture of these two; because of the resultant expected complexity, this has not yet been investigated in detail. However, for CpCp with only 2′,5′ internucleotide linkages the kinetics of the reverse reaction may be readily accounted for by two first order processes; i.e., one of the rings undergoes reversal at a rapid rate and the second at a lower rate. The ratio of the reaction rates is dependent on temperature and varies from 2 to 6 as the temperature is decreased from 40–20°C. The rate constants conform to the Arrhenius equation and the activation energies are 14 kcal/mole for the first ring and 26 kcal/mole for the second.

The photochemical reaction kinetics for a trinucleotide, CpCpCp, containing a mixture of internucleotide linkages, are very similar to those for CpCp, with the exception that the initial quantum yield, 12×10^{-3}, decreases to its limiting value, 7.8×10^{-3}, after about 10 per cent photolysis. The degree of thermal reactivation is also quite high, about 90 per cent.

For highly polymerized poly-C, containing only 3′,5′ internucleotide linkages, the amount of energy absorbed by the individual rings is considerably less than in small oligos because of the very high hyperchromicity. Nonetheless the quantum yield is, on the average, about the same and equals 7.5×10^{-3}. The photochemical reaction may be thermally reversed to the extent of about 85 per cent and proceeds in two stages with rate constants, at 30°C, of 9.5×10^{-4} for the first stage and 1.7×10^{-4} sec^{-1} for the second (see also Table 7-1).

The behavior of polymers of Cp is fully consistent with the principal reaction being that of hydration of cytosine residues. There have been several reports of the isolation of cytosine dimers, but these have not been substantiated. It is, on the other hand, not inconceivable that dimerization occurs to some extent, followed immediately by hydration, and this may be the explanation for the small departure of the kinetics for CpC from a first order reaction.

In the T-even bacteriophage T_4, Drake (1963) reports that the mutagenesis resulting from irradiation by UV is due more to modifications in the base-pair G-HMC (hydroxymethylcytosine) than in the base-pair A-T. On the basis of a personal communication from Wacker, he postulated that the reactions involved are the dimerization of adjacent cytosine residues and also deamination of these residues. It is necessary to point out that there is no evidence whatever for the ultraviolet induced deamination of any 4-aminopyrimidines. It remains to establish whether adjacent 5-hydroxymethylcytosine residues in a polymer chain would form stable dimers. The best way to test this would be by means of a synthetic homopolymer, but such a polymer is as yet unavailable.

(v) *Hetero-oligonucleotides*

The photochemical behavior of several small, hetero-oligonucleotides has already been investigated and the techniques employed described in sufficient detail so that they may be employed on other model oligos (Wierzchowski and Shugar, 1960b; Shugar and Wierzchowski, 1960). The principal results are roughly summarized in Table 7-1, and we shall here dwell only on a few of the salient points involved.

TABLE 7-1. REACTION CONSTANTS FOR PHOTOCHEMISTRY OF URACIL AND CYTOSINE RESIDUES IN POLYNUCLEOTIDES

	Uracil residues in						Cytosine residues in					
	UpUp	poly-U	ApUp	CpU	GpUpA GpApU UpGpA	poly-A poly-U Complex	CpCp	poly-C	ApCp ApApCp	CpU	GpCp GpCpGpCp	poly-I-poly-C Complex
$\phi \times 10^3$	70* 8.0	55* 8.0	7.0	10	6	2.5* 3.5	7.5	7.5	8.2	8.0	5.5	7.5
$k_{20°} \times 10^4$ sec^{-1}	0.01	0.01	0.015	0.015	—	—	0.2* 1.3	[1.7 9.5] 30°	0.6	[1.8]30°	—	—
E kcal/mole	—	—	26.2	—	—	—	14 26	—	18	—	—	—
Per cent TR	85	65	100	90	≥70	100* 80	90	85	100	90	90	≥70

* Reaction proceeds in two stages; upper value is for initial stage.
TR = Thermal reversibility.

For a dinucleotide such as CpU, the kinetics of the photochemical reaction are approximately first order. The uracil ring reacts somewhat more rapidly (quantum yield about 10×10^{-3} as compared to 20×10^{-3} for free Up) than the cytosine residue for which ϕ is about 8×10^{-3} as compared to 13.5×10^{-3} for free C3′p. Hence, as the photochemical reaction approaches completion, the absorption spectrum begins to resemble that for $Cp \cdot H_2O$ alone. As regards the reverse reaction, this initially involves exclusively the cytosine ring with a first order rate constant of 1.8×10^{-4} sec^{-1} at 30C°. For the uracil ring the rate constant at this temperature is 1.5×10^{-6}, almost 100 times lower. Hence, reversal of the uracil residue requires a marked increase in temperature, following which total reversibility for the dinucleotide is about 90 per cent.

The kinetics of the reverse, thermal, reaction for CpU is of additional interest. The proposal has been advanced that cytosine residues may participate in dimer formation much in the same way as uracil or thymine. If, however, this were the case, we should expect the photoproduct of CpU to dissociate in a single step. The fact that each residue undergoes thermal reversal independently of the other furnishes strong evidence against the participation of cytosine residues in dimerization reactions, and supports the contention that each of the aromatic rings in CpU undergoes photochemical hydration.

For dinucleotides such as ApUp and ApCp, it is even simpler to follow both the forward, photochemical, and the reverse, thermal reactions, since the pyrimidine residues may be completely transformed without affecting the purine components at all. Reversibility in both these dinucleotides is 100 per cent. The pertinent kinetic data are tabulated in Table 7-1. The changes in absorption spectrum for the reaction of ApUp are illustrated in Fig. 7-20, and provide an interesting point with regard to the nature of hyperchromicity in small oligonucleotides. The hyperchromicity of ApUp is 13 per cent; as a result of this, simple calculation shows that complete photolysis of the Up component should lead to a decrease in absorption at 2600 Å of 32 per cent; the experimental value from Fig. 7-20 is 31–32 per cent. This therefore provides supporting evidence for the contention that hyperchromicity in small oligonucleotides is due to interaction between aromatic rings and not necessarily to hydrogen bonding.

A somewhat similar conclusion may be arrived at from an examination of the behavior of ApCp at wavelengths longer than 2700 Å. At shorter wavelengths the hyperchromicity of this dinucleotide does not completely disappear, due undoubtedly to the fact that $Cp \cdot H_2O$ possesses partial aromatic character, placed in evidence by its very high extinction at 2400 Å (Fig. 6-9B). The foregoing two examples therefore present an interesting new approach for studing the nature and origin of hyperchromicity in small oligonucleotides, which is deserving of wider application.

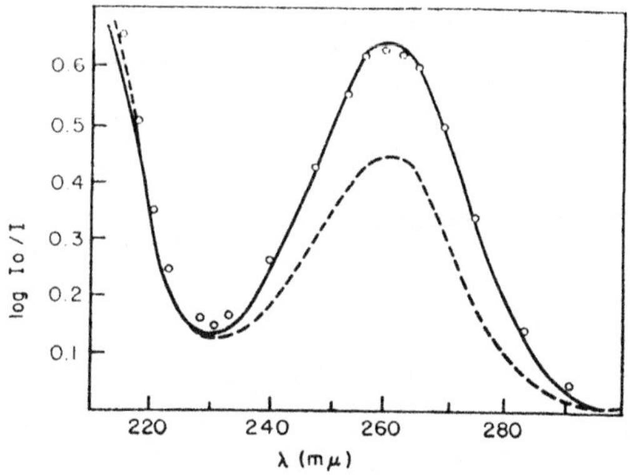

FIG. 7-20. Absorption spectrum at pH 7 of ApUp:

——— prior to irradiation

------ after 70 min irradiation so that the Up component has been completely transformed photochemically to Up · H_2O

○ ○ ○ ○ after heating of irradiated solution 180 min at 80°C.

The extension of the above to the photochemical behavior of a wider variety of model oligos is clearly necessary and desirable for a better understanding of the photochemistry of nucleic acids. This has been carried out for several tri- and tetra-nucleotides (Wierzchowski and Shugar, 1962), some of the results being summarized in Table 7-1, above.

While the foregoing investigations were based on the use of only one wavelength, 2537 Å, much more useful information would be forthcoming from the use of different wavelengths. In particular, it would be desirable to examine the behaviour of some model di- and tri-nucleotides at wavelengths where only one or two of the bases absorb; this would furnish some information as to the relative importance of energy transfer in provoking photochemical reaction in a polymer chain (see below). No such experiments have yet been reported on small oligonucleotides, and initial efforts might profitably be confined to dinucleotides such as ApUp where the adenine residue absorbs at longer wavelengths at which the absorption of the uracil residue falls to zero. Numerous additional combinations of such dinucleotides are possible.

(vi) *Twin-stranded polynucleotides*

The results presented in the foregoing sub-sections are applicable to single stranded polynucleotide chains. While the same general behavior might be expected in native, twin-stranded DNA, some quantitative differences should result from the hydrogen bonding of base pairs and this has, in fact, been verified (Shugar and Wierzchowski, 1960, 1962; Shugar, 1960b) using as models the complexes poly-A-poly-U and poly-I-poly-C.

Some of the data for uracil and cytosine residues in such twin strands are presented in Table 7-1. The most interesting aspect about these models is that the hyperchromicity due to interchain bonding is so high that the decrease in absorption resulting from photochemical transformation of pyrimidine residues is almost compensated for by the increase in absorption due to the accompanying decrease in *interchain hyperchromicity* (see Fig. 7-21). The procedure adopted, therefore, and illustrated in Fig. 7-21 for poly-A-

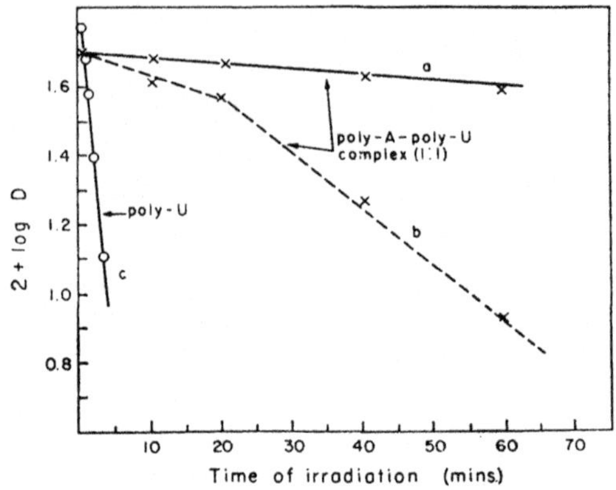

FIG. 7-21. Effect of irradiation on uracil residues in poly-A-poly-U (1:1): Curve *a* shows the experimentally observed decrease in absorption at the maximum vs. time of irradiation; curve *b* is the actual decrease, after correcting for loss of hyperchromicity; and curve *c* is that for poly-U alone. Each point on curve *b* was obtained by enzymatic hydrolysis of the complex after irradiation, as a result of which twin-strand hyperchromicity is eliminated; the absorption of the hydrolysate was then subtracted from that of a control.

poly-U, involves irradiation of the complex with a given dose. The complex is then treated with ribonuclease to hydrolyze the poly-U to the point where all twin-strand hyperchromicity is abolished by dissociation of the complex. The resultant absorption of the solution is then the sum of poly-A plus that portion of poly-U which has not undergone photolysis; this gives us the number of uracil residues which have been transformed, thus providing the necessary information for calculation of the quantum yield. The solution may also now be heated to measure the degree of thermal reversibility. Note in particular the appreciable decrease in ϕ for uracil residues in the complex as compared to poly-U alone (Table 7-1) and how markedly this is reflected in the kinetics for poly-A-poly-U as compared to poly-U alone (Fig. 7-21).

Particularly striking is the fact that, for the initial stage of photolysis, thermal reversibility is 100 per cent; it is only in the later stages of the reaction that thermally non-reversible photolysis occurs. The initial reaction must therefore involve hydration only. This is quite understandable for, in a twin-stranded chain such as poly-A-poly-U, each pyrimidine ring is hydrogen-bonded to a purine, so that rotation about the glycosidic linkages must be severely restricted. Under these conditions, interaction between the adjacent pyrimidine rings in the poly-U strand would be most unlikely.

It is only following partial collapse of the twin-stranded structure that thermally non-reversible photoproducts, probably in part uracil dimers, begin to make their appearance. The gradual collapse of the twin-stranded structure is indeed a logical consequence of the initial hydration of some of the uracil residues. For, as follows from the spectral observations referred to above, the hydration of each uracil residue leads to the dissociation of its twin-strand hydrogen bonds to the corresponding adenine residue. It was, in fact, observed in independent experiments that irradiated poly-U fails to form a twin-stranded complex with poly-A; but it remains to examine this phenomenon quantitatively, i.e. to determine how many uracil residues in poly-U must undergo photochemical transformation to completely inhibit twin-strand formation.

Qualitatively similar results have been obtained for the twin-stranded complex poly-I-poly-C (Wierzchowski and Shugar, 1960, 1962), where dimerization (of cytosine residues) is not to be expected. Somewhat surprising is the fact that neither for this complex, nor for poly-A-poly-U, was it possible to obtain full thermal reversibility of even the initial photochemical transformations without prior dissociation of the complexes. No explanation for this anomaly is as yet forthcoming. Even if, as the evidence indicates, the initial hydration of pyrimidine residues results in partial dissociation of the twin-stranded structure, it is nonetheless puzzling that the strands will not reform on heating. This problem merits further investigation.

Obvious interest attaches to the behavior of a twin-stranded complex in which one of the strands contains only thymine residues, which do not undergo hydration. Such a complex with poly-A is possible by making use of the poly-ribosethymine phosphate of Griffin *et al.* (1959). The properties of poly-ribosethymine phosphate, or poly-rT, and its 1:1 complex with poly-A, have now been elucidated (Shugar and Szer, 1962; Szer *et al.*, 1963) and photochemical investigations on these polymers are in progress. Poly-rT possesses the interesting property that its T_m is about 35°; hence at 20° it is largely in a highly ordered form, while above 40° it is a random coil. At a temperature of about 25°, the behavior of poly-rT is qualitatively that to be expected from the findings for TpT and oligo-T. Bearing in mind the absence of the hydration reaction, the behaviour of the 1:1 complex of poly-rT with poly-A is also in accord with expectations; at 2537 Å, the complex is remarkably resistant to irradiation.

The pronounced decrease in quantum yield for the uracil resides in poly-A-poly-U leads to a rather interesting conclusion. It has been postulated by a number of observers that photochemical effects in nucleic acids are due in part to energy transfer along the chains. There is no question but that energy transfer along such chains does exist. But, in view of the decrease in quantum yield for uracil residues in poly-A-poly-U, and the unchanged quantum yield for cytosine residues in poly-I-poly-C (Table 7-1), it seems unlikely that such energy transfer is photochemically effective. If it were, we should expect higher quantum yields.

2. *Reversal of Biological Inactivation*

It has long been known that ultraviolet-inactivated bacteriophage cannot be photoreactivated with either visible or near ultraviolet radiation unless they are absorbed to their host cells, the implication being that some cellular factor or mechanism is involved in the process. Attempts to photoreactivate isolated ultraviolet-inactivated T-DNA with the 3650 Å band of a mercury arc, under conditions where photoreactivation obtained for ultraviolet-inactivated T1 phage absorbed to *E. coli B* cells, likewise gave negative results. (Zamenhof *et al.*, 1956; Zamenhof, 1957). It would indeed have been surprising if photoreactivation had been observed under these conditions since, like purified phage, DNA and irradiated DNA do not exhibit *specific* absorption either at 3650 Å or at longer wavelengths.

It was later shown by Goodgal *et al.* (1957) and Rupert *et al.* (1958) that T-DNA for Sd *Haemophilus influenza*, following ultraviolet inactivation to the point where the residual activity is about 1 per cent, may be partially reactivated by exposure to a cell-free extract of *E. coli B* in the presence of magnesium ions under the influence of visible light. The effect of visible radiation on the reactivation process is illustrated in Fig. 7-22. The degree of reactivation is proportional to the concentration of the extract, the intensity of illumination with visible radiation and the temperature; the influence of the latter is shown in Fig. 7-23.

The active portion of the extract was found to consist of a dialysable, heat-stable, component and a non-dialysable constituent which could be destroyed by heating for 1 min at 95°C. The degree of reactivation attained with these extracts varied from 10–50 per cent of the activity of a non-irradiated control. It appears difficult to avoid the conclusion that the photorecovery process is catalyzed by some light-activated enzyme. It is of interest in this connection that the active extract is obtained from *E. coli B*, an organism which is itself photoreactiveable, whereas an extract from *H. influenza* (an organism which is not photoreactiveable following ultraviolet irradiation) is inactive in the *in vitro* photoreactivation system.

Attempts have been made to photoreactivate T-DNA from other organisms, but the results have not been as unequivocal, nor as marked as for

H. infl. DNA. Rebeyrotte and Latarjet (1960) have found that transforming DNA from *D. pneumoniae* may be photoreactivated to a small, but definite, degree by an illuminated *E. coli* extract. This has been confirmed by Pakula *et al.* (1960); but negative results have been reported by

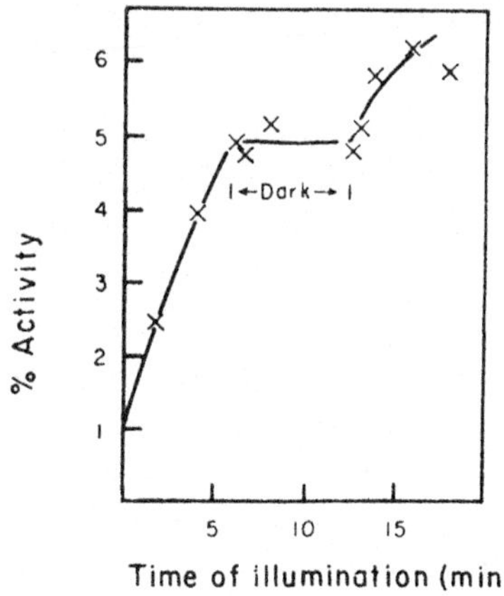

FIG. 7-22. Photoreactivation of ultraviolet inactivated transforming DNA from *Haemophilus influenza* in the presence of a cell-free extract from *E. coli* B, under the influence of visible light. Over the time interval defined by "dark" the light source was removed and is seen to be accompanied by a cessation of photoreactivation (Rupert, Goodgal and Herriot, 1958).

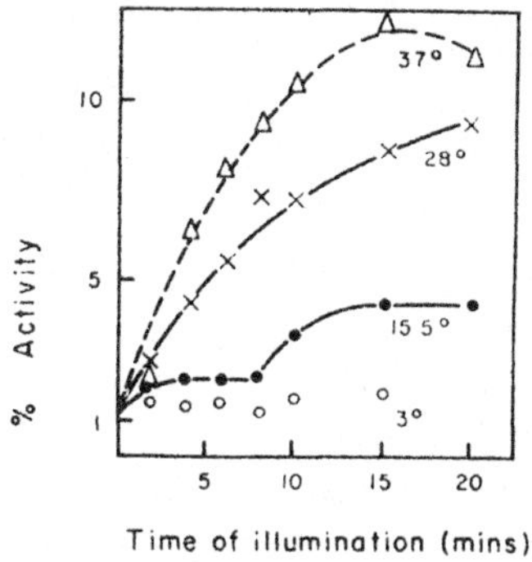

FIG. 7-23. Influence of temperature on rate of photoreactivation of transforming DNA by *E. coli* extract and visible light (Rupert, Goodgal and Herriot, 1958).

Ellision and Beiser (1960). Pakula *et al.* (1960) have also demonstrated slight photoreactivation of streptococcal T-DNA by a *coli* extract.

From the foregoing results it would be premature to draw any conclusions regarding the specificity of the *coli* extract towards T-DNA from different sources. On the other hand, in view of the high efficiency of photoreactivation obtained with the streptomycin resistance marker of *H. infl.* T-DNA, it appears almost logical to examine the activity of the *E. coli* extracts against several different markers of this transforming principle.

An enzyme system has also been found in cell-free yeast extracts which, under the influence of visible light, exhibits similar photoreactivating ability towards *H. influenza* T-DNA as an *E. coli* extract (Rupert, 1958, 1960). Such extracts have, however, been found quite ineffective in attempts to photoreactivate streptococcal T-DNA (Pakula and Walczak, personal communication, 1960).

The *E. coli* extract system has been utilized by Rupert (1958) and by Rupert and Herriot (1958) in an ingenious manner to demonstrate ultraviolet damage, as well as photoreversal of such damage, in non-transforming DNA. The technique is based on the observation that photoreversal of irradiated transforming DNA is not inhibited by non-transforming DNA, but does show definite inhibition in the presence of *irradiated* non-transforming DNA. If, however, the latter is first subjected to photoreactivation, it then no longer inhibits photoreactivation of irradiated T-DNA. This observation is of considerable value in that it demonstrates, rather indirectly it is true, activity of the photoreactivating enzyme system against a wide variety of nucleic acids. It would nonetheless be much more satisfying to obtain more concrete evidence of the ability of the enzyme system(s) to biologically photoreactivate T-DNA samples from bacterial strains other than *H. influenza.*

The mode of action of the photoreactivating enzyme from yeast extracts has been examined in greater detail by Rupert (1960a, 1960b), who has shown that it is bound in the dark by ultraviolet irradiated T-DNA, but not by non-irradiated DNA. An investigation of the kinetics of the reaction, which follow Michaelis–Menten kinetics at constant visible light intensity, suggested that the necessary illumination provokes photolysis of the enzyme-substrate complex with concurrent repair of the DNA substrate.

The problem that now poses itself is that of determining what it is in the irradiated DNA that the photo-enzyme reverses or "repairs." Two concrete possibilities which may be envisaged are removal of a water molecule from the 5,6 saturated bonds of pyrimidine residues and/or the dissociation of pyrimidine dimers. Attempts to demonstrate activity of an *E. coli* extract against hydrated pyrimidine photoproducts have been unsuccessful. On the other hand extensive and convincing evidence has now accumulated which demonstrates that the photo-enzyme does catalyze the dissociation of thymine dimers (Wulff and Rupert, 1962; Setlow and Setlow, 1963; Bollum and Setlow, 1963; Wacker, 1963; Smith, 1963) with the simultaneous partial restoration of biological activity of UV-inactivated transforming and "primer" DNA. The available evidence also indicates that the dimerization reaction which is reversed by the isolated photo-enzyme accounts for about 50 per cent of the activity which is lost as a result of UV-irradiation. These results should stimulate efforts to find some simpler substrates for the photoreactivating enzyme, such as TpT-dimer or irradiated oligo- or poly-T.

Considerable interest likewise attaches to the possibility of reversal of ultraviolet damage, by cell-free enzyme systems, in infectious RNA and in bacteriophages. Pichtl (1960, personal communication) has indeed reported that an *E. coli* extract is effective in inducing photoreactivation in phages; but the degree of photoreactivation obtained was much too low to be significant. No efforts in this direction have as yet been reported with plant or animal viruses.

G. SUMMARY OF PRIMARY EFFECTS OF ULTRAVIOLET LIGHT ON NUCLEIC ACIDS

From the data hitherto accumulated, we can now formulate some general conclusions with regard to the chemical and physico-chemical modifications induced in nucleic acids by doses of radiation in the range responsible for biological effects, i.e. doses of the order of 10^{16}–10^{17} quanta/cm^2 per min, or 10–100 ergs/cm^2 per sec applied for time intervals of from 1 to 30 min. We shall now summarize these briefly.

Physico-chemical studies show that:

(a) There is no rupture of internucleotide linkages

(b) There is no change in viscosity or intrinsic viscosity

(c) There is no measureable change in molecular weight

(d) There is, in the case of single-stranded chains, only a detectable drop in ultraviolet absorption.

(e) For T-DNA there is a slight decrease in the temperature of the transition profile (or denaturation temperature), suggestive of the rupture of a very few twin-strand hydrogen bonds. There is also some evidence for formation of heat-stable cross-links between the twin strands.

From the foregoing, it follows that the damage accompanying biological inactivation must be highly localized in character and confined to only a few residues.

Chemical effects are as follows:

(a) Carbohydrate moieties are unaffected

(b) Purine bases are apparently unaffected

(c) For Cp residues, and for isolated Up residues, the predominant reaction is the formation of 5-hydro-6-hydroxy derivatives which are heat labile and capable of reverting to the parent compound under the influence of heat by elimination of a water molecule

(d) In tracts of Up residues the foregoing hydration reaction is accompanied by dimerization of some pairs of adjacent uracil rings

(e) In tracts of Tp residues the principal reaction is dimerization of neighbouring pairs of thymine rings. There is, however, at least one additional type of transformation undergone by thymine residues; but the nature of this reaction remains to be elucidated.

(f) In twin-stranded chains, like DNA, dimerization of adjacent residues in each strand does not occur until partial collapse of the twin-stranded structure. When this stage has been reached, there is apparently some dimerization of residues between strands

(g) In dried films of nucleic acids, and possibly in nuclei, there is cross-linking, most likely due to pyrimidine dimer formation, between chains

(h) The behavior of 5-substituted cytosine residues, particularly in the T-even phages, remains to be clarified. Available evidence suggests that they undergo some rearrangement which is potentially reversible, by thermal or other means

(i) Reference should be made to Table 7-1 for the influence of base sequences, chain-lengths and Watson–Crick structure on quantum yields and extent of thermal reversibility for Up and Cp residues. Extension of these results to other base sequences is to be desired.

(j) In bacterial cells and in bacteriophages there is some cross-linking induced between DNA and protein; but both the nature and extent of this reaction remain to be clarified.

While the foregoing outline is by no means meant to be exhaustive, it nonetheless indicates that much has yet to be done to clarify the nature of the chemical and physico-chemical modifications accompanying the biological effects resulting from the ultraviolet irradiation of nucleic acids. However, the progress made during the past few years should provide the encouragement necessary to a complete elucidation of all the photochemical reactions involved.

Number of residues affected by inactivation: While it is not possible to determine unequivocally the number of residues transformed when a nucleic acid molecule is inactivated, the number involved may be estimated roughly in several ways if we bear in mind that purines are not affected and, from the experiments on model polynucleotides mentioned above, that radiation absorbed by purine residues is not photochemically effective against pyrimidine residues.

As pointed out above, 90 per cent inactivation of pneumococcal T-DNA is accompanied by a drop of 1.5°C in the temperature of the transition profile (or denaturation temperature, T_m), indicating rupture of a few inter-strand hydrogen bonds (Marmur and Doty, 1959). From Fig. 2-20 on page 76 it will be seen that a decrease in guanine-cytosine content of 2 per cent results in a drop in T_m of about 1°C. Since a G-C base pair involves three hydrogen bonds as compared to two for an A-T base pair, a drop of 1°C in T_m corresponds to a decrease of 2 hydrogen bonds per 100 base pairs and 1.5° to 3 bonds per 100 base pairs or 200 residues. Loss of transforming activity is therefore accompanied by a modification of one to two bases per 100. We have ignored in this calculation the possibility of reformation of some hydrogen bonds by thermal reversal, as well as the considerable error to which the decrease in T_m is subject.

From the 0.7 per cent decrease in ultraviolet absorption of ultraviolet inactivated infectious RNA from TMV (McLaren and Takahashi, 1957) and from the quantum yield data for the same RNA as well as for the genetic markers of various transforming principles, the data of Table 7-2 make it possible to estimate that about 1 base in 200–400 is affected as a result of biological inactivation. These estimates are to be compared with those of Freese (1959) on reverting point mutations in phage T4 induced by chemical agents or base analogous, from which it was concluded that each mutation is due to the modification of a small number, and probably only one, base pair. The positions of the altered bases in the inactivated RNA's constitute one of the most important problems to be solved in molecular biology.

Finally, it is difficult to avoid the conclusion that there is some relationship between biological and physico-chemical reactivation; we shall, however, defer this subject to a later chapter.

H. INDIRECT IRRADIATION EFFECTS

(photosensitization, free radicals)

In a system containing other components in addition to nucleic acids, excitation of the former by radiation in the region in which they absorb may result, as in the case of proteins and enzymes, in modifications of the nucleic acids through photodynamic action, activity of free radicals, etc. Relatively little attention has been devoted to this subject, although it has been known for many years that photodynamic action may be both mutagenic and lethal to living organisms. Furthermore, as will be recalled from the preceding section, the photoreactivation of inactivated T-DNA is largely the end result of the absorption of light by a cellular component (s) other than DNA, in this case by a light-activated enzyme.

It was found by Spiegel-Adolf (1958) that exposure of nucleic acids to a photosensitizer such as fluorescein in the presence of diffuse daylight results in modifications of the absorption spectra of the former. For DNA the absorption at 2600 Å decreases; for RNA it initially increases about 11 per cent, following which a steady decrease sets in. This difference in behavior of the two types of nucleic acids was claimed to parallel that of thymine and uracil. Of the two purine components, guanine was found to be the much more susceptible. On the other hand, the exposure times required were extremely long, up to 67 days for some nucleic acid samples, so that these required sterilization prior to irradiation; consequently, at least insofar as the DNA is concerned, it must have been denatured even before the start of the experiment.

The influence of photodynamically active dyes, such as eosin and methylene blue, and carcinogens (1,2-benzanthracene and 20-methylcholanthrine) on the viscosity of DNA was investigated qualitatively by Koffler and Markert (1951). All of the above substances were found to be effective in reducing the viscosity of the nucleic acid in the presence of visible, near ultraviolet or ultraviolet radiation; in the latter case a suitable correction was introduced for the effect due to direct absorption of ultraviolet by the NA. Unfortunately, as in the studies of Spiegel-Adolf (1958), no quantitative data are provided; it is, nonetheless, evident from the long exposure times required to induce a noticeable reduction in viscosity that the quantum yield for the process must be very low indeed, a rather interesting result in view of the relatively high sensitivity of enzymes to the dyes used (Chapter V).

An experiment of Zamenhof *et al.* (1956) is of some assistance in estimating the biological efficiency of the above process. T-DNA was irradiated in the presence of acriflavine with visible light from a 150 W bulb at a distance of 1.5 cm for a period of 1 hour. Under these conditions no loss in transforming activity could be observed, although the light intensities used were probably even greater than in the experiments of Koffler and Markert (1951). By contrast, acridine readily photosensitizes the inactivation of TMV (Oster and McLaren, 1950), while many dyes are known to be active against a variety of viruses and bacteriophage.

Polymethylacrylic acid is considerably more susceptible than DNA to degradation by visible light in the presence of rose bengal or acriflavine, and Alexander and Fox (1954) therefore investigated this substance in greater detail. It was concluded that the reaction mechanism involves the rupture of carbon–carbon bonds. Since the reaction could be inhibited by addition of reagents which normally counteract the degradation produced by ionizing radiations, it was proposed that the photodynamic action involved the formation of $HO_2\cdot$ radicals as a result of the reaction of the excited dye molecules with dissolved oxygen, as follows:

$$D + h\nu \rightarrow D^*$$
$$D + O_2 \rightarrow D^+O_2^-$$
$$O_2^- + H_2O \rightarrow OH^- + HO_2\cdot$$

A somewhat similar scheme has been proposed by Bellin and Oster (1960) from a study on the mechanism of photodynamic action based on an examination of the kinetics of the dye-sensitized photo-autoxidation of toluenediamine. The light-excited dye is believed to undergo a transition to a long-lived excited state and reacts with dissolved oxygen to produce labile peroxides which, in turn, oxidize the substrate. It was found that only those dyes which are photoreducible can serve as photosensitizers for oxidation, and it is claimed that only such dyes are effective for the photosensitized inactivation of T-DNA. In a parallel investigation (Oster and Bellin, 1960) it was found that, following injection of thionine into mice bearing ascites

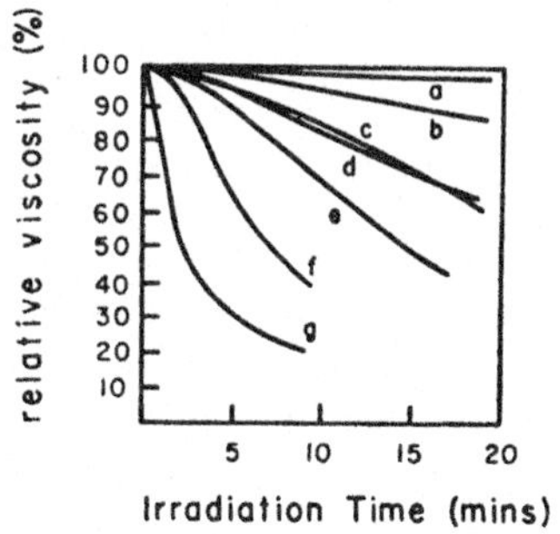

FIG. 7-24. Change of viscosity of 0.1 per cent solution of DNA, irradiated by means of a mercury arc under the following conditions:

(*a*) in glass tube, under N_2 or O_2
(*b*) in quartz tube, under N_2
(*c*) in quartz tube, under O_2
(*d*) in glass tube, with 0.002 M H_2O_2
(*e*) in glass tube, with 0.01 M H_2O_2
(*f*) in quartz tube, with 0.002 M H_2O_2
(*g*) in quartz tube, with 0.01 M H_2O_2
(Butler and Conway, 1953).

tumor, irradiation with visible light provoked tumor regression and increased the life span of the animals.

In a separate investigation on tumours *in vitro* it was observed that the capacity of dyes to act as photosensitizers against tumour cells parallels their capacity to sensitize the photoinactivation of other substances and likewise correlates with their ability to undergo photoreduction (Bellin *et al.*, 1961).

A rather useful approach to the overall problem of photosensitized reactions involving nucleic acids was adopted by Simon and Van Vunakis (1962). A detailed investigation was made of the reaction products resulting from exposure to visible light in the presence of methylene blue of the commonly encountered nucleic acid derivatives and of DNA itself. It was found that guanine residues were selectively attacked, either in the free state or when incorporated in DNA. These results are reminiscent of those found earlier with proteins and enzymes, and are undoubtedly worth following up in more detail. In particular, the use of the biological activity of transforming DNA in such systems should provide valuable information. Acridine orange was not found to exhibit selective activity against guanine residues; the action of this dye on DNA has been studied in considerable detail by Freifielder *et al.* (1961). An extension of this type of approach to the photosensitized inactivation of bacteriophages (Section VIII-D) is highly desirable.

A number of observers have examined the action of photochemically produced radicals on DNA in connection with attempts to interpret the mechanism of action of various ionizing radiations. If DNA is irradiated in the presence of H_2O_2 at wavelengths below 3100 Å so that the hydrogen peroxide is dissociated with very high yield:

$$H_2O_2 + h\nu \rightarrow 2OH\cdot$$

the resulting photochemically produced hydroxyl radicals are much more efficient in degrading nucleotide chains than ultraviolet light alone (Smith and Butler, 1951), as may be seen from Fig. 7-24. The loss of viscosity resulting from the action of such radicals is quantitatively similar to that induced by a dose of X-rays yielding an equal number of hydroxyl radicals, (Butler and Conway, 1953; Butler, 1955), the reactions brought about including deamination and dehydroxylation of the bases, ring destruction, chain breakage, oxidation of carbohydrate moieties, etc.

CHAPTER VIII

Inactivation of Viruses

A. INTRODUCTION

Luria has supplied us with a workable definition: "Viruses are submicroscopic entities, capable of being introduced into specific living cells and of reproducing inside such cells only" (1953). The entry of a single virus particle into a living cell can cause a cell or a cell system to produce more particles essentially like the original invading particle. The process of replication may or may not cause the host to be in a state of pathogenicity. Ultimately, some or all of the newly formed virus is liberated to the environment, whereby the infectious process can spread.

The purification of a virus involves a number of techniques for increasing their amount relative to the amount of non-viral source material present. These include differential filtration, precipitation, and absorptions and centrifugations. These procedures, often in combination, yield preparations of viruses which sometimes crystallize in two (tactoid formation) or three dimensions. In particle size the viruses fall between those of proteins and small bacteria. Chemical analysis together with electron microscopy and X-ray crystallography applied to these entities have shown virsuses to contain protein and one of the two types of nucleic acid; plant viruses contain RNA, bacterial viruses and animal viruses may contain either. The "life cycles" of the viruses share some common features. Following attachment of the virus to the host cells, the nucleic acid is separated from the protein moiety either by "injection" of the nucleic acid into the cell, or else the protein is removed from the nucleic acid by a "digestive" process within the cell. The injection or entry is preceded by some digestive process at the cell membrane (bacteria) or accompanied by pinocytosis, phagocytosis, or some other but not clearly defined energetic process. It is also possible, in numerous cases, to isolate the nucleic acid moiety by physico-chemical means and infect cells with it alone (Gierer and Schramm, 1956; Fraenkel-Conrat, 1956). Presumably the protein moiety protects the virus from disintegration by nucleases in the host, once it is formed. Once in the cell the nucleic acid then directs the synthesis of itself, in the manner, presumably, as does a gene, and of its own specific protein (Zinder, 1960). These two moieties then combine in the cell to produce new virus. A combination of moieties can also take place in a test tube under proper conditions of concentration, pH, and ionic strength, as with e.g. tobacco mosaic virus (TMV).

Virologists early in the field irradiated crude viruses to determine whether they could be inactivated by ultraviolet light and action spectra were found at a few wavelengths to see if they corresponded to protein or to nucleoprotein. The early quantitative work has been reviewed by McLaren (1949), Luria (1951, published 1955) and Pollard (1953). Recent reviews are by Kleczkowski (1957), Gard and Maale (1959) and Shugar (1960). No attempt will be made to cite all examples; instead, a detailed description of the mechanism of inactivation (and reactivation) of viruses will be presented as far as is known, together with some timely qualitative information.

B. INACTIVATION STUDIES WITH A PLANT VIRUS

Shortly after isolation of tobacco mosaic virus in pure form, Stanley (1936) reported that ultraviolet treatment produced inactive products, which, although slightly altered, retained certain chemical and serological properties characteristic of the virus (nucleo) protein. The inactive virus (which had lost its ability to infect susceptible plants) was found to be still soluble but tended to be more easily denatured than active protein. What was remarkable, however, was how subtle must be the inactivation reaction step: the optical rotation of solutions of virus treated with ultraviolet radiation, and the isoelectric point of the virus, were practically unchanged; the molecular weight was not greatly changed and the product could be crystallized; the serum of an animal injected with the product has a neutralizing effect on TMV. The molecules of TMV were found by others to be inactivated according to the usual (pseudo-) first order kinetics, "thus indicating that the absorption of a single quantum is sufficient to bring about inactivation of this gigantic molecule" (Uber, 1941).

Bawden and Kleczkowski (1953) maintain that the first order kinetics observed does not necessarily mean that a single quantum can inactivate the virus particle, because inactivation by heat is also first order, "but no one suggests that heat inactivation occurs because of a single-hit." The authors failed to consider the dependence of the ultraviolet inactivation constant on the initial virus concentration, and the success of the reciprocity law also supports the single-hit theory. Besides, heat inactivation, involving a quantized activation energy, can also be thought of as a "single-hit" or excitation per reacting molecule.

By assembling information from several sources, Uber was able to calculate a quantum yield which, he pointed out, is very small compared to that of urease and other proteins. Spectral data on the virus substantiated the presence of nucleic acid in the protein, and an action spectrum established that only wavelengths less than 3000 Å are appreciably effective (Hollaender and Dugger, 1936).

Irradiation of virus at 37°C results in an increase in turbidity, although warming of the virus to 37°C without irradiation does not give such an increase. Irradiation at room temperature produces inactivation without an increase in turbidity, so the virus is sensitized to heat by ultraviolet light (Oster and McLaren, 1950). Following this report, Kleczkowski (1954) studied the relationship between the loss of activity and the increase in the heat sensitivity caused by ultraviolet light. The change in heat susceptibility did not coincide with the loss of infectivity, because the amount of denatured material forming at 69°C for 30 min following irradiation to 99 per cent inactivation was only slightly greater than that which formed in the non-irradiated solution. Only when irradiation was continued (2537 Å) was the amount of denatured material increased appreciably. The changes in ultraviolet irradiated TMV that increase its susceptibility to heat denaturation probably involve the protein moiety only and indicate that the RNA component is inactivated with higher quantum yield than the yield for denaturation of the protein at this wavelength. The intrinsic viscosity of solutions of ultraviolet inactivated TMV kept at room temperature near pH 7 is the same as that of active TMV, indicating no change of shape or aggregation (Oster and McLaren, 1950). Electron micrographs showed differences in the manner in which particles were distributed only after prolonged irradiation. Whereas in control preparations most particles were separate, heavily irradiated virus showed particles almost all aggregated side-by-side in two-dimensional sheets of various sizes. When the antigenicity of TMV is destroyed at 2537 Å, electron microscopic examination shows complete destruction of the original structure (Kleczkowski, 1962). The ultraviolet spectrum is unchanged following nearly complete inactivation, and inactivation was found to be independent of whether oxygen or oxygen-free nitrogen was bubbled through solutions before and during irradiation. No nucleic acid is released from the virus during inactivation. High doses of ultraviolet light does lead to splitting of RNA from protein (Zech, 1961). TMV does not show any post-irradiation thermal inactivation, as does chymotrypsin (Bawden and Kleczkowski, 1953).

Taken together these results, somewhat different from those with enzymes, suggest that the nucleic acid is the essential target in the inactivation process of the virus. Let us examine the status of structural analysis of TMV following Fraenkel-Conrat (1956) and Ginoza (1958). The elementary particle or molecule, about 150 Å in diameter, is in the form of protein sub-units set in helical array about the long axis of the particle (Franklin, 1956). A hole of radius 19 Å extends down the length of the particle. There are rather deep indentations in the virus surface which follow the pitch of the helix. The nucleic acid in the virus, probably a single-stranded polymer chain (Gierer, 1957) is probably situated 40 Å from the particle axis and is deeply imbedded in the virus protein, Fig. 8-1a. Hydration of the virus is small, perhaps a monolayer of water per rod (McLaren and Rowen, 1951).

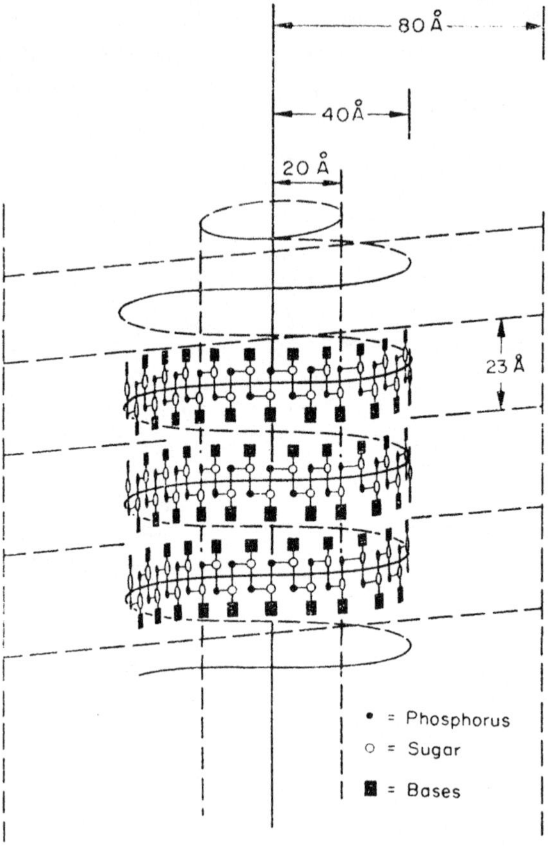

FIG. 8-1a. Schematic representation of a single-strand nucleic acid helix of 40 Å radius in position within the tobacco virus rod. The spacings of the purine and pyrimidine bases, ribose rings and phosphorus atoms with respect to the helix are approximately to scale (Ginoza, 1958).

The protein and ribonucleic acid components may be separated by mild chemical treatment and it has been found that the isolated RNA by itself is infective (Gierer and Schramm, 1956; Fraenkel-Conrat, 1956). When the two components are mixed together under proper conditions they reform infective virus particles. Thus, one is naturally led to a study of the action of ultraviolet light on the infective RNA. Some of the available information has already been cited in the previous chapter, and the subject will now be treated in greater detail.

The RNA from TMV has a molecular weight of about 2.1 to 2.5 $\times$ 10^6, which corresponds to all of the nucleic acid in the TMV molecule. As prepared by the method of Gierer and Schramm it is polydisperse, with about 35 per cent sedimenting under a peak in a sedimentation diagram between 25 and 30 Svedberg units (McLaren and Takahashi, 1957; Friesen and Sinsheimer, 1959). This RNA seems to have a secondary structure consisting of small helical regions involving about half of its nucleotides at room temperature (Doty *et al.*, 1959). The fractional optical densities at 2300 Å, 2537 Å, and 2600 Å, which are due to light scattering are small, *ca.* 0.02 to 0.04 (cf. Hopkins and Sinsheimer, 1955). Intrinsic viscosities of RNA preparations are as high as 270.

During most of the course of inactivation by irradiation at 2537 Å, one-hit kinetics are observed. It is interesting to note that if the initial preparation of RNA consists of a mixture of active, i.e. infective, and inactive molecules, as may well be the case with the RNA, equation (1-8a) still holds, provided that all molecules are of the same size and optical properties. Here

$$[A_0] = [N_i] + [N_0]$$

where $[N_i]$ is the initial concentration of inactive RNA at zero time of irradiation, and $[N_0]$ is the initial concentration of infective nucleic acid. This may be deduced from the method of derivation of the equation (McLaren, 1951). In other words, only the total concentration of nucleic acid need be known, together with the per cent loss of activity with a given dose.

Unlike TMV itself, photoreactivation of ultraviolet inactivated nucleic acid is moderately efficient (Bawden and Kleczkowski, 1959; Rushizky *et al.*, 1960) and must be taken into account in interpreting quantum yield data. The quantum yields for RNA with or without *in vivo* photoreactivation after irradiation and inoculation are essentially constant between 2300 and 2800 Å, the range of wavelengths studied; the yields being 1.8×10^{-3} and 3.6×10^{-3} respectively. The figure 1.8×10^{-3} is obtained either in water or in M/100 phosphate, pH 7.1, and at 6°C or 13°C. It is reduced by a factor of three to five if the concentration during irradiation is reduced four to five-fold (i.e. to 10 μg/ml.). Between concentrations of about 40 and 100 μg/ml., little or no difference in quantum yields was observed at 2537 Å.

Concentration dependence is difficult to interpret; and this renders it difficult to compare the yields with those of model compounds with assurance. Perhaps the best procedure is to utilize quantum yields at infinite dilution since we do not know the possible influence of hydration, dimerization, or mutual photosensitization (Chessin, 1960) on the results. For comparison with RNA, uridylic acid was irradiated at optical densities of 0.25 and 1.25 at 2537 Å (1 cm path length) and the quantum yields for loss of characteristic absorption spectrum were 0.017 and 0.019. With polyuridylic acid irradiated under the same conditions, quantum yields were 0.032 and 0.039, respectively. We thus find a twofold increase in ultraviolet sensitivity of the uridylic acid polymer over that of the monomer, but no concentration dependence of the quantum yield.

Rushizky and Pardee (1960) calculated some quantum yields, based on optical density changes (Table 8-1).

TABLE 8-1. SOME QUANTUM YIELDS FOR NUCLEIC ACIDS AND CONSTITUENTS

Substance	Φ
Uridylic acid	0.02
Cytidylic acid	0.01
Deoxycytidylic acid	0.003
{RNA (From TMV)	0.005*
{Ribonucleotide mixture	0.014*
{DNA (thymus)	<0.002*
{Ribonucleotide mixture	0.004

* Based on pyrimidine bases reacting and absorbing.

By comparison, a mixture of free bases in the proportions found in RNA behaved photochemically as one would expect for the summation of the reactivity of the separate monomers; by contrast, the quantum yield for infectious RNA, based on optical density changes, is less than expected. A hyperchromic effect is probably obtained when the pyrimidine bases are modified and the hydrogen-bonded purines are released in RNA, and tends to cancel the anticipated optical density decrease (cf. Shugar and Wierzchovski, 1960; Wierzchowski and Shugar, 1962). A quantum yield for RNA based on the optical density change and the total quanta absorbed is about that found for loss of biological activity. This probably means, then, that uridylic and cytidylic acid residues in the RNA were altered during inactivation. There is no reduction of intrinsic viscosity of the NA during irradiation; therefore inactivation proceeds without depolymerization of the RNA (McLaren and Takahashi, 1957). The intrinsic viscosity of RNA extracted from TMV inactivated with ultraviolet light does not differ from RNA extracted from active TMV (Buzzell *et al.*, 1959). A reduction of intrinsic viscosity to one-half its initial value would be expected for a random break in each RNA rod should such a break be involved in inactivation (Lauffer *et al.*, 1956). At present we are unable to decide from the irradiation data alone whether some particular base must be altered with a high quantum yield, or whether any one may be altered but with a low quantum yield (more than half of all nucleotides are necessary for activity (Schuster and Schramm, 1958)). It is clear that the inactivation step with ultraviolet light must be rather local on the RNA chain and does not involve chain

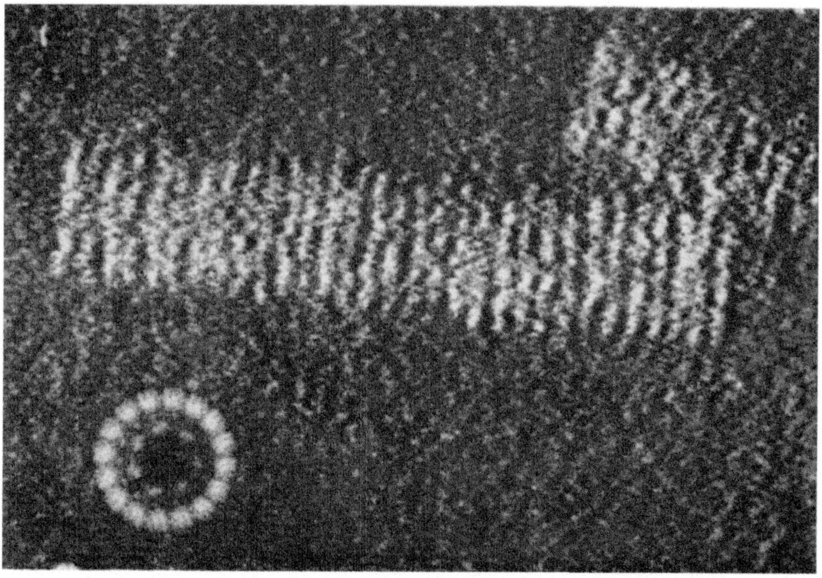

FIG. 8-1b. The stacked disc configuration of TMV "*X*" protein (1000 000×) (R. Markham *et al.*, 1963).

breakage. In fact, inactive TMV can be reconstituted from ultraviolet inactivated RNA and virus protein (Chapter VII).

If uracil residues dimerize at all in the irradiation of TMV-RNA, such dimerization either does not affect infectivity or is not photoreversible (Kleczkowski, 1963).

It may be noted that the ultraviolet sensitivities of free poliovirus ribonucleic acid and TMV-RNA differ but little, if any. The two RNAs are of similar molecular weight and base composition (Norman, 1960).

The action spectra for a number of biological changes induced by ultraviolet light implicate the nucleic acids as the primary absorbers (Gates, 1928, 1930). This has been the basis for interpreting, e.g., ultraviolet induced mutations (Stadler and Uber, 1942; Loofbourow, 1948; Blum, 1950). The action spectrum of TMV does not, however, look very much like the absorption spectrum of RNA (Hollaender and Duggar, 1936) because the inactivation Φ depends on the wavelength. When TMV is placed on a plant, presumably the protein must be stripped from the RNA in order for the latter to initiate the replication mechanism of the plant cells. At long wavelengths (2537 Å and 2804 Å) the RNA may be inactivated in the TMV and the virus appears dead by bio-assay. At shorter wavelengths (e.g. at 2260 Å) the protein may be denatured and the RNA may be trapped in inert protein, though still infective. If the RNA cannot "get out" of the denatured protein on a plant leaf the virus will appear to be "dead". The action spectrum will not match the absorption spectrum. This seems to be the case (Fig. 8-2; cf. Fig. 8-3) (Siegel and Norman, 1958). On the other hand, the action spectrum of RNA from TMV looks very much like the absorption spectrum (Fig. 8-4).†

In turning again to the irradiation of TMV *per se*, we have a more difficult problem in attempting to calculate a quantum yield. Because of the large size and shape of TMV, an appreciable fraction of the apparent ultraviolet absorption by a solution is scattered (Schramm and Dannenberg, 1944), Chapter I. Consequently, irradiation studies must take this into account. A rough correction for scatter has been obtained by means of the Raleigh rule for turbidity, τ. The absorption of a TMV preparation was measured at 3500 Å–4000 Å, where nucleic acid and protein have little or no absorption and the scattering constant K was calculated (eqn. 2-3). It is 5.3×10^{-19} cm^3 for a TMV solution containing 15.6 μg P per ml (Reddi, 1957). From the constant K, Reddi calculated the scatter below 3500 Å and subtracted the turbidities from the ultraviolet spectrum to obtain a spectrum for

† The apparently satisfactory comparison can only be considered tentative in view of the concentration dependence of the quantum yield at a given wavelength. However, the quantum yields were essentially independent of wavelength, i.e. were roughly constant, under the condition of only two-fold variations in concentration; i.e. D was maintained at about 1 between 2300–2800 Å.

TMV corrected for scattering. This is not altogether sound, since the turbidity depends on a lower power of the wavelength than 4 when the virus approaches λ_0 in length (Oster, 1948).

A more direct way of estimating the scattering is to compare the optical density of two solutions, one of which consists of a low molecular weight protein and the other contains the protein polymerized into rods of the length

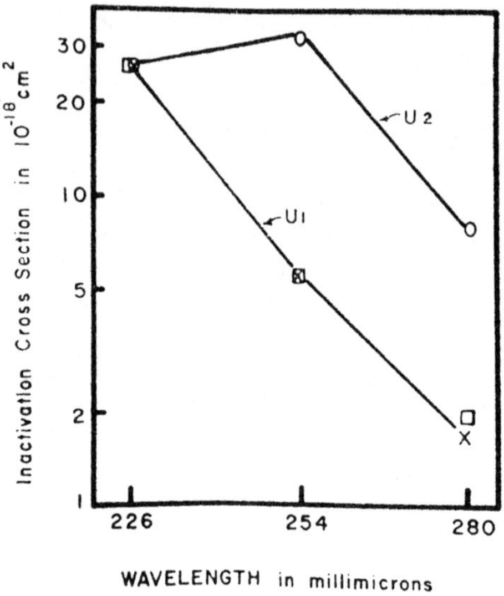

FIG. 8-2. Action spectra for two strains of TMV (Siegel and Norman, 1958).

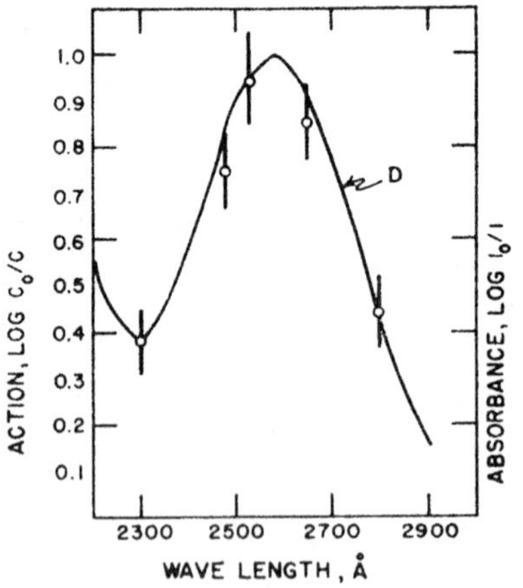

FIG. 8-3. Action and absorption spectra of infectious nucleic acid from tobacco mosaic virus (McLaren and Claesson, 1961). The equation utilized is (1-10).

and width of TMV itself. The difference represents the turbidity due to scattered light, provided re-absorption of scattered light is neglected and provided the refractive index of the rods is the same as that of the virus. As will be seen from the results to be described, these provisos are reasonable (McLaren and Takahashi, 1958).

A suitable protein for study is X-protein from infected tobacco plants. It is monomeric at pH 7.3 and polymerized at pH 5.3 as rods 150 Å in diameter and with lengths more or less commensurate with TMV itself;†

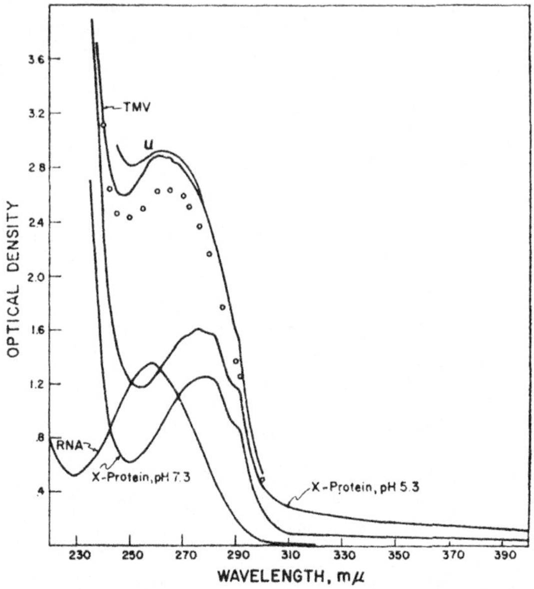

FIG. 8-4. Ultraviolet absorption spectra of tobacco mosaic virus (1.054 mg/ml); X-protein at pH 7.3 and at pH 5.3 (1.0 mg/ml); RNA (0.054 mg/ml). Circles are sums of points on the RNA curve and the X-protein (pH 5.3) curve. Uppermost curve, U, sums that of RNA and a hypothetical curve for X-protein fully polymerized to rods of 2710 Å length (McLaren and Takahashi, 1958).

cf. Fig. 8-1b. The absorption spectra of X-protein in monomeric and polymeric forms are shown in Fig. 8-4. As may be seen by the difference between the two curves, the scattering is appreciable at all wavelengths below 4000 Å. Electron micrographs of the pH 5.3 solutions revealed that only a few per cent of the polymeric rods were as long as those of TMV, and in order to estimate the scattering which would have resulted from the X-protein if all of

† Rod formation is also possible at pH 7, if the solution is exposed to ultraviolet light (Kleczkowski, 1959).

this protein were of uniform rods equal in length to those of TMV, use was made, as a first approximation, of the relationship between turbidity and weight-average molecular weight, $\bar{M}_w = \Sigma c_i M_i / \Sigma c_i$, namely eqn. (2-2). c_i denotes the weight concentration of the ith component (in g/cm^3). From the proportionality constants for the weight-average molecular lengths (proportional to weights) obtained from the electron micrographs, and the observed turbidity at given wavelengths (eqn. 2-2), the turbidity for a monodisperse system of particles of 2710 Å length (the length of TMV, Doty and Edsall, 1951) was calculated. With this information it is of interest to see how well the spectrum of TMV can be accounted for. In the figure we see the sum of the spectra of ribonucleic acid (RNA) from TMV, present to the extent of 5.1 per cent in TMV (Knight and Woody, 1958), and polymerized X-protein (the circles) and the sum of the spectrum of RNA from TMV (the small scatter by RNA has been neglected) and of a calculated spectrum for X-proteins polymerized to the full length of TMV (the uppermost curve). The agreement between the spectrum of TMV and the calculated spectrum is satisfactory above 2600 Å wavelength. The scattering of ultraviolet light by the polymerized protein is clearly enormous. It is difficult to correct for secondary scattering and re-absorption of scattered light, but re-absorption may be very high and depends on the geometry of the system (Claesson, 1956).

In calculating the quantum yield for inactivation of the virus, the intensity of the light absorbed by the virus, I_{abs}, is (cf. equation 1-8a)

$$I_0 \times \text{opacity} \times \frac{\varepsilon_a[A_0]}{\varepsilon_a[A_0] + \varepsilon_s[S]} \left(1 - \frac{\varepsilon_B}{\varepsilon_a}\right) \qquad (8\text{-}1)$$

where the opacity is the fraction of I_0 absorbed by the solution, $\varepsilon_B/\varepsilon_a$ is the fraction of the opacity due to light scattering; ε_a and ε_s are the extinction coefficients of virus and solvent respectively and ε_B is the scattering coefficient of virus. Taking the fraction $(1 - \varepsilon_B/\varepsilon_v)$ as about 0.64 at 2537 Å, the quantum yield for tobacco mosaic virus is 3.4×10^{-5} (McLaren and Takahashi, 1957).

The quantum yield for the RNA is higher than that for TMV at the same temperature. A quantum yield for RNA incorporated in TMV, based on the RNA content of TMV and assuming that the protein is only an absorber and does not transfer energy to the RNA (Shore and Pardee, 1956) is 4.6×10^{-5}. We may thus conclude that the naked RNA is more ultraviolet sensitive than is RNA incorporated in intact virus, Table 8-2.

On a per milligram basis, the energy required to inactivate 50 per cent of the naked RNA is 0.3 J, and for RNA incorporated in the TMV molecule the corresponding figure is 2.8 J. In TMV the protein may protect the RNA somewhat by preventing nucleic acid from undergoing reaction with solvent

TABLE 8-2. ENERGY FOR INACTIVATION OF TMV AND INFECTIOUS RNA FROM TMV (McLAREN AND TAKAHASHI, 1957)

Substance	Einsteins per mole $1/\Phi$	Joules per mg (50% inactivation)
RNA, free	0.33×10^4	0.43
RNA in TMV	2.1×10^4	2.8
TMV	2.3×10^4	0.16
Enzymes, antibodies	$M/450$	0.75

or from unfolding. In the form of TMV, for each milligram of RNA inactivated, requiring 2.8 J/mg, a corresponding amount of protein, namely 18.6 mg, is involved, giving an energy value for the inactivation of whole TMV of only about 0.16 J/mg. The energy required to inactivate a milligram of a typical enzyme is 0.75 J. It would seem likely, however, that the greater ultraviolet sensitivity of TMV as compared to enzymes is only apparent, for if inactivation involves only the nucleic acid moiety of TMV at 2537 Å, the energy requirement is 2.8 J/mg, not 0.15 J/mg for the biologically active component.

We come again to the question of the influence of protein on the action spectrum of RNA. This may be studied by irradiation of the intact virus. Quantum yield calculations are complicated by the necessity of correcting apparent light absorption by the virus for light scattered by the virus. Also, in dilute solution re-absorption of scattered light by the virus will be small, whereas in concentrated solutions, re-absorption may be appreciable. Neglecting the re-absorption, for which we have no means of evaluation, we can correct absorbances for scattering by means of the experimental scattering determinants of McLaren and Takahashi, 1958, and compute yields for the inactivation of TMV from the rate data of Rushizky *et al.* As may be seen from Table 8-3, quantum yields for TMV are not independent of wavelength and therefore the action spectrum cannot parallel the absorption spectrum (McLaren and Moring Claesson, 1961). To find the quantum yields, Φ_s for RNA incorporated in the virus particle, assuming that only light absorbed by the RNA moiety leads to inactivation, Φ must be corrected further for the fraction of the absorbed light absorbed by RNA. These values are also tabulated. It becomes clear that from 2483–2804 Å the sensitivity of RNA in the virus is fairly independent of wavelength, although much less than that of the free RNA (devoid of protein). The quantum yield for free RNA varies from 3.4×10^{-3}–3.8×10^{-3} in the range between 2804 and 2300 Å; that is, it is essentially constant within the experimental variability of the tests. On the other hand, Φ_s is much larger at 2300 Å than in the range of 2483–2804 Å, and we suspect that the quantum yield calculated in this way is without physical meaning at 2300 Å. In other words, since Φ for free RNA is independent of wavelength (in the wavelength range under consideration), we can conclude that the high value of Φ at 2300 Å really means that the virus is rendered non-viable because light absorbed by both the protein and the RNA moieties can lead to inactivation (at 2300 Å most of the light is absorbed by protein). That some protein denaturation in the virus occurs during irradiation of TMV even at 2537 Å is indicated by the thermal sensitivity of irradiated virus. Apparently the

TABLE 8-3. QUANTUM YIELDS FOR THE INACTIVATION OF TOBACCO MOSAIC VIRUS, Φ, AND FOR THE RNA INCORPORATED IN THE VIRUS, Φ_s

Wavelength, Å	Rate constants for inactivation $\times 10^5$	$\Phi \times 10^5$ (virus)	$\Phi_s \times 10^5$ (RNA in virus)
2804	1.8	2.3	9.3
2652	3.9	4.9	11
2537	4.3	5.8	12
2483	4.0	5.9	14
2300	8.7	~12	~200

host plant cannot extract active RNA from a coat of protein which has become denatured by ultraviolet light, and therefore the quantum yield is higher than would be expected if the RNA was equally available to the plant after irradiation at any wavelength.

The work described above on infectious RNA and TMV involved "common TMV" or strain Ul (Siegel and Wildman, 1954). However, strains of TMV differ markedly in sensitivity to ultraviolet inactivation in suspension (Siegel *et al.*, 1956). For example, strain U2 differs nearly six fold from U1 in sensitivity to inactivation at 2537 Å and 2804 Å (Siegel and Norman, 1958). (Both strains have the same ultraviolet sensitivity at 2537 Å when irradiated dry (Siegel, 1957).) The nucleic acids from both U1 and U2 TMV have the same sensitivity, and reconstituted U1-TMV has the same sensitivity as U2-TMV.† The detailed type of bonding existent in the natural U1 strain partially protects the RNA and this bonding does not reappear during reconstitution, a very important observation. It is

† It was also reported that the RNA from both strains have the same sensitivity as U2 virus. Since the inactivation data were not corrected for extinction coefficients or for light scatter, eqn. (8-2), this statement is of doubtful validity. Even with optically thin solutions these terms cannot be ignored in quantatitive comparisons.

relevant to note that the RNA is more easily released from U2 than fromU1 virus by the heat detergent method of Fraenkel-Conrat.

Siegel *et al.* (1956, 1957) have utilized ultraviolet light to study the course of infection of tobacco plants with RNA and TMV. At room temperature the survival of virus in irradiated leaves as a function of dose is exponential for approximately 5 hours after inoculation, after which the appearance of new intracellular infectious particles is indicated by a change to a multi-target survival curve. During the 5 hour period the infection is characterized by three discernible phases of ultraviolet sensitivity, i.e. a period during which the infection has essentially the same sensitivity as the virus *in vitro*, a period during which resistance to ultraviolet light increases, and a period during which a secondary level of resistance is maintained. In contrast, infective centers caused by the nucleic acid display little or no lag period before they begin to increase in resistance to ultraviolet light. Siegel, Ginoza, and Wildman conclude that when infection is initiated with the intact virus, the nucleic acid moiety of the virus is freed from the protein moiety soon after infection is established and before virus multiplication begins. Unfortunately, some technical problems involving the host plants were not appreciated by Siegel and these results may be not altogether correct (Bawden and Kleczkowski, 1960; Schramm and Engler, 1958).

Bawden and Kleczkowski (1955) tested seven plant viruses for photo-reactivation, i.e. to see if plants exposed to visible light, after inoculation with virus preparations partially inactivated by ultraviolet light, produce more local lesions than plants kept in darkness. TMV did not exhibit photoreactivation but potato X, tomato bushy stunt, tobacco necrosis, cabbage black ringspot, cucumber mosaic, and tobacco ringspot viruses did. A certain time is required after inoculation before a state is reached in which the ability of the virus to infect the host is affected by light. The time is 30 min at 20°C with potato virus X. It has been suggested that the critical light intensity is between 80 and 380 foot candles (Chessin, 1958). The phenomenon involves some light-sensitive mechanism in the host cells; exposing irradiated viruses to visible light does not affect their infectivity. (Luria, 1955, defines an inactive virus particle as one that is unable to parasitize a host cell or which, upon parasitizing a host cell under conditions where no reactivation occurs, fails to give rise to the production of active virus.)

It is not surprising to learn, after considering the difference in sensitivities of U1 and U2 TMV, that the sensitivity of plant viruses is not a direct function of their content of nucleic acid (Bawden and Kleczkowski, 1955).

C. PHOTOCHEMICALLY PRODUCED VACCINES

Since Stanley's original observation (1936) that ultraviolet inactivated TMV retains the serological properties of the original virus, the separation of infectivity from the antigenic properties of viruses by radiation has been frequently recognized as a fairly general and feasible phenomenon and has been taken advantage of by many workers in attempts to prepare improved vaccines. The earlier literature on this subject will be found summarized in two papers by Levinson *et al.* (1944, 1945). The principle in all cases has been the same, viz. to completely destroy infectivity with as little damage as possible to the protein component, so that as much of the original antigenicity as possible is retained (Wierzchowski and Shugar, 1958; Shugar, 1960). This is frequently possible because of the high selective absorption of the nucleic acid components with respect to the protein portion. In some instances, however, there is only a small margin between the radiation dose required to inactivate the virus and that which will affect antigenicity, so that careful control of conditions is often an important prerequisite.

The theoretical basis for these procedures is perhaps best illustrated by the action spectra for loss of infectivity and of hemagglutination ability of influenza virus, Fig. 8-5, as well as the dose-effect curves for loss of agglutination ability of influenza virus, which are of the "multiple-hit" type, Fig. 8-6,

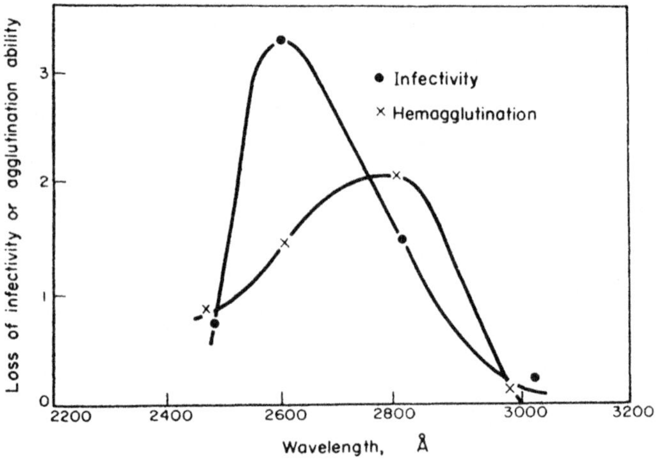

FIG. 8-5. Action spectra for influenza virus (Tamm and Fluke, 1950).

as compared to loss of infectivity, which follows a semi-log course (Tamm and Fluke, 1951). It would undoubtedly be desirable to devote more attention to the use of different wavelengths, with a view to finding the maximum possible margin between loss of infectivity and antigenicity. The failure to do so in the past is perhaps due in large part to the desire to avoid the additional cost likely to be associated with the use of such installations on a large scale. With the advent of high-power quartz xenon lamps, giving a high-intensity continuous spectrum in the ultraviolet, it may now be possible, in conjuction with suitable filters, to conduct such investigations on a practical scale.

By very short exposures of continuously flowing 1 mm thick films of virus solution to the radiation from a specially developed source with appreciable output over the entire ultraviolet region, including that below 2000 Å, Levinson *et al.* (1944, 1945) produced rabies and St. Louis encephalitis vaccines as well as poliomyelitis vaccine (Milzer *et al.*, 1944, 1945) and lymphocytic choriomeningitis vaccine (Milzer and Levinson, 1949).

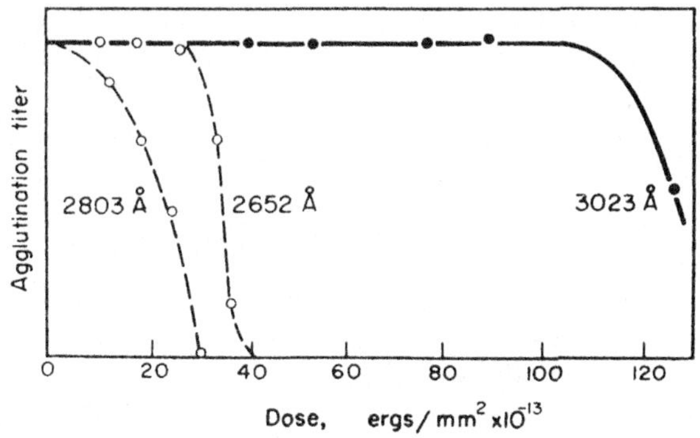

FIG. 8-6. Dose-effect curve for influenza virus (Tamm and Fluke, 1951).

It was subsequently shown by Habel and Sockrider (1947) that the characteristics of the source were of less importance as compared to the technique used for irradiating thin films, and that commercial germicidal lamps were equally effective if due allowance is made for the lower intensity output.

Similar procedures were used by McKinstry and Reading (1944) for SK-mosaic poliomyelitis vaccine and by Andrews *et al.* (1948) for vaccinia virus. The latter, while reported to immunize mice against ectromelia, offered little protection to rabbits and guinea-pigs against vaccinia. This finding was ascribed by Collier *et al.* (1955) to the fact that immunizing doses were administered intraperitoneally rather than subcutaneously, that the time interval between doses was too short, and that excessive radiation doses were applied with the type of irradiation technique used.

Collier *et al.* (1955) have used the Habel and Sockrider (1947) irradiation device to produce ultraviolet inactivated vaccinia virus which produced good immunity in rabbits and monkeys, and suggested that such irradiated virus may be useful in special cases where vaccination is normally contra-indicated.

More recently the American authors have extended their original procedure to produce a poliomyelitis vaccine which is claimed to compare favorably with formalin-inactivated vaccines. The principle employed involved a combination of irradiation, followed by warming of the irradiated virus to 37–40°C (Shaughnessy *et al.*, 1957). In view of the generally increased susceptibility of enzyme activity and virus infectivity to heat, following a period of irradiation sufficient to only partially destroy activity, this technique should undoubtedly find wider application.

A somewhat similar principle has been used by Taylor *et al.* (1957a, b), also for poliomyelitis vaccine. This consisted of a combination of ultraviolet irradiation with formalin treatment and gave consistently good killing of infectivity without materially affecting antigenicity. It provided a vaccine which exhibited satisfactory results in clinical trials and was subsequently reported to possess certain advantages over other vaccines (Beardmore *et al.*, 1957; McLearn and Taylor, 1958). The latter reference should be consulted for a comprehensive review of the theoretical and practical aspects involved. The use of ultraviolet light for preparation of polio vaccines has been investigated by a number of other workers (Verlinde and Klavenbeck, 1952; Roizman *et al.*, 1959).

The resistance of some viruses to the photodynamic action of sensitizing dyes has been taken advantage of in a procedure to rid live polio vaccines of other potentially harmful organisms. The method, details of which are not at hand at the time of writing (Hiatt, *7th International Congress of Biological Standardization, London,* 1961), involves the use as a sensitizing agent of toluidine blue, to which polio virus is highly resistant (see below).

Ultraviolet irradiation has also been successfully applied to the freeing of plasma products, intended for clinical use, from the viruses causing hepatitis and jaundice (McCall *et al.*, 1957). LoGrippo and Hartmann (1958) have demonstrated that eastern equine encephalmyelitis virus may be completely eliminated from blood plasma products by a procedure involving first addition of 500 mg/ml. of β-propiolactone, and then irradiation at an ultraviolet intensity of 2 mW/cm^2. The electrophoretic pattern of the plasma proteins was unaffected by this treatment.

Vaccines are, on the other hand, not considered to be the ultimate answer to the problem of virus infections, since they are too specific in their effects. For example, influenza vaccines made a few years ago have offered no protection against the new Asian viruses. In addition, there are too many virus species to make it feasible to develop a vaccine against each of them. One possible new approach is suggested by the phenomenon of "virus interference" (Isaacs and Burke, 1958). It has been known for more than two decades that infection by one virus may subsequently prevent infection by a second, e.g. Rift valley virus "protects" monkeys against the fatal yellow fever virus. "Killed" virus may also be used as the interfering agent, and by far the best general method of virus inactivation is by means of ultraviolet irradiation. If inactivated, interfering virus is mixed with tissue cells and incubated, a product called "interferon" may subsequently be isolated.

Interferon shows all the interfering activity of the original virus, i.e. it prevents virus multiplication in cells. It appears to be proteinoid in character. Influenza virus irradiated for a short time produces maximum yield of interferon; excessively irradiated virus does not produce any. Curiously enough, when fresh live virus is added to cells treated with interferon, more of the latter is produced.

While considerable speculation has been devoted to the mode of action of interferon, recent evidence indicates that it acts by uncoupling nuclear oxidative phosphorylation and inhibits virus production, but not interferon production (Isaacs *et al.*, 1961).

D. BACTERIOPHAGES

It is probably no exaggeration to state that ultraviolet radiation has been more extensively employed than any other of the wide variety of agents which are capable of inactivating bacteriophages. Indeed, in recent years phage photochemistry has become firmly established as a highly developed and specialized field of its own. In large part this is due to the fact that, apart from inactivation of infectivity, ultraviolet light provokes a number of additional and even more interesting physiological and genetic effects. These readily lend themselves both to qualitative and quantitative investigations because of the remarkable convenience with which the phage-host relationship may be studied experimentally. Ultraviolet light has likewise proved to be a powerful tool for following the intracelluar course of phage development, for investigations on the induction of phage development in lysogenic bacteria, etc.

In addition to the phenomenon of photoreactivation, exhibited also by some plant and animal viruses, ultraviolet inactivated bacteriophage are subject to partial reactivation by means of other mechanisms, of which the best known are multiplicity reactivation and cross reactivation. Since our ultimate aim is to correlate biological effects with the chemical modifications provoked by irradiation, it is to be expected that the existence of reversible photobiological effects will prove of considerable value for this purpose.

It has also become fairly widespread practice to combine the result of phage photochemistry with those due to the effects of ionizing radiations, including decay of incorporated ^{32}P, with a view to extracting the maximum possible information. An excellent review of this subject, with special reference to phages T2 and T4, has appeared (Stahl, 1959).

1. *DNA Content and UV Lethality*

While the behavior of different phage strains to ultraviolet light varies considerably, there is no conclusive evidence from existing data of a strict correlation between ultraviolet sensitivity of infectivity and phage taxonomy, such as appears to prevail for photosensitized inactivation (Sect. E). The accumulation of additional data may, of course, lead to some modification of this situation, but there are facts which argue rather against it. A case in point is that for the T-even bacteriophages T2, T4, T6, the former of which is twice as sensitive as the other two; such variations in a given group lead to overlapping sensitivities between different groups. The source of the difference in ultraviolet sensitivities of T2 and T4 is, on the other hand, of considerable theoretical interest (Streisinger, 1956; Harm, 1961a, 1961b) and will be discussed in a special section below.

There is no apparent correlation between the DNA content of a phage and its radiation sensitivity, although such a correlation does exist for inactivation by X-rays or by ^{32}P-decay (Garen and Zinder, 1955). Of the two phages C16 and S13, for example, the DNA content of the former is only about 3 per cent that of the latter (Sinsheimer, 1957), but C16 is almost 3 times as sensitive to ultraviolet light as S13 (Latajet and Walh, 1945). The reproductive ability of phage P22 is much more resistant to ultraviolet light as compared to T2 on the basis of relative DNA content per particle (Garen and Zinder, 1955). Phage λ is inactivated at only 12 per cent of the rate for T5, notwithstanding that both phages are of the same size (Weigle and Delbruck, 1951).

It has been variously suggested that the DNA of some phages may be intrinsically more sensitive than that of others. This is undoubtedly a conceivable possibility; e.g. the relative contents of the base pairs guanine–cytosine and adenine–thymine in a given phage DNA would most likely influence the ultraviolet sensitivity of the phage. Although the base compositions of a number of phage DNA's are fairly well established, quantitative data with regard to their relative UV sensitivities are not as well known. It would be surprising if a *strict* correlation of this type existed; e.g. T2 is twice as sensitive as T4, but both contain the same quantity and type of DNA. An additional factor must, however, be taken into account, viz. the base sequences; the importance of this has already been demonstrated quite clearly in studies on the photochemistry of model oligonucleotides. It is perhaps also worth recalling that the higher the A-T content of an isolated transforming DNA, the more extensive is the amount of cross-linking resulting from ultraviolet irradiation (Marmur and Grossman, 1961). It should also be recalled that the UV sensitivity of bacterial cells has been reported as being dependant on their A-T (or G-C) content. The difficulty of adequately testing for any such relationship in the bacteriophages is due in large part to the complex and varied effects of the host cells employed for the measurement of phage activity. This point will become clearer when we come to discuss the various mechanisms by means of which inactivated phages are susceptible to reactivation by their host cells.

2. *Action Spectra*

It is, on the other hand, perhaps somewhat naive to consider the susceptibility of a given phage DNA as an entity divorced from the integrated morphological structure of the phage and its protein component. We have already seen, above, how important these factors are in determining the ultraviolet sensitivity of a given strain of TMV. It is true that most action spectra which have been obtained for phages exhibit the characteristic maximum at 2600–2650 Å shown by nucleic acid absorption spectra. But in almost all instances the number of points on the action spectra have been insufficient to exclude with certainty some participation of the protein components. In at least one instance where particular attention was devoted to this matter, the action spectrum (of phage M5) definitely suggests in-

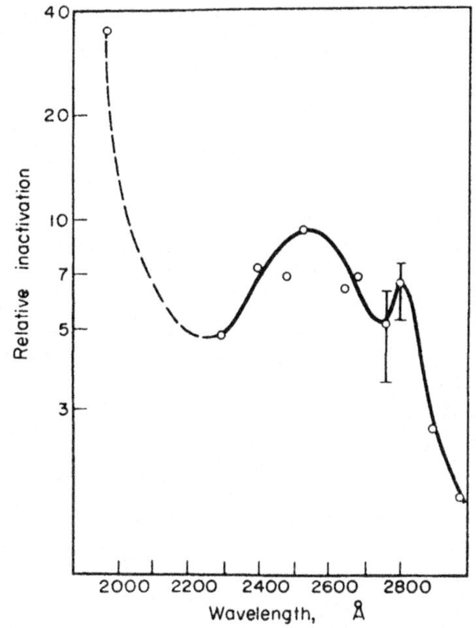

FIG. 8-7. Action spectrum of *Bact. megaterium* bacteriophage M5, the secondary maximum of which at 2800 Å indicates involvement of the protein component (Franklin *et al.*, 1953).

volvement of the protein moiety, Fig. 8-7. A similar qualitative finding has been reported by Fluke for phage T1 (Fluke, 1956). A point undoubtedly deserving of comment in this connection is the reported ultraviolet-induced cross-linking between DNA and protein both *in vitro* and *in vivo*, as well as the extreme sensitivity of this type of reaction (Smith, 1963). The possibility of such a reaction in the bacteriophages is certainly worth investigating.

3. *Quantum Yields*

Luria's (1955) emphasis on the value of quantum yield determinations, instead of relative sensitivities, at different wavelengths for action spectra determinations has not lost its validity, but the experimental difficulties involved have hitherto been a rather strong deterring factor.

Quantum yields have been measured over a wide range of wavelengths for phages T1 and T2 by Zelle and Hollaender (1954), the results being tabulated in Table 8-4. These values cannot be considered as fully established, since, according to the authors, the number of active phage particles per plaque-forming unit was not known accurately. Furthermore, corrections for scattering were not sufficiently good to formulate the ultraviolet absorption spectra, hence the amount of absorbed light, with suitable accuracy. The authors believe their values are correct to within a factor of two. Reference to the original paper is recommended because of the excellent discussion of the experimental techniques and possible pitfalls involved in such measurements.

It is rather curious that the quantum yields vary so little over such a wide range of wavelengths. We have seen, in the studies already reported on dimerization in thymine oligonucleotides (Johns *et al.*, 1962; Sztumpf and Shugar, 1962), that the extent of dimerization of adjacent pairs of thymine

TABLE 8-4. QUANTUM YIELDS FOR INACTIVATION OF PHAGES (ZELLE AND HOLLAENDER, 1954)

Particle	λ in Å	$\Phi \times 10^4$	Particle	λ in Å	$\Phi \times 10^4$
T1 phage	2220	5.0	T2 phage	2220	2.7
	2378	5.0		2378	3.2
	2537	6.3		2537	3.1
	2650	6.5		2650	3.4
	2804	6.0		2804	3.5
	3022	7.3		3022	1.8

residues is markedly dependent on the wavelength. If dimerization were a factor in inactivation, one might expect this to be reflected in a variation of quantum yield with wavelength. Furthermore, at the shortest wavelength used, 2220 Å, protein absorption is quite high and nucleic acid components undergo reactions other than hydration and dimerization; despite this the quantum yields are apparently unaltered. This certainly would bear closer examination.

4. *Kinetics of Inactivation of Infectivity*

In assaying the infectivity, a surviving particle is regarded as one which gives rise to a visible plaque under the standard conditions employed. However, in view of the existence of the phenomenon of photoreactivation, Luria (1955) defines an inactive phage particle as one that is unable to parasitize a host cell or which, upon parasitizing a host cell under conditions where no reactivation occurs, fails to give rise to the production of active virus particles. The use of the term "inactive phage" is somewhat misleading, since phages which are incapable of reproduction are still capable of being adsorbed to their host cells and of killing the host cell. Furthermore, phages which do not multiply under conditions of single infection of the host may do so when more than one phage particle is adsorbed to the host cell, a phenomenon known as multiplicity reactivation (Luria, 1947).

Although there are numerous examples where inactivation of infectivity follows the expected first order course (Campbell-Renton, 1937; Latarjet and Wahl, 1945), the number of exceptions is so large as to raise some doubts as to whether an observed logarithmic survival curve is not the result of some artifact. Deviations from exponentiality are frequently dependent on the conditions under which the phages have been irradiated. But there are also many instances where the method of assay is involved and, particularly in these latter cases, much useful information about the phage-host interactions may be obtained from an examination of the survival curves.

For the T-type bacteriophages which have been most extensively investigated, the survival curves are rarely strictly exponential, although the deviations may sometimes be small. For T2, T4, T5 and T6 the curves are usually concave downward, while for T1, T3 and T7 they are concave upward (Benzer, *et al.*, 1950, in Delbruck, 1950; Dulbecco, 1950). According to Tessman (1954, quoted in Stahl, 1959), the downward concavity exhibited by the survival curve of T4 is a function of the wavelength used for inactivation. (Cf. with results of Wetzed *et al.* (1960) for T1.) While no interpretation has been offered for this anomaly, Doerman and Chase (quoted in Stahl, 1959) report exponential survival curves for highly purified T4 which has been taken to indicate that the ultraviolet light may provoke a "reactivating effect" on some of the phage particles which have been inactivated by some "impurity" in the preparations normally used. If one accepts this interpretation, it becomes necessary to assume that the "impurity" is associated with the phage DNA, since irradiation of the latter after it has been injected into the host cell may also show the initial downward concavity.

The downward concavity, or "multi-hit" nature, of the survival curves for T2 had been earlier noted by Laterjet and Morenne (1951), using as irradiation sources commercial daylight lamps which were found to emit traces of radiation to wavelengths as low as 2894 Å. The departure of the inactivation curves from exponentiality was, at the time, ascribed to a possible dose-rate effect, since the intensity of the ultraviolet light emitted by these lamps is very low and long exposure times were required. This possibility has, however, not been futher explored. It has been claimed by Eckart (1954) that the reciprocity relationship does not hold for irradiation of T1 at 2650 Å, but this is in contradiction with the results of Fluke and Pollard (1949), and of Fluke (1956) who found that an 8-fold variation in intensity at 2652 Å was without effect.

Hill and Rossi (1954a) investigated in some detail the influence of irradiation conditions on the survival curves of T1 in the presence and absence of photoreactivation. While their subsequent attempts to interpret the findings were largely speculative (Hill and Rossi, 1954b), the experimental observations are nonetheless of considerable interest. For example, irradiation in acetate buffer at neutral pH in the "dark" (i.e. under conditions where no photoreactivation could occur) gave a survival curve concave upwards and 2.5 times higher than that in phosphate buffer; on the other hand, the survival curves following photoreactivation were almost coincidental. In 95 per cent ethanol the light and dark survival curves were exponential to residual activities of 10^{-5} with fairly high rate constants. With a decrease in ethanol concentration, the curves became complex again (concave upwards) while the rate constants (or sensitivity) decreased, Fig. 8-8. Irradiation of the phage in frozen medium also resulted in rather large variations in ultraviolet sensitivity, with an increase in phosphate buffer and a comparable decrease in acetate (cf. Levine and Cox, 1960). Irradiation of the dried phage still gave complex inactivation curves, while the inactivated particles exhibited no photoreactivation; but this finding is contrary to that of Fluke (1956) who reported that, whereas the survival curve for T1 in solution is complex, irradiation of dried phage gave a strictly exponential curve to low residual activities, while the inactivated phage exhibited photoreactivation. The source of this discrepancy undoubtedly lies in the difference in experimental conditions employed (see Fluke, 1956).

It follows that the effects of irradiation of extracellular phage are profoundly dependent on the conditions of irradiation and it is highly probable that the type of photochemical modifications induced are to some extent dependent on the nature of the medium in which irradiation is carried out. This is also evident from the fact that the extent of photoreactivation is dependent on the conditions under which ultraviolet inactivation was carried out. It should be recalled that irradiation of nucleic acids and model polynucleotides in the dried state is accompanied by a high degree of cross-linking. It may well be worth examining the influence of the various factors outlined above on the photochemistry of some model polynucleotides to determine what modifications they induce in hydration and/or dimerization.

Another observation of some interest is due to Kellenberger and Weigle

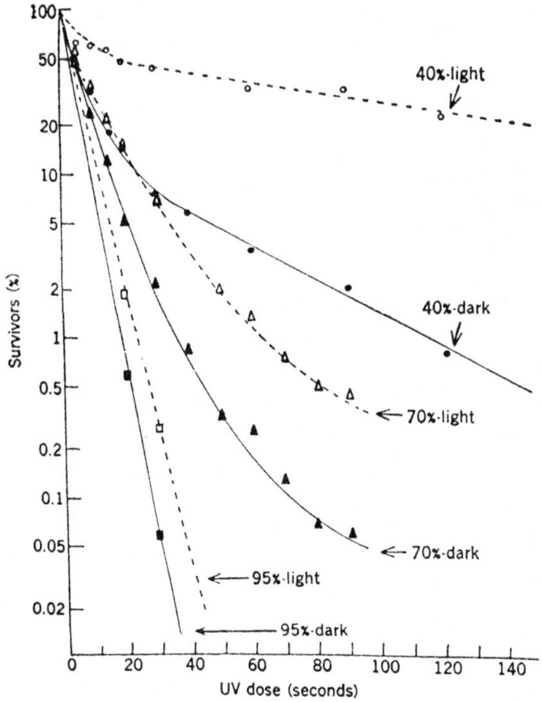

FIG. 8-8. Effect of ethanol concentration on the dark and light survival of phage T1 at 2537 Å (Hill and Rossi, 1954a).

(1958), who found that the ultraviolet sensitivity of phage λ increased 25 per cent between 0°C and 37°C. Changes in temperature prior and subsequent to irradiation were without effect. Obviously some dark reaction is involved and the one that comes most readily to mind is that involving the photohydration of cytosine residues, a reaction which is subject to thermal reversibility in this temperature range. The actual situation must, however, be somewhat more complex than this since, if only photohydration were involved, an increase in temperature following irradiation should show some effect.

5. *Phage-host Interactions*

Whereas an ultraviolet inactivated bacteriophage is, by definition, unable to undergo reproduction, it is nonetheless by no means biologically inert, and can still exhibit some of the properties of the non-irradiated phage. This is further evidence that the injuries provoked by the radiation are highly localized in character, such that only a given region, or regions, of the phage genome are affected.

Ultraviolet inactivated phages are still capable of killing the host cell, thus indicating that the phage may still adsorb to the cell and inject its DNA. For phage T2 the host-killing property is enormously more resistant to radiation than is infection. For an ultraviolet dose which results in a survival of infectivity of 10^{-6}, almost 35 per cent of the initial phage particles still retain their host-killing property (Luria and Delbruck, 1942).

Many of the other properties of ultraviolet inactivated phages are similarly retained. For example, following infection of *E. coli* cells with irradiated T2, the rate of synthesis of deoxycytidylate hydroxymethylase is decreased, but the enzyme appears without any lag. The synthetase activity of the phage is considerably more radiation resistant than infectivity (Delikas, 1961).

The phenomenon of interference is also encountered with bacteriophages, the first such example being reported by Luria and Delbruck (1942), who found that the multiplication of normal T1 in the host cells is prevented by simultaneous infection with ultraviolet inactivated T2. On the other hand, ultraviolet inactivated T1 does not inhibit multiplication of T2.

An important observation is that of Luria (1944) who found that a considerable increase occurs in the latent period of phages which have survived moderate ultraviolet doses. The significance of this observation lies in the fact that the phage particle has undergone some type of damage which did not destroy infectivity. The magnitude of this growth-delaying effect varies for different phages and increases with increasing dose. That the effect is due to some damage in the DNA of the phage was demonstrated by Setlow *et al.* (1955) who found that the action spectrum is nucleic acid-like in nature, at least for phage T1, which was the only one studied.

Some observations of Sauerbeier (1960) on phage T1 are pertinent in relation to the foregoing. The reaction of formaldehyde with non-hydrogen bonded amino groups was utilized to demonstrate that the number of amino groups accessible to formaldehyde in irradiated T1 increases with the dose, indicating rupture of some hydrogen bonds as a result of irradiation. Presumably these are the hydrogen bonds between complementary strands, since the single-stranded phage ϕX174 did not show this effect. Of even more importance was the finding that not all hydrogen bond-breaks led to inactivation of phage T1. The hydrogen bonds broken by ultraviolet irradiation apparently do not rejoin during storage of the phage extracellularly. It is, however, possible that this would occur through some intracellular reaction following injection of the phage DNA into the host cell. Hydrogen bond breakage between complementary strands may result from hydration of pyrimidine rings (Shugar, 1960; Wierzchowski and Shugar, 1962). It was suggested by Sauerbeier (1962) that such breakage would result from the transition of thymine, cytosine or guanine from a keto to an enol form, but no direct evidence for this exists. One method of testing the foregoing hypotheses would be an examination of irradiated T2; this phage contains no cytosine, but only hydroxymethylcytosine, which would not be expected to undergo hydration. Hence if phage T2 behaved like T1 in its reaction with formaldehyde following irradiation, this would most likely exclude hydration as the source of the presumed hydrogen bond breakage.

(a) *Kinetics of survival:* As has already been pointed out, the complex survival curves for phage irradiated extracellularly are due in part to the conditions of irradiation and to some extent to the interaction of the host cell with the irradiated phage; the latter effect is a necessary consequence of the fact that the assay of phage infectivity requires adsorption of the phage to the host cell. There is, however, additional and even more convincing evidence for such interaction between the host and irradiated phage obtained from studies of survival of infectivity on host cells the physiological state of which has been modified.

Phages such as T2, T4, and T6 are considerably more ultraviolet sensitive than T1, T3, and P22. However, if the reproductive ability of P22 is assayed on irradiated host cells, its ultraviolet-sensitivity is decreased and approaches that of T2. The apparently higher radiation resistance of P22 on normal host cells is consequently proposed as being due to partial genetic homology between the phage and host genomes: after infection the host material can replace ultraviolet-damaged homologous phage material (Garen and Zinder, 1955).

It was found by Kellenberger and Weigle (1958) that the survival of phage λ is dependent on the physiological state of the cells, from which it was concluded that some of the ultraviolet-induced lesions can be "repaired" by the cells. Such lesions were referred to as "sublethal." It was concluded, as above, that at least part of the genome of the temperate phage is more or less homologous with part of the bacterial genome so that crossing between the host genome and undamaged portions of the phage genome was possible.

There are many other experiments which suggest that crossing of phage and host genomes takes place. This must, of course, involve chromosome fragmentation with subsequent recombination of the fragments to produce homologous phage. Direct evidence for such processes has recently been obtained by Meselson and Weigle (1961) in experiments in which the phage and/or genomes were labelled with heavy isotopes and the presence of the labels sought in the progeny.

Some rather striking effects of the host cells on phage survival have been reported by Ellison *et al.* (1960). The survival of ultraviolet irradiated T1, T3, and T7 was appreciably lower when assayed on strain B_s, a radiation-sensitive mutant of the normal host cell, *coli B*. The survival curves were complex when assayed on strain B, but were exponential with B_s as the host, Fig. 8-9. Since T1 was found to undergo the same degree of photoreactivation either with B or B_s, the differences in the survival curves are not due to lack of adsorption to, or penetration of, the host cells. The similarity in behavior of T1, T3, T7, as compared to T2, T4, and T5, is in agreement with other evidence that these two sets of phages differ with respect to host interaction.

(b) *Photoreactivation:* Following adsorption of an ultraviolet-inactivated phage to the host cell, irradiation with visible light usually provokes partial

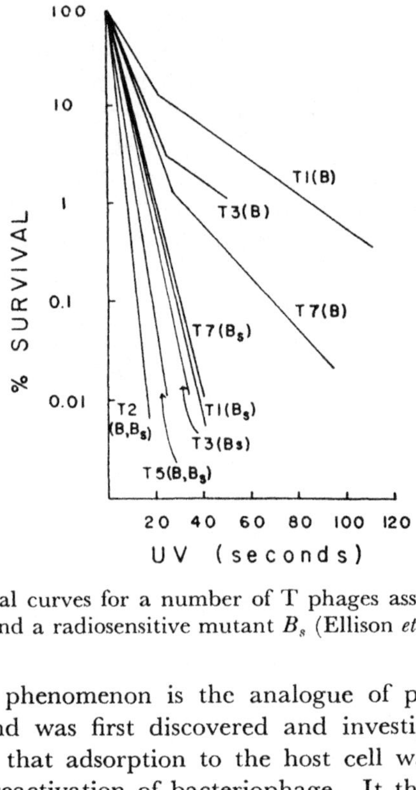

FIG. 8-9. Survival curves for a number of T phages assayed with *E. coli* strains *B* and a radiosensitive mutant B_s (Ellison *et al.*, 1960).

reactivation. This phenomenon is the analogue of photoreactivation in micro-organsims and was first discovered and investigated by Dulbecco (1950), who found that adsorption to the host cell was a necessary prerequisite for photoreactivation of bacteriophage. It therefore follows that some intracellular factor is necessary in addition to visible light and it is most likely that this factor is an enzyme or enzyme system present in the bacterial cell or synthesized following injection of the irradiated phage. In all probability this enzyme is similar to that extracted from *coli* and yeast cells and is capable of photoreactivating transforming DNA, probably by direct reversal of ultraviolet lesions. This may be inferred indirectly from the observation of Lennox *et al.* (1954) who found that the ultraviolet sensitivity of photoreactivated phages was the same as that of the nonirradiated.

Indirect evidence for such a hypothesis has recently been presented by Harm (1962), who demonstrated that the photoreactivation of phage T4*v* (a T4 mutant which behaves similarly to T2 with respect to ultraviolet-sensitivity and photoreactivation) adsorbed to *E. coli* is appreciably inhibited if the cells are simultaneously infected with several phages of the same type which have been so heavily irradiated that their contribution to the survival can be neglected. The heavily irradiated phage, which are not susceptible to photoreactivation, compete with normally irradiated phage for the intracellular PR-enzyme.

Pichl (1960) has actually reported photoreaction of ultraviolet-inactivated T2 phage by means of an illuminated extracellular *E. coli* extract; but the amount observed was so small that it can hardly be considered as established. Furthermore, the system used was very complex. Attempts to demonstrate photoreaction for bacteriophages by means of extracellular systems, as for transforming DNA, might be expected to be accompanied by difficulties, since the photoreactivation of host-adsorbed phage probably involves the naked phage DNA following its injection into the host. In extracellular systems the phage DNA would most likely be "screened" by the phage protein from the action of any PR-like enzyme. It should also be noted that attempts to use Pichl's system for photoreactivation of transforming DNA have been unsuccessful (Pakula, personal communication, 1960).

As for other organisms, photoreactivation of bacteriophages is not a 100 per cent phenomenon. Dulbecco (1950) has coined the term "photoreactivable sector" as an index of the degree to which a phage is susceptible of photoreactivation. The absorption of ultraviolet light by a phage has a probability *a* of producing a photoreactiveable inactivation and a probability *b* of producing a non-photoreactiveable inactivation, where $a - b$ is the probability of producing any inactivation, and is proportional to the absorption cross-section of the phage for ultraviolet light. Taking $a - b$ as 1, *a* is the photoreactiveable sector of the cross section and *b* the non-photoreactiveable sector; *a* then varies between 1 (100 per cent photoreactivation) and 0 (no photoreactivation) and is the index of photoreactivibility. The PR sectors for the T phages are shown in Table 8-5.

TABLE 8-5. PHOTOREACTIVEABLE SECTORS OF T-PHAGES

Phage	T1	T2	T3	T4	T5	T6	T7
PR sector	0.68	0.56	0.39	0.20	0.20	0.44	0.35

While the PR sector for a given phage is fairly reproducible, it depends on the conditions of irradiation (Hill and Rossi, 1954a) as well as on the type of host cell used for adsorption of the phages prior to irradiation with visible light (Dulbecco, 1955).

It is clear that the PR sectors are not correlated with the other known properties of the T phages. However, as will be shown below, the differences in the PR sectors for T2 and T4 are directly related to their ultraviolet-sensitivities which, under suitable condition, may even be made to coincide.

(c) *Multiplicity reactivation:* In the case of photoreaction, we are dealing with the direct reversal of a photochemically induced lesion in a nucleic acid molecule by some light-activated process which is apparently an enzymatic reaction. Other mechanisms are, however, known which may lead to the partial reactivation of ultraviolet-inactivated phage. These include, amongst others, multiplicity reactivation (MR) and cross-reactivation(CR), which we shall now consider briefly in turn.

When some ultraviolet-inactivated phage are assayed under conditions such that more than one inactivated phage particle is adsorbed per host cell, it is found that there is an increased probability that the multiple-infected host cell will lyse with the concomitant release of viable phage progeny. If the ultraviolet inactivating dose is barely sufficient to inactivate the phage, the probability of MR may even approach unity. With increasing doses the probability decreases. On the other hand, since the probability of MR also increases, as might be anticipated, with increasing multiplicity of infection, this can offset the effect of even heavy radiation doses.

The phenomenon was first extensively investigated by Luria (1947) and Luria and Dulbecco (1949) and, on the basis of the demonstration by Delbruck and Bailey (1946) of genetic recombination between related phages in the same host cell, was interpreted by them as being due to the recombination of undamaged segments of individually inactivated phages to generate a fully active bacteriophage. This is also consistent with what has been repeatedly emphasized in this text with regard to the highly localized character of the lesions produced in nucleic acid chains by ultraviolet radiation. In fact, both multiplicity reactivation (and cross reactivation) provide additional supporting evidence for the highly localized nature of ultraviolet lesions, as will be seen from what follows.

It was further suggested by Luria (1947) that the undamaged segments of the inactivated DNA molecules are capable of independently reproducing themselves prior to assembly to form active phage.

On the basis of several simplifying assumptions, Luria and Dulbecco (1949) calculated the expected frequency of reproductive infections. These assumptions were:

(i) that each phage contains a definite number n of genetic units of equal radiation sensitivity;

(ii) that one of each undamaged genetic unit is required in order to give rise to viable phage progeny;

(iii) that the number of phages adsorbed to a given host cell follows a Poisson distribution.

If y is the fraction of multiply-infected bacteria giving rise to phage production; x the average multiplicity of infection; h the average number of lethal damages or hits per phage particle (the number of phage-lethal hits is the negative logarithm of the surviving ultraviolet-irradiated fraction and represents the average number of total lethal damages per particle due to a given ultraviolet dose); k the absolute number of phage particles adsorbed to individual cells; then the frequency of productive infection ν is given by

$$\nu = \frac{\sum_{k=2}^{\infty} \frac{x^k e^{-x}}{kl} [1 - (1 - e^{-h/n})^k]^n}{1 - (x + 7)e^{-x}} \tag{8-2}$$

from which n can be calculated directly since all the other quantities are measurable.

Some controversy has centered around the validity of the assumption that the adsorption of the phages to the host cells follows a Poisson distribution, but this appears to have been effectively disposed of by Harm (1956), and by the results of Bernstein on the Vi-phage II of *Salmonella typhi.* On the other hand, it is most unlikely that the various genetic units of a given phage particle are equally radiation sensitive. Wide variations in radiation sensitivity are normally encountered for different genetic markers in transforming DNA; and all the experimental results hitherto obtained for the radiation sensitivity of model oligo- and polynucleotides demonstrate the importance of base sequences and secondary structure on quantum yields for the various photochemical reactions involved.

For radiation doses such that h attains values of 10–20, the above equation was found to adequately represent the experimental data for phages T2, T4, and T6 with values of n equal to 25, 15, and 30 respectively. For considerably higher doses, on the other hand, serious divergences are encountered (Dulbecco, 1952). These have stimulated a series of extensive experimental and theoretical investigations of the phenomenon, the results of which are summarized in some detail by Barricelli (1956, 1960), who has extended the foregoing theoretical treatment to obtain data for T2 and T4 phages up to high radiation doses.

The modified treatment is based on the assumption, also arrived at from experimental considerations by Harm (1956), of the existence of two different types of lethal damage due to ultraviolet light. One type of damage, at sites which involve ordinary genetic loci, is produced with a probability small as compared to that for phage inactivation as a whole. The other types of damage involve a few specific sites, which are believed not to exceed three in phages T2 and T4, with a considerably higher probability of inactivation; these are denoted as "vulnerable centers" and are defined as follows: "Any structure, whether genetic or not, whose functioning is required in at least one of the infecting phages to permit of the reproduction of any of the phage present in the bacterium, is a vulnerable center if its probability of inactivation represents an appreciable fraction of the inactivation probability of a phage particle" (cf. Krieg, 1959). For full details of the theoretical treatment, reference should be made to the original publications (Barricelli, 1956, 1960), but the approximate form of the pertinent equation is

$$\log W_m = r \log x - 0.434(gh + xg_0L^{x-1}h^x) \quad (8\text{-}3)$$

where W_m is the probability of multiplicity reactivation
r is the number of vulnerable centres
x is the average multiplicity of infection
h is the radiation dose in phage-lethal hits
g is the fraction of inactivations in vulnerable centres
g_0 is the fraction of inactivation in ordinary genes

$$L = L'g_0$$

where L' is the fractional length of ordinary chromosome damages relative to the total length of the ordinary genome.

The foregoing equation gives a very good fit to the experimental data for phages T2 and T4 for various multiplicities of infection and for high radiation doses. However, the fact that the theoretical formulation fits the experimental data is by no means conclusive proof that the detailed mechanism proposed is necessarily correct in all its details. As pointed out by Bernstein (1957), Barricelli's equation contains three arbitrary parameters as compared to the single one in Luria's, thus lending itself to a better fit with the experimental data. It is important to specify the significance of these parameters in more definite form; and attention is indeed drawn by Barricelli (1960) to the fact that two genetic structures in the rII region of the second linkage group of phage T4 appear to be identifiable as "vulnerable centers" (Krieg, 1959). Furthermore, the above formula also adequately fits the multiplicity reactivation data of Dulbecco (1952) for phage T2 both in the case of darkness and under conditions of maximum photoreactivation, two of the three presumed vulnerable centres thus being photoreactiveable. This latter fact thus provides a supplementary clue which may prove of utility in further attempts to delineate the nature of the vulnerable centres.

It would, nonetheless, be somewhat premature to generalize the theoretical formulation of Barricelli at the present time. While extremely useful as a working model, its range of application must obviously be considerably extended before this can be done. It is necessary to point out in this connection that, for the Vi-phage 11 of *Salmonella typhi*, the original equation of Luria (1947) fully accounts for the degree of multiplicity reactivation observed, even at high doses, an analysis of the data indicating the presence of four ultraviolet sensitive units per phage particle (Bernstein, 1957).

It is, of course, also conceivable that no single model will adequately represent the behavior of all phages, or even the behavior of a given phage when different host cell strains are employed for multiplicity reactivation. It should be recalled that MR appears to be non-existent for phages T3 and T7, and is normally barely detectable for T1. However, Tessman and Ozaki (1957) have shown that MR of ultraviolet-inactivated T1 is measurable with the use of suitable irradiated host cells; the results are interpreted as being due to masking of MR in normal cells by some, as yet undefined, host cell reactivation phenomenon.

(d) *Cross reactivation.* In cross reactivation (CR) the host cells are simultaneously infected with a single irradiated particle from a homogeneous phage strain and a phage particle from a non-irradiated strain. The two strains are such that their genotypes differ by one or more mutant markers. Under these conditions the majority of the phage progeny possess genotypes derived principally from the parent non-irradiated strain; but the progeny may also include members which contain genetic marker(s) of the irradiated phage. The inactivated phage, is, of course, under these conditions not reactivated as a separate entity. This "rescue" of markers of the irradiated phage *via* the multiplication of non-irradiated phage has led to the introduction of the term "marker rescue" in place of cross reactivation. The importance of this phenomenon in investigations on the mechanism of phage reproduction and genetics is immediately obvious.

Cross reactivation experiments differ from multiplicity reactivation in one important respect. In the latter we must differentiate between several types of ultraviolet damage in addition to that which may interfere with the injection of the phage DNA into the host cells. It is necessary to distinguish between lesions which interfere with the replication of the damaged site (genetic damages) and those which may interfere with the ability of a phage particle to effect some heterocatalytic function on behalf of all the phages infecting a host cell (functional damage). We may therefore have genetic lesions, functional lesions, and also mixed genetic and functional lesions (cf. Symonds and McCloy, 1958).

In CR experiments, on the other hand, any functional deficiency provoked by ultraviolet irradiation will be compensated for by the non-irradiated infecting phage, so that it becomes possible to analyze the damage in the irradiated phage exclusively for their genetic characteristics.

Two experimental techniques have been applied to studies on CR (Doermann, 1961). The first makes use of the single-burst technique which is restricted to doses not exceeding 50 phage-lethal hits. The second utilizes the rII-K12(λ) system of Benzer (1957) and is based on the fact that rII mutants of phage T4 cannot produce plaques on K12(λ) while rII+ mutants can. Wild-type phage is therefore irradiated and mixed with *E. coli B* so that only one or two phage particles are adsorbed per host cell.

The host cells are simultaneously infected with two or three phage particles of non-irradiated carrier phage of some rII genotype. Following removal of non-adsorbed virus with anti-T4 serum, the infected host cells are plated on K12(λ) before they have had an opportunity to lyse. Only a host cell which liberated wild-type phage will make a plaque on K12(λ). It is consequently possible to measure the probability that an irradiated wild-type phage will transmit to the progeny the rII$^+$ allele of the rII mutant gene in the carrier. In this way the chance of rescuing the rII$^+$ allele of any rII marker put into the carrier can be determined quite accurately. Experimentally it is possible to have as many as 10^6 rII-producing cells on a single plate without interfering with the assay of the r$^+$ phage; hence a low fraction of rescues can be readily detected. This permits use of ultraviolet doses attaining 1000 or more phage-lethal hits.

The general results obtained with the single-burst technique have been summarized by Doermann (1961), as follows:

> A given marker of the irradiated phage may be completely absent from the burst even when another appears in a nearly normal number of progeny phage particles, indicating that a given portion of the phage genome may be damaged without affecting another region. If, however, two markers are linked they may be inactivated together, the probability of such an event being greater the more closely linked the markers are. Distantly linked markers behave as though they are unlinked. If one of the markers of a linked pair is damaged, but the second appears in the burst, the latter tends to appear in a smaller number of phages than another, unlinked, marker contributed by the irradiated parent phage to the same burst. The general conclusion from these results is that a fraction of the UV-induced lesions are discretely distributed along the phage genome and that surviving markers are rescued by means of genetic recombination.

The results obtained at higher doses by use of the rII method substantiate the foregoing and provide some additional useful data. In single rII$^+$ rescue, the survival curves are convex to the origin, indicative of a gradual shrinkage in size of the surviving target with increasing dose. For any given marker the probability of rescue decreases further as additional lesions are formed between it and the nearest lesion already present. Ultraviolet damage genomes multiply only to a very limited extent, if at all. The subject has been examined in detail from a theoretical point of view (Barricelli and Doerman, 1961) but, as for multiplicity reactivation, it is necessary to point out that the range of experimental data available is as yet too limited to permit generalization.

(e) *Gene-controlled reactivation.* The difference in ultraviolet sensitivities of phages T2 and T4, to which repeated reference has been made above, has provided some interesting information about the genetic mechanism involved in the ultraviolet inactivation and photoreactivation of these phages. It was first noted by Luria (1947) that, at 2537 Å, T2 is twice as sensitive as T4, notwithstanding that the DNA contents of the two phages, as well as their size and morphology, are similar. It was also reported that the T2- and T4-like progeny of a genetic cross between the two phages exhibit relative sensitivities corresponding to those of the parental types; i.e. the ultraviolet sensitivity level behaves genetically as a unit character. Some very elegant experiments of Streisinger (1956) and Harm (1961a, 1961b) have both confirmed and extended these observations in such a way as to permit of the genetic location of the factor controlling these levels in the two phages, as well as their relation to the photoreactiveable sector in both phages. These latter studies are, in fact, typical of the modern approach to the use of ultraviolet (and ionizing radiations) as a tool for the elucidation of the genetic structure of bacteriophages, a field which is expanding so rapidly as to render it difficult to summarize the literature.

The genetic locus controlling the ultraviolet sensitivities of the foregoing phages, referred to by Streisinger (1956) as *u*, is closely linked to the rII locus of T2 and controls also the degree of photoreactivation (PR) in such a way that a high ultraviolet sensitivity (in T2) corresponding to the u^+ allele is accompanied by a high degree of PR and, conversely, a low sensitivity (in T4) corresponding to the *u* allele is accompanied by a low level of PR. Under conditions where PR is allowed to take place after ultraviolet irradiation, the survival curves for both phages are almost identical.

Two hypotheses may be formulated to interpret the foregoing: The primary photochemical transformations induced in the phage DNA are dependent on the presence of the *u* or u^+ allele in the irradiated phage genome. *Or* the primary photochemical modifications are similar in nature and the *u*-gene controls some intracellular processes which determine whether a primary lesion results in lethality or not; the *u*-gene might directly reverse the radiation damage, or do so indirectly by modifying the intracellular state of the cell, e.g. by directing the formation of an enzyme which carries out the reactivation.

The second alternative has been shown to be the most plausible (Harm, 1961a, 1961b) in a series of experiments which profited from the fact that *u*-gene function is much more resistant to ultraviolet light than infectivity. The survival of T2 in the dark (i.e. in the absence of PR) was examined in the absence and presence of the *u*-gene from T4.

In Fig. 8-10, curve A represents the survival of T2, singly infecting normal *E. coli B* cells. Curve B is the survival curve for T2, singly infecting *E. coli B* cells which were simultaneously infected with heavily irradiated T2 phage at a multiplicity of 3. The small difference between A and B is due to multiplicity reactivation between the phage under study and the heavily irradiated phage. Curve D is the survival for T4 singly infecting *E. coli B* cells, and E when the host cells were simultaneously infected with heavily irradiated T4 at a multiplicity of 3; again the slight difference is due to multiplicity reactivation. If one now irradiates T2, singly infecting the host cells which are simultaneously infected with heavily irradiated T4 at a multiplicity of 3, one obtains curve C which is close to that for T4 under similar conditions (curve D). Hence in the presence of the *u*-gene of T4, the survival of T2 is similar to that of T4.

Since under conditions of PR, the survival of T2 is also similar to that of T4, it follows that the *u*-gene in irradiated T4 performs the same, or an

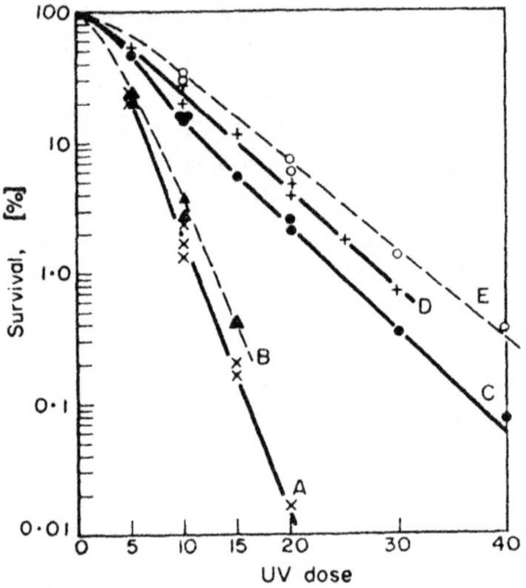

FIG. 8-10. Showing reactivation of ultraviolet inactivated phage T2 by *u*-gene of phage T4. For details see text. (Harm, 1961a).

equivalent, function as does PR on irradiated T2. In other words, *u*-gene reactivation leads to reversal of the same ultraviolet lesions as photoreactivation (Harm, 1961, see also Harm, 1963).

The foregoing has derived additional support from the finding that the T2 survivors are still genotypically u^+. Furthermore, a T4 mutant was also isolated which was found to be genotypically u^+, using as criteria ultraviolet sensitivity and photoreactivation, whereas the host range properties were those of T4.

It follows that the lesions reversed by PR and the *u* allele must be similar. It remains to establish the nature of these lesions. Indirect evidence has shown that photoreactiveable lesions in transforming DNA involve, *to a partial extent*, cross-links between complementary strands. Stahl *et al.* (1961) found that the ultraviolet sensitivities of T2-u^+ and a mutant T2-*u*, in both of which almost all of the thymine had been substituted by 5-bromouracil, were identical and that neither showed any PR. This has been interpreted as evidence in favour of thymine dimerization as the source of the lesions which are reactivated in the normal phages by PR or the *u*-gene (Harm, 1961b). Such an interpretation is, however, by no means unequivocal. It has been shown by means of the density gradient technique, for example, that only a fraction of the lethality of various phages can be due to twin-strand cross-linking in the DNA components (Wierzchowski and Glisin, personal communication). Furthermore, experimental data must obviously be acquired to elucidate the nature of the lesions.

E. PHOTOSENSITIZED INACTIVATION

Although the photosensitized inactivation of viruses has come in for considerable attention, relatively little is as yet known regarding the sites involved in this type of reaction. This problem undoubtedly merits closer attention since, as will appear from some of the results to be presented, there is a considerable degree of specificity involved both as regards the sensitizer and the acceptor (i.e. the virus particle) as has already been shown to prevail for the photosensitized inactivation of enzymes and proteins. Furthermore, as for these latter, the presence of oxygen is a necessary pre-requisite for photodynamic action.

Of the numerous early observations made in this connection, undoubtedly the most significant are those of Burnet (1933), who demonstrated that serologically related phages exhibit similar susceptibility to visible light in the presence of a given dye, but that there are marked differences in sensitivity from one serological grouping to another. Subsequent investigations have provided confirmation of this important finding (see below), but its particular significance remains to be elucidated. Indirect evidence, to be presented below, suggests that it is the external phage structure which is the controlling factor. An additional observation reported by Burnet (1933) and Burnet *et al.* (1937) is that, in the presence of methylene blue, infectivity may be destroyed with only slight loss of antigenicity. This could be interpreted to mean that loss of infectivity results from photodynamic action of the dye on the nucleic acid component, since protein antigenicity is relatively resistant to photosensitization; e.g. the enzyme activity of a protein is much more susceptible to photodynamic action than the antigenic properties. But conclusive evidence for the involvement solely of the nucleic acid component is still lacking.

In addition to the foregoing, it has also been noted that the nature of the dye employed as a sensitizer may be of considerable importance. For example, Herzberg (1953) observed that thiazine, acridine and thioxanthone dyes all sensitized vaccinia virus to visible light, but that methylene blue and thiopyronine were most active, at a dilution of 10^{-6}. Galston and Baker (1949) found riboflavin to be quite active as a photosensitizer of phage T2r to visible light; while Galston (1950) reported that riboflavin is also effective against phage T6r, the inactivation curve being of the "multiple-hit" type as compared to the first order reaction normally found for photosensitization processes involving enzymes. But the activity of riboflavin against viruses is by no means general, as shown by the results of Welsh and Adams (1954).

A fairly comprehensive study of the photodynamic inactivation of bacteriophages is that of Welsh and Adams (1954), who investigated 14 different strains from several serological groups. Neither riboflavin nor eosin, which are known effective sensitizers for some enzymes, as well as other systems including phages T2r and T6r (see above), were effective, but methylene blue was active in all cases. Except for a short initial lag period (which has also been reported by a number of investigators), the survival curves were logarithmic for all strains examined. From the influence of dye concentration and pH on the reaction kinetics for inactivation, it was concluded that the rate-determining factor was the amount of dye adsorbed to the phage particles. But by far the most interesting finding was that demonstrating, from kinetic studies, the correctness of the earlier conclusion of Burnet (1933), according to which serologically related phages exhibit similar photodynamic sensitivity, with marked differences from one serological group to another. In the case of T2 phage, it was also noted that destruction of the host-killing property occurred at only one-third the rate of loss of plaque-forming ability, a finding which is undoubtedly deserving of further attention. Attempts to demonstrate photoreactivation and multiplicity reaction were completely unsuccessful; this is in agreement with Galston's findings for T6r phage and shows that the mechanism involved must be quite different from that prevailing with ultraviolet irradiation.

The behavior of a number of viruses towards the action of toluidine blue (Hiatt *et al.*, 1960) is illustrated in Table 8-6 in which the first order rate constants are listed for inactivation of different viruses under analogous conditions. In practically all instances the survival curve exhibited an initial small lag period which was ignored in the calculation of the first order rate constants. Qualitatively similar results were obtained with the related thiazine dye, methylene blue. The relatively high resistance of polio-virus, which contains RNA rather than DNA, was regarded by the authors (Hiatt *et al.*, 1960) as a confirmation of Yamamoto's (1958) thesis that the action of thiazine dyes is due to their combination with bacteriophage DNA. This argument is open to considerable doubt, even if Yamamoto's proposal is regarded as valid. Thiazine dyes, such as toluidine blue and methylene blue, are strongly bound by both RNA and DNA, and are widely used cytochemically for both nucleic acids. Indeed the binding of methylene blue by RNA has formed the basis of a technique for measurement of ribonuclease activity (Shugar, 1953).

It should once more be emphasized that some direct evidence as to the extent of participation of virus nucleic acids in the photosensitization process

TABLE 8-6. FIRST-ORDER RATE CONSTANTS (k) FOR PHOTODYNAMIC INACTIVATION OF VARIOUS VIRUSES IN THE PRESENCE OF 2×10^{-5} M TOLUIDINE BLUE IRRADIATED WITH POLYCHROMATIC LIGHT, 0.39 W/CM²

Virus	$k(\text{sec}^{-1})$	Rate relative to that for the coliphage T3
T3 coliphage	0.44	1.0
Vaccinia	0.44	1.0
Adenovirus 3	0.081	0.18
Adenovirus 4	0.074	0.17
Adenovirus 7	0.074	0.17
ECHO-1	0.0024	0.006
Coxsackie A9	0.0015	0.003
Poliovirus I, II or III	0.0005	0.001
Coxsackie B2	0.005	0.001

is to be desired. Free vesicular exanthemo virus (an RNA virus) suspended in an acridine orange solution is not light sensitive but the simultaneous introduction of the dye and virus to the host results in progeny that are sensitive to light. Apparently the dye is unable to penetrate the protein coat of the virus *in vitro* (Hackett, 1962).

The relatively high resistance of poliomyelitis virus to photosensitized inactivation has been taken advantage of by Hiatt as a means of eliminating potentially harmful organisms from live polio vaccines (Hiatt, *7th International Congress of Biological Standardization, London,* 1961).

Oster and McLaren (1950) had earlier shown that TMV is inactivated by blue light in the presence of acriflavine and that binding of the dye by the virus was probably involved since, in the presence of salt, which reduced the extent of binding of the dye, the loss in activity was considerably reduced. No conclusions were drawn as to the site of possible adsorption of the dye. Contrary to results of most observers with other viruses, the survival curve for TMV in the presence of acriflavine was found to be first order for the entire reaction, with no initial lag period. Furthermore, partial removal of oxygen from the system markedly reduced the loss in activity, as expected if the reaction involved photo-oxidation.

Additional investigations with TMV should be most promising if one bears in mind the ease with which infectious RNA may be isolated from this virus. It would be interesting, for example, to photodynamically inactivate TMV, and then examine the activity of the RNA isolated from the inactivated virus. The photosensitizing dye would, of course, have to be first removed by dialysis. Such an experiment would inform us to what extent the nucleic acid component of the intact virus is involved in the virus inactivation. Furthermore, if it is indeed largely the nucleic acid component which is most susceptible in the intact virus, then a comparative investigation of the photodynamic effect on the intact virus and on the isolated RNA component might provide some useful information as to the extent to which the reaction on the intact virus is influenced by dye diffusion through the protein coat to the nucleic acid component. RNA from TMV causes a color shift when combined with acridine orange, and visible light causes inactivation in the presence of small amounts of dye (Chessin, 1960).

The relationship between the chemical structure and photodynamic activity of various dyes against the T-group bacteriophages has been extensively investigated by Yamamoto (1956, 1958). Thiazine, oxazine and acridine dyes were found to be very active and the following features were listed as common to all of them: (a) active dyes are basic hetero-tricyclic compounds possessing two chargeable N atoms; (b) the distance between these two N atoms is almost equal in the various dyes and a conjugated double-bond system lies between them; (c) all of the dyes possess an ionic affinity for nucleic acids. It is, of course, to be expected that adsorption of the dye to certain sites will facilitate its activity as a photosensitizer but, since the relative concentration of dye to phage molecules is always quite high, it is also likely that the ability of the dye to form an excited long-lived (triplet) state is of equal, if not of greater, importance.

In Table 8-7 are presented some of Yamamoto's (1958) results for the activity of a variety of dyes against the T-series of bacteriophages belonging to four serological groups. The results are seen to be in accord with the conclusions of Burnet (1933) and Welsh and Adams (1954) regarding a correlation between serological grouping and photodynamic sensitivity. The relatively high resistance of the T-even bacteriophages is attributed by Yamamoto (1958) to their supposedly smaller pore-size, since they are more susceptible to osmotic shock than the other T-coliphages; this would make penetration of dye molecules to the nucleic acid components more difficult.

Some indirect evidence in favour of such a conception has been provided by Helprin and Hiatt (1959) in a study of the inactivation of $T2r^+$ and T3 coliphages by the photodynamic action of toluidine blue, which demonstrated that these phages differ widely in their rate of uptake and tenacity of binding of photosensitizing dyes. The inactivation of $T2r^+$ was found to be very markedly dependent on the time of pre-incubation with the dye in the dark; but this effect was absent for T2. In common with other observers, it was assumed that combination of virus and dye was a necessary prerequisite for photodynamic action. Pre-incubation of T2r with dye at 37°C resulted in a higher degree of inactivation, in agreement with the assumption of the binding of a larger amount of dye. The high temperature coefficient of this

TABLE 8-7. PHOTODYNAMIC ACTIVITY OF VARIOUS DYES AT NEUTRAL pH AGAINST THE T BACTERIOPHAGES; DYE CONCENTRATION M/60,000; PHAGE TITERS, 5×10^7/ml

Dye	First order inactivation rate constants (min^{-1}) (Serological groups)						
	Strain						
	T1	T2	T4	T6	T3	T7	T5
Methylene blue	0.24	0.01	0.02	0.01	0.55	0.72	0.77
Toluidine blue	0.12	0.00	—	—	0.36	0.57	0.40
Brilliant cresyl blue	0.07	0.00	—	—	0.18	0.33	0.66
Neutral red	0.01	0.01	0.01	—	0.06	—	0.01
Crystal violet	—	0.00	—	—	0.04	—	0.06
Methyl green	—	0.00	—	—	—	—	0.01
Pyronine	—	0.01	—	—	0.02	—	0.05
Acridine orange	0.06	0.00	0.00	0.00	0.97	1.15	0.35
Acriflavine	0.04	0.01	—	—	0.29	0.46	0.28
Rivanol	—	0.01	—	—	—	—	0.10

process, a Q_{10} of about 4, is similar to that for diffusion across a membrane. If the dye-phage mixtures were suddenly diluted, the sensitivity of the T3 phage decreased, as might be expected; but T2r$^+$ under such treatment was still inactivated at an appreciable rate, in agreement with the hypothesis that the dye was bound at the interior of the phage from which its removal by diffusion was restricted (cf. also Kaufman and Hiatt, 1959).

Triphenylmethane dyes, which were themselves inactive against the T phages, were found by Yamamoto (1958) to competitively inhibit active dyes; but this does not provide any more concrete information as to the nature or number of sites involved.

Attention should be drawn to the fact that a number of phages have been found to exhibit sensitivity to visible light, even in the absence of a specific photosensitizer, with an action spectrum exhibiting a maximum in the near ultraviolet. This phenomenon was examined in some detail by Wahl and Latarjet (1946, 1947) for phages S13 and C16, the action spectra for which are quite different. The effect was verified by Welsh and Adams (1954) for a number of phages. It is, however, quite small compared to that prevailing in the presence of methylene blue, of the order of 1 per cent or less. This phenomenon could conceivably be due to the presence of some pigment in the phages, present either as an impurity or as a trace constituent; and it would undoubtedly be of value to carefully re-examine the behavior of some phages which may now be isolated in much higher purity than in the period 1946–1954. The possible importance of the foregoing in relation to phage structure is further emphasized by the fact that bacteriophages, in which most of the thymine has been replaced by 5-bromouracil, exhibit marked photosensitivity at wavelengths longer than 3100 Å, of the order of about 1 per cent of that at 2537 Å. This phenomenon, which has been variously reported from a number of laboratories, is not as puzzling as appears at first sight if it is recalled that halogenated uracils and their glycosides absorb farther to the red than the non-halogenated derivatives, so that they may exhibit absorption to the red of 3100 Å (for tabulation and review of spectra of all halogenated uracils and their glycosides, see Berens and Shugar, 1963). In fact high molecular weight polymers of 5-bromouracil have been found to undergo degradation by exposure to near ultraviolet radiation (Michelson, personal communication).

Boyce and Setlow (1963) have examined quantitatively the influence of incorporated bromouracil on the photochemical sensitivity of bacteriophage T4. The phage preparation used was such that 80 per cent of the thymine was replaced by bromouracil. Compared to the normal phage, the sensitivity was increased 7-fold at 2300 Å and 4-fold over the range 2400–2900 Å. This was considered to arise in part from impairment of enzymatic reactivation of photochemical lesions (cf. Howard–Flanders *et al.*, 1962; see Section D-5) and in part from the nature and content of the photoproducts induced in the brominated phage DNA (see, for example, below). But, above 3000 Å, the sensitivity went up by a factor of 88, and this could be largely accounted for by the greater ultraviolet absorption of the phage due to its bromouracil content as compared to the normal thymine-containing phage.

The foregoing is also of practical interest to the phage worker, since a number of commercial "daylight" lamps are known to emit small quantities of radiation down to 3000 Å and even to 2900 Å (Latarjet amd Morenne, 1951). It was, in fact, the use of such lamps which led to the observation that brominated phages are photosensitive in the near ultraviolet. It had been observed much earlier by Latarjet and Morenne (1951) that phage T2 is slowly inactivated by lamps of this type, the survival curve being of the multiple-hit type and not exponential. Since the lamp emission was found to include traces of radiation down to 2890 Å, this can account for the slow inactivation, but the complex kinetics observed suggest a different inactivation mechanism than at 2537 Å, if we assume that the reciprocity law prevails at both wavelengths.

Some insight into the nature of the reaction involved has been furnished by Wacker *et al.* (1962) by a study of the behaviour of isolated DNA under similar conditions. A DNA preparation containing incorporated ^{14}C-bromouracil was irradiated with a GE 20-watt fluorescent lamp. Hydrolysis of the irradiated DNA, followed by chromatography, revealed the presence of ^{14}C-uracil. No uracil was obtained on irradiation of free bromouracil under identical conditions, although dehalogenation is of frequent occurrence in the irradiation of halogeno substituted aromatic rings.

CHAPTER IX

Some Selected Problems in Photobiology

A. RADIATION RECEPTORS IN BIOLOGICAL SYSTEMS: ACTION SPECTRA

One of the major problems to be resolved in relation to the photochemistry of a biological system is that of determining the component(s) responsible for the absorption of the biologically effective radiation. The component(s) in question may not necessarily be that which is transformed by the incident radiation; the latter may, instead, be transferred to some other cell constituent which then undergoes the reaction leading to the ultimate modification observed experimentally. An examination of the response of a biological system to light of different wavelengths is theoretically capable of providing a clue as to the nature of the constituent responsible for the initial absorption, provided certain conditions are fulfilled. Actually what we obtain by means of such a study is the response (or action) of the sytem to different wavelengths, the resultant plot of which, as we have already stated (Chapter I), is commonly referred to as an "action spectrum". As we shall now show, such an action spectrum should, under defined conditions, correspond point by point with the absorption spectrum of the component(s) responsible for the absorption of the incident energy producing the biological response. From a knowledge of the absorption spectra of the various components of the system, that which is involved may often be identified by simple comparison.

The theory of action spectra was developed by the Warburgs in the nineteen twenties and constitutes one of the finest developments in experimental biology (Warburg, 1949). For biological systems the assumptions are

(a) the photochemical quantum yield for the photoreceptor is the same through the breadth of the spectrum investigated;

(b) there is no important absorption by substances other than that of the compound photochemically activated and light scatter is negligible;

(c) the absorption spectrum of this compound in the living system must not differ essentially from its absorption spectrum in the solvent in which it is determined;

(d) only a small fraction of the incident light is absorbed by the compound in question (cf. Chapter I).

If Φ is the quantum yield for the biological effect produced, then

$$-\mathrm{d}C/\mathrm{d}t = \Phi I_0 \beta C \qquad (9\text{-}1)$$

where C is the concentration of the light sensitive substance in moles/cm^3 and β is the absorption coefficient in cm^2/mole (cf. Chapter I). In a complex system we actually measure, not dC/dt but some effect such as inactivation, mutation, etc. which is proportional to this differential.

By equation (9-1), for a given wavelength we have

$$\beta = \ln (C_0/C)/I_0\Phi t \tag{1-8b}$$

and at two wavelengths, λ_1 and λ_2, we can find the ratio

$$\beta_1/\beta_2 = \frac{[\ln C_0/C]_{\lambda_1}}{[\ln C_0/C]_{\lambda_2}} \frac{(I_0 t)_{\lambda_2}}{(I_0 t)_{\lambda_1}} \tag{9-2}$$

where β_1/β_2, for a given dose, is proportional to the ratios of action at each wavelength. Alternatively, if Φ is independent of wavelength, we have for a given amount of action and exposure time

$$\frac{\beta_2}{\beta_1} = \frac{(I_0)_1}{(I_0)_2} \tag{9-3}$$

The ratio β_1/β_2 also equals σ_1/σ_2 where, as before, σ is the absorption cross section. Similarly, all ratios of β_2 to β_1 can be determined with light of all wavelengths available from the light source and for which I_0 in mole quanta/cm^2 per sec can be measured. For an equal amount of photochemical change at two wavelengths, it can be shown also that

$$\beta_2 = \frac{(\text{energy incident at } \lambda_1)}{(\text{energy incident at } \lambda_2)} \frac{\lambda_1 \beta_1}{\lambda_2} \tag{9-4}$$

so that β_2 is proportional to $1/I_0\lambda_2$—consequently a plot of the reciprocal of the energy required to produce a given biological effect versus wavelength, normalized at λ_1, will give a plot coincident with the absorption spectrum, provided all assumptions are met (cf. equation 1-10).

In some instances the agreement between absorption and action spectra is remarkably close, as illustrated by the action spectrum for scotopic vision (Fig. 9-1). Such close correspondence is, however, not the general rule, as might be expected from an examination of the assumptions on which the above derivation is based. (See Reviews by Loofbourow (1948), Blum (1950) and Setlow (1957).)

The requirement that the system under study be in a stationary state is an obvious one. For example, one would hardly expect the above relationships to hold for a bacterial suspension in a culture medium at 37°C or, for that matter, even at room temperature if the time of irradiation is appreciably by comparison with the rate of metabolism of the system. The simplest solution, in doubtful cases, is to shorten the irradiation time by the use of higher intensities, or to lower the temperature of the system so as to decrease the rate of metabolism during the period of irradiation. For isolated systems, such as purified enzymes, biologically active DNA or viruses, such problems naturally do not exist.

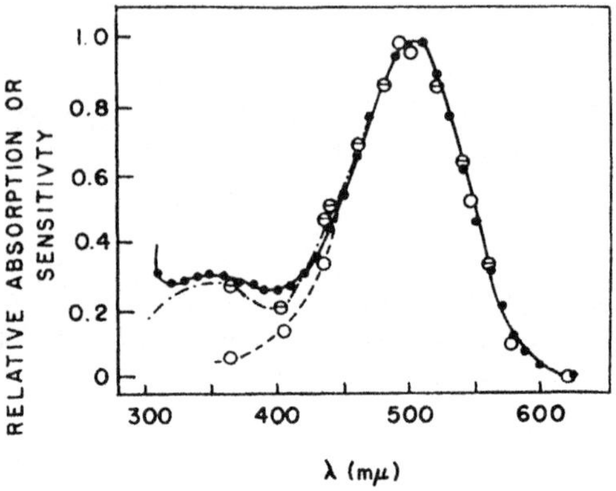

FIG. 9-1. Action spectrum for scotopic vision as compared to absorption spectrum of visual purple:

●●●●● absorption spectrum of rhodopsin
○○○○ sensitivity of the aphakic (lensless) eye
⊖⊖⊖⊖ photosensitivity of rhodopsin

(from Wald (1947) as given in Blum (1950)).

The reciprocity law should hold for all the absorbing components of the system and this is probably true in most instances. For inactivation of enzymes by ultraviolet light, reciprocity has been shown to prevail for a 10^5 variation in light intensities up to 10^9 ergs/cm^2 per sec (McLaren *et al.*, 1952). While comparable observations are apparently not as yet available for nucleic acids or nucleoproteins, it has been reported that the quantum yield for photochemical transformation of Up is unaffected by a 50-fold variation in the incident light intensity (Wierzchowski and Shugar, 1959). The law even holds for damage to mouse skin over a 10^7-fold intensity range (Claesson *et al.*, 1958).

In the derivation of the above relationships it was tacitly assumed that each of the absorbing components absorbs much in the way it does in the isolated state. In practice this is not always so, the absorption spectrum of a given constituent being frequently modified by light scattering and/or by complex formation with other components of the system. Another possible source of error may be the dichroism associated with the selective orientation of some molecules *in vivo* (Commoner, 1949; Lipkin and Commoner, 1949), the net result of which would be an appreciable decrease in extinction coefficients. Several controversies have revolved around this subject, largely because of its possible significance in microspectrophotometric methods for estimation of nucleic acids in intact cells, but the experimental evidence appears to negate its existence to any appreciable extent in all but one of the various cellular materials in which it has been looked for (Thorell and Ruch, 1951; Ruch, 1957).

In most experimental work on action spectra with living organisms, scattering has usually been ignored. It is, however, not difficult to approximately correct for in the case of an optically thin suspension of micro-organisms (Mitchell, 1950) by procedures analogous to those already described for viruses. For example, the absorption spectrum of a suspension of *M. pyogenes* may be experimentally corrected for scattering by varying the refractive index of the medium, Fig. 9-2. We have found that this

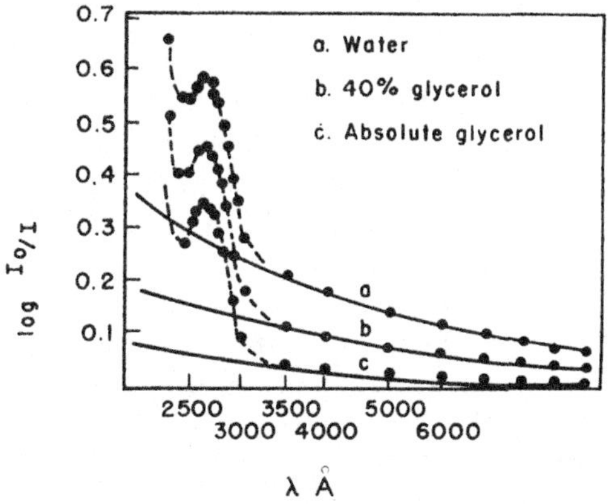

FIG. 9-2. Effect of refractive index of medium on absorption and scattering spectra of *M. pyogenes*. The continuous lines represent scattering, which is measured to 3500 Å and then extrapolated to shorter wavelengths (from Mitchell, 1950).

procedure is applicable to a wide variety of organisms and, once the extent of the scattering has been calculated, it may be applied to the aqueous suspensions usually employed for irradiation purposes. A variety of other methods has now been developed for the correction for, or elimination of, scattered light in measurements of absorption spectra of micro- as well as some higher, organisms. The simplest and most convenient of these is that of Shibata *et al.* (1954), theoretical and experimental details of which have been outlined by Shibata (1958). Reference should also be made to the papers of Duysens (1956), Keileen and Hartree (1958), Latimer and Eubanks (1962) and Amesz *et al.* (1961), the latter of which treats in detail from a theoretical point of view the various methods for correction of the absorption spectra of scattering suspensions.

To the foregoing must be added the further requirements that the attenuation of the incident radiation, prior to reaching the sensitive substance in the cell, should not be a function of the wavelength or, better still, that it be sufficiently small so that it may be entirely ignored. The extent to which this requirement prevails is, in practice, not easily established. Examination of Fig. 7-2, above, shows that in the case of micro-organisms, the nucleic acid content of which is fairly high, the nucleic acid spectrum completely obscures that of proteins. In addition the principal absorption in this case is due to the RNA of the cytoplasm. Supplementary criteria must therefore be used to distinguish between an action spectrum due to RNA and DNA; usually this is based on the nature of the response measured such as mutation, which is known to be due to DNA. But there are instances where, although an action spectrum has clearly established the involvement of nucleic acids, it has not been possible to determine unequivocally whether it is due to RNA or DNA (see below).

A brief reference to a detail of purely technical interest may be pertinent at this point. The running of an action spectrum usually requires the use of a monochromator, which may not always be accessible. When it is desired simply to determine whether a given action spectrum is nucleic acid-like in character or not, a modification proposed by Epstein and Schiff (1961) may prove useful. The sample under investigation is irradiated with a high-pressure mercury lamp, and then with the same lamp in conjunction with a Corning No. 9700 filter, which cuts off at 2800 Å (a suitable microscope cover glass may also be used as the cut-off filter, see Appendix). The same procedure is applied to the photoinactivation of the bacteriophage T2, for which the action spectrum is known to be that of DNA. If protein absorption is of importance in the action spectrum of the test sample, then the ratio for inactivation with unfiltered and filtered radiation will be reduced as compared to the same ratio for T2 phage.

1. *Wavelength Dependence of Quantum Yields*

Another one of the fundamental assumptions inherent in action spectra techniques is the independence of quantum yields with respect to wavelength. It becomes of interest, therefore, to examine to what extent this is so for proteins and nucleic acids.

(a) *Nucleic acids.* At the moment there is not much data available on the effects of different wavelengths on the photochemical transformation of nucleic acid derivatives, a point already stressed in Chapter VIII and, to some extent, we must rely on indirect observations for any conclusions. This situation should, however, shortly be improved. Setlow (1961) has begun an extensive investigation on the wavelength dependence of quantum yields for the main nucleic acid derivatives, the results of which have not yet been published. Preliminary data for the photochemical transformation

TABLE 9-1. QUANTUM YIELDS FOR PHOTOCHEMICAL TRANSFORMATION OF UP AT DIFFERENT WAVELENGTHS IN AQUEOUS MEDIUM AT NEUTRAL pH, BASED ON LOSS OF ABSORPTION OF PRINCIPAL MAXIMUM AT 2600 Å (SETLOW, 1961)

λ(Å)	Φ	Remarks
2800	0.016	Reaction reversible in dark
2650	0.017	Reaction reversible in dark
2480	0.019	Reaction reversible in dark
2380	0.017	Reaction reversible in dark
2300	0.022	Per cent reversibility reduced
2250	0.038	Initial value only since other dark reactions also observed

of Up is shown in Table 9-1; it should be recalled that, in aqueous medium, the principal reaction in this case is addition of a water molecule to the 5,6 double bond of the pyrimidine ring. It will be seen that at wavelengths below 2350 Å other reactions occur in addition to water uptake. At wavelengths longer than 2350 Å the reaction is apparently independent of wavelength. The marked increase in quantum yield below 2350 Å, if confirmed and if it applies to other nucleic acid derivatives, will undoubtedly lead to complications in extension of action spectra to this wavelength region and may necessitate a revision of some conclusions which have already been based on a comparison of wavelength effects above and below 2350 Å (e.g. Siegel and Norman, 1958).

The situation with regard to dimer formation is somewhat more complex. Although reliable data are lacking, there appears to be very little doubt that the dimerization reaction is wavelength dependent. It will, however, be recalled that the quantum yield for thymine dimer photodissociation is almost independent of wavelength (Section VI-2); by analogy the same may be expected to hold for uracil dimers. Since, at any given wavelength, an equilibrium is established between dimer formation and dissociation in DNA, the action spectrum would be expected to be representative of the dimerization reaction. With preparations of RNA the situation may be complicated by the fact that photodissociation of uracil dimers is necessarily accompanied by at least partial hydration of the liberated uracil residues. Some useful information might be gained by a quantitative comparison of the action spectra for the photochemical transformation of the uracil and thymine residues in UpU and TpT on the one hand, and poly-U and poly-T on the other, due precautions being taken that both polymers are either in the form of random coils or with similar degrees of secondary structure (see Section II. B-3).

The action spectrum for the ultraviolet-induced cross-linking of dry films of DNA is exhibited in Fig. 9-3, and is in fairly good agreement with the absorption spectrum of DNA, suggesting that the quantum yield is wavelength independent. This finding may, on the other hand, be objected to on the grounds that the property measured was a physico-chemical, and not a biological one. It is consequently worth noting that the quantum

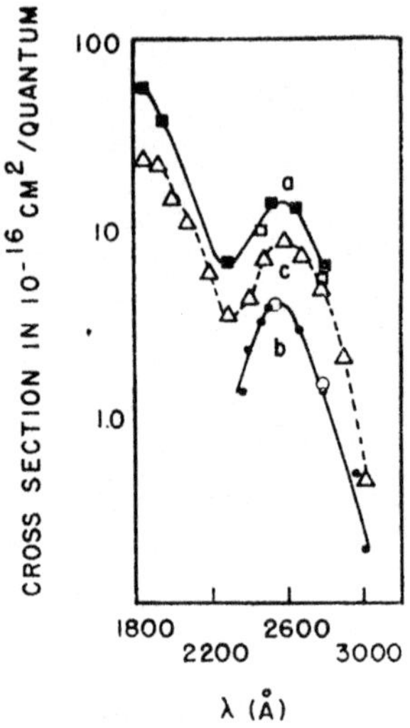

FIG. 9-3. Action spectra for cross-linking of dry films of DNA. Films were irradiated and the resulting loss of solubility in water or salt solution measured:

(a) Action spectrum based on loss of solubility in 0.2 M NaCl.
(b) Action spectrum based on loss of solubility in water
(c) Absorption spectrum of DNA in aqueous solution

(from Setlow and Doyle, 1954).

yield for cross-linking of dried DNA was found to be unaltered at low temperatures (90°K), whereas that for inactivation of transforming DNA in solution is reduced to one-half at 77°K (Lerman and Tolmach, 1959). By contrast, the yield for dried T1-phage at 3547 Å has been reported to be unchanged between room temperature and 90°K (Setlow and Doyle, 1953a).

The quantum yield for inactivation of infectious RNA from TMV has been reported as independent of wavelength over the range 2300 Å to 2800 Å (Rushizkey *et al.*, 1960), but the virus ϕX174 has been found to show a definite dependence of quantum yield on wavelength (Setlow and Boyce, 1960). Finally the yields for inactivation of the coliphages T1 and T2, for both of which the action spectra are approximately nucleic acid-like in character, exhibit more than minor variations over the range 2300 to

3100 Å (Zelle and Hollaender, 1954). We can only conclude at the moment that existing evidence favors the view that quantum yields for nucleic acids are not entirely independent of wavelength but are sufficiently so to permit their being identified by action spectra (Stadler and Uber, 1942). Additional measurements would be highly desirable.

(b) *Proteins.* The wavelength dependence of quantum yields for proteins also varies for different proteins. To some extent this has been correlated with the amino-acid content, and, in particular, cystine (Chapter V). The action spectra for some proteins are illustrated in Fig. 9-4. Although the

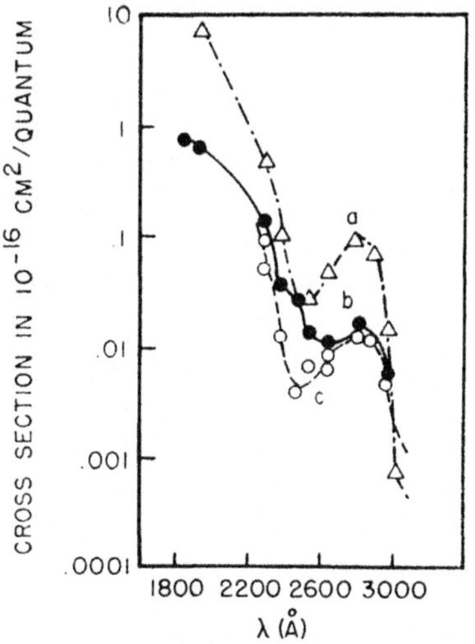

FIG. 9-4. Action spectra for inactivation of (a) aldolase in aqueous solution; (b) chymotrypsin in the dried state; (c) gramicidin in alcoholic solution (from Setlow and Doyle, 1953, 1957).

absorption spectra of all three are similar, it will be noted that the action spectra exhibit net differences, most pronounced in the case of chymotrypsin. For some proteins, such as trypsin in the dry state, the departure of the action from the absorption spectrum is even more marked. There is, however, one feature characteristic of all proteins hitherto examined and which serves as a basis for action studies, viz. the action spectrum exhibits a distinct maximum in the neighborhood of 2800 Å and a relatively steep shoulder to the red of this maximum; in addition, if there is a minimum to the violet of 2800 Å, it is located between 2450 and 2550 Å whereas for nucleic acids this minimum is in the range 2300–2400 Å.

This principle is well illustrated by the action spectrum for the destruction of paramecin, the toxic substance released into the culture medium by *Paramecia* and the chemical composition of which has not yet been established.

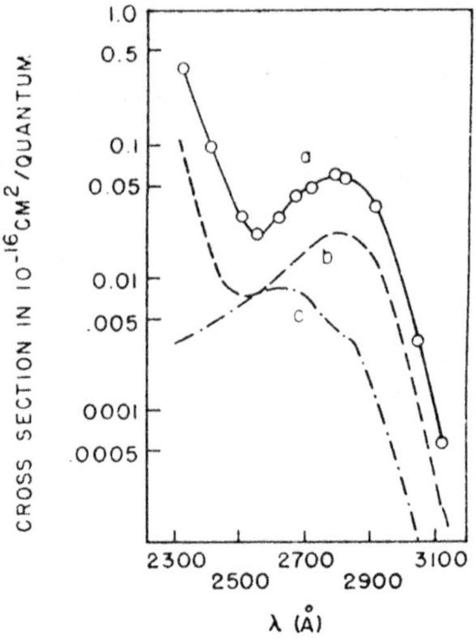

FIG. 9-5. Action spectrum for destruction of paramecin, the killing substance of *Paramecium:*

(*a*) Action spectrum of paramecin
(*b*) Absorption spectrum of trypsin (16 μg/ml)
(*c*) Absorption spectrum of DNA (0.5 μg/ml)
(from Setlow and Doyle, 1956).

The action spectrum is shown in Fig. 9-5, along with the absorption spectrum for DNA and that for a characteristic protein, trypsin. It will be seen that the action spectrum exhibits fairly good correspondence with that for the protein.

2. *Examples of Action Spectra Studies*

Because of the variety of biological effects resulting from ultraviolet irradiation, probably more action spectra studies have been carried out on micro-organisms than with any other material. The similarity between nucleic acid absorption and bactericidal effects has been known for many years and it is because of this that special types of mercury resonance lamps (familiarly known as bactericidal or germicidal lamps) have been so widely used in microbiological and clinical work. The action spectrum for "killing" of micro-organisms is very similar to the absorption spectrum shown in Fig. 9-2 (see also Figs. 9-7 and 9-11, below) (cf. Gates, 1928).

Numerous illustrations of the application of action spectra to attempted identification of the components of biological systems responsible for specific biological effects will be found in some of the reviews cited above. We shall refer here only to a few recent, special, examples.

Several attempts have been made to identify the cellular component(s) responsible for protein synthesis in micro-organisms, the proteins elected for study being those which exhibit enzymatic activity, since this property can be readily and specifically estimated quantitatively. What is actually measured is the relative destruction at different wavelengths of the ability to induce enzyme formation. For the galactozymase of yeast such an action spectrum was first reported by Swenson (1950) to be nucleic acid-like in character. Rajewsky *et al.*, (1959) also found that the action spectrum for velocity of lysine decarboxylase synthesis in *Bacterium cadaveris* resembled the absorption spectrum of nucleic acid (Fig. 9-6), but the final enzymatic

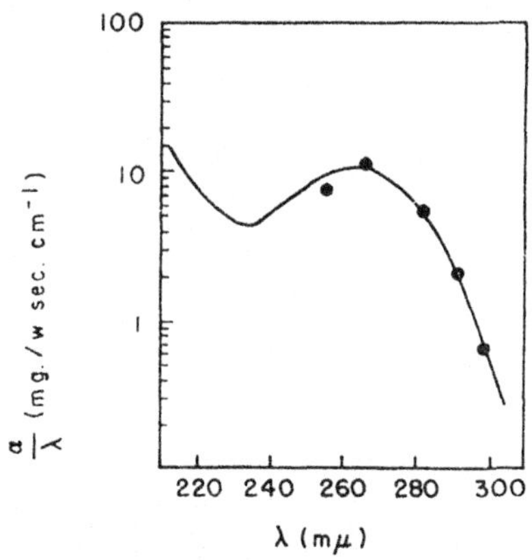

FIG. 9-6. Action spectrum for inactivation of induction rate of lysine decarboxylase in *Bacterium cadaveris* (dots are experimental points) compared with absorption spectrum of nucleic acid (continuous line) (from Rajewsky *et al.*, 1959).

activity appeared to be determined by some other mechanism, the action spectrum for which resembled that of a high cystine-containing protein. The target volume for rate of enzyme formation was estimated by X-ray inactivation and found to be of the order of 18×10^6 in molecular weight units, which would exclude soluble RNA. There are, however, some doubts as to the validity of such target volume calculations, paticularly for intracellular material (Alexander, 1960). It is, on the other hand, unlikely that the error could be so great as to invalidate the conclusion that S-RNA is not involved in the process (see below).

A more extensive investigation has been made of the action spectra for formation in *E. coli* of the inducible enzymes β-galactosidase and tryptophanase, with the results shown in Fig. 9-7. It should be noted that not only the action spectra for formation of both enzymes, but likewise that for viability, all resemble the nucleic spectrum and are quite dissimilar to that for enzyme inactivation. Making use of the known values for quantum yields for uracil and cytosine residues, the magnitude of the functional unit responsible for enzyme induction was estimated to be of the order of $3-7 \times 10^5$ molecular weight units. Notwithstanding the approximate nature of the calculations (e.g. no account was taken of the fact that the quantum yield for thymine in DNA is higher than that for free thymine), they support the

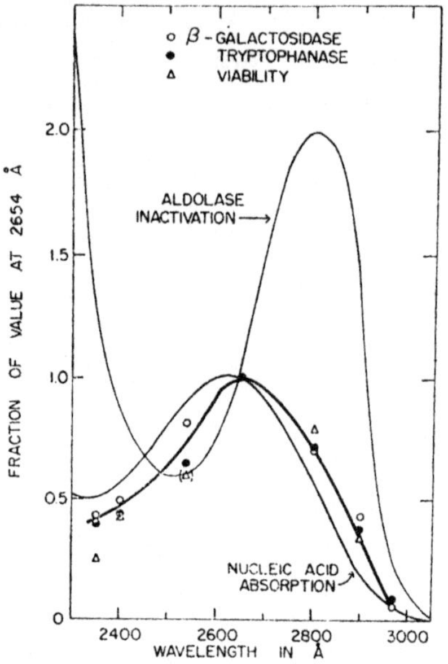

FIG. 9-7. Action spectra for viability and for inactivation of ability for formation of the induced enzymes β-galactosidase and tryptophanase in *E. coli*, as compared to nucleic acid absorption and action spectrum for inactivation of a typical protein, aldolase. Ordinates are reciprocals of doses required to reduce viability to 5 per cent (care being taken to avoid photoreactivation), β-galactosidase formation to 20 per cent, and tryptophanase formation to 10 per cent, of the control (non-irradiated) rates, divided by comparable values obtained to 2650 Å (from Rushizky *et al.*, 1960).

conclusion of Rajewsky *et al.* (1959) that S-RNA is not the functional unit involved (Rushizky *et al.*, 1960). Attention should be drawn to the fact that the action spectra shown in Fig. 9-7 were checked for reciprocity and found to be unaffected by a several-fold variation in incident light intensity.

The interesting analysis made by Setlow (1960), Setlow and Setlow (1960) and Setlow and Boyce (1960) of the action spectra for infectivity of bacteriophages containing single and double stranded DNA components has already been cited (Chapter VIII). The experimental basis for this was the observation that phage T2 exhibits an action spectrum, for loss of infectivity, with a minimum at 2300 Å; that of phage ϕX174, the DNA constituent of which is known to be single stranded (Sinsheimer, 1960) possesses a minimum at 2300 Å. Furthermore, the action spectrum for ϕX174 was found not to be parallel to that of the absorption spectrum of either the intact phage or its isolated DNA; and computations of the quantum yield, based on absorption of light by the nucleic component of the virus,

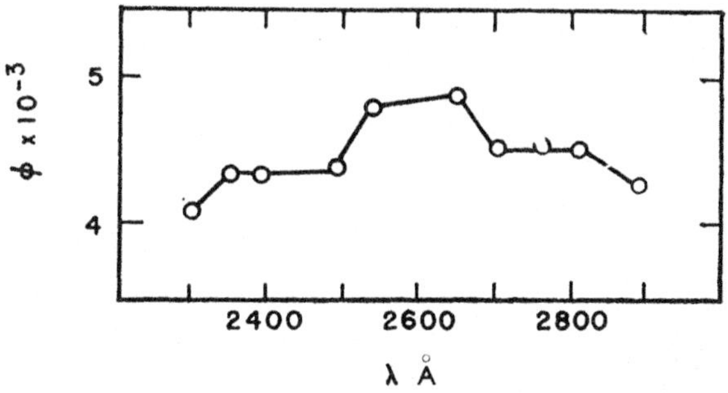

Fig. 9-8. Quantum yield for destruction of infectivity of phage ϕX174 at pH 7 as a function of wavelength (from Setlow and Boyce, 1960).

showed that it varied with the wavelength (Fig. 9-8). In view of the deduction by Shore and Pardee (1956a, 1956b) that energy transfer from protein to nucleic acid does not occur in TMV, it was concluded that the action spectrum could not be interpreted in terms of that for a nucleoprotein, but rather through effects on specific bases of the DNA component of the virus. Bearing in mind the established higher radiation sensitivity of pyrimidines as compared to purines, an explanation for the action spectrum was sought in terms of the action of radiation on the cytosine and thymine constituents. An analysis of the assumptions made for this purpose would be too extensive to enumerate here and reference should be made to the original publication for details (Setlow and Boyce, 1960). The variation in action spectrum with pH was then used to test this hypothesis. At acid pH the action spectrum for ϕX174 was indeed modified in the direction to be expected from the appreciable change in absorption exhibited by the cytosine component at this pH (Fig. 9-9).

The suggestion is made by Setlow and his group that the action spectrum for a bacteriophage such as T2 more clearly resembles that of the absorption spectrum of DNA because of the twin-stranded nature of its DNA component; this is presumed to facilitate energy migration so that all the absorbed energy is utilized for inactivation. There is really very little experimental evidence to assess the correctness of this argument; it should be recalled that quantum yield measurements for pyrimidine residues in both single and twin-stranded

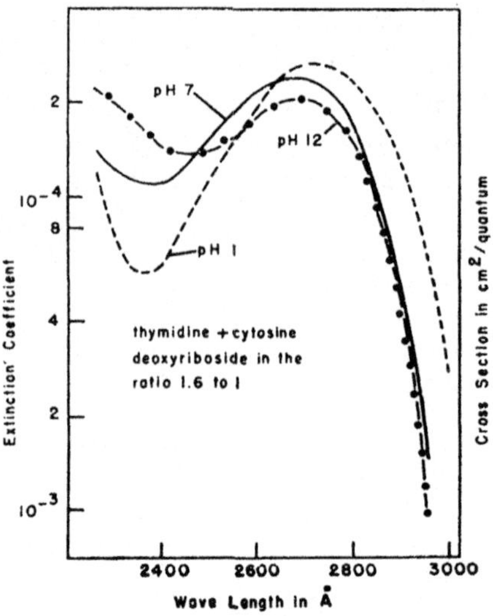

FIG. 9-9A. Calculated absorption spectrum of a mixture of thymine and cytosine deoxyribosides in the molar ratio 1.6 : 1 at pH 1, 7 and 12.

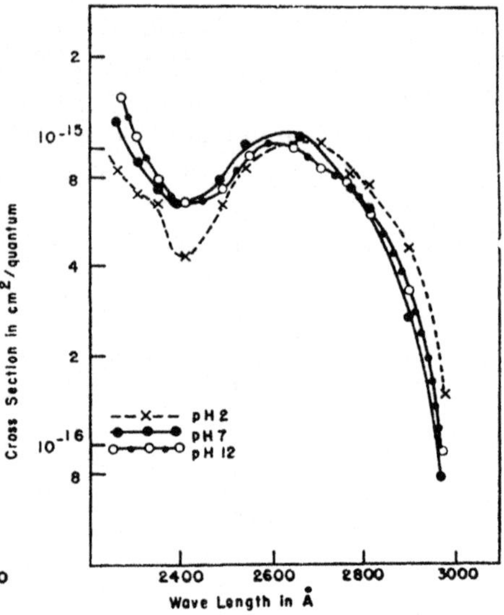

FIG. 9-9B. Action spectra for destruction of infectivity of phage ϕX174 at pH 2, 7 and 12.

Note the similar changes at acid pH in both action and absorption spectra (Setlow and Boyce, 1960).

model polynucleotides provide a rather potent argument against energy transfer being involved in the photochemical transformation of the pyrimidines (Shugar and Wierzchowski, 1960; Wierzchowski and Shugar, 1961b). Attention should also be drawn to the fact that a comparison between phage T2 and phage ϕX174 is perhaps not fully justified inasmuch as the base components are not the same; phage T2 contains, not cytosine, but 5-hydroxymethylcytosine and glucosylated 5-hydroxymethylcytosine, and the photochemical behavior of the glycosides of these latter differs profoundly from that for cytosine glycosides (Shugar, 1960a; Wierzchowski and Shugar, 1960a, 1960c).

Notwithstanding the foregoing, the observation of a shift in the minimum of the action spectrum as between a single and double stranded DNA has been applied to an examination of the state of phage T2 during replication in its host cell (cf. Chapter VIII). Phage T2r$^+$ and its complex with *E. coli* B were irradiated at different wavelengths and at different times during the latent period. The action spectrum for the phage alone exhibited a minimum at 2350 Å which, 3 min following infection, shifted to 2400 Å; and at 11 min following infection shifted back to 2350 Å. This was interpreted as a change in viral DNA from a twin-stranded structure to a single strand at 3 min and back to a twin-strand at 11 min. Particular emphasis is placed by the authors (Setlow and Doyle, 1960) on the accuracy with which the *shift* in the minimum of the action spectrum may be determined, as compared to the lower accuracy of measurement of the entire action spectra; and the observation of such a shift for single phage during a period when it is known to be undergoing replication most certainly is more convincing than the comparison between two different phages with different base compositions.

In view of the importance of the foregoing, Setlow and Setlow (1961) have reported in preliminary form an experiment designed to test directly the shift in action spectrum between single and double-stranded DNA. *Hemophilus influenzae* transforming DNA was irradiated at 85°C, where it is in the normal, two-stranded form; and at 98°C where the inter-chain hydrogen bonds were broken so that it may be considered effectively single-stranded. The action spectrum for inactivation at the higher temperature was, in fact, found to shift 25 Å to the red, in agreement with expectation.

An additional, important, application has been reported in preliminary form from the same laboratory. Because of its single-stranded DNA component, the method of replication of the virus ϕX174 has posed some difficult theoretical problems. Setlow and McNeal (1961) therefore examined the action for inactivation of the phage-*E. coli* C complex during the early replication stage. From the shape of the action spectra, it was concluded that ϕX174 DNA is involved in a double-stranded configuration during replication; a finding which, if confirmed, is of considerable theoretical significance.

Action spectra techniques have been extensively applied to the problem of photoreactivation (Jagger, 1958), but without great profit to date. We shall revert to this subject in the following section (p. 339).

3. *Partial Cell Irradiation*

A rather interesting field of development in recent years is the revival of a technique introduced many years ago for the selective ultraviolet irradiation of fractions of intact individual cells. The details of this procedure, now commonly referred to as Partial Cell Irradiation (PCI), will be found briefly described in the Appendix. The methods now in current use make it possible to irradiate areas of the order of several microns in diameter or to irradiate an entire cell with the exception of a small, predetermined, area. In essence, as emphasized by Zirkle (1957), PCI complements microsurgery. The use of a microbeam makes it possible to act on a given portion of a cell with a view to analyzing its normal function and/or the relation of the latter to the functioning of the cell as a whole. The applicability of the technique is, of course, limited to large cells. Its scope has, on the other hand, been considerably enhanced by the use of microbeam sources of ionizing radiations (see Zirkle, 1957, for review).

In addition, monochromatic microbeams of various wavelengths have been employed, although practically no quantitative action spectra have as yet been reported. For example, wavelengths between 2250 and 3000 Å are all effective in causing disappearance of the spindle in newt cells (even when the cytoplasm alone is irradiated) but, although 2250 Å appeared to be the most effective, the data were not sufficiently quantitative to establish an action spectrum.

A phenomenon which has received considerable attention is chromosome "paling." If a portion of the chromosome content of a cell is submitted to an ultraviolet microbeam, its refractive index is decreased and the irradiated portion becomes pale when viewed under phase-contrast. This effect does not occur when the total cell is irradiated and the extent of paling is increased, with the same incident energy per unit area, the smaller the area of the chromosome irradiated. This paling is accompanied by a decrease in ultraviolet absorption (Perry, 1957) and a diminished or even negative reaction with basophilic dyes and the Feulgen stain. Curiously enough the action spectrum appears to be protein-like in nature, although the material which disappears is DNA (Zirkle, 1957; Zirkle and Uretz, 1960). Further studies have indicated that the action spectrum corresponds to that for a tyrosine-containing protein and that the paling is due to loss of both DNA and protein (Zirkle and Uretz, 1963).

In a study on the effects of selective ultraviolet irradiation of the cytoplasm of different types of cells, Montgomery and Hundley (1960) were able to show that nuclear shielding during irradiation of the cytoplasm (a) confers no protection on the cytoplasm or on the longevity of the cell and (b) prevents nuclear manifestations of ultraviolet induced nuclear damage.

In at least one instance an attempt has been made to establish the possible existence, and site, of photoreactivation of ultraviolet induced damage. Irradiation of the nucleus of the egg of the wasp *Habrobracon*, leads to the early arrest of development, whereas irradiation of the cytoplasm results in different embryonic abnormalities. The nuclear damage was found to be susceptible to photoreactivation, but not the cytoplasmic damage (von Borstel and Wolff, 1955).

B. REVERSAL OF INITIAL PHOTOBIOLOGICAL EFFECTS: PHOTOREACTIVATION

The reversal of ultraviolet induced modifications or damage in biological systems by subsequent exposure to a variety of agents, in particular visible and near ultraviolet light, had been observed sporadically over a number of years; but it was Kelner (1949a) who first established unequivocally the reality of the phenomenon by his demonstration that visible light could provoke partial recovery in ultraviolet inactivated spores of *Streptomyces griseus*. Once experimentally established, the significance of photoreactivation (PR) was almost immediately appreciated, the more so in that subsequent investigations have established the existence of this effect in almost all species of biological organisms. A wide variety of ultraviolet induced effects may be "reversed" or photoreactivated, including "killing," mutations, division delay, the infective ability of bacteriophage, etc. (see reviews by Dulbecco, 1955 and Jagger, 1958, 1960 for details).

In what follows we shall concern ourselves principally with the primary effects of ultraviolet light on nucleic acids and their constituents. A few words of explanation are perhaps in order before proceeding further.

1. *Ultimate Effects of Ultraviolet Irradiation*

Relatively little is as yet known about the final effects of irradiation, particularly when we are dealing with the intact cell. This is perhaps best exemplified by mutagenesis in micro-organsims, where phenotypically mutant cells are formed only after a certain delay following irradiation. During this lag period, which precedes what may be regarded as the "sub-cultivation" period leading to the formation of stable mutants, the yield of the latter may be considerably modified by a variety of treatments, of which the best known is photoreactivation or photorestoration. Cellular metabolism appears to be an essential prerequisite for mutant formation, which has been variously linked with DNA replication as the terminal event in the process, although there is some evidence that RNA and protein synthesis are involved as well (Witkin, 1956; Witkin and Thisl, 1960; Doudney and Haas, 1960; Weatherwax and Landman, 1960) Once cell metabolism has been allowed to take over, following the application of ultraviolet radiation, the ability to induce reversibility disappears. What actually happens during this period, when the ultraviolet induced effects are "fixed," is at the moment largely a matter of speculation, and we can do no more here than refer the reader to the foregoing references, particularly that of Doudney and Hass (1960) which contains a brief review of the literature on the subject up to 1960. Some ultraviolet-induced mutations have also been reported which are quite inert to post-irradiation treatment. Finally, it is necessary to emphasize that mutagenesis by ultraviolet light in extracellular RNA and DNA is unknown (e.g. Litauer and Ephrussi-Taylor, 1959). It is also normally an extremely rare event in bacteriophage (Krieg, 1959). A notable exception is the X-phage of *Serratia*, where up to 2 per cent mutations to clear-plaque types has been reported by irradiation of the free phage either with ultraviolet or X-rays (Ellmauer and Kaplan, 1959); independent confirmation of this would be extremely useful since, if true, it would provide a useful model for investigation of the mechanism of mutation induction.

2. *Characteristics of Photoreactivation*

Entirely apart from its biological significance, the phenomenon of PR is of special importance in relation to attempts to pin down at the molecular level the sites of action of ultraviolet radiation as well as the nature of the photochemical transformations involved, since a reversible system obviously limits the variety of cellular constituents likely to be involved as well as the nature of the accompanying chemical modifications. PR consequently provides an important tool for identification of the primary effects of radiation which, in turn, must be clarified before we can expect to elucidate the subsequent steps referred to above. Before proceeding to a consideration of the possible mechanisms responsible for PR, we shall recapitulate briefly some of its salient features:

(a) PR is so widespread that failure to observe it can by no means be construed as evidence for its non-existence; reference should be made to Jagger (1958) for numerous illustrations. An extremely instructive example is that of TMV which for a number of years defied attempts at photoreactivation, notwithstanding that other plant viruses exhibited the phenomenon. It was finally demonstrated by Bawden and Kleczhowski (1959) and Rushizky *et al.* (1960) that it is the nature of the nucleic acid-protein linkage which prevents PR following ultraviolet inactivation, and that ultraviolet-inactivated, isolated infectious RNA from TMV is photoreactiveable. A similar explanation may well apply to the absence of PR in some bacteria, where the intracellular state of binding of DNA may inhibit the phenomenon.

(b) The type of ultraviolet damage or modifications which can subsequently be reversed is that caused by wavelengths in the range 2300 to 3100 Å (e.g. Zelle *et al.*, 1958; Eschke *et al.*, 1960). Shorter wavelengths usually give rise only to irreversible effects. This is quite understandable in the light of what has already been stated in earlier sections of this text on the effects of radiation below 2300 Å on the photochemical transformation of pyrimidine derivatives.

(c) Action spectra studies demonstrate that most of the damaging effects of ultraviolet radiation result principally from primary absorption by nucleoproteins and, in many instances, by the nucleic acid component alone (see next paragraph). One instance exists where nucleic acid precursors have been postulated as the primary receptors (Hass and Doudney, 1957), but there is really no concrete evidence to support this; on the other hand, the photochemistry of nucleic acid derivatives is well enough known now as to permit a direct, experimental test of such a possibility.

(d) Most photoreactiveable phenomena include reproduction, mutation or transformation, all of which involve the nucleus and, consequently,

DNA-protein. There are also at least two authenticated examples of cytoplasmic PR (Pierce and Giese, 1957; Skeb and Errera, 1957), presumably involving RNA-protein. In the case of bacteriophage there is hardly room for doubt but that PR operates exclusively on DNA-protein and the evidence appears to favor DNA as the PR site. There is, of course, no question as to what is being photoreactivated in transforming DNA. The ability of plant viruses to undergo PR points to RNA-protein and the demonstration of photoreactivation for isolated infectious RNA is fairly

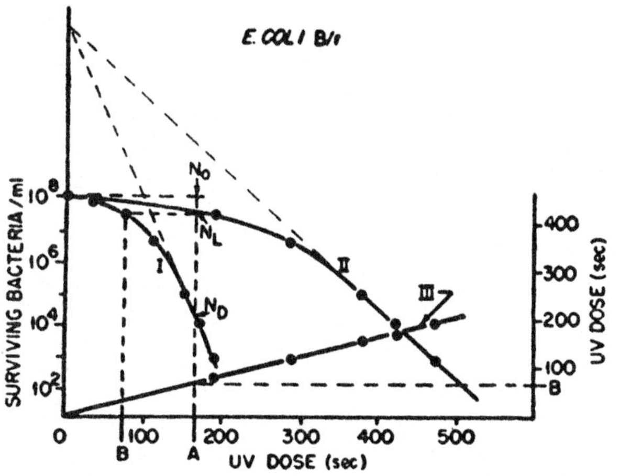

FIG. 9-10. Illustrating survival curves for *E. coli B*/r (I) after UV irradiation and (II) after maximum photoreactivation. Curve III depicts the ratio of the UV dose for a given survival (right ordinates) to the UV dose which would give the *same survival after photoreactivation* (abscissa). N_0 is the initial number of viable cells, N_D is the number of viable cells after application of a given UV dose and N_L the number of viable cells after max. PR. See text for additional details (from Novick and Szilard, 1949, as modified by Jagger, 1958).

convincing evidence that it is the RNA of the other plant viruses as well that is implicated. There is, consequently, reasonably good evidence for the contention that the sites at which ultraviolet radiation induces the damage subsequently reversed by visible light involve nucleic acids.

(e) PR is not a 100 per cent phenomenon; it varies for different organisms and, for any given organism or biologically active phage or nucleic acid, the degree of photoreactivation attainable depends on the ultraviolet dose which has been used for inactivation. For a description of the various methods employed for calculating percent PR, reference should be made to Novick and Szilard, (1949), Dulbecco, (1955), Jagger (1958). The general principles involved, as applied to PR in *E. coli B*/r are shown in Fig. 9-10.

In the figure,

N_0 is the initial number of viable cells

N_D is number of surviving cells after ultraviolet irradiation (dark survival)

N_L is number of surviving cells after PR (light survival) from which

$$\text{Per cent light survival is } (N_L/N_0) \times 100$$

$$\text{Per cent PR is } (N_L - N_D)/N_0 - N_D) \times 100$$

Another useful formulation is the

$$\text{Dose-reduction factor} = \frac{\text{ultraviolet dose for given survival before PR}}{\text{ultraviolet to give same survival after max. PR}}$$

originally suggested by Kelner (1949b) and represented by curve III in Fig. 9-10. For any given survival the numerator of the above fraction is taken from the right-hand ordinate in Fig. 9-10 (e.g. dose at B) and the denominator from the abscissa (e.g. dose at A). It will be seen that in this particular case the dose-reduction factor is constant, so that the overall effect of PR is to reduce the effective ultraviolet dose by a constant factor. If the dose-reduction factor is constant, then tangents to the light and dark survival curves (curves II and I, respectively in Fig. 9-10) at a given survival level should intersect at the ordinate. PR may be quite high for low doses of ultraviolet radiation but usually decreases appreciably with increasing doses of ultraviolet radiation. A PR of 50 per cent represents fairly good recovery but a value for percent PR may be quite deceptive if it is not referred to the percent survival resulting from ultraviolet irradiation. It is even more useful to exhibit percent PR for different ultraviolet survival values, as in Fig. 9-10, but this is not always done.

In a metabolizing system, exposure to visible light must take place shortly after ultraviolet "damage" has occurred, but this period may be increased by lowering the rate of metabolism of the system, e.g. by decreasing the temperature. On the other hand, UV-inactivated *isolated* T-DNA may be kept for weeks without impairment of its ability to undergo PR. Consequently the damage (or lesions) susceptible of reversibility involves some modification which is reasonably stable outside the cellular environment.

It remains to be established whether the existence of less than 100 per cent PR is due to some irreversible ultraviolet damage (this is, of course, certainly true for high ultraviolet doses) or whether the PR process itself is not a fully efficient one.

(f) The PR process itself apparently requires the presence of some cellular factor. For living cells this might be concluded only by inference. But PR for irradiated bacteriophage, T-DNA and infectious RNA has generally been found possible only in the presence of some cellular environment. Only one exception has been reported, viz. for reversion of c-mutations in the X-phage of *Serratia* (Kaplan *et al.*, 1960), which takes place in extracellular phage, although the PR for "killing" apparently did require prior adsorption of the phage to the host cell. The significance of this observation, if confirmed, can hardly be overestimated. It should be recalled that this is the only phage reported to undergo appreciable mutation when irradiated with UV in the extracellular state. Should PR for the free phage be confirmed, it would be interesting to investigate what wavelengths are responsible, and what is the relationship of per cent PR to "survival" values for different ultraviolet doses.

(g) Most of the available evidence appears to favour the view that PR involves a direct reversal of the damage or lesions resulting from UV irradiation and not the induction of some new metabolic pathway which bypasses the radiation-formed lesions. This is supported by the fact that irradiated phage and infectious RNA may be photoreactivated; and that T-DNA, although requiring some cellular factor, may be photoreactivated outside the cell.

3. *Radiation Receptors in Photoreactivation*

From the brief summary given in the previous section, it may be concluded that PR involves the repair or reversal of some type of modification(s) induced in nucleic acid molecules by ultraviolet light. One might therefore expect that a knowledge of the nature of the constituents absorbing the photoreactivating visible light would be of assistance in elucidating the nature of the mechanism involved.

Unfortunately, despite numerous efforts based principally on action spectra studies (Setlow, 1957; Jagger, 1958), the nature of the photoreceptors operative in PR is unknown. The range of wavelengths effective in promoting PR embraces the interval 3100–5000 Å, but the action spectra, two typical examples of which are shown in Fig. 9-11, are in no instance sufficiently characteristic to permit of positive identification. In fact, from the data hitherto accumulated on this subject it must be concluded that the receptors for PR in different organisms are not even the same. It is important to recognize that any suggestion to the effect that nucleic acids themselves are the receptors of the radiation responsible for PR must cope with the fact that neither nucleic acids nor their photoproducts exhibit selective absorption above 3100 Å.

On the other hand, the fact that wavelengths as low as 3100 Å still photoreactivate leaves open the question as to whether PR operates at even lower wavelengths. This is not likely to be an easy problem to resolve, since wavelengths below 3100 Å, where proteins and nucleic acids absorb, are precisely those which produce damage. It should nonetheless be recalled that photochemically formed pyrimidine dimers are subject to dissociation at wavelengths below 2900 Å; hence if inactivation is linked in any way with dimerization, photoreversal at wavelengths below 2900 Å is quite conceivable from a qualitative point of view.

Undoubtedly a more effective system for action spectra studies of the nature of the absorbing component in PR would be the *in vitro* PR system of Rupert *et al.* (1958) and Rupert (1960) for T-DNA. But, in the long run,

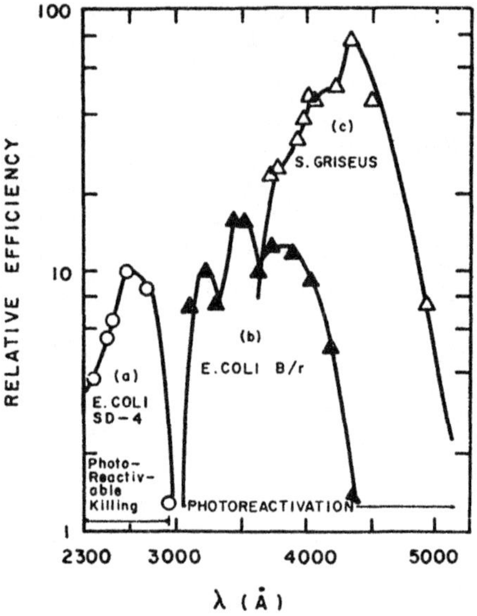

FIG. 9-11. Action spectra for: (a) photoreactiveable "killing" in *E. coli* SD-4 (Zelle *et al.*, 1958); (b) Photoreactivation of "killing" in *E. coli* B/r (Jagger and Latarjet, 1956); (c) Photoreactivation of "killing" in spores of *Streptomyces griseus* (Kelner, 1951). The ordinate scales of curves *b* and *c* bear no relation to that of curve *a* (Taken from Jagger, 1960).

it may prove simpler to achieve this by further purification and, eventually, isolation of the active component.

4. *Physical Models as Analogies of Photoreactivation*

A number of physical models have been advanced as possible mechanisms for PR, all of a rather general nature, but sharing in common the basic assumption of a direct reversal of ultraviolet-induced damage. All of these models are based on analogies from solid state physics and, although they now appear to be unnecessary, it is nonetheless instructive to briefly examine them.

It has been found that the germination of some seeds is stimulated by red light (5250–7000 Å), and this effect may be reversed by radiation farther to the red (7000–8200 Å) (Borthwick *et al.*, 1952; Evenari and Stein,

1953; Evenari *et al.*, 1953), so that we have here a biological counterpart of PR. In some respects the process bears a marked resemblance to phenomena exhibited by certain crystalline phosphors and it has been suggested that they result from the trapping of electrons in a semi-conducting system under the influence of red light and subsequent release of the trapped electrons under the influence of light of longer wavelength. By analogy Stein (1954) has proposed a similar mechanism for PR.

An almost identical scheme has been advanced by Reid (1957), based on a process in photographic emulsions. On exposure to light ($h\nu_1$), electrons are excited from halide ions in a silver halide crystal into a "conduction band" *C* where they move freely throughout the crystal until they are "trapped" at an impurity *T* (or a point at which the regular lattice structure is interrupted). Silver ions attracted by the negative charge move through the lattice to the impurity *T* where they are neutralized with resultant formation of a microscopic speck of silver, giving rise to the "latent image." The latter may be "erased" by irradiation of the film with infrared light ($h\nu_2$), due to the fact that infrared radiation has sufficient energy to release electrons from the trap *T* into the conduction band *C*, and insufficient energy to liberate electrons from the halide ions. The initial process is therefore reversed by a return of the electrons to the halide ions while the Ag^+ ions diffuse away from the now positively-charged impurity or trap *T*. It was suggested that in a macromolecule or a crystalline aggregate of molecules, where essentially solid-state conditions exist, electrons might readily be released and then retained in "traps" until released by photoreactivating light.

The foregoing mechanism of electron capture in crystals had earlier been proposed by Stein (1954) in a general discussion of possible PR mechanisms, including also the possible formation of metastable states. This latter hypothesis was elaborated upon by Duchesne and Garson (1957) and Duchesne and Rosen (1959), who postulated that ultraviolet irradiation of a biological system was accompanied by the internal conversion of the first excited state of some constituent (unspecified) of a DNA molecule, followed by its transfer to a lower, metastable, state. The return to the original, or ground, state under the influence of visible radiation would then take place in the reverse direction via the initial excited state. Apart from the improbability of the existence of metastable states with such enormous lifetimes as required by PR in nucleic acid chains, the model is more formal than real. The *in vitro* PR of T-DNA (Rupert *et al.*, 1956), regarded by Duchesne and his collaborators as supporting evidence for their hypothesis, is interpreted by them as due to some effect of the medium which facilitates (in the physical sense) the transitions between the various excited states. Such an interpretation is obviously contrary to the experimental facts, even if we ignore the fact that DNA does not absorb the PR light. Another argument against the formation of metastable states as a mechanism for PR is forthcoming from the finding that photoreactivation is not the only means by which ultraviolet lesions may be reversed (see below). The foregoing authors also draw an analogy between PR and the phenomenon of photochromism (Hirshberg, 1956) where an ultraviolet induced change in colour of some compounds (spiropyrans) may subsequently be reversed by irradiation with visible light. Here too the analogy is rather a formal one since (a) in photochromism the substance affected by ultraviolet light is also the one which absorbs the visible radiation resulting in eradication of the color change, and (b) this effect is observed only in solid glasses at low temperatures where the excited states responsible for the color change are "frozen" or "trapped."

5. *Alternative Methods for Reversal of Ultraviolet Effects*

PR of ultraviolet induced "damage" is an impressive and exciting phenomenon which captured the imagination of many research workers. But visible light is by no means the only agent which can provoke the dark reversal of ultraviolet induced effects. It is merely the one which has been most extensively investigated. A variety of other factors may also reverse ultraviolet induced damage (see Jagger, 1958 for review), the best-known of these being an increase in temperature following ultraviolet irradiation which we shall refer to as thermal reactivation or thermal restoration (TR).

We shall confine our attention here to TR, not because the other methods of reversal are of minor importance, but only because they have been less extensively investigated and the paucity of data renders it somewhat difficult to draw any concrete conclusions. Even TR has by no means received the attention it deserves, as will be clear from what follows.

TR was first demonstrated in ultraviolet irradiated micro-organisms by Stein and Meutzner (1950) and Anderson (1951). The former authors showed that, following "killing" of *E. coli* by ultraviolet light, an increase in temperature to 43° led to an increase in survival comparable in magnitude to that resulting from PR. The phenomenon is not as general as PR; Anderson (1951) found that, out of 10 *coli* and 7 yeast strains, only 2 of the *coli* strains showed TR. It must, on the other hand, once more be emphasized that, by comparison with PR, TR has not been widely investigated. Furthermore, many micro-organisms do not exhibit PR; it would, in fact, be extremely interesting to test some of these for TR.

Whether TR represents the same reversal mechanism as PR is yet to be finally established; existing evidence indicates that, in at least some instances, it is, but it would be premature at the moment to draw any final, general, conclusion. For *E. coli B* and *S. cerevisae* it appears to be so. For the former (Stein and Laskowski, 1956) PR raises the survival level to that of TR (at 44°), following which both survivals remain parallel; for the latter strain, TR raises the survival level to that of PR, following which both levels are parallel (Giese *et al.*, 1957).

TR has also been reported for T1 phage adsorbed to the host cell (Bresch, 1950) and this has been both confirmed and extended to a number of coliphages adsorbed to their host cells and at a temperature of 45° (Lukianiec and Shugar, 1961; Shugar, 1960a). It is of some significance that no TR could be obtained without prior adsorption of the phage to their host cells. The use of temperatures above 45°C on the free phages gave negative results because of complications resulting from heat denaturation. It would therefore appear from these experiments that some intracellular factor is as important for TR as for PR, a fact of considerable significance in any attempt to define the mechanism of TR and its relation to PR. On the other hand, Kaplan *et al.* (1960) report appreciable TR for ultraviolet-induced c-mutations in the *free* X-phage of *Serratia*; we have already referred above to this phage, which has been reported by the same authors to mutate under the influence of ultraviolet light in the free state as well as to exhibit extracellular PR, and it is even clearer now that independent confirmation of these findings is of considerable importance.

In general, more quantitative data on TR, particularly in relation to PR on the same organisms, are much to be desired. It would likewise be of interest to examine whether the *in vitro* system of Rupert *et al.* (1958) which, like PR of micro-organisms, exhibits a temperature-dependent dark reaction, could give a TR effect.

Attention should be drawn to the detailed study of Buzzell (1956) on the kinetics of TR for *E. coli B.* Using cells in the logarithmic phase of growth, which are most homogeneous with respect to radiation sensitivity, she demonstrated three types of ultraviolet "killing" which differed in being either rapidly, slowly, or not at all susceptible to TR. The kinetics of the slow reaction were regarded as indicating that more than one damage must be nullified. A comparison with corresponding data for PR under identical conditions would have been of considerable value in this instance. On the other hand, if such a comparison is contemplated, it is well worth considering whether it might not be more fruitful to execute it with phages rather than with micro-organisms because of the greater simplicity of defining the ultraviolet-induced lesions.

6. *Potential Reversible Photochemical Transformations in Nucleic Acids*

From the evidence outlined in the foregoing sections, to the effect that PR (and TR) involves the direct reversal of ultraviolet-provoked lesions in essential nucleic acid molecules, we are provided with an additional and, what is more, a particularly precise criterion by means of which to attempt to define the primary effects of irradiation. From the data presented in previous chapters, we shall now briefly recapitulate what is known about the modifications resulting from irradiation of polynucleotide chains with doses not exceeding the order of magnitude of those inducing biological effects. Under these conditions the pyrimidine rings are the ones to react initially, as follows:

(a) Water addition across the 5,6 double bonds, with resultant saturation of the latter, for uracil and cytosine residues. The water molecule may subsequently be eliminated in the dark, most conveniently by heating at neutral pH. Thymine residues do not undergo water addition.

(b) Dimer formation between pyrimidine rings *via* formation of a cyclobutane ring between the 5,6 double bonds, again with saturation of the latter. This may occur between thymine or uracil residues. No positive evidence exists for such dimer formation between cytosine residues; and what evidence does exist from reaction kinetics is rather against it. In addition to dimer formation between adjacent residues in single-stranded chains, there is indirect evidence for such a reaction between residues in the twin strands of a double helix. It should be recalled that photochemical dissociation of dimers may occur in the same wavelength region in which they are formed; hence at any given wavelength the degree of dimerization will represent an equilibrium between dimer formation and dissociation.

(c) Several typical examples will now be cited for purposes of illustration (see also Table 7-1, page 265):

(i) When a uracil or cytosine residue is sandwiched between two purines, the only reaction undergone is that of hydration, and this is fully reversible in the dark by heating or acidification.

(ii) In a sequence such as CpUp, each residue undergoes only hydration; there is no dimerization. This is shown by the fact that the reverse reaction takes place in two stages, the Cp residue first eliminating its water molecule at 30–40°C, followed by the Up residue at a much higher temperature.

(iii) In polymer tracts of Up residues, dimerization varies from 15–20 per cent for a dinucleotide to over 35 per cent for long chains (seven or more residues); the remaining residues undergo hydration. This is for a wavelength of 2537 Å. At longer wavelengths the proportion of dimerization will be greater. In tracts of Tp, where hydration cannot occur, 34 per cent dimerization has been reported for a dinucleotide, while 9–18 per cent of the total thymine of the DNA of various bacterial strains has been isolated as dimers following irradiation at 2537 Å. The quantum yield for thymine dimer formation in single-stranded polymer chains is fairly high compared to the effect of radiation on isolated thymide residues. In a poly-U chain the quantum yield for dimerization is apparently greater than for hydration, but more precise quantitative data are required. It is most likely, but remains to be confirmed, that dimerization does not occur in tracks of Cp, the only reaction being hydration.

(iv) In air-dried films of nucleic acids, cross-linking occurs between strands, most likely through dimerization of pyrimidine rings, with very high quantum yield. Since such films contain considerable moisture, such cross-linking may occur in nuclear DNA; and it is quite conceivable that it will take place in solution as well.

(v) On the other hand, in a twin-stranded chain such as poly-A-poly-U, the dimerization reaction appears to be initially inhibited, possibly due to steric factors, and hydration appears to be the only reaction until 25 per cent of the uracil rings have reacted.

For twin-stranded chains, photochemical transformation of the pyrimidine residue is accompanied by dissociation of the hydrogen bonds, linking it to the purine residue of the corresponding base pair.

(d) For 5-substituted cytosine residues, which are confined principally to the T-even bacteriophages, the photochemical transformation involves ring opening and/or rearrangement. The precise mechanism remains to be elucidated; but experiments with 4-aminopyrimidine analogues, which exhibit similar photochemical transformations, suggest that this reaction is subject to reversibility.

Summing up it will be seen that the principal characteristic feature of the initial photochemical transformations which occur in polynucleotide chains is that they are, with the possible exception of the cross-linking of air-dried films, *potentially* reversible. Dehydration and dimer dissociation may be two of the functions of the light-activated enzyme system which photoreactivates transforming DNA. Attempts to demonstrate enzymatically catalyzed dehydration have been entirely negative. But the evidence for enzymatic reversal of thymine dimerization is fully established (for recent, extensive review see Rupert, 1963).

It is, of course, not yet possible to conclude that the photochemical transformations outlined above are the only ones which actually occur. But, if there are any additional reactions, they are most likely limited in nature in view of the fact that they must be reversible for such important biological phenomena as mutation, division delay, "killing," etc. When doses of the order of 10^4 ergs/mm^2 or higher are employed, the resulting reactions are much more far-reaching in scope, but they are also less interesting from a biological point of view.

We are also not yet in a position to specify the precise modifications resulting in a polynucleotide chain following hydration or dimerization of a particular residue(s), beyond the fact that in a twin-strand chain the resulting dissociation of base-pair hydrogen bonds will provoke a deformation in the chain as well as an apparent change in base sequence. This is experimentally reflected in a change in the temperature profile and a small decrease in the "melting" temperature T_m (Marmur and Doty, 1959). Additional physico-chemical studies on irradiated nucleic acids, using biological activity as a criterion of radiation effects, and on model polynucleotides, should materially assist in elucidating the nature of the overall modifications; and density gradient centrifugation should be a useful tool in such studies.

It must once more be emphasized that biological reactivation, whether it is PR or TR, appears to require the participation of some cellular factor activated by light and/or heat or ionic strength, etc. and which, in the case of transforming DNA, appears to be an enzyme. The only exception to the foregoing, and to which we have had occasion to refer to more than once above, is the reversal of *c*-mutations in the X-phage of *Serratia marcesens* by light or heat without adsorption of the phage to the host cells. Until independent confirmation of this is forthcoming, it would be premature to formulate any detailed hypothesis as to the mechanism involved, however tempting it may appear in the light of the ready reversal of hydration by heat alone or of dimerization by ultraviolet light alone.

One word of caution is perhaps necessary with regard to the reported PR; the light source used for this purpose was not a well-defined one and it would be advisible to employ at least some well-defined wavelength band, if not monochromatic radiation, bearing in mind that irradiated phage does not normally absorb in the near ultraviolet.

If we exclude the foregoing exception, the widespread existence of PR, not only in micro-organisms but in higher organisms as well, raises some interesting questions as to the origin and real function of the PR enzyme. Is it a relic of the past, at a time when ultraviolet radiation more readily penetrated the atmosphere, so that it came into being as one of the defense mechanisms against harmful radiations? Or is it an enzyme which normally performs some other function, PR being merely incidental to its normal operation in the cellular environment?

C. RADIATION INDUCED SYNTHESES: ORIGIN OF LIFE

We have heretofore been dealing with the physico-chemical and biological modifications resulting from the irradiation of biologically important macromolecules and their constituent components. An additional and related subject of special interest which has begun to attract widespread attention not only from a speculative, but to a large extent from an experimental point of view is that of the origin of life on the primitive earth and, consequently, the mechanisms by which the first organic compounds were synthesized under primitive earth conditions. The interest in this field has been further accentuated by the rapid development of space satellites and interplanetary space vehicles by means of which we may well expect to learn something in the very near future as to the possible existence of organic material and, perhaps, some forms of micro-organisms in outer space and on the planet Mars.

Of the many factors which may have played some role in the evolution of life on the primitive earth, there is little doubt but that various types of radiation and, in particular, ultraviolet radiation must have been of special importance, particularly during the earliest stages when only primitive organisms existed. There is, on the other hand, considerable indirect evidence that ultraviolet radiation may have been of primary importance in the initial formation of life by inducing the synthesis of some of the primary organic molecules by non-biological processes. A brief discussion of the present status of this subject is therefore not out of place, the more so in that some experimental evidence in support of such a hypothesis has already been forthcoming from a number of laboratories. This should, however, not be construed as indicating that ultraviolet radiation was the sole, or even the most important factor involved in the synthesis of the primitive biological building blocks. It is nonetheless worth drawing attention to the fact that laboratory investigations designed to test such possibilities have been increasing our knowledge of photochemical reactions in some fields, e.g. in the far ultraviolet, in the absorbed state, etc.

In order to examine a given system for potential production of some of the biological "building blocks," one must necessarily make some assumptions as to the nature of the starting materials available on the primitive earth and of the nature of the primitive atmosphere which is, in turn, directly linked to the mode of formation of the earth. One can then simulate these conditions on a laboratory scale and, with some simple substances, investigate the formation of complex organic molecules under the influence of an energy source.

One of the principal points, which has been the subject of much discussion, is whether the primitive atmosphere was an oxidizing or reducing one. The comprehensive treatment of this subject by Oparin (1936) was based largely on the assumption that the primitive atmosphere was a reducing one containing mainly hydrogen, methane, nitrogen and ammonia; as well as smaller amounts of carbon dioxide and carbon monoxide, along with trace quantities of substances such as higher hydrocarbons, hydrogen sulfide, phosphine, etc. This view has come to be generally accepted and has served as the basis for most experimental work in the field (Lanham, 1952; Bernal, 1951; Miller and Urey, 1959; Calvin, 1959, 1960). A number of attempts have indeed been made to synthesize more complex organic molecules under oxidizing conditions (Rabinowitch, 1945; Garrison *et al.*, 1951) but the results have either been negative or unsatisfactory from the point of view of yields. Miller and Urey (1959) point out that cosmic dust clouds, from which the earth may have been formed, contain a large excess of hydrogen; while the planets Jupiter, Saturn, Uranus and Neptune are known to possess atmospheres of methane and ammonia. These substances are readily observed today because of lower temperatures and high gravity on these planets, as a result of which their hydrogen has not had sufficient time to escape. Furthermore meteorites, which from the closest approach within our reach of the building material from which the earth may have been formed, are highly reduced, their iron content being mainly in the free form or as ferrous sulfide, the phosphorus as phosphides and the carbon as elemental carbon or iron carbide.

The carbon content of meteorites is not very high, usually of the order of 2 per cent. More significant is the fact that some of it is in the organic form. Calvin and Vaughan (1960) have extracted carbon from some meteorites and demonstrated by spectral methods that a good deal of it is in the organic form. One compound was isolated by column chromatography; it appeared to be a heterocyclic substance and exhibited an absorption spectrum in the quartz ultraviolet, the spectral titration of which in acid medium revealed the presence of a functional group with a p*K* in the neighborhood of pH 4–5. The conclusion of the authors that this compound is cytosine is, in our opinion, of doubtful validity from the spectral data presented. The approach used by the authors is valuable, however, since an analysis of various meteorites undoubtedly will provide us with extensive data regarding the synthesis of organic compounds in the various stages of evolution of the astral bodies from which the meteorites must originally have come. These, and related problems, have formed the subject of extensive discussions (Abelson, 1961; Sagan, 1961; Bernal, 1961; Urey, 1962; Oro, 1963; Gilvarry and Hochstein, 1963).

Assuming, therefore, that the original atmosphere was a reducing one containing principally hydrogen, methane, nitrogen and ammonia, the next question relates to the nature of the agent or agents responsible for producing transformations to more complex molecules. These have been listed by Miller and Urey (1959), see (Table 9-2).

It is clear from the following table that ultraviolet radiation must have been an important, if not the principal, source of energy for the possible

TABLE 9-2. PRESENT SOURCES OF ENERGY AVERAGED OVER EARTH'S SURFACE

Source of energy	Cal/cm^2 per year
Total radiation from sun	260,000
Ultraviolet light <2500 Å	570
<2000 Å	85
<1500 Å	3.5
Electric discharges	4
Cosmic rays	0.0015
Radioactivity (to 1 km depth)	0.8
Volcanoes	0.13

transformation of the components of a primitive reducing atmosphere. The threshold wavelengths for absorption of these components are presented in Table 9-3, which also gives some measured values of the quantum yields for photolysis, as compiled by Terenin (1957).

TABLE 9-3. THRESHOLD ABSORPTION WAVELENGTHS AND QUANTUM YIELDS FOR DISSOCIATION OF VARIOUS GASES (TERENIN, 1957))

Gas	Threshold wavelength for absorption (Å)	Quantum yield for dissociation
H_2	900	—
N_2	1100	—
CH_4	1450	0.4
CO	1545	1(1295 Å)
CO_2	1690	—
H_2O	2000	0.3 (1650 Å)*
NH_3	2250	0.5 (1470 Å)

* For liquid water Barret and Baxendale (1960) report the quantum yield for photodissociation to be 0.6 at 1849 Å.

In a mixture of gases it is, of course, not essential that all of them be excited by absorption. For example, irradiation of a mixture of NH_3 and CO at wavelengths above 1850 Å gives rise to compound formation even though only ammonia absorbs in this region. It has also been observed that low yields of formaldehyde and amino-acids result from the irradiation of CH_4, NH_3, and H_2O with the Hg 1849 Å line although only the latter two absorb at this wavelength (Miller, 1957). The active carbon intermediates are apparently formed through free radical reactions. Of even greater significance is the influence of adsorption of some of the gases on oxides and other surfaces, to which we shall return below.

For earlier references to photochemical transformations in various gases under the influence of ultraviolet radiation, reference should be made to Noyes and Leighton (1941) and Dauvillier (1958), among others. The first extensive successful laboratory synthesis of organic compounds from mixtures of gases such as the above made use of electrical discharges (Miller, 1953, 1955, 1957). A considerable number of experiments have now been performed with both ultraviolet and ionizing radiations, with positive results. Before going on to a discussion of some of these experiments, it is of interest to consider the possible sites of synthesis under primitive earth conditions. Since the absorption of the gases referred to above is largely in the wavelength region below 2200 Å, most of the photochemical reactions in which they could be involved in the *free state* would have been in the upper layers of the atmosphere, since shorter wavelengths would have been partially absorbed and, to some extent, attenuated by Rayleigh scattering before reaching the earth's surface. The resulting photoproducts, on the other hand, would absorb at longer wavelengths. The problem then arises as to whether they could have reached the surface of the oceans prior to undergoing extensive photodecomposition. This is, indeed, one of the principal objections which can be raised against the hypothesis of photochemical synthesis in the upper atmosphere as the source of primitive organic substances, but it is by no means an unequivocal one.

An attempt to treat this problem quantitatively has been made by Hull (1960), and while some of the assumptions on which it is based are of questionable validity, the approach used is certainly a useful one; it should stimulate the planning of future experiments along more quantitative lines. For example, assuming that glycine is formed in the upper atmosphere, that it is sensitive to radiation up to 3000 Å and that the atmosphere was transparent to 2250 Å, it was suggested that decarboxylation of activated (excited) glycine would occur with a quantum yield of unity. With an absorption coefficient of about 0.15 cm^{-1} $(moles/l.)^{-1}$ at 2500 Å and an ultraviolet intensity below 3000 Å of about 10^{-2} cal/cm^2 per min (or 10^{15} $quanta/cm^2$ per sec), its half-life would be about 30 days. This is to be compared with the half-time for transport from the stratosphere to the earth's surface, which was calculated from present fall-out data to be about 3 years. Consequently, about 97 per cent of the glycine formed in the upper atmosphere would undergo photodecomposition before it could reach the earth's surface. The rate of formation of glycine was estimated from some experiments of Groth and van Weyssenhoff (1957) in which a mixture of ammonia, methane, and water vapour were irradiated in the wavelength range 1165–1470 Å, with the resultant formation of only detectable quantities of glycine and alanine after exposure to 10^{21} quanta. From these results Hull estimates an upper limit for the quantum yield of formation of the amino acids of 10^{-6}. Taking the flux of solar radiation below 1500 Å as 3.5 cal/cm^2 per year (Miller and Urey, 1959) or 3×10^{11} $quanta/cm^2$ per sec, it was calculated that the rate of formation of glycine would be about 3×10^5 $molecules/cm^2$ per sec. During the average life-time in the atmosphere this could be built up to 10^{12} $molecules/cm^2$. Assuming that 3 per cent reaches the surface of the sea, and a half-life in the sea of 1000 years, the maximum concentration in the sea was estimated to be about 10^{-12}M and, consequently, too low as starting material for the spontaneous generation of life. The situation becomes much more hopeless when one attempts to apply the above line of reasoning to the formation of more complex molecules and, eventually, to polymers such as proteins.

Hull emphasizes that the foregoing calculations are not in conflict with the results of laboratory experiments, where the amounts of organic compounds synthesized have been appreciable, since in the latter the overall yield is increased by the simple expedient of continuous removal of the reaction, or photoproducts. But, as Bernal (1960) correctly points out, this is precisely the crux of the argument overlooked by Hull in presenting the above calculations. Actually a variety of mechanisms could have operated to remove the photoproducts from their zone of formation, at the same time concentrating them to the point where reactions between them would be possible and lead to the formation of more highly organized systems (Bernal, 1951, 1957), and it has been proposed by Bernal (1960) that such processes might be susceptible of investigation on a laboratory scale, e.g. by using large volumes of dilute solutions of organic substances and adsorption of the photoproducts of the latter on clay surfaces under controlled conditions.

There exist, on the other hand, other possible mechanisms by means of which primitive synthesis might have taken place under conditions much more favourable than those indicated above. For example, Terenin (1957) has drawn attention to the fact that the threshold wavelengths for absorption and excitation of H_2O and NH_3 may be extended several hundred Å to the red when they are adsorbed on suitable surfaces or catalysts such as alumina and silicates (Figs. 9-12). The greater facility with which water molecules undergo photodissociation under such conditions is due to the formation,

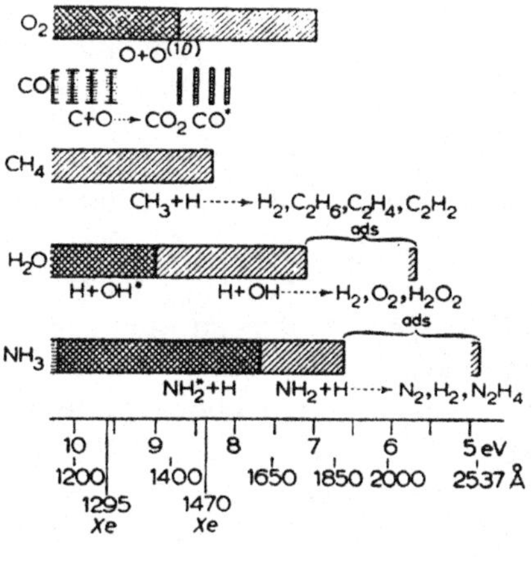

FIG. 9-12. Spectral regions of absorption and photochemical dissociation of some gases. The shaded regions are those in which the gases normally absorb with resultant dissociation into the atoms and radicals shown. The hatched regions are those in which some of the radicals exhibit luminescence, which has been utilized to study their properties. The arrows indicate the primary products formed by the subsequent reactions of the radicals. The regions enclosed by the brackets and designated by plus signs indicate extension of the normal region of spectral absorption when the gases are absorbed on a suitable surface or catalyst such as oxides (Terenin, 1957).

on the surface of porous silicates, oxides and alumino-silicates, of hydrates such as —Si—OH (with single bonds above and below the Si). The splitting of such hydroxyls requires quanta of lower energy than the dissociation of OH from H_2O.

Various homogeneous and heterogeneous catalysts have also been employed in aqueous media in attempts to increase photoproduct yields with impressive results in a number of instances, Table 9-4. It has been possible in this way to even extend some syntheses into the visible region of the spectrum.

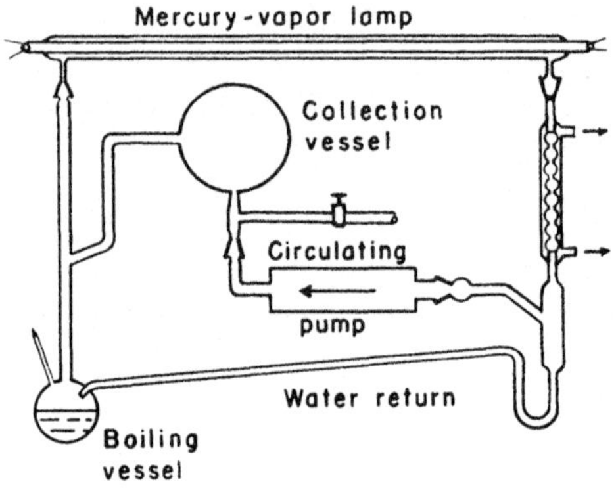

FIG. 9-13. Apparatus for simulating primitive earth conditions of synthesis under influence of ultraviolet radiation (Groth, 1960).

Figure 9-13 illustrates schematically the type of equipment employed for the study of syntheses in the gaseous phase. The reaction takes place in the mantle surrounding the mercury vapour lamp. Water from the boiler enters the circulating system and is condensed together with the reaction products. For shorter wavelengths, using Xenon and Krypton sources, LiF and CaF_2 windows must be inserted between the source and the reaction system, and such types of equipment are also described by Groth (1960). The experimental facilities are, of course, considerably simplified in those instances where irradiations are carried out in the aqueous phase.

Experiments have now been reported using widely varying conditions, such as different wavelengths, varying pressures for various gases in the gas phase, temperature, type of catalyst, etc. A cursory outline of most of this work is presented in Table 9-4. The usual procedure has been to isolate and identify the reaction products by paper or column chromatography. But it will be noted from Table 9-4 that relatively little emphasis has hitherto been placed on estimation of quantitative yields. A notable exception is the work of Miller (1955, 1957), using electrical discharges as the energy

TABLE 9-4. SUMMARY OF TYPES OF LABORATORY EXPERIMENTS CONDUCTED TO DEMONSTRATE PHOTOCHEMICAL SYNTHESIS OF ORGANIC COMPOUNDS UNDER PRIMITIVE EARTH CONDITIONS

Starting products	Radiation	Photoproducts	Remarks	Reference
Various ternary organic compounds in 10% aqueous solution with 0.1% KNO_3 at pH 6.5	Hg lamp	Glycine, serine, and alanine	0.1–1% yield	Caltrera and Ferrari (1959) Ferrari and Caltrera (1961)
Paraformaldehyde and KNO_3 (in presence of $FeCl_3$ as catalyst)	Sunlight	12 amino-acids	—	Bahadur (1954, 1957)
Paraformaldehyde and colloidal molybdenum oxide in H_2O	500 watt bulb	7 amino-acids	Nitrogen source was nitrate or NH_3	Bahadur *et al.* (1958)
H_2 vapour (100 mm) CH_4(500 mm) NH_3(100 mm) CO^3(500 mm)	Schumann ultraviolet	Alanine, amino-butyric acid, valine, norleucine	—	Dodonova *et al.* (1959)
0.1 M aqueous solutions of NH_4 salts and organic carboxylic acids	Hg resonance lamp (2537 Å)	Alanine, glycine aspartic acid	—	Deschreider (1958, 1959)
H·CHO and NH_4 salts in aqueous medium with and without adsorbents and catalysts	Hg lamp	Variety of compounds including 10 amino-acids	Yields about 10–4 M in amino-acids	Pavlowskaya and Passynsky (1957, 1958) Pavlowskaya *et al.* (1960)
CH_4, NH_3, CO in contact with liquid H_2O, in presence or absence of adsorbents	Schumann ultraviolet	Alanine and other amino-acids	Longer wavelengths effective in presence of adsorbents	Terenin (1957)
Suspension of FeS in aq. NH_4 Cl in CH_4 atmosphere	Hg arc	Phenylalanine, methionine, valine	—	Ellenbogen (1958)
CH_4, NH_3, H_2O	Xenon arc (1295, 1470 Å)	Amines and amino-acids	Negative results with Hg 1849 Å	Groth and Weyssenhoff (1958, 1959)
C_2H_5, NH_3, H_2O	Hg arc (1849 Å)	Amines and amino acids	—	
Various inorganic substances with and without catalysts in aqueous medium	1849 Å or full Hg arc	Amino-acids and polymeric material	Investigated effect of various catalysts	
Glycine	Hg arc	Glycylglycine ?	Many unidentified products	Reid (1957)
Cystine	Hg arc	Various amino-acids		

source; and it would obviously be desirable to apply such quantitative techniques to photochemically induced syntheses.

From Table 9-4 it is apparent that the photochemical formation of complex organic molecules, including amino-acids and peptides, may occur from simple starting products under a wide variety of conditions. While those occurring in the gas phase are readily interpretable *via* the radicals formed through photochemical dissociation of the starting products, unquestionably the most interesting are those which have been conducted in the liquid or adsorbed phases, since the wavelengths required are usually in the neighborhood of, or well above, 2200 Å. They could therefore have taken place on the earth's surface or in the oceans under conditions where the photoproducts could be readily concentrated and protected from the effects of shorter wavelengths. This removes many of the difficulties associated with theories based on synthesis in the stratosphere (see above).

Not too much is known about the nature of the reactions in the liquid and sorbed phases, the isolation of some of the intermediates being a necessary prerequisite for this purpose. An interesting example is the work of Reid (1957) on the irradiation of alkaline aqueous solutions of CH_2O, NH_3 and CO_2 with the Hg 1849 Å line which resulted in the formation of formamine (hexamethylenetetramine) in high yield:

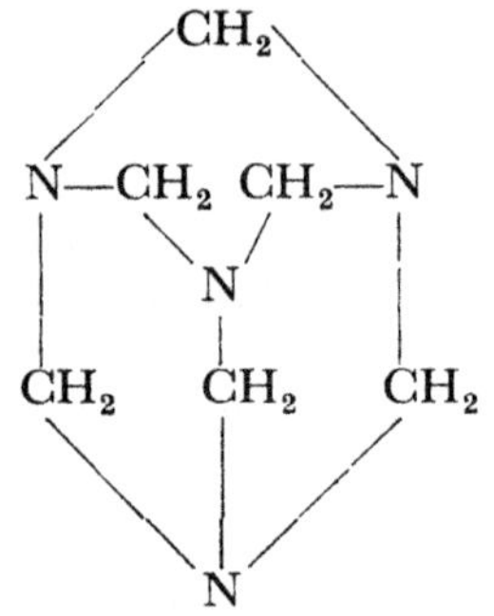

This compound, in turn, gave rise to glycine, alanine and histidine (?) in alkaline medium; and to glycine, alanine, formic acid, glutamic acid (?) and leucine (?) at pH 6.

Considerably more is known about simple thermal syntheses of amino-acids which have also been proposed as a source of primitive earth syntheses. For example, Oro' *et al.* (1959) have proposed the following general sequence of reactions, based on the synthesis of a number of amino-acids

$$HCOH + NH_2OH \xrightarrow{20\text{–}100^\circ C} HCH{:}NOH \xrightarrow[-H_2O]{\text{acid or base catalysis}} HCN$$

The resulting products then undergo condensation reactions, first to amides and nitriles, and then to hydroxy- and amino-acids. This reaction sequence has been proposed as an explanation for the finding many years ago (Baly *et al.*, 1922) that amino-acids are formed upon the ultraviolet irradiation of aqueous solutions of formhydroxamic acid and formaldehyde; hydrolysis of the formhydroxamic acid yields hydroxylamine, which will combine with formaldehyde to initiate the above sequence. The same reaction sequence

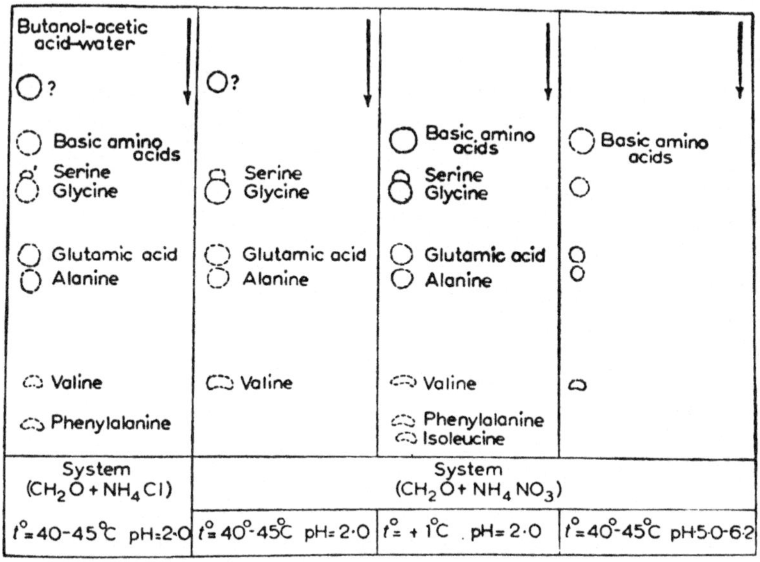

FIG. 9-14. Reaction products isolated by chromatography from aqueous system irradiated with mercury arc (Pavlovskaya and Passynski, 1957).

has been suggested as applicable to the results of Bahadur and collaborators (see Table 9-4).

Finally, mention should be made of the thermal methods which have been successfully developed for polymerization of peptides with the formation of products resembling polypeptides (Fox and Harada, 1958; Kovacs and Nagy, 1961).

Ionizing radiations have also been used to synthesize amino-acids and related compounds from simple starting materials (Dose and Rajewsky, 1957; Hasselstrom *et al.*, 1957; Paschke *et al.*, 1957; Dose and Ettre, 1958) but with results which are considerably less impressive than with ultraviolet radiation. This is perhaps not surprising since the radiation products, once formed, are not as readily shielded from the effects of ionizing radiation. Particularly interesting is the recent observation that γ-irradiation results in polymer formation from sugars, hydroxy-, and amino-acids in aqueous medium; the polymers are formed by C—C additions (Barker *et al.*, 1959).

It will be noted, from the foregoing, that relatively little attention has been devoted to the problem of the primitive synthesis of purine and pyrimidine analogues. Miller (1955), in his experiments with electrical discharges, sought but found no evidence for the formation of such compounds; i.e. he found no substances with absorption maxima corresponding to purines or pyrimidines. It does not, however, follow that photochemical reactions would necessarily lead to the direct formation of nucleic acid derivatives; these might be formed as the result of secondary reactions. For example, it has been proposed (Shugar, 1960) that initial syntheses of pyrimidines might proceed via the dihydropyrimidines; these exhibit only end-absorption in the quartz ultraviolet and would be resistant to radiation to the red of 2400 Å. It has been shown that polynucleotide chains of such polymers may be formed (Janion and Shugar, 1960). Dihydropyrimidines might at some subsequent stage have undergone oxidation leading to formation of the corresponding pyrimidines.

A possible route for dihydropyrimidine formation may be envisaged from the observation of Fox (1957) that ureidosuccinic acid can be obtained by thermal pathways. The ring closure of this compound, which proceeds readily in acid medium, would then lead to the formation of dihydro-orotic acid. It remains to be shown how this could be oxidized to orotic acid. A finding of interest in this connection is the demonstration by Barszcz and Shugar (1961) that γ-irradiation of the photoproduct of 1,3-dimethyluracil, i.e. of 1,3-dimethyl-5-hydro-6-hydroxyuracil (which is an analogue of 1,3-dimethyl-dihydrouracil) leads to the formation in appreciable yield of uracil and 3-methyluracil. It remains then to establish whether this reaction would also proceed photochemically under the appropriate conditions.

On the other hand, purines (including adenine and guanine) and amino-pyrimidines have been synthesized by thermal means (100–220°) from mixtures of NH_4CO_3-serineformamide, from urea and glycine with formamide or ammonium formate or alanine, and from urea and serine (Oro and Kimball, 1960), as well as from HCN via imidazole (Oro and Kimball, 1961; cf. Kliss and Matthews, 1962).

Hence, while a good deal of information has now been accumulated as to how amino-acids and some peptides may have been synthesized under primitive earth conditions from simple starting materials, we as yet have very little to go on insofar as nucleic acid "building blocks," not to speak of polynucleotides, are concerned.

The very fact that these approaches are being made bespeaks an optimism which may be only emotional. That they are heuristic is obvious but the gap between the *quick and the dead* is not bridged by them. Bohr spoke as follows:

> In every experiment on living organisms there must remain some uncertainty as regards the physical conditions to which they are subjected, and the idea suggests itself that the minimal freedom we must allow the organism will be just large

> enough to permit it, so to say, to hide its ultimate secrets from us. On this view, the very existence of life must in biology be considered as an elementary fact, just as in atomic physics the existence of the quantum of action has to be taken as a basic fact that cannot be derived from ordinary mechanical physics. Indeed, the essential non-analyzability of atomic stability in mechanical terms presents a close analogy to the impossibility of a physical or chemical explanation of the peculiar functions characteristic of life (1932).

Can biological phenomena be explained in terms of the known properties of micro- and macromolecules and a synergism involving form as well as substance? Clearly the form of life substance can only be knowable when, if ever, we discover the proper questions. Progress in bridging the gap can only be realized by keeping the, perhaps eternal, gap in mind.

Appendix. Some Techniques in Photochemistry

A. LIGHT SOURCES

1. *Mercury Arcs*

Most of the sources used for photochemical investigation in biology are commercial types of mercury lamps; there is an enormous variety on the market, each designed for some specific purpose. Of the various factors which determine the output and wavelength distribution of such lamps, the two principal ones are the mercury vapor pressure and the nature of the outer lamp envelope. Extensive descriptions of mercury vapor lamps are to be found in the manufacturers' literature, in various handbooks such as that of the Illuminating Engineering Society (New York), and in various texts and articles, e.g. those by Noyes and Leighton (1941), Ellis, Wells and Heyroth (1941), Luckiesh (1946), Bourne (1948), Koller (1952), Withrow and Withrow (1955).

(a) *Medium pressure* (up to a few atm) mercury arcs possess the advantage, aside from stability, of providing a number of discrete, intense, lines in the ultraviolet and visible which may be isolated by means of various filter combinations (Section C), or monochromators, if the areas to be irradiated are small.

(b) *High pressure* mercury arcs operate at pressures of a few up to almost 300 atm. At the higher pressures air or water cooling must be employed; in the event that water is used as the coolant, its purity may be of some significance when the lamp is to be used at wavelengths below 2800 Å. On the other hand, water-cooled lamps possess the advantage that a good deal of the infrared is absorbed by the water.

As the pressure in a mercury arc is increased, the background emission becomes more and more prominent until, at a pressure of about 285 atmospheres, the emission is almost entirely continuous from 2700 Å upward. Such sources are not always suitable for use with a monochromator because of the broadness of the individual lines in the spectrum. Some of the commercial high pressure mercury arcs are enclosed in a glass outer envelope which may be removed for work in the ultraviolet, although at the expense of shortening the effective life of the lamp. For industrial photochemical purposes there are models of high pressure lamps incorporating continuous flow systems.

2. *Carbon Arcs*

Carbon arcs are considerably less efficient than mercury arcs, as regards effective light output with respect to power input, but this is of relatively little importance in biological work where the prime considerations are total intensity and wavelength distribution of the emitted light. This is perhaps best illustrated in connection with the design of what is probably the largest monochromator hitherto constructed exclusively for biological investigation (Ehret and Monk, 1955; Monk and Ehret, 1956). It was found more suitable in this case to use a carbon arc, taking advantage of the increased stability and varied spectral distribution of the emitted light resulting from the introduction into the electrode cores of various salts and metals.

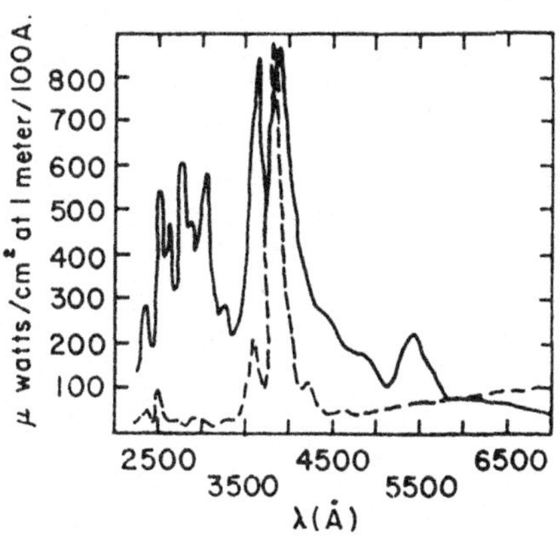

FIG. A-1. Spectral distribution of light from National Carbon arcs:

(a) National D soft carbon core, 50 V, 60 A, -------;
(b) National W polymetallic-cored carbon, 60 V, 80 A (from National Carbon Co. catalogs, and Koller, 1952).

Figure A-1 exhibits the enormous difference in ultraviolet emission of two arcs with different electrode core compositions.

3. *Xenon Arcs*

A relatively new development of considerable significance for photochemical work is the high pressure xenon arc (Schulz, 1947; Baum and Dunkelman, 1950; Anderson, 1951) in which the discharge takes place between two tungsten electrodes spaced several mms apart in an atmosphere of xenon at pressures ranging from 20 to 50 atm. A source of this type is as concentrated as a carbon arc and considerably more stable. Figure A-2 exhibits the ultraviolet emission of a 22.8 A 0.5 kW xenon arc, compared with that from a 10 mm² carbon crater operating at 12 A. The xenon arc is seen to be continuous and 3–4 times as intense as a carbon arc of similar area. Its intensity compares favorably with that of some high pressure mercury arcs. According to Monk and Ehret (1956) a 10 kW xenon arc, in course of development, is expected to give a better emission than a commercially available carbon arc operated at 78 V, 180 A. It is

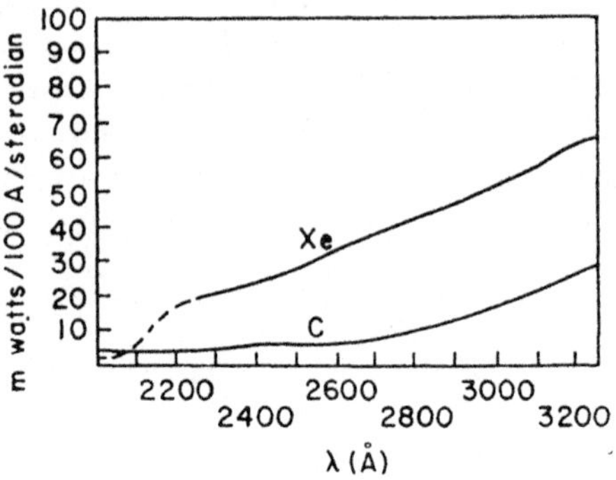

FIG. A-2. Radiant energy emission from a high pressure 500 W xenon arc (Xe) as compared to that from a 10 mm² carbon arc crater operated at 12 A (C) (Baum and Dunkelman, 1950).

quite possible that such sources will become commercially available in the near future.

A xenon arc has been constructed for use in the vacuum ultraviolet (Wilkinson and Tanaka, 1955). The gas is contained in a pyrex tube and is excited by microwave energy at a frequency of 2450 Mc/s, while the radiation, which emerges through a fluorite window, exhibits a strong continuous emission from 1500–2250 Å.

4. *Low-pressure Mercury Resonance Lamps*

These are undoubtedly the most widely used sources of ultraviolet radiation for photochemical work in the field of biology, since they can be constructed so that more than 90 per cent of the total emission is at 2537 Å. Such lamps may be made in the laboratory or obtained commercially in various shapes and sizes; some firms will construct resonance lamps according to requirements. Other types of resonance lamps include the germicidal or bactericidal lamps so widely used for disinfection.

The following table gives the emission as a function of wavelength of a typical commercial resonance lamp and a typical germicidal lamp.

A typical germicidal lamp such as the G30T8, the diameter of which is 2.5 cm and the effective length 80 cm, emits a total of 7 W in the ultraviolet. This corresponds to 10,800 $\mu W/cm^2$ at the surface of the lamp *or*, since almost all of this is at 2537 Å, 84 $\times$ 10^{16} quanta/cm² per min.

Relative emission (per cent)

λ(Å)	British Thermal Syndicate resonance lamp T/M5/369*	Typical germicidal lamp†
2482	0.025	—
2537	100.0	100.0
2652	0.06	0.14
2753–2893	0.06	0.12
2967	0.19	0.37
3022	0.06	0.17
3126–3132	0.61	1.43
3342	0.034	—
3650–3663	0.55	1.30
4047	0.40	—
3906–4077	—	1.60
4339–4358	1.0	3.4
5461	0.90	2.25
5770–5791	0.14	0.60

* Taken from British Thermal Syndicate Bulletin. Multiply figures in this column by 1.2 to get emission in watts/cm^2 at 100 cm horizontally from lamp.

† From Buttolph (1955).

According to the above table, the output of the 2537 Å line of the BTS lamps represents more than 95 per cent of the total emission. This is in agreement with the measurements of Peel (1937). However, Parker (1953) reports that a lamp of this type in his laboratory gave an emission of only 85 per cent at 2537 Å.

Heidt and Boyles (1951) have examined the effect of several variables on the emission of resonance lamps used for photochemical purposes, in which the lamp, in the form of a spiral, was immersed in water. Particularly significant was the effect of temperature, the intensity passing through a maximum at about 45°C. The output of "germicidal" lamps is similarly dependent on the ambient temperature (Koller, 1952; Buttolph, 1955). Account must also be taken of the drop in intensity with age of the lamp and, in particular, the effect of any coating of dust that may form on the lamp surface.

The emission of various germicidal lamps as a function of distance from the lamp axis is given in a chart by Buttolph (1955). At short distances from the lamp axis the intensity is inversely proportional to the distance from the axis, whereas at distances greater than the length of the lamp the intensity is roughly inversely proportional to the square of the distance from the lamp. For practical purposes, where the lamp intensity has been determined at a given distance, it is permissible to use Buttolph's chart to calculate the intensity at any other distance.

Of considerable importance from a photochemical point of view is the emission of all types of resonance lamps at 1849 Å, one of the other resonance lines of mercury. While the output of this line is usually small by comparison to that at 2537 Å, due to its absorption by the lamp envelope, its biological effectiveness is frequently considerably greater than that of the 2537 Å line and may therefore be quite troublesome. Furthermore, some of the commercial germicidal lamps (cold-cathode Sterilamps) are made in a number of different glasses so that, although the intensity of the 2537 Å line is unchanged, that of the 1849 Å line varies. The purpose of such lamps is to produce varying amounts of ozone for disinfection, but for photochemical work it is best to avoid the use of such sources. Santer (1957) has described a procedure for measuring the 1849 Å output of resonance lamps. Attention must also be directed to the possible influence of ozone produced by such lamps on a photochemical reaction, particularly where very long exposure times are used (Shugar, Adamiec and Sztumpf, 1959).

Further details on the operating characteristics and wavelength distribution of resonance and "germicidal" lamps may be found in the manufacturers' literature or by consulting Ellis, Wells and Heyroth (1941), Koller (1952), Buttolph (1955).

5. 1849 Å *Resonance Lamp*

Even the best fused quartz absorbs appreciably at 1850 Å so that only a fraction of the emission of the 1849 Å resonance line of low pressure mercury lamps is transmitted. This difficulty has been circumvented to some extent by Dacey and Hodgins (1950), who constructed a resonance lamp with a sapphire window through which the output intensity over an area of several cm^2 is about 3.5×10^{17} quanta/min. The effective area over which such a source may be used could be considerably enlarged by use of a sapphire tube (Nelson, 1956) for construction of the lamp. On the other hand, for biological purposes such sources suffer from the serious disadvantage that the 2537 Å resonance line is also present. The utility of such sources could be considerably enhanced if some means were available for filtering out the 2537 Å line.

6. *Simple Laboratory Sources of* 2537 Å *Radiation for Inactivation Studies and Preparative Photochemistry*

A simple type of source suitable for the irradiation of sizeable quantities of material, and which can easily be put together in any laboratory, is exhibited in Fig. A-3A.

G is a germicidal lamp surrounded by a glass tube T with a diameter

2 mm (or more) greater than that of G. T is tapered at the bottom and cemented to G by means of a plastic ring and plastic cement or shellac, at W. To maintain T concentric with G a few small plastic spacers are inserted at the upper tapered rim, at P. The solution to be irradiated is introduced at the upper end. To T is sealed the second tube J, thus providing a water

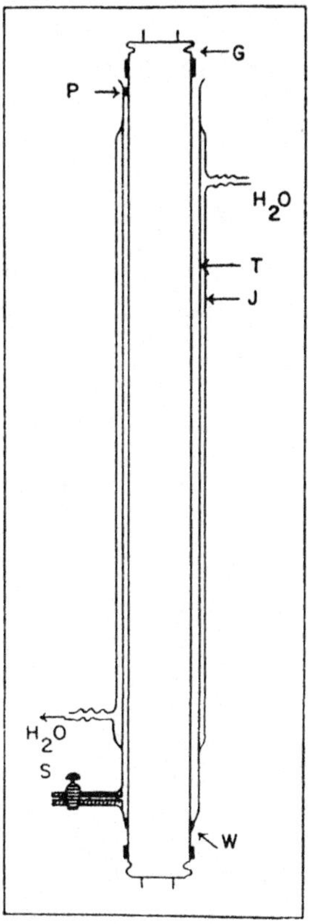

FIG. A-3A. Simple laboratory source of 2537 Å for preparative or other purposes (Shugar, 1960).

jacket for temperature control. During irradiation the solution may be stirred by a stream of oxygen, nitrogen, or air through the capillary tube and stopcock S. Samples may be drawn off during irradiation either at P by use of a fine pipette, or via the stopcock S; if the latter, it is more convenient to use a two-way stopcock.

The above may be used for the irradiation of up to 250 mg of some nucleic acid derivatives. By closing off the upper end at P (or by introducing a capillary tube below P) it may also be used as a continuous flow equipment. Finally, should there be any undesirable radiation emitted by the lamp (e.g. below 2537 Å or in the visible region), this may be eliminated by introducing an additional tube (quartz) between G and T into which a chemical filter layer may be introduced.

The total output of a germicidal lamp is utilized in this way. The temperature range over which the source may be used, however, is necessarily limited by the effect of temperature of the water jacket on the operational characteristics of the lamp. According to Buttolph (1955) the reduction in intensity of a germicidal lamp from that at room temperature (presumably 20°C) is about:

10%	at	10°C	or	37°C
20%	at	5°C	or	42°C
30%	at	2°C	or	47°C

A simple set-up for accurate, quantitative measurements on enzymes, viruses, peptides, etc., is illustrated in detail in Fig. A-3B (Oster and McLaren,

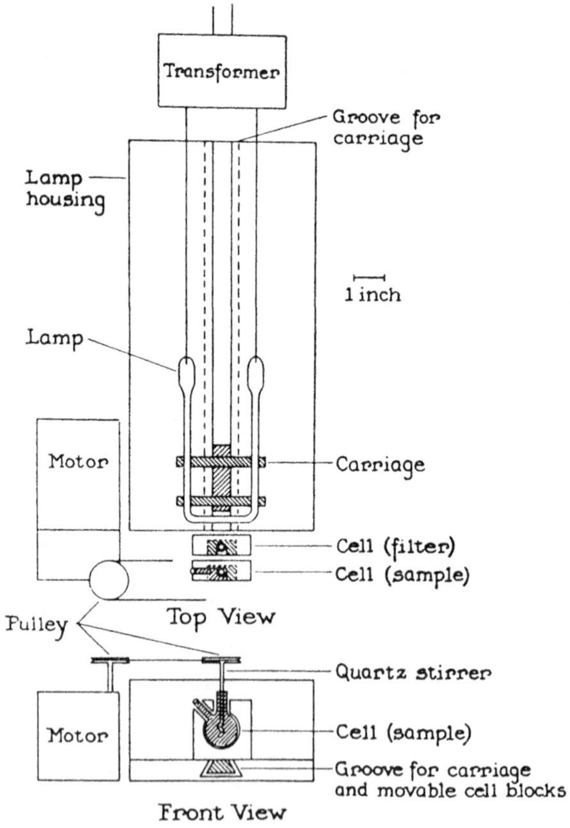

FIG. A-3B. Precision ultraviolet irradiation with an inexpensive 2537 Å UV source (Oster and McLaren, 1950).

1950). By using an actinometer solution in the same or nearly identical way as used for the solution containing enzyme or virus, one may accurately estimate the number of quanta which fall on the solution, without the elaborate care necessary to obtain parallel light, as is required with a thermopile or other ultraviolet meter. The correction for reflection from cell walls is also eliminated, provided that the refractive index is the same for both solutions. The filter cell, with 2–5 per cent acetic acid, cuts out any 1849 Å line from the mercury resonance lamp, and a uranyl actinometer lets through the same visible light as does the protein solution, so the former "sees" the same quanta as the latter.

7. *Monochromators*

The use of monochromatic radiation is becoming more and more a pre-requisite in studies on the photochemistry of proteins and nucleic acids. A wide variety of monochromators is now available commercially, although a number of them are intended largely for investigations in the field of microspectrophotometry. Some of the commercially available spectro-graphs may, in conjunction with suitable sources, be profitably employed as monochromators for photochemical studies.

In most instances, however, it has been necessary to construct one's own monochromator. The requirements for such an instrument are not nearly as rigorous as for spectroscopy since spectral purity and dispersion need not be of primary quality. The simplest instrument for this purpose uses a water prism, which may be quite large, and the original design of which is due to Harrison (1934). Modifications of this type of instrument have been described by Cannon and Rice (1942) and Fluke and Setlow (1954). An excellent description of Harrison's design will be found in the text by Strong (1939), which includes other useful data on monochromators. Reference may also be made to the article by Withrow and Withrow (1955).

Grating monochromators specially designed for photobiological investiga-tions have been described by French *et al.* (1947), and Monk and Ehret (1956). The latter instrument represents the ultimate in monochromators for photochemical investigations in the field of biology. Although too elaborate and expensive for most laboratories, its description merits perusal by any prospective purchaser or constructor of a monochromator, because of the useful quantitative data presented with regard to light sources and the energies available for various types of experiments.

8. *Flash Discharge Sources*

These are modifications of the gas-discharge lamps originally developed for high-speed photography. The energy of a large bank of condensers is discharged through a tube containing a rare, or other, gas or a mixture of gases, so that the light energy is emitted as a single flash of short duration

(Norrish and Porter, 1949). Porter's equipment delivers single flashes via a condenser bank of 100–500 μF capacity. An inexpensive source rated at 20 μF capacity has also been described (McLaren *et al.*, 1952).

Such sources are coming into widespread use and serve a twofold purpose:

(a) The amount of useful light energy emitted in one flash is of the order of 10^4 times, or more, than those obtainable with conventional light sources, thus considerably extending the range of light intensities over which a photo-chemical reaction may be studied. In addition, the course of a given reaction may be investigated under markedly different conditions because of the fact that a high concentration of photochemical intermediates is produced.

(b) The rapid formation of a high concentration of intermediates makes possible a study of their nature and properties, principally by optical methods (Porter, 1950).

The emission of such lamps is usually continuous over the entire spectral range, occasionally with some superimposed lines which depend on the nature of the gas used. The spectral distribution of the radiation emitted by one such source is shown in Fig. A-4, the determinations having been made by ferrioxalate actinometry.

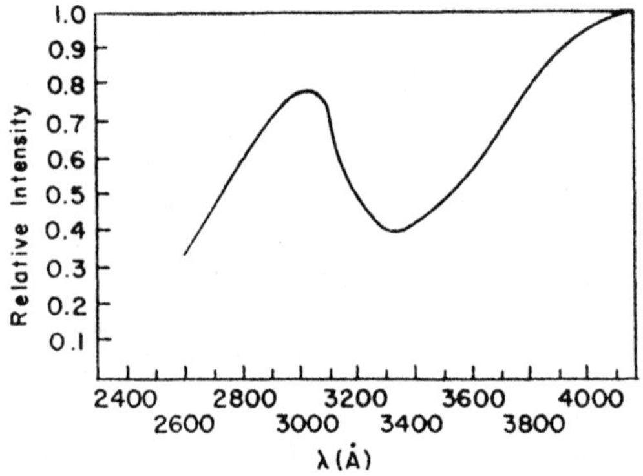

FIG. A-4. Distribution of light emission from a flash lamp filled with pure oxygen at 5 mm Hg. The energy of the flash was obtained from 800 μF, at 5 kV (Claesson and Wettermark, 1957).

The total emission of such sources is a function of the capacity and charging voltage of the condenser bank. The duration of the flash is dependent on the capacity and total inductance of the circuit. Approximately, the flash duration time is

$$\tau = \pi\sqrt{LC}$$

For the study of short-lived intermediates it is obviously desirable that the flash time be as short as possible. Most of the sources described in the literature have flash times of the order of 100–150 μsec. Ramsay (1957) has described a source in which a 40 μF condenser bank is discharged at 8 kV with flash times of about 10–20 μsec. By decreasing the total inductance in the circuit (including that of the lamp itself) to a minimum, Claesson and Lindqvist (1957b) were able to obtain very high energies with flash times of 5–25 μsec.

The mechanical strength of the lamp is one of the factors which limit the amount of energy in the discharge. Two typical electrodes for flash photolysis lamps are shown in Fig. A-5.

Mains *et al.* (1955) have constructed a flash lamp with magnesium electrodes which gives the line spectrum of magnesium and, in particular, the

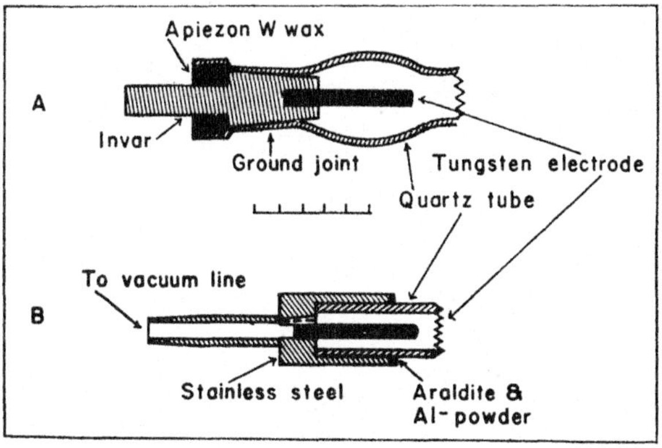

FIG. A-5. Two types of photolysis flash lamp electrodes: A, with conical ground joint; B, with cylindrical cemented joint (Claesson and Lindqvist, 1957a).

2800 Å line with such a high intensity that the lamp may for many purposes be regarded as a monochromatic source. Another innovation is that of Nelson and Ramsay (1956) and Nelson (1956) who prepared a flash lamp with a sapphire window, thus extending the emission intensity in the far ultraviolet region. Intensity measurements with high intensity flash equipment may still be made with uranyl oxalate actinometry with the same quantum yield data used at low intensities (Christie and Porter, 1952).

For additional literature references on such sources, see following section on flash-photolysis flash-spectroscopy.

B. FLASH-PHOTOLYSIS FLASH-SPECTROSCOPY

The experimental procedure used for investigating photochemical intermediates produced by irradiation with a flash photolysis lamp is shown schematically in Fig. A-6. The reaction mixture in the cuvette R is irradiated by the flash from the lamp F connected to the condenser bank C. At a given predetermined interval following the flash, the smaller flash lamp S is triggered. The light from S passes through a filter W and then through the reaction vessel R (which has plane parallel windows) and on to the slit of the spectrograph P. The entire spectrum may then be recorded on a photographic plate; photographs may also be taken at different time intervals following flash irradiation. A more convenient and accurate procedure, which may be applied once the spectrum of the intermediates

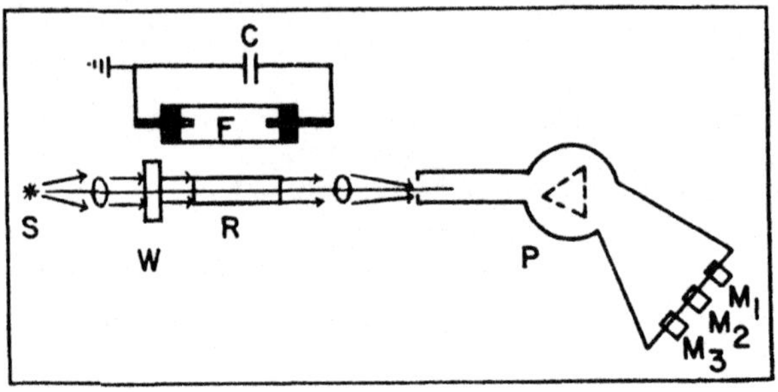

FIG. A-6. Schematic layout of equipment for flash-photolysis flash-spectroscopy (see text for details).

is known, is to replace the photographic plate by one or more photomultipliers ($M_{1,2,3}$) set at appropriate wavelengths and to use a continuous source at S. The formation and decay of intermediates can then be followed kinetically with much greater ease.

The various designs for such equipment differ only in minor details from that originally proposed by Porter (1950). For details of construction reference should also be made to the publications of Gibson (1956), Stron, Chien, Graf and Willard (1957), Bridge and Porter (1958), Claesson and Lindqvist (1957a), and McLaren *et al.* (1952), bearing in mind that each was developed for some specific purpose. Norrish and Thrush (1956) have reviewed a good deal of the literature on this subject, including the type of results obtained. An additional review, dealing in particular with application of flash photolysis in irradiation studies, is that of Porter (1959).

Herzberg and Shoosmith (1956) and Thrush (1959) have considerably extended the field of application of these techniques by the construction of equipment for recording of transient spectra in the vacuum ultraviolet, with good recording sensitivity to as low as 1250 Å. Another innovation of considerable importance is that of Marlmberg (1957), who constructed a flash source which emits flashes of a few mμsec duration with an energy of several μJ per flash. A short, concise article by Nelson (1962) describes various types of flash discharge lamps and their applications.

C. SPECTRAL FILTERS

When a suitable monochromator is not available, or if the intensity obtainable from one is insufficient or does not embrace a sufficiently large area, various filters or filter combinations may be used to cut off at either short or long wavelengths or to isolate some spectral line or group of lines, or more generally, bands. For many photochemical experiments the fine resolution offered by a monochromator is, indeed, unnecessary.

1. *Solid Filters*

A wide variety of filters is available commercially, principally for wavelengths in the near ultraviolet and visible regions of the spectrum. These include colored glass filters; gelatin filters (these tend to fade with continued use and must also be kept away from moisture) and interference filters. The latter are characterized by high transmission at the peak of the band transmitted (in the visible region) and by their ability to isolate a relatively narrow band of wavelengths. Operating as they do on the interference principle, they transmit also integral sub-multiples of the transmitted wavelengths and due account must be taken of this. Such filters have recently been made available for the ultraviolet by Barr and Stroud in England with peak transmissions of up to 20 per cent in the range 2200–3200 Å and up to 25 per cent for the region 3200–3900 Å; as well as by Schott, Jena, G.D.R.

Interference filters are normally offered as 1 in. diameter discs or, upon request, up to a maximum diameter of 2 in. An auxiliary glass filter is incorporated for elimination of unwanted orders, cemented to the interference filter for wavelengths above 2800 Å; for shorter wavelengths, the lack of suitable optical cements makes it necessary to mount the two components in a holder. The transmission characteristics of a typical 2nd order filter suitable for isolation of the 2537 Å line is shown in Fig. A-7 along with the transmission characteristics of a glass filter suitable for elimination of the unwanted 1st order. It will be observed from the figure that the efficiency of the filter will be lower than the 20 per cent shown because of the only gradual cut-off of the OX7 glass filter. In this particular case it might be more advantageous to replace the OX7 by an acetic acid filter (see Fig. A-9). On the other hand for many experiments it may be unnecessary to remove the 1st order. These interference filters are quite stable to prolonged exposure to ultraviolet light, but the auxiliary filters are not and it is therefore best to order them uncemented.

The lower wavelength limit of such filters is limited by the lack of suitable auxiliary filters for the region below 2550 Å. This difficulty may be partially overcome by individual users by ordering only the interference filter and using a chemical filter for cut-off at the short wavelength end.

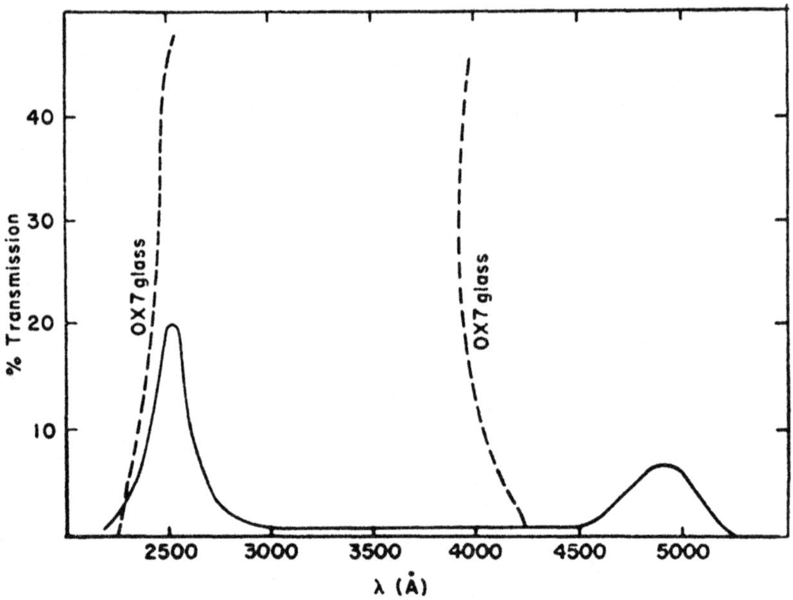

FIG. A-7. Typical 2nd order interference filter for isolation of the 2537 Å line, as well as transmission characteristics of glass for elimination of unwanted 1st order (Barr and Stroud pamphlet No. 1238).

Excellent, and very inexpensive, cut-off filters for the region to the red of 2600 Å are provided by commercial microscope cover glasses. These usually come in thicknesses of 0.08–0.11 mm and are normally available in sizes ranging from 20 × 20 mm to 100 × 100 mm; but can be obtained in even larger sizes on request. The cut-off wavelengths and transmission characteristics of these cover glasses vary from one manufacturer to another, but this is rather advantageous since one may then select the type most suitable for a given purpose by trial. Several representative transmission curves are given in a paper by Sztumpf and Shugar (1962b) and their application to specific problems in nucleic acid photochemistry described in two other communications (Sztumpf and Shugar, 1962a; Fikus *et al.*, 1962). The transmission characteristics may be further varied by using two or more thicknesses in one filter arrangement. Following one or two hours exposure to high doses of UV, the characteristics may undergo slight modification, following which they remain stable This stability is an important factor in preparative work where the high intensities normally required are frequently injurious to other types of filter combinations

Some of the principal suppliers of the large variety of commercially available solid filters include Baird Associates, Farrand Optical Co., Bausch and Lomb, Eastman Kodak and Corning Glass in the U.S.A.; Barr and

Stroud and Chance Bros. in England; the Zeiss Works, Jena, German Democratic Republic. Extensive tables of transmission characteristics of American and Schott Jena filters, and of Wratten filters, will be found in the *Handbook of Physics and Chemistry* (Chemical Rubber Publishing Co., Cleveland, Ohio).

2. *Chemical Filter Combinations*

The spectral absorption characteristics of many chemical compounds, both organic and inorganic, offer a convenient means of isolating spectral bands or arc lines in almost any region of the spectrum. A large variety of such filters, or filter combinations, have been devised, some of them including commercially available glass, gelatin or interference filters as one of the components.

Only a few of the known combinations will be listed here. Much work remains to be done in this field and for those interested in the principles involved in the development of such filters, the excellent article by Kasha (1948) should be consulted. A few words of caution are, however, necessary when employing such filters:

(a) The degree of purity of the chemical compounds used may occasionally be of primary importance.

(b) Some of the compounds listed in the literature undergo slow photodecomposition with resultant modification of transmission characteristics, so that they must be renewed from time to time. For example, Bass (1948) proposes a solution of thiophene in iso-octane as a cut-off filter for the ultraviolet, but the sensitivity of this compound to irradiation is such that the author himself recommends changing the solution every 20–30 min. In practice, if the transmission characteristics of a given compound are highly desirable and the compound in question not too expensive, it may be quite practical to use it in a continuous flow cell.

(c) Fluorescence of some compounds may be troublesome and should always be watched for.

In any event, it is recommended that published data on chemical filter combinations be accepted rather as a guide and that each observer check the absorption characteristics of all components both prior to, and during, their use. It is obviously necessary that the cross section of the filter be greater than that of the irradiation vessel.

Naturally, the characteristics of the source must always be taken account of in estimating the potential efficiency of a projected filter. A good illustration of this is shown in Fig. A-8, which exhibits the transmission characteristics of a filter combination of nickel sulfate and a Zeiss (Jena) WG7 glass filter in the 3130 Å region, as well as the actual transmission of such a filter with two different types of sources.

(a) *Short wavelength cut-off filters.* For elimination of all wavelengths below

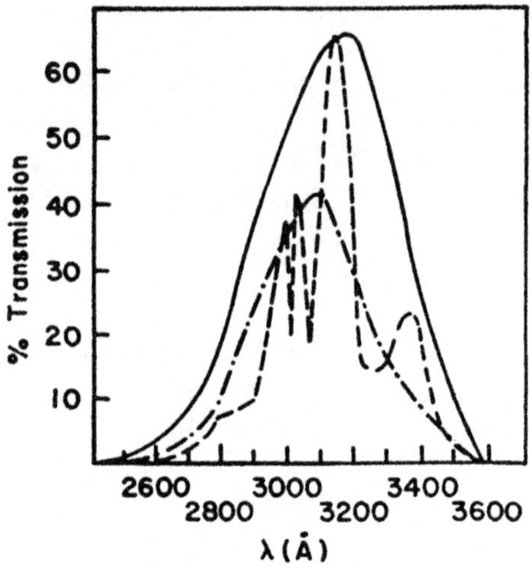

Fig. A-8. Per cent transmission of filter consisting of (2 mm Jena WG7 *and* 10 mm $NiSO_4 \cdot 6H_2O$, 500 g/l.) ———; and products of transmission of filter and intensity of source for flash photolysis lamp, –·–·–·, and for Philips SP 500 W mercury arc, -------. The dashed curves have been normalized to have the same area (Claesson, Juhlin and Wettermark, 1958).

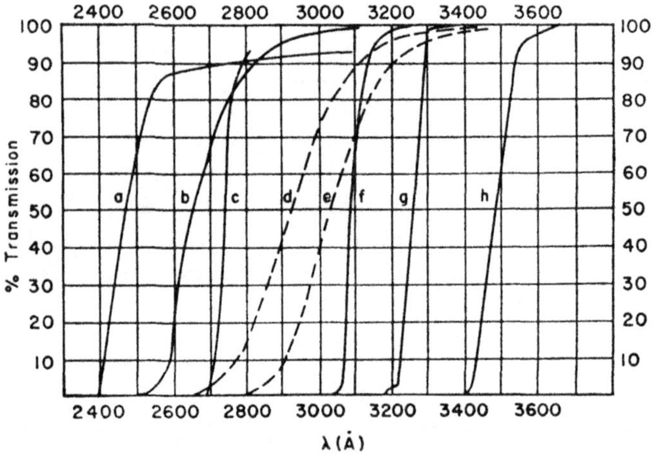

Fig. A-9. Ultraviolet cut-off filters:

(a) 20 per cent aqueous acetic acid, 10-mm layer.
(b) Carbon tetrachloride, 5-mm layer (Kasha, 1948).
(c) Cumene in iso-octane, 10 mg/ml, 10-mm layer, (Bass, 1948).
(d) 0.5 per cent $CuSO_4$, 10 mm layer.
(e) 1.5 per cent $CuSO_4$, 10 mm layer.
(f) Potassium acid phthalate, 5 mg/ml, 10-mm layer (Kasha, 1948).
(g) Napthalene in iso-octane, 12.8 mg/ml, 10-mm layer (Kasha, 1948).
(h) 2,7-dimethyl-3,6-diazacyclohepta-1,6-diene iodide (see Schwarzenbach and Lutz, *Helv. Chim. Acta* 23, 1139, 1940 for preparation of the perchlorate), in water, 0.2 mg/ml (Kasha, 1948).

a given value, undoubtedly the best results are to be obtained with chemical filters, since the cut-off is much sharper than for any other type. Such cut-off filters frequently form one of the components of a band-pass filter.

Figure A-9 exhibits the cut-off characteristics of a number of substances in the range 2400–3400 Å. Additional compounds for cut-off in the ultraviolet are listed by Bass (1947) but some of them are unstable to light, while several exhibit some structure at the cut-off wavelength itself and must therefore be used in conjunction with a Corning 9863 transmitting only in the region 2350–4000 Å.

(b) *Band-pass filters.* A useful system of band-pass filters for the ultraviolet has been developed by Kasha (1948) and these are presented in Fig. A-10,

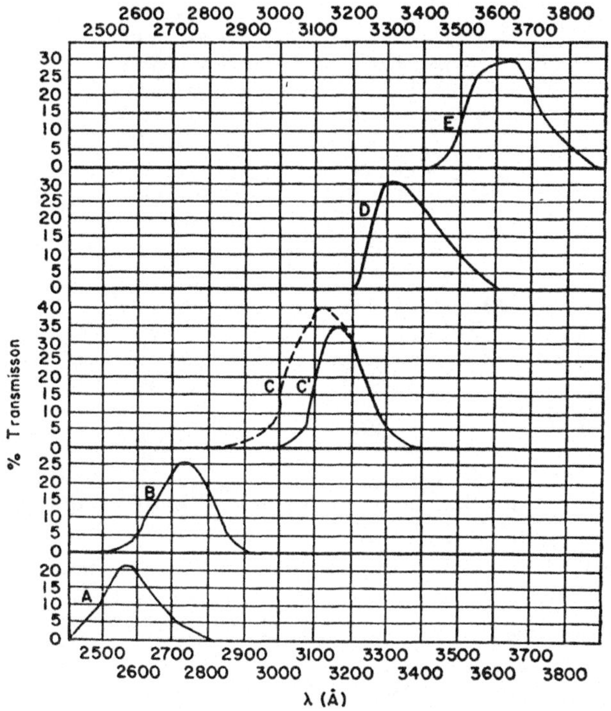

FIG. A-10. Ultraviolet band-pass filters, each consisting of following components, details of which are given in table below (Kasha, 1948).

A: 5 + 14

B: 4 + 8 + 9 + 13

C: 3 + 8 + 11

C′: 3 + 8 + 11 + 12

D: 2 + 7 + 10

E: 1 + 6 + 9

as well as the accompanying table. As mentioned above, the original publication may be profitably consulted for an idea of the methods used in the development of such filters, stability of compounds, etc.

COMPOSITION OF COMPONENTS FOR BAND-PASS FILTERS ILLUSTRATED IN FIG. A-10

Component number	Compound	Concentration	Optical path
1	$CuSO_4 \cdot 5H_2O$	100 g/l. water	5 cm
2	$NiSO_4 \cdot 6H_2O$	100 g/l. water	5 cm
3	$NiSO_4 \cdot 6H_2O$	200 g/l. water	5 cm
4	$NiSO_4 \cdot 6H_2O$	300 g/l. water	5 cm
5	$NiSO_4 \cdot 6H_2O$ and $CoSO_4 \cdot 7H_2O$	240 g/l. water and 45 g/l. water	5 cm
6	Corning filter C.S. 7–37(5860 or 586)		5 mm
7	Corning filter C.S. 7–51(5970 or 597)		4 mm
8	Corning filter C.S. 7–54(9863 or 986)		3 mm
9	*	200 mg/l. water	1 cm
10	Naphthalene	12.8 g/l. *iso*-octane	1 cm
11	K_2CrO_4	200 mg/l. water	1 cm
12	$KHC_8H_4O_4$	5.0 g/l. water	1 cm
13	CCl_4	—	5 mm
14	1,4-diphenylbutadiene	4.24 mg/100 ml ether	1 cm

* 2,7-dimethyl-3,6-diazacyclohepta-1,6-diene iodide.

A large variety of other filter combinations, both for the ultraviolet and visible regions of the spectrum, are presented by Noyes and Leighton (1941), Bowen (1936) and Masson, Boekelheide and Noyes (1956). An extensive bibliography of the literature on filters is given by Scott and Sinsheimer (1955). The textbooks of Wood (1934) and Strong (1939) also contain considerable useful information.

Several more recently developed filter combinations are worthy of mention:

Hunt and Davis (1947) have made an extensive investigation of all filter combinations proposed for isolation of the 3130 Å group of the mercury arc and propose the following:

50 mm 0.178 M nickel chloride
50 mm 0.005 M potassium chromate
10 mm 0.0245 M potassium bi-phthalate
5 mm Corning 9863

This combination is said to transmit 25 per cent of the 3130 Å group and less than 0.01 per cent in the rest of the ultraviolet. Chilas (1961) has described a combination of interference and chemical filters for the isolation of relatively broad bands located at 2210 Å, 2600 Å and 2680 Å.

Koda, Haradona and Kamiyama (1958) separated the 2537 Å mercury line with 10 mm of α-naphthol in iso-octane or methyl alcohol (0.004 wt. per cent) combined with an ordinary filter of aqueous nickel or cobalt sulphate. This filter is said to be transparent to 3100 Å.

Buraway and Roach (1958) isolated the 4358 Å mercury line by means of crystal violet perchlorate and 9:10-dibromoanthracene in ethanol-

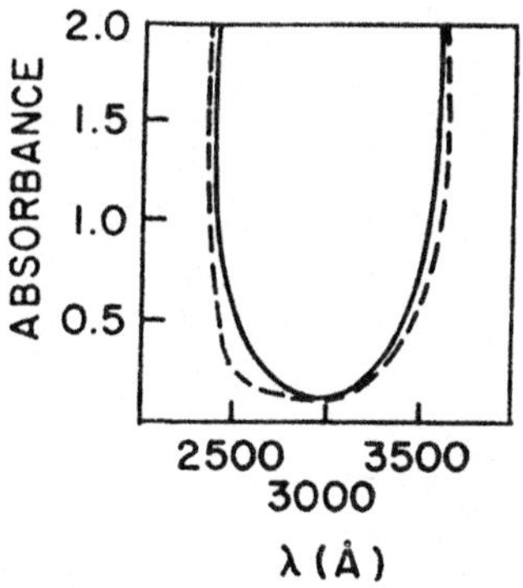

FIG. A-11. Absorption of nickel sulfate–sorbitol sandwich
(a) between fused quartz plates, -------;
(b) between a quartz plate and a Corning 9863 plate, ——— (from Strait, Goyan and Kumler, 1956).

toluene. The transmission is 90 per cent at 4358 Å and almost zero at 4047 Å and 5461 Å.

A rather novel, semi-solid, band-pass filter for the ultraviolet has been devised by Strait *et al.* (1956) and is prepared as follows: 2 moles sorbitol and 1 mole $NiSO_4 \cdot 7H_2O$ are dissolved by heating in fourteen moles water and then further purified. Subsequent gentle heating at 60°C to remove water yields a highly viscous liquid which, on cooling, congeals to a semi-solid complex and is "sandwiched" between two discs of silica or between one disc of silica and another of Corning 9863 glass. Filters such as this have been made up to 15 cm diam and can withstand temperatures of solid CO_2 or 70°C for several hours. Such a filter transmits 60–80 per cent at 2800–2900 Å and cuts off at 2400 Å and 3600 Å. Figure A-11 exhibits the absorption characteristics of such a filter. Reference should be made to the original publication for complete directions.

D. MEASUREMENT OF LIGHT INTENSITY

1. *Physical Methods*

Characterization of a photochemical reaction, by quantum yield or otherwise, requires a knowledge of the number of quanta absorbed by the reacting system and hence measurement of the incident light intensity. The fundamental method for measurement of radiation intensity is by use of a thermopile; this instrument is calibrated against a true "black body" which radiates according to Stefan's law. In practice this is rarely done because of the difficulty of setting up a truly black body and, when thermopiles are used, they are usually calibrated against standard lamps purchased from some government standardizing bureau. Such lamps are usually supplied with a certificate indicating the dependence of intensity on lamp current. By varying the lamp current, the filament temperature and consequently the wavelength distribution of the emitted light will vary (de Vos, 1954). If the response of the thermopile is directly proportional to lamp intensity, this is a reasonably good indication that the thermopile response is independent of wavelength, as it must be. An excellent, brief, outline of the construction and use of thermopiles is given by Strong (1939) and Withrow and Withrow (1955).

The more commonly used methods for measurement of light intensity involve the use of (a) photoelectric cells, (b) a photochemical reaction the quantum yield for which has been established, in other words, chemical actinometry.

The photoelectric cells used include the "barrier-layer" type and phototubes. The latter are generally preferred since their sensitivity is much higher, they are less subject to "fatigue," and they are now available commercially with spectral sensitivities to suit almost every need. Photoelectric cells suffer, on the other hand, from the disadvantage that their response is usually not linear and also varies with wavelength; as a result of this they must be used only for monochromatic radiation and a calibration curve must be drawn for each wavelength.

A wide variety of instruments have been devised for the measurement of light intensities with phototubes. For general details reference should be made to a standard text or to the *Standard Handbook for Electrical Engineers* (McGraw-Hill, N.Y.). Typical arrangements are described by Taylor and Haynes (1947) Pal, (1948), Launer (1949), and Mellors (1952). The latter author describes a portable instrument in which a gas-filled cell is used at low voltages to give a linear response; the instrument has an adjustable range from 0–1000 $\mu W/cm^2$ for light of wavelength 2537 Å. He writes:

> It is, of course, always possible to extend the range of any such instrument by the interposition between it and the source of an optical or neutral filter so that the intensity incident on the cell is within the range of the instrument. It is then

only necessary to multiply the instrument reading by the appropriate filter transmission factor to get the true intensity.

Although barrier-layer cells are much less sensitive than phototubes, are subject to fatigue and are relatively insensitive in the ultraviolet, Ponzio (1952) and Latarjet, Morenne and Berger (1953) have shown how to use such cells for routine measurements of the ultraviolet emission of germicidal lamps at 2537 Å, based on a principle suggested by Beese (1939). Such a cell may be readily constructed in the laboratory or purchased at cost from the Service de Radiologie et Cancerologie of the Institut Pasteur, Paris, France. It is shown schematically in Fig. A-12. The incident beam, the

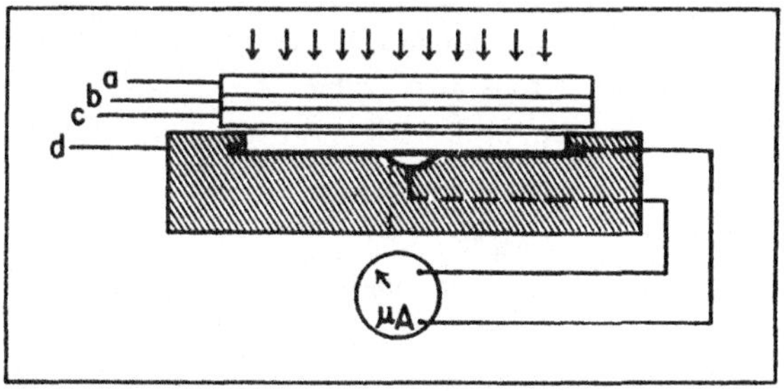

FIG. A-12. Application of selenium barrier-layer cell to measurement of radiant energy at 2537 Å (Latarjet, Morenne and Berger, 1953).
(a) Corning filter No. 9863 *or* Schott UG5.
(b) Willemite screen.
(c) Katathermic glass filter to eliminate infrared.
(d) Selenium cell.
(μA) Microammeter.

intensity of which is to be measured, passes through a Corning 9863 filter (a Schott UG5 filter may also be used) which removes visible radiation and gives a reasonably good transmission at 2537 Å. The transmitted ultraviolet radiation causes the Willemite screen to fluoresce. The resultant green fluorescent light passes through the Katathermic glass filter (which absorbs the infrared) and strikes the selenium photocell which is connected to a direct current microammeter with an internal resistance of 2000 ohm and a total scale deflection of 5 μA.

Earlier models of such meters did not incorporate the filter systems used in the above, so that it was necessary to take two readings, one of the total lamp output and the other with a glass or filter screen to remove the ultraviolet; the difference between the two readings gave the 2537 Å emission (Luckiesh and Taylor, 1940; Kerr, 1947).

Jagger (1961) has recently described a simple direct reading instrument based on the use of a high quality ultraviolet-sensitive photovoltaic cell developed by the General Electric Co. A new type of direct reading meter for the ranges 2400–3200 and 2700–3200 Å has also just been made available by Philips of Holland.

Although in the following section special emphasis will be laid on the value of chemical actinometry, it should be noted that a light meter such as the above (or the phototube light meters supplied by some commercial firms) is extremely useful for mapping the distribution of intensity of radiation from resonance and germicidal lamps.

"Solar blind" detectors. The photoemission properties of caesium and rubidium tellurides render these substances of particular interest for measurements of ultraviolet energy. They exhibit a high photoelectric yield in the region 2000–3000 Å and are relatively insensitive to the near ultraviolet and visible radiation. Photocells using these materials have been proposed for ultraviolet photometry (Taft and Apker, 1953) and for measurement of germicidal radiation (Dunkelman, 1957).

2. *Chemical Actinometry*

Chemical actinometry possesses considerable advantages for light intensity measurements, particularly where the source does not involve the use of a monochromator. A chemical actinometer gives the average dose over a give period of time and over a given area. By carrying out actinometry in the reaction vessel itself, one may also eliminate complications resulting from the geometry of the irradiating system.

(a) *Uranyl oxalate actinometer.* This is still the most widely used chemical actinometer (Leighton and Forbes, 1930; Forbes and Heidt, 1934), the reaction utilized being the uranyl ion photosensitized decomposition of oxalic acid, as follows:

$$H_2C_2O_4 \rightarrow H_2O + CO_2 + CO$$

Actinometer solutions: Uranyl oxalate is prepared by mixing hot solutions of uranyl nitrate and oxalic acid. The uranyl oxalate, which separates out on cooling, is dried under vacuum and then for 3 hr in a 110°C oven.

Solution a: 0.01 mole (3.75 g) uranyl oxalate and 0.05 mole (6.3 g) oxalic acid made up to 1 liter in water. This solution is for use in the wavelength region 3300–4350 Å for a cell optical depth of 10 mm.

Solution b: Solution *a* diluted 10-fold with water, for use in the wavelength range 2400–3300 Å for a 10 mm cell depth.

Actinometry: following irradiation, the actinometer solution is made strongly acid by the addition of about 1/4 of its volume of 9–10 N H_2SO_4. The solution is then warmed to over 70°C to remove CO and titrated with 0.01–0.1 N permanganate. A non-irradiated control is titrated in the same way. If the titre difference, expressed in terms of 0.1 N permanganate, is *t* ml., the quantum yield Φ, and the fractional light absorption (for a 10 mm optical path) f_a, then

$$\text{Number of } Nh\nu = \frac{t}{2\Phi f_a} \times 10^{-4}$$

Division of this amount of energy by the time of exposure and the volume of solution in cm^3 gives I, the intensity falling on the solution in the cell. For further details see Claesson, 1956.

For those cases where there is high light absorption it is advisable to stir the solution during irradiation. It is also best to carry the reaction to approximately 50 per cent completion for good results. The following table gives the values of quantum yield and fractional light absorption for 10 mm layers of the actinometer solution:

λ(Å)	Solution	Φ	f_a
4350	a	0.58	0.242
4050	a	0.56	0.284
3660	a	0.49	0.342
3350		0.53	
3130	b	0.56	0.658
3020	b	0.57	0.82
2780	b	0.58	
2650	b	0.58	1.0
2537	b	0.60–0.64	1.0
2080	b	0.47–0.50	1.0

A more detailed description of the uranyl oxalate actinometer is given by Noyes and Leighton (1941) and by Masson *et al.* (1956) for those cases where actinometry is applied to the light beam emerging from the reacting system.

Undecomposed oxalate may also be titrated with ceric sulphate (Forbes and Heidt, 1934) or the titration may be replaced by spectrophotometry according to Pitts, Margerum, Taylor and Brim (1955). The ceric sulphate is added to the actinometer solution so that there is a slight excess in the blank. The difference in residual concentrations of the oxidizing agent in the blank and irradiated samples is then determined by difference spectrophotometry at 3200 Å, the maximum of absorption of ceric sulfate. The principal advantages of this procedure are that spectrophotometry is much more convenient than titration and shorter exposure times may be used. The claim of the authors that the results are more accurate than permanganate titration cannot, however, be considered as established.

Claesson and Lindqvist (1957a) titrate the excess ceric ion with arsenic solution thereby obtaining a sharper end point which they claim makes possible a reproducibility of 0.01 per cent and an overall accuracy of 1 per cent. The actinometer solution used was 0.01 M in uranyl oxalate and 0.04 M in oxalic acid. The actinometer solution (25 g) is first freed of CO by bubbling nitrogen through it for 20 min at 70°C. After cooling it is mixed with 28 g of a solution of 0.01 M $Ce(SO_4)_4 \cdot 2(NH_4)_2SO_4$ in 2 M H_2SO_4. The ceric ions are then estimated by titration with 0.0025 M arsenic solution, using osmium tetroxide as catalyst and ferroin as indicator. Sharp indicator changes were obtained by addition of a slight excess of arsenic and back titration with 0.0025 M ceric sulphate solution.

Zill (1955) has applied a manometric procedure to the uranyl oxalate actinometer, measuring the CO and CO_2 liberated during decomposition of oxalic acid according to the equation given above.

The liberation of CO, which takes place with a quantum yield of 0.17 as compared to 0.60 for oxalate decomposition, has been made use of by Porter and Volman (1962) for actinometry on a micro scale. CO released was determined by a flame ionization detector after catalytic hydrogenation to methane. The gas was separated on a gas chromatographic column. While the procedure is reported to give an absolute sensitivity three-fold better than that of the malachite green leucocyanide actinometer (see below), its complexity is such that it is suitable only for special applications.

N.B. One of the principal attributes of the uranyl oxalate actinometer for measurements of light intensity in the ultraviolet region most used for photochemical studies on enzymes, nucleic acids and viruses, is that the concentration of the actinometer solution used is such that it is relatively non-absorbing above 3100 Å, and almost entirely transparent to visible light. When the source used is a resonance lamp, for example, no special precautions must be taken against shielding from daylight or artificial light.

(b) *Malachite Green Leucocyanide actinometer.* For relatively low light intensities the uranyl oxalate actinometer becomes rather inconvenient because of the extremely long exposure times required. Under such circumstances the malachite green leucocyanide actinometer (Harris, Kaminsky and Simard, 1935; Calvert and Rechen, 1952) may be used. Irradiation of a solution of malachite green leucocyanide in ethyl alcohol transforms the former into an intensely colored dye with absorption maximum at 6200 Å. The quantum yield is 1.0 ± 0.03 over a concentration range of 10^{-3}–10^{-4} M, is independent of wavelength over the range 2480–3300 Å, and the temperature coefficient is zero over the range 7–37°C. The amount of dye formed may be followed by simple spectrophotometry, the optical density at 6200 Å being proportional to the number of quanta absorbed. The method of preparation and purification of the leucocyanide is given by Calvert and Rechen (1952).

The light intensities to be measured should not exceed about 10^{14} quanta/cm^2 per sec since the quantum yield appears to then drop below unity as a result of the high absorption of the photoproduct. This does not necessarily preclude its use for measuring intensities of powerful sources, since a suitable filter may be interposed to cut the incident intensity down to the required level.

One inconvenient feature about this actinometer is that it must be kept in the dark except when exposed to the source to be measured. Since the quantum yield is unity over a wide range of wavelengths, this means also that in measuring the output of a resonance lamp, the visible and near ultraviolet components must be eliminated by means of a suitable filter; or, two measurements may be made, one with the full light from the source and the other with the 2537 Å line eliminated by means of a short wavelength cut-off filter.

(c) *Ferrioxalate actinometer.* The potassium ferrioxalate actinometer (Parker, 1953; Hatchard and Parker, 1956) is applicable over a wide range of wavelengths and intensities, and only spectrophotometry is involved in the measurement of light intensities. A serious disadvantage for routine work is that all manipulations must be conducted either in the dark or by means of a Kodak OB safelight (or other suitable lamp) with exposure only to the source being measured.

On the other hand, the characteristics of the actinometer have been very extensively investigated over a wide range of operating conditions (Hatchard and Parker, 1956) and its accuracy checked by independent observers (Baxendale and Bridge, 1955; Claesson and Lindqvist, 1957a; Claesson, Juhlin and Wettermark, 1958).

Actinometer solutions: Mix, with vigorous stirring, 3 vols of 1.5 M potassium oxalate and 1 vol. of 1.5 M ferric chloride (using analytical grade reagents). The precipitated potassium ferrioxalate, $K_3Fe(C_2O_4)_3 \cdot 3H_2O$, is recrystallized 3 times from warm water and the crystals sucked dry on a filter in a current of air at about 45°.

0.006 M *actinometer solution:* Dissolve 2.947 g of the crystals in 800 ml water. Add 100 ml 1N H_2SO_4 and water to a final volume of 1 liter.

0.15 M *actinometer solution:* Made up in same way as above.

Both the above solutions may be stored in amber bottles in the dark for long periods.

Calibration curve for Fe^{++} ion. The following solutions are prepared:

(a) 0.4×10^{-6} mole/ml Fe^{++} in 0.1N H_2SO_4. This should be prepared fresh by dilution of standard 0.1M $FeSO_4$ in 0.1N H_2SO_4.

(b) 0.1 per cent 1:10-phenanthroline monohydrate in water.

(c) Buffer solution: 600 ml N sodium acetate and 360 ml N H_2SO_4 plus water to a final volume of 1 liter.

To a series of 20-ml calibrated flasks add:

0, 0.5, 1.0, · · · 4.5, 5.0 ml of (a)

and 10, 9.5, 9.0, · · · 5.5, 5.0 ml 0.1N H_2SO_4

To each flask add 2 ml of (b) and 5 ml of (c), with mixing after each addition, then make up to mark and allow to stand for *at least ½ hr, in the dark.* Optical densities are then read in 10 mm cuvettes at 5100 Å and a calibration curve drawn, the value for the sample free from Fe^{++} ion being first subtracted from all the others. The optical density of the blank should be about 0.01 and 1.81×10^{-6} mole Fe^{++}/20 ml should give an optical density of 1.0. The calibration curve is linear and accurately reproducible. The "difference optical density" is independent of the presence of potassium ferrioxalate and of the volume of added buffer, provided the latter is equal to at least ½ the volume of 0.1N H_2SO_4 present (final pH is 3.5).

Procedure for actinometry. The actinometer solution is stirred during irradiation with a stream of oxygen-free nitrogen and exposed to the source long enough to produce 0.005×10^{-6} to 3×10^{-6} mole Fe^{++}/ml. An aliquot of the solution is then pipetted into a 20 ml calibrated flask, to which is then added 2 ml of solution (b) and a volume of (c) equal to ½ the volume of the irradiated aliquot. The flask is then made up to mark with water and, after ½ hr (or longer) the optical density is read at 5100 Å in a 10 mm cuvette. The optical density of a control (non-irradiated) solution is subtracted from that of the irradiated solution. From the corrected optical density the amount of Fe^{++} ion formed is read off the calibration curve and, from the known quantum yield (see Table below) the number of *Nhν* absorbed is calculated. From the known absorption of the solution (see Fig. A-13), the incident intensity in *Nhν* is calculated. In practice it is

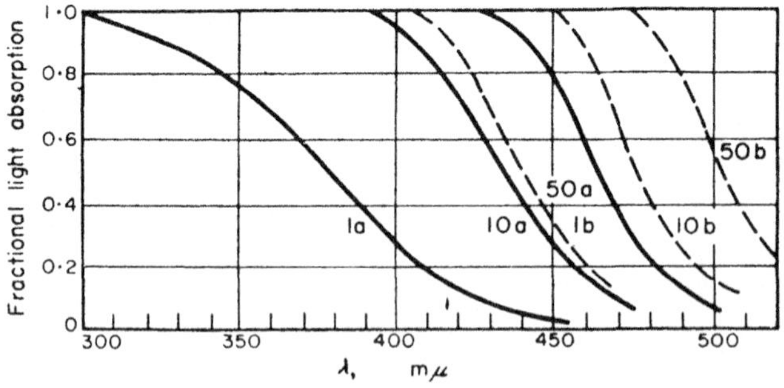

FIG. A-13. Fractional light absorption by potassium ferrioxalate actinometer solutions, 0.006 M (———) and 0.15 M (------) in 0.1 N H_2SO_4. Figures refer to optical depths of solutions in mm (from Hatchard and Parker, 1956).

perhaps best to measure directly the absorption of the actinometer solution. The authors state that the amount of ferrioxalate decomposed should correspond to a reduction in concentration of not more than 0.005 M.

QUANTUM YIELD FOR FERRIOXALATE ACTINOMETER AS A FUNCTION OF WAVELENGTH AND TEMPERATURE (from Hatchard and Parker, 1956)

λ(Å)	Ferrioxalate conc.	Φ(22°C)	Temp. coeff. (22–32°C) (per cent 10°C)*
5460	0.15 M	0.15	—
5090	0.15 M	0.36	—
4800	0.15 M	0.93	—
4680	0.15 M	0.92	—
4360	0.006 M	1.11	0
4050	0.006 M	1.14	+5
3650	0.006 M	1.21	+1
3340	0.006 M	1.23	−2
3130	0.006 M	1.24	+2
2970/3020	0.006 M	1.24	+2
2540	0.006 M	1.25	—

* For most practical purposes, the temperature coefficient may be neglected.

(d) *Other actinometer systems.* Several other actinometer systems have been proposed for use in the ultraviolet and vacuum ultraviolet (see Masson *et al.*, 1956) but these have hitherto not found any wide application in biological studies.

Latarjet *et al.* (1952), Lee and Puck (1958) and Lerman and Tolmach (1959) have made use of the known inactivation rates of bacteriophages T2 (or C16) for checking intensities of 2537 Å sources. A more convenient procedure involves the use of pyrimidine derivatives such as dimethyluracil and uridylic acid (Chapter VI), since it is necessary to measure only a decrease in absorption over a given time interval under the influence of the source of radiation (Shugar and Wierzchowski, 1958).

A relatively simple actinometer system has been described by Warburg and Schocken (1949) for the visible region of the spectrum up to 6600 Å. It is based on the chlorophyll (or protoporphyrin) sensitized oxidation of allylthiourea (or thiourea), the quantum yield for which is unity over a wide range of wavelengths. The reaction involves measurement of oxygen disappearance in the Warburg manometer. The actinometer solution is contained in a rectangular manometer vessel and consists of 200 mg thiourea in 5 ml pyridine, plus 2 mg crystalline ethyl chlorophyllid (for red, yellow and blue light) *or* 10 mg crystalline protoporphyrin (for green and blue light). Attention is drawn to the fact that, although this actinometer can be used to the blue end of the visible spectrum, its value in the ultraviolet remains to be investigated.

An interesting method for measuring the 1849 Å emission of resonance lamps has been devised by Santer (1957), based on the measurement of the ionization produced in water vapor in air at constant relative humidity.

E. MICROBEAM, OR PARTIAL CELL, IRRADIATION

The use of ultraviolet microbeams for the irradiation of localized areas of living cells was originated and first applied by Tschachotin (1912) and his collaborators (see Tschachotin, 1938). The principle of this technique consists in bringing to a microscopic focus a beam of ultraviolet light, using quartz optics. The position of focus was first located by means of a fluorescent solution, following which the specimen to be irradiated was introduced into this position. This obviously involved considerable technical difficulties.

With the advent of the compound microscope using reflecting objectives, it became possible to introduce considerable simplification into the above equipment and a more modern design is exhibited in Fig. A-14; the following description of its operation is taken from Zirkle (1957), and Uretz (1954).

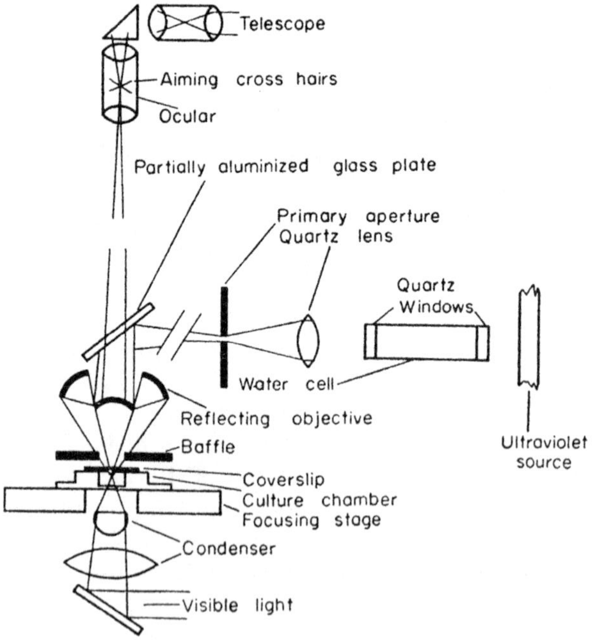

FIG. A-14. Schematic representation of equipment for ultraviolet irradiation of portion of living cell (for explanation see text) (from Zirkle, 1957).

Observation and location of the specimen to be irradiated is achieved by means of the compound microscope with reflecting objective in which only the focussing stage is moveable by means of a vertical adjustment for focussing. The image of the specimen to be irradiated is brought into focus in the plane of the aiming cross hairs in the ocular. By observing the target image through an adjustable horizontal telescope focussed on the aiming cross hairs, variations in eyesight amongst different observers are compensated for, so that the target specimen is always brought to focus at the same optical plane below the objective.

For irradiation purposes, an image of a suitable ultraviolet source is passed through a water cell to remove a good deal of the heat and is focussed by means of a quartz lens on a primary aperture. The light emerging from this aperture is directed into the reflecting objective by means of an aluminized mirror, so that a reduced image of the beam is formed below the reflecting objective. To focus this ultraviolet micro-image (or "hot spot") on to the same plane as the viewing microscope, visible light is focussed onto the primary aperture and the latter moved along the beam axis until the visible micro-image is focussed on a mirror placed in the object focal plane of the viewing microscope. The primary aperture is then moved perpendicularly to the beam axis until the micro-image is centered under the cross hairs. Finally, the target is brought into focus in the microscope and located horizontally by means of a mechanical stage so that the required area is in the path of the microbeam.

By means of the above equipment microbeams with a diameter down to about 7μ may be obtained. The size (and shape) of the microbeam may also be varied by means of a set of apertures of different shapes and sizes. Both monochromatic and heterochromatic radiation may be used. Beam energy is measured by means of a photomultiplier tube previously calibrated against a thermopile. Studies by Uretz (1954) and Perry (1957) provide good illustrations of the application of this equipment.

An entirely different scheme has been developed more recently by Montgomery and Bonner (1958), based on the use of the ultraviolet flying spot television microscope (Montgomery, Roberts and Bonner, 1958; see Zworykin, 1957 for review article on television microscopy). The rectangular pattern traced out on the face of an ultraviolet emitting cathode ray tube by a scanning spot of ultraviolet light is directed into a compound microscope consisting of a quartz ocular and reflecting objective. The image of the cathode ray tube pattern is thus demagnified and projected onto the specimen to be irradiated in the focal plane of the reflecting objective. In effect, then, the cathode ray tube pattern replaces the conventional ultraviolet source used in the equipment described above. The light transmitted through the target specimen is, in this equipment, collected by a quartz condenser and directed into an ultraviolet sensitive photomultiplier tube which is in turn connected to a video amplifier driving in synchrony a monitor tube on which a magnified visible image of the specimen appears. By varying the intensity of the scanning spot at any desired position on the ultraviolet emitting cathode ray tube, any given area on the target specimen down to 1μ in diameter may be intensely irradiated while the remainder of the specimen is subjected only to non-damaging levels of irradiation for the simultaneous recording of the ultraviolet absorption images of the specimen by time-lapse motion picture photography. It is also possible by means of this set-up to intensely irradiate the entire target with the exception of a given area down to 1μ in diameter.

Although the complexity of this equipment is such as to preclude its coming into widespread use, its versatility is indeed astonishing and should lead to a considerable extension of our knowledge of the effects of radiation on living cells. A modification of this equipment, with applications, has been described by Freed *et al.* (1959).

Finally, mention should be made of the technique involving mechanical division of living cells into nucleated and enucleated portions, each of which may then be irradiated with conventional sources, thus making possible a study of the potential effects involving the nucleus and/or cytoplasm. Cells susceptible of being mechanically divided include *amoeba* and *acetabularia*. A typical study of this nature is that of Skreb and Errera (1957).

Bibliography

ABELSON, P. H. (1961). Extra-terrestrial life. *Proc. Nat. Acad. Sci.* (U.S.), **47,** 575–581.

ALBERT, A. (1958). Additions to double bonds in *N*-heteroaromatic six-membered rings, in *Heterocyclic Chemistry* (Albert, A., ed.), Butterworth, London. pp. 20–29.

ALECHINSKY, A. and REUSE, J. (1946). Nicotinamide and ultraviolet irradiation of histidine. *Compt. rend. soc. biol.* **140,** 1224.

ALEXANDER, A. and MOROSON, H. (1960). Effect of ultraviolet light on the shape and size of DNA. *Nature,* **185,** 678.

ALEXANDER, P. (1953). The combination of protamine with desoxyribonucleic acid. *Biochim. Biophys. Acta,* **10,** 595.

ALEXANDER, P. (1960). Protection of molecules *in vitro* against damage by ionizing radiations, in *Radiation Protection and Recovery* (HOLLAENDER, A., ed.), Pergamon Press, London, chap. 1.

ALEXANDER, P. and FOX, M. (1954). Photodynamic degradation of macromolecules sensitized by dyes and carcinogens. *1st Intern. Photobiol. Congr.* Amsterdam.

ALEXANDER, P. and STACY, K. A. (1959). Cross-linking of deoxyribonucleic acid in sperm heads by ionizing radiations. *Nature,* **184,** 958.

ALLEN, A. J., STEIGER, R. E., MAGILL, M. A. and FRANKLIN, R. G. (1937). Amino Acids, Acyl Amino Acids, Dipeptides, Acyl Dipeptides and Derivatives of these Compounds. II. *Biochem. J.* **31,** 195.

ALPER, T. GILLIES, N. E. and ELKIND, M. M. (1960). The signoidal survival curve in radiobiology. *Nature,* **186,** 1062.

AMESZ, J. DUYSENS, L. N. M. and BRANDT, D. C. (1961). Methods for measuring and correcting the absorption spectrum of scattering suspensions. *J. Theor. Biol.* **1,** 59.

ANDERSON, E. H. (1951). Heat reactivation of ultraviolet-inactivated bacteria. *J. Bacteriol.* **61,** 389.

ANDERSON, T. F. (1945). On a bacteriolytic substance associated with a purified bacterial virus. *J. Cell. Comp. Physiol.* **25,** 1.

ANDERSON, W. T. (1951). Xenon compact arc lamps. *J. Opt. Soc. Amer.* **41,** 385.

ANDERSON, W., HAYES, D., MICHELSON, A. M. and TODD, A. R. (1954). Deoxyribonucleosides and related compounds. IV. The configuration of the glycosidic centre in deoxyadenosine and deoxycytidine. *J. Chem. Soc.,* 1882.

ANDREWS, C. H., ELFORD, W. J. and NIVEN, J. S. F. (1948). Vaccinia and electromelia in the mouse. *Brit. J. Exp. Pathol.* **29,** 329.

ANFINSEN, C. B. (1957). Structural basis of ribonuclease activity. *Federation Proceedings,* **16,** 783.

ANFINSEN, C. B., HARRINGTON, W. F., HVIDT, A., KINDERSTROM-LANG, K., OTTESEN, M. and SCHELLMAN, J. (1955). Studies on the structural basis of ribonuclease activity. *Biochim. Biophys. Acta,* **17,** 141.

APPLEYARD, R. K. (1952). The irradiation of dried hemoglobins by fast charged particles. II. *Arch. Biochem. Biophys.* **40,** 111.

APRISON, M. H. (1958). Effect of ultraviolet radiation on serum oxidase activity. *Arch. Biochem. Biophys.* **78,** 260.

ARNOLD, W. and OPPENHEIMER, J. R. (1950). Internal conversion in the photosynthetic mechanism of blue-green algae. *J. Gen. Physiol.* **33,** 423.

Arnow, L. E. (1937). The formation of dopa by the exposure of tyrosine solutions to ultraviolet radiation. *J. Biol. Chem.* **120,** 151.

Arnow, L. E. (1938). On the kinetics of tyrosine destruction and dopa formation by ultraviolet rays. *J. Phys. Chem.* **42,** 415.

Arnow, L. E. (1942). Destruction of phenylalanine by ultraviolet radiant energy. *Proc. Soc. Exptl. Biol. Med.* **49,** 578.

Atwood, K. C. and Norman, A. (1949). On the interpretation of multi-hit survival curves. *Proc. Nat. Acad. Sci.* **35,** 696.

Augenstein, L. (1959). Structural correlation between esterase and protease activities of trypsin. *Science,* **129,** 718.

Augenstein, L. G. and Drist, K. L. (1962). On the nature of enzyme inactivation by radiation in *Biological Effects of Ionizing Radiation at the Molecular Level.* Intl. Atomic Energy Agency, Vienna, pp. 263–276.

Augenstein, L. G. and Ghiron, C. A. (1961). The inactivation of trypsin by ultraviolet light. *Proc. Nat. Acad. Sci.* **47,** 1530.

Augenstein, L. G. and Ray, B. R. (1957). Trypsin monolayers at the air–water interface. II. Effect of X-ray and ultraviolet radiation upon enzymatic activity. *J. Phys. Chem.* **61,** 1380.

Bahadur, K. (1954). Photosynthesis of amino acids from formaldehyde and potassium nitrate. *Nature,* **173,** 1141.

Bahadur, K. (1957). The reactions involved in the formation of compounds preliminary to the synthesis of protoplasm and other materials of biological importance, in 1*st Intern. Symp. on Origin of Life on the Earth,* Moscow, 1957; Academy of Sciences, U.S.S.R., 1957; Pergamon Press, London, 1959.

Bahadur, K., Ranganayaki, S. and Santamaria, L. (1958). Photosynthesis of amino acids from paraformaldehyde involving the fixation of nitrogen in the presence of colloidal molybdenum oxide as catalyst. *Nature,* **182,** 168.

Balazs, E. A., Laurent, T. C., Howe, A. F. and Varga, L. (1959). Irradiation of mucopolysacharides with ultraviolet light and electrons. *Rad. Res.* **11,** 149.

Balls, A. K. and Jansen, E. F. (1952). Stoichiometric inhibition of chymotrypsin. *Advances in Enzymol.* **13,** 321.

Baly, E. C. C., Heibron, I. M. and Hudson, D. P. (1922). Photocatalysis. II. Photosynthesis of nitrogen compounds from nitrates and carbon dioxide. *J. Chem. Soc.* **121,** 1078.

Baranowska, J. and Shugar, D. (1960). Photochemistry of model oligo- and polynucleotides. III. Cross-linking and staining properties of UV irradiated films of nucleic acids and oligonucleotides. *Acta Biochim. Polon.* **7,** 505.

Baricelli, N. A. (1956). A "chromosomic" recombination theory for multiplicity-reactivation in phages. *Acta Biotheoret.* **11,** 107.

Barkdoll, A. E. and Ross, W. F. (1944). Synthesis of tyrosyltyrosyltyrosine and tyrosyltyrosyltyrosyltyrosine. *J. Amer. Chem. Soc.* **66,** 951.

Barker, S. A., Grant, P. M., Stacey, M. and Ward, R. B. (1959). Effects of γ-radiation. I. Polymer formation from sugars, hydroxy acids and amino acids. *J. Chem. Soc.,* 2648.

Barrett, J. and Baxendale, J. H. (1960). Photolysis of liquid water. *Trans. Faraday Soc.* **56,** 37.

Barrett, J. and Mansell, A. L. (1960). Ultraviolet absorption spectra of the molecules of H_2O, HDO and D_2O. *Nature,* **187,** 138.

Barricelli, N. A. (1960). An analytical approach to the problem of phage recombination and reproduction. I. Multiplicity reactivation and the nature of radiation damage. *Virology,* **11,** 99.

Barricelli, N. A. and Doermann, A. H. (1961). An analytical approach to the problems of phage recombination and reproduction. III. Cross reactivation. *Virology*, **13,** 460.

BARSZCZ, D. and SHUGAR, D. (1961). Radiation chemistry of nucleic acids. I. Some pyrimidines, dihdropyrimidines and hydrated pyrimidines. *Acta Biochim. Polon.* **8,** 455.

BASS, A. M. (1948). Short wavelength cut-off filters for the ultraviolet. *J. Opt. Soc. Amer.* **38,** 977.

BATT, R. D. MARTIN, J. K., PLOESER, J. MCT. and MURRAY, J. (1954). Chemistry of the dihydropyrimidines: ultraviolet spectra and alkaline decomposition. *J. Amer. Chem. Soc.* **76,** 3663.

BATTISTO, J. R., PRINGLE, R. B., and NUNGESTER, W. J. (1953). The effect of UV irradiation on Immune serum. *J. Infect. Dis.* **92,** 85.

BAUM, W. A. and DUNKELMAN, L. (1950). Ultraviolet radiation of the high pressure xenon arc. *J. Opt. Soc. Amer.* **40,** 782.

BAWDEN, F. C. and KLECZKOWSKI, A. (1953). The behavior of some plant viruses after exposure to ultraviolet radiation. *J. Gen. Microbiology*, **8,** 145.

BAWDEN, F. C. and KLECZKOWSKI, A. (1955). Studies on the ability of light to counteract the inactivating action of ultraviolet radiation on plant viruses. *J. Gen. Microbiol.* **13,** 370.

BAWDEN, F. C. and KLECZKOWSKI, A. (1959). Photoreactivation of nucleic acid from tobacco mosaic virus. *Nature*, **183,** 503.

BAWDEN, F. C. and KLECZKOWSKI, A. (1960). Some effects of ultraviolet radiation on the infection of *Nicotiniana glutinosa* leaves by tobacco mosaic virus. *Virology*, **10,** 163.

BAXENDALE, J. H. and BRIDGE, N. K. (1955). The photoreduction of some ferric compounds in aqueous solution. *J. Phys. Chem.* **59,** 783.

BAYZER, H. and SCHAUENSTEIN, E. (1955). Uber die quantitative berucksichtung der tyndallsabsorption im UV-absorptionspektrum von proteinen. *Mikrochimica Acta*, 490; *J. polymer Sci.* **16,** 45.

BEARDMORE, J. B., HOOK, A. E., SARBER, R. W., MCLEAN, I. W. and TAYLOR, A. R. (1957). Poliomyelitis vaccine safety testing by the tissue culture method. *J. Immunol.* **79,** 489.

BEAVEN, G. H. and HOLIDAY, E. R. (1950). Some practical aspects of the effect of stray radiation in spectrophotometry. *Photoelectric Spectrophotometry Group, Bull. No.* 3.

BEAVEN, G. H. and HOLIDAY, E. R. (1952). Ultraviolet absorption spectra of proteins and amino acids. *Adv. Protein Chem.* **7,** 320.

BEAVEN, G. H., HOLIDAY, E. R. and JOHNSON, E. A. (1955). Optical properties of nucleic acids and their components, in *The Nucleic Acids* (CHARGAFF, E. and DAVIDSON, J. N., eds.), Academic Press, New York, vol. I, chap. 14.

BEAVEN, G. H., HOLIDAY, E. R. and JOPE, E. M. (1950). The ultraviolet absorption spectra of the aromatic amino acids in proteins and related compounds. *Dis. Faraday Soc.* **9,** 406.

BECKER, J. P. (1933). Photochemical changes of hydroxproline and proline caused by X- and ultraviolet radiation. *Strahlentherapie*, **48,** 296.

BECKER, J. P. (1935). Ein vergleich der photochemischen veranderungen von l-Asparaginsaure, l-Asparagin, und Glyzylglyzin durch rontgenstrahlen und ultraviolettes licht. *Strahlentherapie*, **52,** 531.

BECKER, J. P. and SZENDRÖ, P. (1931). Spectroanalytische und chemische untersuchungen uber den abbau von eiweiss durch ultraviolette strahlen. *Arch. ges. Physiol.* **228,** 755.

BEERS, R. F., JR. and STEINER, R. F. (1957). Titration and spectrophotometric studies upon polyadenylic acid. *Nature*, **179,** 1076.

BEERS, R. F., JR. and STEINER, R. F. (1958). Spectrophotometric, light scattering and titration studies upon the interaction of polyadenylic and polyuridylic acids. *Nature*, **181,** 30.

BEESE, N. C. (1939). The response of several fluorescent materials to short wavelength ultraviolet radiation. *J. Opt. Soc. Amer.* **29,** 278.

BEISER, S. M. and ELLISON, S. A. (1958). The effect of ultraviolet irradiation on pneumococcus transforming factor. *7th Intern. Congr. Microbiol.*, Stockholm, communication No. 4f.

BELITSER, V. O. and KOTKOVA, K. I. (1960). Photo-oxidation of fibrinogen and of fibrin monmer. *Ukrain. Biokhim. Zhur.* **32,** 3.

BELLIN, J. S., MOHOS, S. C. and OSTER, G. (1961). Dye-sensitized photoinactivation of tumour cells *in vitro*. *Cancer Research*, **21,** 1365–1371.

BELLIN, J. S. and OSTER, G. (1960a). Mechanism of photodynamic action as illustrated by *in vitro* systems. *3rd Intern. Photobiol. Congr.*, Copenhagen.

BELLIN, J. S. and OSTER, G. (1960b). Photodynamic inactivation of transforming principle. *Biochim. Biophys. Acta*, **42,** 533.

BENDICH, A. (1955). Chemistry of purines and pyrimidines, in *The Nucleic Acids* (Chargaff, E. and Davidson, J. N., eds.), Academic Press, New York. vol. I, chap. 3.

BENDICH, A. and CLEMENTS, G. C. (1953). A revision of the structural formulation of vicine and its pyrimidine aglucone, divicine. *Biochim. Biophys. Acta*, **12,** 462.

BENESCH, R. and BENESCH, R. E. (1955). The acid strength of the —SH group in cysteine and related compounds. *J. Amer. Chem. Soc.* **77,** 5877.

BENZER, S. (1957). The elementary units of heredity, in *The Chemical Basis of Heredity* (MCELROY, W. D. and GLASS, B., eds.), Johns Hopkins Press, Baltimore.

BERGMANN, E. and SAMUEL, R. (1941). On the mechanism of the primary photodissociation of organic molecules. *J. Org. Chem.* **6,** 1.

BERGMANN, W. and BURKE, D. C. (1955). Marine products XXXIX. The nucleosides of sponges. III. Spongothymidine and spongouridine. *J. Org. Chem.* **20,** 1501.

BERL, W. G. and FEAZEL, C. E. (1954). Sugar analysis: ultraviolet spectra of sugars in alkaline solution. *Agr. Food Chem.* **2,** 37.

BERNAL, J. D. (1951). *The Physical Basis of Life.* Routledge & Kegan Paul, London.

BERNAL, J. D. (1957). Stages in biopoiesis, in *1st Intern. Symp. on Origin of Life on the Earth*, Moscow, 1957; Academy of Sciences U.S.S.R., 1957; Pergamon Press, London, 1959, pp. 23–35.

BERNAL, J. D. (1960). Thermodynamics and kinetics of spontaneous generation. *Nature*, **186,** 694.

BERNAL, J. D. (1961). Significance of carbonaceous meteorites in theories on the origin of life. *Nature*, **190,** 129–131.

BERNHART, F. W. (1939). Denaturation and molecular splitting of egg albumin by ultraviolet radiant energy. *J. Biol. Chem.* **128,** 289.

BERNSTEIN, A. (1957). Multiplicity reactivation of ultraviolet-irradiated Vi-phage II of *Salmonella typhi*. *Virology*, **3,** 286.

BERTHELOT, A. and AMOREUX, G. (1938). Formation of indole-3-acetic acid by the action of ultraviolet light on tryptophan. *Compt. rend.* **206,** 699.

BEUKERS, R. and BERENDS, W. (1960a). The effect of paramagnetic substances on the conversion of some pyrimidines by ultraviolet irradiation. *Biochim. Biophys. Acta*, **38,** 573.

BEUKERS, R. and BERENDS, W. (1960b). Isolation and identification of the irradiation product of thymine. *Biochim. Biophys. Acta*, **41,** 550.

BEUKERS, R. and BERENDS, W. (1961). The effects of UV-irradiation on nucleic acids and their components. *Biochim. Biophys. Acta*, **49,** 181.

Beukers, R., Ijlstra, J. and Berends, W. (1958). The effect of ultraviolet light on some components of the nucleic acids. II. In rapidly frozen solutions. *Rec. trav. chim.* **77**, 729.

Beukers, R., Ijlstra, J. and Berends, W. (1959a). The effect of ultraviolet light on some components of the nucleic acids. III. Apurinic acid. *Rec. trav. chim.* **78**, 247.

Beukers, R., Ijlstra, J. and Berends, W. (1959b). The effect of ultraviolet light on some components of the nucleic acids. IV. The influence of oxygen on the extent of the first irreversible reaction, shown by uracil and orotic acid. *Rec. trav. chim.* **78**, 879.

Beukers, R., Ijlstra, J. and Berends, W. (1959c). The effect of ultraviolet light on some components of the nucleic acids. V. Reversibility of the "first irreversible reaction" under special conditions. *Rec. trav. chim.* **78**, 883.

Beukers, R., Ijlstra, J. and Berends, W. (1960). The effect of ultraviolet light on some components of the nucleic acids. VI. The origin of the ultraviolet sensitivity on DNA. *Rec. trav. chim.* **79**, 101.

Bigelow, C. C. (1959). Difference spectra of ribonuclease and two ribonuclease derivatives. *Compt. rend. Lab. Carlsberg*, **31**, 305.

Bitny-Szlachto, S. and Shugar, D. (1958). Quantitative staining with radioactive indicators. Preparation of ^{35}S-labelled thizine dyes. *Bull. Acad. Polon. Sci.*, Cl. II, **6**, 129.

Bitny-Szlachto, S. and Shugar, D. (1959). Quantitative staining with radioactive indicators. Preparation of ^{14}C-labelled crystal violet and methyl green. *Bull. Acad. Polon. Sci.*, Cl. II, **7**, 293.

Blix, U., Iland, C. N. and Stacey, M. (1954). The serological activity of deoxyribonucleic acids. *Brit. J. Exp. Path.* **35**, 241.

Blout, E. R. and Assadourian, A. (1954). The ultraviolet absorption spectra of desoxypentose nucleic acid. *Biochim. Biophys. Acta*, **13**, 161.

Blum, H. F. (1941). *Photodynamic Action and Diseases Caused by Light.* Reinhold, New York.

Blum, H. F. (1950). Action spectra and absorption spectra, in *Biophysical Research Methods* (Uber, F. M., ed.), Interscience, New York. pp. 417–450.

Blumenfeld, O. O. and Levy, M. (1958). Spectrophotometric and stoichiometric titrations of ribonuclease in eight molar urea. *Arch. Biochem. Biophys.* **76**, 97.

Bogle, D. S., Burgess, V. R., Forbes, W. F. and Savige, W. E. (1962). Photolysis and Photo-oxidation of amino acids and peptides. V. *Photochem. Photobiol.* **1**, 277–290.

Bohr, N. (1932). Light and life. *2nd Intern. Congr. on Light*, Copenhagen.

Bollum, F. J. and Setlow, R. B. (1963). Ultraviolet inactivation of DNA primer-activity. I. Effects of different wavelengths and doses. *Biochim. Biophys. Acta*, in press.

Borthwick, H. A., Henricks, S. B., Parker, M. W., Tooke, E. H. and Toole, V. (1952). Reversible photoreaction controlling seed germination. *Proc. Nat. Acad. Sci. U.S.* **38**, 662.

Bothner-By, C. T. and Balazs, E. A. (1957). Effect of ionizing radiations on glucose solutions. *Rad. Res.* **6**, 302.

Bourne, H. K. (1948). *Discharge Lamps for Photography and Projection.* Chapman & Hall, London.

Bovey, F. A. and Yanari, S. S. (1958). Disulfide bonds of pepsin in relation to physical properties and enzymatic activity. Abstracts of papers, 134th Meeting of American Chemical Society, Chicago.

Bowen, E. J. (1946). *The Chemical Aspects of Light.* Oxford, London.

BOWEN, E. J. (1947). The absorption of light by chemical compounds. Fourth John Mercer lecture. *J. Soc. Dyers and Colourists*, **63,** 281.

BOWEN, E. J. (1950). Light absorption and photochemistry. *Quart. Rev.* **4,** 236.

BOYCE, R. and SETLOW, R. (1963). The action spectrum for ultraviolet light inactivation of systems containing 5-bromouracil-substituted DNA. *Biochim. Biophys. Acta*, **68,** 455–461.

BRAUDE, E. A. (1945). Intensities of light absorption. *Nature*, **155,** 753.

BRAUDE, E. A. (1949). Studies in light absorption. Part VIII. Dibenzyl and stilbene derivatives. Interaction between unconjugated chromophores. *J. Chem. Soc.*, 1902.

BRAUDE, E. A. (1955). Ultraviolet and visible light absorption, in *Determination of Organic Structures by Physical Methods* (Braude, E. A. and Nachod, F. D., eds.), Academic Press, New York. chap. 4.

BRAUDE, E. A., SONDHEIMER, F. and FORBES, W. F. (1954). Steric effects in the electronic spectra of organic compounds. *Nature*, **173,** 117.

BRAUN, W. (1953). *Bacterial Genetics.* W. B. Saunders, Philadelphia.

BRAUNER, L. (1952). The pH changes in the photolysis of heteroauxin. *Naturwiss.* **39,** 282.

BRESCH, C. Z. (1950). ZUR reaktivierung von bakteriophagen. *Z. Naturforsch.* **5b,** 420.

BRIDGE, N. K. and PORTER, G. (1958). Primary photoprocesses in quinones and dyes. II. Kinetic studies. *Proc. Roy. Soc.* (*London*) **A244,** 276.

BRIGHENTI, L. and FALASCHI, A. (1962). Effect of radiation on heart lactic dehydrogenase and pancreas ribonuclease and protective action of a sulfanilamide disulfide. *Biochim. Biophys. Acta*, **59,** 376–388.

BRODE, W. R. (1939). *Chemical Spectroscopy.* John Wiley, New York.

BRODSKIJ, A. (1952). *Chimia Izotopow.* U.S.S.R. Academy of Sciences, Moscow.

BROUWER, D. M. (1957). Ondersoekingen in verband met de rol van een imidazoolgroep bij de werking van chymotrypsine. Thesis, University of Leiden.

BROWN, D. M., COHRAN, W., MEDLIN, E. and VARADARAJAN, S. (1956). Nucleotides. XXXIX. 5′deoxy-5′-iodo-O^2:2′cyclourindine: an X-ray study. *J. Chem. Soc.*, 4873.

BROWN, D. M., TODD, A. R. and VARADARAJAN, S. (1956). Nucleotides. XXXVII. The structure of uridylic acids *a* and *b* and a synthesis of spongouridine (3-β-D-arabofuranosyluracil). *J. Chem. Soc.*, 2388.

BUCHER and KASPER. (1947). Photochemische spaltung des kohlen oxydmyoglobins durch ultraviolette strahlung. *Biochim. Biophys. Acta*, **1,** 21.

BUELL, M. V. and HANSEN, R. E. (1959). Photoelectric technique for spectrophotometry between 194 and 225 mμ. *Anal. Chem.* **31,** 878.

BULL, H. B. (1951). *Physical Biochemistry.* John Wiley, New York. chap. 10.

BURAWAY, A. and ROACH, A. G. (1958). Color filters for the 4358 Å line of the mercury discharge spectrum. *Nature*, **181,** 762.

BURNETT, F. M. (1933). The classification of dysentry-coli bacteriphages. III. A correlation of the serological classification with certain biochemical tests. *J. Pathol. Bacteriol.* **37,** 179.

BURNETT, F. M., KEOGH, E. V. and LUSH, D. (1937). Immunological reactions of filtrable viruses. *Australian J. Exp. Biol. Med. Sci.* **15,** 227.

BURTON, M. (1948). Radiation chemistry: effect of molecular size. *J. Phys. Colloid Chem.* **52,** 810.

BUTENANDT, A., FRIEDRICH-FREKSA, H., HARTWIG, S. and SCHEIBE, G. (1942). Beitrag un feinstrukturen des tabakmosaikvirus. *Z. Physiol. Chem.* **274,** 276.

BUTLER, J. A. V. (1955). Effect of ultraviolet on nucleic acids and nucleoproteins and other biological systems. *Experientia*, **11,** 289.

BUTLER, J. A. V. and CONWAY, B. E. (1900). The action of photochemically generated radicals from H_2O_2 on deoxyribonucleic acid and simple model substances. *Proc. Roy. Soc.*, **B141,** 562.

BUTLER, J. A. V., GILBERT, V. and JAMES. D. W. F. (1952). The action of ionizing radiations and of radiomimetric substances on DNA. Part VI. Physicochemical measurements of the action of bis-chloroethylmethylamine. *J. Chem. Soc.*, 3268.

BUTTOLPH, L. S. (1955). Practical applications and sources of ultraviolet energy, in *Radiation Biology* (Hollaender, A., ed.), McGraw-Hill, New York, vol. 2, chap. 2.

BUZZELL, A. (1956). The interrelationships between ultraviolet irradiation and the thermal reactivation of the *Bacterium E. coli* B. *Arch. Biochem. Biophys.* **62,** 97.

BUZZELL, A., TRKULA, D. T. and LAUFFER, M. A. (1959). Action of ultraviolet light on ribose nucleic acid in tobacco mosaic virus. *Proc. First Nat. Biophysics Conference, Columbus,* 1957, New Haven, Yale University Press. pp.254–255.

CALTERA, I. R. and FERRARI, G. (1959). Photochemical synthesis of amino acids from inorganic nitrogen compounds and ternary organic compounds. *Ann. chim. (Rome)*, **49,** 1639.

CALVERT, J. G. and RECHEN, H. J. L. (1952). Precision actinometry at low light intensities with malachite green leucocyanide. *J. Amer. Chem. Soc.* **73,** 2101.

CALVIN, M. (1959). Round trip from space. *Evolution*, **13,** 362.

CALVIN, M. (1962). The origin of life on earth and elsewhere. *Ad. Biol. Med. Phys.*, **8,** 315–342.

CALVIN, M. and VAUGHN, S. K. (1960). Extraterrestrial life; some organic constituents of meteorites and their significance for possible extraterrestrail biological evolution. *1st Intern. Space Science Symp.*, Nice, France.

CAMPBELL-RENTON, M. L. (1937). Radiation of bacteria with ultraviolet light. *J. Pathol. Bacteriol.* **45,** 237.

CANNON, C. V. and RICE, O. K. (1942). A monochromator using a large water prism. *Rev. Scient. Inst.* **13,** 513.

CANZANELLI, A., GUILD, R. and RAPPORT, D. (1951). Ammonia and urea production and changes in absorption spectra of nucleic acid derivatives following ultraviolet irradiation. *Amer. J. Physiol.* **167,** 364.

CARPENTER, D. C. (1940). Splitting the CONH linkage by means of ultraviolet light. *J. Amer. Chem. Soc.* **62,** 289.

CARPENTER, D. C. (1941). Seminar on the irradiation of proteins. *J. Franklin Inst.* **232,** 76.

CARTER, C. E. (1950). The decomposition of DPN and ATP by ultraviolet light. *J. Amer. Chem. Soc.* **72,** 1835.

CASPAR, D. L. (1956). Radial density distribution in the tobacco mosaic virus particle. *Nature*, **177,** 928.

CASTILLE, A. and RUPPOL, M. (1928). Spectres ultraviolets de quelques produits biologiques purs et melangés avec des alkaloides. *Bull. acad. roy. med. Belge* (5), **6,** 263.

CAVALIERI, I. F. (1953). Studies on the structure of nucleic acids. VII. On the identification of the isomeric cytidylic and adenylic acids. *J. Amer. Chem. Soc.* **75,** 5268.

CAVALIERI, L. F. and BENDICH, A. (1950). Ultraviolet absorption spectra of pyrimidines and purines. *J. Amer. Chem. Soc.* **72,** 2587.

CAVALIERI, L. F., BENDICH, A., TINKER, J. F., and BROWN, G. B. (1948). Ultraviolet absorption spectra of purines, pyrimidines and triazolopyrimidines. *J. Amer. Chem. Soc.* **70,** 3875.

CAVALIERI, L. F. and STONE, A. L. (1955). Studies on the structure of nucleic acids.

VIII. Apparent dissociation constants of desoxypentose nucleic acid. *J. Amer. Chem. Soc.* **77,** 6499.

CHEESMAN, D. F. and DAVIES, J. T. (1954). Physiochemical and biological aspects of proteins at interfaces. *Adv. Protein Chem.* **9,** 439.

CHA, C. Y. and SCHEREGA, H. A. (1960). Structural studies of ribonuclease. I. Hydrogen ion equilibria in a denaturing solvent. *J. Amer. Chem. Soc.* **82,** 54.

CHARGAFF, E. and ZAMENHOF, S. (1948). The isolation of highly polymerized desoxyribonucleic acid from yeast cells. *J. Biol. Chem.* **173,** 327.

CHERKASOV, A. S. and VEMBER, T. M. (1959). Effect of oxygen on the photochemical transformation and the concentration quenching of fluorescence of some anthracene derivatives. *Optike i Spektroskopiya,* **6,** 503.

CHESSIN, M. (1958). Light quality and photoreactivation of plants and viruses. *Ann. Applied Biol.* **46,** 388.

CHESSIN, M. (1960). Photodynamic inactivation of infectious nucleic acid. *Science,* **132,** 1840.

CHILAS, C. B. (1961). Broad-band ultraviolet filters. *J. Opt. Soc. Am.,* **51,** 895.

CHRISTENSEN, E. and GIESE, A. C. (1954). Changes in absorption spectra of nucleic acids and their derivatives following exposure to ozone and ultraviolet irradiation. *Arch. Biochem. Biophys.* **51,** 208.

CHRISTIE, M. I. and PORTER, G. (1952). Photochemical studies with the gaseous flash discharge. Uranyloxalate actinometry at high intensities. *Proc. Roy. Soc.,* **A, 212,** 390.

CLAESSON, I. M. (1956). The effect of ultraviolet light (λ 2537 Å) on *Helix pomatia* hemocyanin and bovine serum albumin as studied by changes in the ultraviolet spectrum and sedimentation behaviour. *Arkiv for Kemi,* **10,** 1.

CLAESSON, S., JUHLIN, L. and WETTERMARK, G. (1958). The reciprocity law of ultraviolet irradiation effects. *Acta Dermato-Venereologica,* **38,** 123.

CLAESSON, S. and LINDQVIST, L. (1957a). A flash photolysis apparatus for photochemical studies at very high light intensities and energies. *Arkiv for Kemi,* **11,** 535.

CLAESSON, S. and LINDQVIST, L. (1957b). A fast photolysis flash lamp for very high light intensities. *Arkiv for Kemi,* **12,** 1.

CLAESSON, S. and WETTERMARK, G. (1957). The spectral distribution of oxygen-filled photolysis flash lamps as determined by chemical actinometry. *Arkiv for Kemi,* **11,** 561.

CLARK, J. H. (1936). The effects of radiation on proteins, in *Biological Effects of Radiation* (Duggan, B. ed.), McGraw-Hill, New York. pp. 303–322.

CLARKE, V. M., TODD, A. R., and ZUSSMAN, J. (1951). Nucleotides. VIII. Cyclonucleoside salts. A novel rearrangement of some toluene-*p*-sulphonylnucleosides. *J. Chem. Soc.,* 2952.

COHN, W. E. (1955). The separation of nucleic acid derivatives by chromatography on ion-exchange columns, in *The Nucleic Acids: Chemistry and Biology* (Chargaff, E. and Davidson, J. N. eds.), Academic Press, New York, vol. I.

COLLIER, L. H., McLean, D. and VALLET, H. (1955). The antigenicity of ultraviolet irradiated vaccinia virus. *J. Hygiene,* **53,** 513.

COMMONER, B. (1949). On the interpretation of the absorption of UV light by cellular nucleic acids. *Science,* **110,** 131.

COMMONER, B. HEISE, J. J., LIPINCOTT, B. B., NORBERG, R. E., PASSONNEAN, J. V. and TOWNSEND, J. (1957). Biological activity of free radicals. *Science,* **126,** 57.

COMMONER, B. and LIPKIN, D. (1949). The application of the Beer–Lambert law to optically anisotropic systems. *Science,* **110,** 41.

CONRAD, W. E. (1954). The isolation of oxamide and parabanic acid from the products of ultraviolet irradiation of uracil. *Rad. Res.* **1,** 523.

COULTER, C. B., STONE, F. M. and KABAT, E. A. (1936). Structure of the ultraviolet absorption spectra of certain proteins and amino acids. *J. Gen. Physiol.* **19,** 739.

COWGILL, R. W. (1963). Fluorescence and the structure of Proteins. I. Indole and phenol compounds. *Archives Biochem. Biophys.* **100,** 36.

CRAMMER, J. L. and NEUBERGER, A. (1943). The state of tyrosine in egg albumin and insulin as determined by spectrophotometric titration. *Biochem. J.* **37,** 302.

CRAMPTON, C. F. (1957). Studies on the dissociation of histones from the nucleohistone of calf thymus. *J. Biol. Chem.* **227,** 495.

CRAMPTON, C. F., LIPSHITZ, R. and CHARGAFF, E. (1954). Studies on nucleoproteins. I. Dissociation and reassociation of the desoxyribonucleohistone of calf thymus. *J. Biol. Chem.* **206,** 499.

CRAMPTON, C. F., STEIN, W. H., and MOORE, S. (1957). Comparative studies on chromatographically purified histones. *J. Biol. Chem.* **225,** 363.

CRICK, F. H. C. (1963). In *Progress in nucleic acid research* (J. N. DAVIDSON and W. E. COHN, eds.), Academic Press, N.Y.

CRICK, F. H. C. and WATSON, J. D. (1954). The complementary structure of desoxyribonucleic acid. *Proc. Roy. Soc.* **A 223,** 80.

DACEY, J. R. and HODGINS, J. W. (1950). Mercury 1850 Å and Xenon 1470 Å resonance lamps as ultraviolet sources for photochemical studies. *Can. J. Research B* **28,** 90.

DANIELLI, J. F. and DAVIES, J. T. (1951). Reaction at interfaces in relation to biological problems. *Adv. Enzymology,* **11,** 35.

DANIELS, M. and GRIMISON, A. (1963). Photochemistry of thymine. *Nature,* **197,** 484.

DANNENBERG, H. (1951). Uber die ultraviolettabsorption der proteine. *Angew. Chem.* **63,** 208.

DAUVILLIER, A. (1958). *L'origine Photochimique de la vie.* Masson et Cie., Paris.

DAVIES, D. R. and RICH, A. (1958). The formation of a helical complex between polyinosinic acid and polycytidylic acid. *J. Amer. Chem. Soc.* **80,** 1003.

DAVISON, P. F. and BUTLER, J. A. V. (1956). The chemical composition of calf thymus nucleoprotein. *Biochim. Biophys. Acta,* **21,** 568.

DEBYE, P. and EWARDS, J. O. (1952a). A note on the phosphorescence of proteins. *Science,* **116,** 143.

DEBYE, P. and EDWARDS, J. O. (1952b). A note on the phosphorescence of proteins. *J. Chem. Phys.* **20,** 236.

DECOURCY, S. J., ELY, J. O. and ROSS, M. H. (1953). Effect of ultraviolet light on deoxyribonucleic acid in rat thymocytes. *Nature,* **172,** 119.

DEDEKEN, R. H., BROEKUYSEN, J., BECHET, J and MORTIER, A. (1956). Étude spectrophotometrique de la dissociation de la fonction sulphydrile et structure moléculaire de la cysteine. *Biochim. Biophys. Acta,* **19,** 45.

DEERING, R. A. and SETLOW, R. (1963). Effects of ultraviolet light on thymidine dinucleotide and polynucleotide. *Biochim. Biophys. Acta,* in press.

DEGARILHE, M. P. and LASKOWSKI, M. (1956). Optical changes occurring during the action of phosphodiesterase on oligonucleotides derived from desoxyribonucleic acid. *J. Biol. Chem.* **223,** 661.

DEKKER, C. A. (1960). Nucleic Acids: Selected topics related to their enzymology and chemistry. *Ann. Rev. Biochem.,* **29,** 453–474.

DELBRUCK, M. (1950). *Viruses 1950.* California Institute of Technology Press, Pasadena.

DELBRUCK, M. and BAILEY, W. T. (1946). Induced mutations in bacterial viruses. *Cold Spring Harb. Symp. Quant. Biol.* **11,** 33.

DELIKAS, N. (1961). The ability of irradiated bacteriophage T2 to initiate the synthesis of deoxycytidylate hydroxymethylase in *E. coli*. *Virology*, **13,** 242.

DELZENNE, G. (1959). Les sensibilisateurs de photopolymerization. *Ind Chim. Belge*, **24,** 739.

DENFFER, D. V. (1954). Photoperiodismus und wuchsstoffe. *1st Intern. Congress Photobiol.*, Amsterdam.

DENFFER, D. V. and FISCHER, A. (1952). Paper chromatographic determination of 3-indole-carboxaldehyde in photolytically decomposed 3-indole-acetic acid solution. *Naturwiss.* **39,** 549.

DESCHREIDER, A. R. (1952). The action of ultraviolet rays on flour. VI. The effect on L-cysteine. *Bull. ecole meunerie belge*, **14,** 64. *Chem. Abs.* **46,** 11477c.

DESCHREIDER, A. R. (1958). Photosynthesis of amino acids. *Nature*, **182,** 528.

DESCHREIDER, A. R. (1959). Amino acid synthesis by ultraviolet irradiation (2537 Å) from a mixture of carboxylic acids and ammonium salts. *Rev. fermentations et inds. aliment.* **14,** 113.

DESCHREIDER, A. R. and RENARD, M. (1955). Degradation products of L-tryptophan by ultraviolet light. *Bull. inst. agron. et stas. recherches Gembloux*, **23,** 269. *Chem. Absts.* **50,** 3901f.

DESNUELLE, P. (1953). The general chemistry of amino acids and peptides, in *The Proteins* (Neurath, H. and Bailey, K., eds.), Academic Press, New York.

DEVOE, H. and TINOCO, I. (1962a). The hypochromism of helical polynucleotides: Base contributions. *J. Mol. Biol.*, **4,** 500–517.

DEVOE H. and TINOCO, I. (1962b). The hypochromism of helical polynucleotides. *J. Mol. Biol.*, **4,** 518–527.

DEVREUX, S. JOHANNSON, M. and ERRERA, M. (1951). Affinite du vert de methyle pour l'acide desoxyribonucleique. *Bull. soc. chim. biol.* **33,** 800.

DE VOS, J. C. (1954). The tungsten striplamp as a standard source of radiation *Proc. 1st Intern. Photobiol. Congr.*, Amsterdam, pp. 273–275.

DIXON, G. H., KAUFFMAN, D. and NEURATH, H. (1958). Amino acid sequence at the active center of trypsin. *Abstracts of 133rd meeting of Amer. Chem. Soc.*, San Francisco, April 13–18, 1958. p. 26C.

DODONVA, N. Y., SIDOROVA, A. I. and TERENIN, A. N. (1959). Photochemical synthesis under the action of the Schumann (short ultraviolet) radiations. *Vestnik Leningrad Univ.* **14,** No. 16, *Ser. Fiz i Khim.*, No. 3, 33.

DOERMANN, A. H. (1961). The analysis of ultraviolet lesions in bacteriophage T4 by cross reactivation. *J. Cellul. Comp. Physiol.* **58,** Suppl. 1, 79.

DOERMANN, A. H., CHASE, M. STAHL, F. W. (1955). Genetic recombination and replication in bacteriophage. *J. Cellul. Comp. Physiol.* **45,** 51.

DONOVAN, J. W., LASKOWSKI, M., JR. and SCHEREGA, H. A. (1958). Influence of ionization of carboxyl groups on the ultraviolet absorption spectrum of lysozyme. *Biochim. Biophys. Acta*, **29,** 455.

DOSE, K. and ETTRE, K. (1958). Radiochemical synthesis of amino acids and related compounds. *Z. Naturforsch.* **13b,** 784.

DOSE, K. and RAJEWSKY, B. (1957). Strahlenchemische bildung von aminen und aminocarbon sauren. *Biochim. Biophys. Acta*, **25,** 225.

DOSE, K. and RAJEWSKY, B. (1962). Photochemistry of sulphur containing amino acids and peptides. *Photochem. Photobiol.* **1,** 181–191.

DOTY, P. (1957). The physical chemistry of desoxyribonucleic acids. *J. Cell. Comp. Physiol.* **49,** Suppl. i, 27.

DOTY, P., BOEDTKER, H., FRESCO, J. R., HALL, B. D. and HASELKORN, R. (1959).

Configuration studies of polynucleotides and ribonucleic acids. *Ann. N.Y. Acad. Sci.* **81,** 693.

Doty, P., Boedtker, H., Fresco, J. R., Haselkorn, R. and Litt, M. (1959). Secondary structure in ribonucleic acids. *Proc. Nat. Acad. Sci.* **45,** 482.

Doty, P. and Edsall, J. R. (1951). Light scattering in protein solutions. *Ad. Prot. Chem.* **6,** 35.

Doty, P. and Geiduschek, E. P. (1953). Optical properties of proteins, in *The Proteins* (Neurath, H. and Bailey, K., eds.), Academic Press. vol. I, pp. 393–460.

Doty, P., Marmur, J. and Sueoka, N. (1959). The heterogeneity in properties and functioning of deoxyribonucleic acids, in *Structure and Function of Genetic Elements,* Brookhaven Symp. in Biol., No. 12, 1.

Doty, P., McGill, B. and Rice, S. A. (1958). The properties of sonic fragments of DNA. *Proc. Nat. Acad. Sci.* **44,** 432.

Doudney, C. O. and Hass, F. L. (1960). Some biochemical aspects of the post-irradiation modification of ultraviolet-induced mutation frequency in bacteria. *Genetics,* **45,** 1481.

Downes, A. and Blunt, T. P. (1879). On the influence of light upon protoplasm. *Proc. Roy. Soc.* IV **28,** 199.

Drake, J. W. (1963). Properties of UV-induced rII mutants of bacteriophage T4. *J. Mol. Biol.* **6,** 268–283.

Duchesne, J. and Garson, J. (1957). Photo-restauration en biologie. *J. Chim. Phys.* **54,** 789.

Duchesne, J. and Rosen, B. (1959). Sur le role de la conversion interne dans l'effet de photorestauration en biologie. *J. Chim. Phys.* **56,** 76.

Duggan, E. L., Giese, A. C. and Luck, J. M. (1956). The combinations of organic anions with serum albumin. *Arch. Biochem.* **28,** 174.

Dulbecco, R. (1950). Experiments on photoreactivation of bacteriophages inactivated with ultraviolet radiation. *J. Bact.* **59,** 329.

Dulbecco, R. (1952). A critical test of the recombination theory of multiplicity reactivation. *J. Bacteriol.* **63,** 199.

Dulbecco, R. (1955). Photoreactivation, in *Radiation Biology* (Hollaender, A. ed.), McGraw-Hill, New York. vol. II.

Dunkelman, L. (1957). Spectrally selective middle ultraviolet photodetectors. (soar blind detectors). *J. Opt. Soc. Amer.* **47,** 1048.

Duysens, L. N. M. (1956). The flattening of the absorption spectrum of suspensions, as compared to that of solutions. *Biochim. Biophys. Acta,* **19,** 1.

Eager, J. E. and Savige, W. E. (1963). Photolysis and Photo-oxidation of amino acids and peptides. VI. *Photochem. Photobiol.* **2,** 25–38.

Eckart, W. (1954). Ultraviolet inactivation of bacteriophage. *Strahlentherpie,* **94,** 60.

Edwards, B. G. (1949). Ultraviolet spectra of some indole drivatives, including tryptophane and gramicidin. *Arch. Biochem.* **21,** 103.

Egan, R., Michel, H. O., Schlueter, R. and Jandorf, B. J. (1957). On the mechanism of demerization of chymotrypsin. *Arch. Biochim. Biophys.* **66,** 366.

Ehrert, C. F. and Monk, G. S. (1955). Design and performance of the Argonne biological spectrograph. *Rad. Res.* **3,** 223.

Ekert, B. and Monier, R. (1958). Essais de photorestauration du diphosphopyridine nucleotide prealablement soumis aux rayons ultraviolets. *Bull. soc. chim. biol.* **40,** 793.

Ekert, B. and Monier, C. (1959). Structure of thymine hydroperoxide produced by X-irradiation. *Nature,* **184,** B.A. 58.

Ellenbogen, E. (1958). Abstract *Amer. Chem. Soc.* Meeting, Chicago, 1958, p. 47C.

Ellis, C., Wells, A. A. and Heyroth, F. F., eds. (1941). *The Chemical Action of Ultraviolet Rays.* Reinhold, New York.

Ellison, S. A. and Beiser, S. M. (1960). The ultraviolet inactivation of pneumococcus tranforming factor. *Rad. Res.* **12,** 381.

Ellison, S. A., Feiner, R. F. and Hill, R. F. (1960). A host effect on bacteriophage survival after ultraviolet irradiation. *Virology,* **11,** 294.

Ellmauer, H. and Kaplan, R. W. (1959). Auslosung von Klarplaque-Mutationen durch UV-Bestrahlung des freien phagen χ von *Serratia marcescens. Naturwiss.* **46,** 150.

Elmore, D. T. and Todd, A. R. (1952). Nucleotides. XVIII. The synthesis and properties of adenosine-5′-uridine-5′-phosphate. *J. Chem. Soc.,* 3681.

Elson, D. (1958). Evidence for hydrogen bonds in a ribonucleoprotein. *Biochim. Biophys. Acta,* **27,** 207.

Elson, D. (1959). Prepraration and properties of a ribonucleoprotein isolated from Escherichia coli. *Biochim. Biophys. Acta,* **36,** 362.

Engelhard, H. and Eikenberg, K. (1955). Uber die einwirkung von ultraviolettem licht auf serum-albumin. *Z. f. Naturfors.* **10b,** 622.

Ephrussi-Taylor, H. and Latarjet, R. (1955). Inactivation par les rayons X d'un facteur transformant du pneumocoque. *Biochim. Biophys. Acta,* **16,** 183.

Epstein, H. T. and Schiff, J. A. (1961). A simple and rapid method for estimating an action spectrum where nucleic acids (or, in principle, other materials) are the chromophores involved. *Exp. Cell Res.* **23,** 623–624.

Errera, M. (1952a). Étude photochimique de l'acide desoxyribonucleique. I. mesures energetiques. *Biochim. Biophys. Acta,* **8,** 30.

Errera, M. (1952b). Étude photochimique de l'acide desoxyribonucleique. II. Étude des produits de la photolyse. *Biochim. Biophys. Acta,* **8,** 115.

Errera, M. (1952c). Mecanismes de l'action des radiations sur le noyau cellulaire. *Ann. Soc. Royale. Sc. Med. et Natur. de Bruxelles,* **5,** 65.

Errera, M. (1953). Mechanisms of biological action of ultraviolet and visible radiations. *Progress in Biophysics,* **3,** 88.

Eschke, W., Eckart, W. and Wetzel, R. (1960). Uber den einfluss der inaktivierenden wellenlange der UV-strahlung auf die reaktivierung von bakteriophagen. *Naturwiss.* **47,** 521.

Estermann, E. F., Luse, R. A. and McLaren, A. D. (1956). Some observations on the photolysis of acelylanine at 2537 Å. *Rad. Res.* **5,** 1.

Estermann, E. F. and McLaren, A. D. (1962). Photochemistry of proteins XXIII. *Photochem. Photobiol.* **1,** 109.

Evans, J. C. and Ellman, G. L. (1959). The ionization of cysteine. *Biochim. Biophys. Acta,* **33,** 574.

Evenari, G. and Stein, G. (1953). The influence of light upon germination. *Experientia,* **9,** 94.

Evenari, M., Neumann, G. and Stein, G. (1953). Factors modifying the influence of light on germination. *Nature,* **172,** 452.

Falk, M., Hartman, K. A. and Lord, R. C. (1963). Hydration of DNA. III. A spectroscopic study of the effect of hydration on the structure of DNA. *J. Am. Chem. Soc.* **85,** 391–394.

Felix, K. (1953). *The Chemical Structure of Proteins.* J. A. Churchill, London.

Fellig, J. (1954). Production of triuret from uric acid by ultraviolet irradiation. *Science,* **119,** 129.

Felsenfeld, G., Davies, D. R. and Rich, A. (1957). Formation of a three-stranded polyribonucleotide molecule. *J. Amer. Chem. Soc.* **79,** 2023.

FELSENFELD, G. and RICH, A. (1957). Studies on the formation of two- and three-stranded polyribonucleotides. *Biochim. Biophys. Acta*, **26,** 457.

FERRARI, G. and CULTRERA, R. (1961). Photochemical synthesis of amino acids and a new transamination process by transfer of free amino radicals. *Nature*, **190,** 326.

FERRI, M. G. (1951). Fluorescence and photoinactivation of indole acetic acid. *Arch. Biochem. Biophys.* **31,** 127.

FERRI, R. G. and GUIDOLIN, R. (1951). Fluorescence and inactivation of snake poisons. *Science*, **113,** 300.

FEUGHELMAN, M., LANGRIDGE, R., SEEDS, W. E., STOKES, A. R., WILSON, H. R., HOOPER, C. W., WILKINS, M. H. F., BARCLAY, R. K. and HAMILTON, L. D. (1955). Molecular structure of desoxyribose nucleic acid and nucleoprotein. *Nature*, **175,** 834.

FIALA, S. (1949). Polarographic and spectrophotometric studies on photodynamic action by fluorescein dyes. *Biochem. Zeit.* **320,** 10.

FIKUS, M., WIERZCHOWSKI, K. L. and SHUGAR, D. (1962). Photochemistry of cytosine nucleosides and nucleotides. III. Mono- and dimethylamino analogues. *Photochem. and Photobiol.* **1,** 325–335.

FINKELSTEIN, P. (1949). Thesis, Polytechnic Institute of Brooklyn.

FINKELSTEIN, P. and MCLAREN, A. D. (1950). pH dependence of quantum yield and ultraviolet absorption spectrum of chymotrypsin. *J. Polymer Sci.* **4,** 573; **5,** 267.

FLEMMING, K. (1959). Mechanism of histamine formation in histidine solution irradiated with ultraviolet light. *Arch. exptl. Pathol. Pharmakol.* **236,** 27. *Chem. Absts.* **53,** 1035e.

FLUKE, D. J. (1956). The effect of low temperature and drying on the ultraviolet action spectrum of T1 bacteriophage. *Rad. Res.* **4,** 193.

FLUKE, D. J. and POLLARD, E. C. (1949). Ultraviolet inactivation spectrum of T1 bacteriophage. *Science*, **110,** 274.

FLUKE, D. J. and SETLOW, R. B. (1954). Water-prism ultraviolet monochromator. *J. Opt. Soc. Amer.* **44,** 327.

FORBES, G. S. and HEIDT, L. J. (1934). Optimum composition of uranyl oxalate solutions for actinometry. *J. Amer. Chem. Soc.* **56,** 2363.

FORBES, W. F., RIVETT, D. E. and SAVIGE, W. E. (1962). Photolysis and photo-oxidation of amino acids and peptides. IV. *Photochem. Photobiol.* **1,** 217–230.

FORBES, W. F. and SAVIGE, W. E. (1962). Photolysis and photo-oxidation of amino acids and peptides. I. *Photochem. Photobiol.* **1,** 1–14.

FORD, R. A. (1957). Photochemical processes in solids: the kinetics of free radical reactions in nylon. *J. Colloid Sci.* **12,** 271.

FÖRSTER, T. (1960). Transfer Mechanisms of electronic excitation energy. *Rad. Res.*, Suppl. 2, 326.

FOX, J. J., CAVALIERI, L. F. and CHANG, N. (1953). The identification of cytidylic acids *a* and *b* by spectrophotometric methods. *J. Amer. Chem. Soc.* **75,** 4315.

FOX, J. J. and LAMPEN, I. (1959). Pyrimidine nucleosides. *Advances Carbohydrate Chemistry*, **14,** 283.

FOX, J. J. and SHUGAR, D. (1952a). Absorption spectra and structure of barbituric acid derivatives as a function of pH. *Bull. soc. chim. belge*, **61,** 44.

FOX, J. J. and SHUGAR, D. (1952b). Spectrophotometric studies of nucleic acid derivatives and related compounds. II. Natural and synthetic pyrimidine nucleosides. *Biochim. Biophys. Acta*, **9,** 369.

FOX, J. J., YUNG, N. and WEMPEN, I. (1957). Spectrophotometric studies of nucleic acid derivatives and related compounds. IV. On the structure of orotidine. A study of *N*-methylated orotic acids. *Biochim. Biophys. Acta*, **23,** 295.

FOX, S. W. (1957). A chemical theory of spontaneous generation, in *1st Intern. Symp. on Origin of life on the earth.* Academy of Sciences, U.S.S.R., Moscow, 1957; Pergamon Press, London, 1959.

FOX, S. W. and HARADA, K. (1958). Thermal copolymerization of amino acids to a product resemebling protein. *Science,* **128,** 1214.

FRAENKEL-CONRAT, H. (1956a). Rebuilding a virus. *Scientific American,* **194,** 42.

FRAENKEL-CONRAT, H. (1956b). The role of the nucleic acid in the reconstitution of active tobacco mosaic virus. *J. Amer. Chem. Soc.* **78,** 882.

FRAENKEL-CONRAT, H. (1957). Degradation of tobacco mosaic virus with acetic acid. *Virology,* **4,** 1.

FRAENKEL-CONRAT, H. and SINGER, B. (1959). Reconstitution of TMV. III. Improved methods and the use of mixed nucleic acids. *Biochim. Biophys. Acta,* **33,** 359.

FRAENKEL-CONRAT, H., STAEHELIN, M. and CRAWFORD, L. V. (1959). Tobacco mosaic virus reconstitition using inactivated nucleic acid. *Proc. Soc. Exp. Biol. Med.* **102,** 118.

FRANCK, J. and PLATZMAN, R. (1954). Physical principles underlying photochemical, radiation-chemical and radiobiological reactions, in *Radiation Biology* (Hollaender, A. ed.), McGraw-Hill, New York. vol. I, pp. 191–254.

FRANKLIN, R. E. (1956). Location of the ribonucleic acid in the tobacco mosaic virus particle. *Nature,* **177,** 929.

FRANKLIN, R. M., FRIEDMAN, M. and SETLOW, R. B. (1953). The ultraviolet action spectrum of a *Bacillus megatherium* bacteriophage. *Arch. Biochem. Biophys.* **44,** 159.

FREDERICQ, E. (1955). Modifications of the ultraviolet absorption spectrum of insulin. *Arch. intern. physiol. et biochim.* **63,** 261.

FREDERICQ, E. (1956). Fractionation of insulin. Ciba Foundation Colloqium (*Endocrinology*) **9,** 89.

FREED, J. J., EIGLE, J. L., RUDKIN, G. T. and SCHULTZ, J. (1959). UV radiation effects on Ehrlich ascites tumor cells: Observations using a flying spot UV microscope. *J. Biophys. Biochem. Cytol.* **5,** 205.

FREIFIELDER, D., DAVISON, P. F. and GEIDUSCHEK, E. P. (1961). *Biophys. J.* **1,** 389.

FRENCH, C. S., RABIDEAU, G. S. and HOLT, A. S. (1947). The construction and performance of a large grating monochromator with a high energy output for photochemical and biological investigations. *Rev. Sci. Inst.* **18,** 11.

FRIESEN, B. S. and SINSHEIMER, R. L. (1959). Partition cell analysis of infective tobacco mosaic virus nucleic acid. *J. Mol. Biol.* **1,** 321.

FROMAGEOT, C. and SCHNEK, G. (1950). Le spectre ultraviolet du lysozyme; avec des considerations sur le spectre ultraviolet de divers acides amines et de quelques-uns de leurs peptides. *Biochim. Biophys. Acta,* **6,** 113.

FUJIWARE, T., TAKANO, N., MOROKUMA, T., and YAMADA, M. (1954). Pigment formation by ultraviolet decomposition of histidine. *Nagasake Igakkai Zassi,* **24,** 726. *Chem. Absts.* **49,** 2538f.

GALSTON, A. W. (1950). Riboflavin, light and the growth of plants. *Science,* **111,** 619.

GALSTON, A. W. and BAKER, R. S. (1949). Inactivation of enzymes by visible light in the presence of riboflavin. *Science,* **109,** 485.

GARAY, K. and GUBA, F. (1954). Die reversible veranderung des adenosintriphosphates infolge von strahlungswirkungen. *Acta Physiol. Acad. Sci. Hungar.* **5,** 393.

GARD. S. and MAALPE, O. (1959). Inactivation of viruses, in *The Viruses* (Buruet, F. M. and Stanley, W. M., eds.), Academic Press, New York. vol. I, pp. 359–428.

GARDY, W. (1959). Electron spin resonance of free radicals in irradiated biochemicals. *Rad. Res.*, Suppl. 1, 491.

GAREN, A. and ZINDER, N. D. (1955). Radiobiological evidence for partial genetic homology between bacteriophage and host. *Virology*, **1,** 347.

GATES, F. L. (1928). On nucleic derivatives and the lethal action of ultraviolet light. *Science*, **68,** 479–480.

GATES, F. L. (1930). On study of the bactericidal action of ultraviolet light. *J. Gen. Physiol.* **14,** 31–41.

GEHRMANN, G. and OKADA, S. (1957). The amino acid composition of desoxyribonuclease I. *Biochim. Biophys. Acta*, **23,** 621.

GEIDUSCHEK, E. P. and HERSKOVITS, T. T. (1961). *Arch. Biochem. Biophys.*, **95,** 114.

GELLERT, M. (1961). On the assignment of $n \rightarrow \pi^*$ transitions in polynucleotides. *J. Am. Chem. Soc.* **83,** 4664–4665.

GERDIL, R. (1961). The crystal structure of thymine monohydrate. *Acta Crystallog.* **14,** 333.

GIBSON, Q. H. (1956). An apparatus for flash photolysis and its application to the reaction of myoglobin with gases. *J. Physiol.* **134,** 112.

GIBSON, Q. H. (1959). The photochemical formation of a quickly reacting form of haemoglobin. *Biochem. J.* **71,** 293.

GIBSON, Q. H. and AINSWORTH, S. (1947). Photosensitivity of haem compounds. *Nature*, **180,** 1416.

GIERER, A. (1957). Structure and biological function of ribonucleic acid from tobacco mosaic virus. *Nature*, **179,** 1297.

GIERER, A. and SCHRAMM, G. (1956). Infectivity of ribonucleic acid from tobacco mosaic virus. *Nature*, **177,** 702.

GIESE, A. C. (1947a). Radiations and cell division. *Quart. Rev. Biol.* **22,** 253. (see also *Anat. Record*, **99,** 116).

GIESE, A. C. (1947b). Sensitization of nucleoproteins to heat by UV radiations. *Anat. Record*, **99,** 672.

GIESE, A. C. (1950a). Action of ultraviolet radiation on protoplasm. *Physiol. Rev.* **30,** 431.

GIESE, A. C. (1950b). Is UV photolysis of proteins increased by subsequent heating? *Anatomical Record*, **108,** 209.

GIESE, A. C. (1953). Protozoa in photobiological research. *Physiol. Zool.* **26,** 1.

GIESE, A. C., IVERSON, R. M. and SANDERS, R. T. (1957). Effect of nutritional state and other conditions on ultraviolet resistance and photoreactivation in yeast. *J. Bacteriol.* **74,** 271.

GIESE, A. C. and LEIGHTON, P. A. (1935). Quantitative studies on the photolethal effects of quartz ultraviolet radiation upon paramecium. *J. Gen. Physiol.* **18,** 557.

GIESE, A. C., LEIGHTON, H. L. and BAILEY, R. (1952). Changes in the absorption spectra of proteins and representative amino acids induced by ultraviolet irradiation and ozone. *Arch. Biochem. Biophys.* **40,** 71.

GILVARRY, J. J. and HOCHSTEIN, A. R. (1963). Possible role of meteorites in origin of life. *Nature*, **197,** 624–625.

GINOZA, W. (1958). Kinetics of heat inactivation of ribonucleic acid of tobacco mosaic virus. *Nature*, **181,** 958.

GINOZA, W. and GUILD, W. R. (1961). On the inactivation of transforming DNA by temperatures below the melting point. *Proc. Nat. Acad. Sci.* (*U.S.*) **47,** 633.

GINOZA, W. and ZIMM, B. H. (1961). Mechanisms of inactivation of deoxyribonucleic acids by heat. *Proc. Nat. Acad. Sci.* (U.S.) **47,** 639.

GLAZER, A. N. and SMITH, E. L. (1960). Effect of denaturation on the ultraviolet absorption spectra of proteins. *J. Biol. Chem.* **235.** PC 43.

GLISIN, V. and DOTY, P. (1962). *Biochim. Biophys. Acta*, **61,** 458–460.

GOLDENBERG, H. (1950). Thesis, Polytechnic Institute of Brooklyn.

GOLDENBERG, H. and MCLAREN, A. D. (1951). Effect of ultraviolet light on the specific activity of chymotrypsin and trypsin. *J. Amer. Chem. Soc.* **73,** 1131, 5929.

GOLDFARB, A. R. (1953). Absorption spectrum of the peptide bond. II. Influence of chain length. *J. Biol. Chem.* **201,** 317.

GOLDFARB, A. R. and HOFFMAN, E. (1959). The ultraviolet absorption spectra of benzoyl polyglycine anilides and benzoyl polyalanine anilides. *Arch. Biochem. Biophys.* **81,** 493.

GOLDFARB, A. R., HOFFMAN, E. and GUTSTEIN, N. (1958). Absorption spectrum of the peptide bond: basic properties of peptides and proteins. *Arch. Biochem. Biophys.* **76,** 161.

GOLDFARB, A. R. and SAIDEL, L. J. (1951). Ultraviolet absorption spectra of proteins. *Science,* **114,** 156.

GOLDFARB, A. R., SAIDEL, L. J. and MOSOVICH, E. (1951). The ultraviolet absorption spectra of proteins. *J. Biol. Chem.* **193,** 397.

GOLDING, R. M. (1960). Ultraviolet absorption of water solutions. *Nature,* **188,** 308.

GONNARD, P. (1953). Oxidation by ultraviolet irradiation of *o*- and *m*-hydroxyphanylalanine. *Ann. pharm. franc.* **11,** 283. *Chem. Abs.* **47,** 8788b.

GOODGAL, S. H., RUPERT, C. S. and HERRIOT, R. M. (1957). Photoreactivation of *Hemophilus Influenza* transforming factor for streptomycin resistance by an extract of *E. coli,* in *The Chemical Basis of Heredity* (McElroy, W. D. and Glass, B. eds.), Johns Hopkins Press, Baltimore.

GOODMAN, M. and LISTOWSKY, I. (1962). Conformational aspects of synthetic polypeptides. VI. Hypochromic spectral studies of oligo-methyl-l-glutamate peptide. *J. Am. Chem. Soc.* **84,** 3770–3771.

GOODWIN, T. W. and MORTON, R. A. (1946). The spectrophotometric determination of tyrosine and tryptophane in proteins. *Biochem. J.* **40,** 628.

GORIN, G. (1956). The ionization of cysteine. *J. Amer. Chem. Soc.* **78,** 767.

GREEN, M. and COHEN, S. S. (1958). Studies on the synthesis of bacterial and viral pyrimidines. III. Derivatives of dihydrocytosine. *J. Biol. Chem.* **228,** 601.

GREEN, N. M. and NEURATH, H. (1954). Proteolytic enzymes, in *The Proteins* (NEURATH, H. and BAILEY, K.), Academic Press, New York, vol. 2, pp. 1057–1198.

GREENE, F. D., MISROCK, S. L. and WOLFE, J. R. (1955). The structure of anthracene dimers. *J. Amer. Chem. Soc.* **77,** 3852.

GREENSTEIN, J. P. and JENRETTE, W. V. (1941). Physical changes in thymonucleic acid induced by proteins, salts, tissue extracts and ultraviolet irradiation. *Cold Spring Harb. Symp. Quant. Biol.* **9,** 236.

GRIFFIN, B. E., TODD, A. R. and RICH, A. (1958). A synthesis of ribothymidine 5-pyrophosphate and its conversion to polyribothymidylic acid. *Proc. Nat. Acad. Sci.* (*U.S.*) **44,** 1123.

GROSSMAN, L. (1962). The effects of ultraviolet-irradiated poly-U in cell-free protein synthesis in *E. coli*. *Proc. Nat. Acad. Sci.* (U.S.), **48,** 1609–1614.

GROSSMAN, L. and DELBRUCK, M. (1963). The effects of UV monochromatic light on the structure and function of poly-U. *Fed. Proc.* **22,** 2526.

GROSSWEINER, L. I. (1956). Metastable states of photoexcited ovalbumin and constituents. *J. Chem. Phys.* **24,** 1255.

GROSSWEINER, L. I. and MULAC, W. A. (1959). Primary processes in the flash photolysis of ovalbumin and constituents. *Rad. Res.* **10,** 515.

GROTH, W. (1960). Photochemical production of some organic compounds under possible primitive earth conditions, in *Photochemistry in the Liquid and Solid States,* John Wiley, New York.

GROTH, W. and WEYSSENHOFF, H. V. (1958). Photochemical formation of amino acids from mixtures of simple gases. *Naturwiss.* **44,** 510.

GROTH, W. and WEYSSENHOFF, H. V. (1959). Photochemical formation of organic compounds from mixtures of simple gases. *Ann. Physik.* **4,** 69.

GRUNBERG-MANAGEO, M., ORTIZ, P. and OCHOA, S. (1955). Enzymatic synthesis of nucleic acid-like polynucleotides. *Science,* **122,** 907.

GUIDOLIN, R. and FERRI, R. G. (1950). Acao de substancias fluorescentes sobre venenos ofidicos. *O Hospital (Brazil)* **38,** 737.

GULLAND, J. M. and HOLIDAY, E. R. (1936). Constitution of the purine nucleosides. IV. Adenosine and related nucleotides and coenzymes. *J. Chem. Soc.* 765.

HAAS, F. L. and DOUDNEY, C. O. (1957). A relation of nucleic acid synthesis to radiation induced mutation frequency in bacteria. *Proc. Nat. Acad. Sci. (U.S.)* **43,** 871.

HABEL, K. and SOCKRIDER, B. T. (1947). A continuous flow method of exposing antigens to ultraviolet radiation. *J. Immunol.* **56,** 273.

HACKETT, H. J. (1962). The photodynamic effects of acridine orange on an RNA virus (*vesicular exanthema*). *Photochem. Photobiol.* **1,** 147.

HALWER, M. (1951). The photochemistry of riboflavin and related compounds. *J. Amer. Chem. Soc.* **73,** 4870.

HALY, A. R., FENGHELMAN, M. and GRIFFITH, J. C. (1957). Supercontraction of wool irradiated with ultraviolet light or iodinated. *Nature,* **180,** 1064.

HAM, J. S. and PLATT, J. R. (1952). Far ultraviolet spectra of peptides. *J. Chem. Phys.* **20,** 335.

HAMMETT, L. P. (1940). *Physical Organic Chemistry.* McGraw-Hill, New York.

HAMOIR, G. (1951a). Fish tropomyosin and fish nucleotropomyosin. *Biochem. J.* **48,** 146.

HAMOIR, G. (1951b). Further investigations on fish tropomyosin and fish nucleotropomyosin. *Biochem. J.* **50,** 140.

HARGREAVES, A. B. (1955). Acetylcholinesterase response to ultraviolet radiation and its relation with sulfhydryl groups. *Arch. Biochem. Biophys.* **57,** 41.

HARM, W. (1956). On the mechanism of multiplicity reactivation in bacteria. *Virology,* **2,** 559.

HARM, W. (1961a). Gene-controlled reactivation of UV-inactivated bacteriophage. *J. Cellul. Comp. Physiol.* **58,** Suppl. 1, 69.

HARM, W. (1961b). Reactivation of phage from lethal ultraviolet damage in DNA. *J. Chim. Phys.* **58,** 1058.

HARM, W. (1963). Mutual competition between UV-inactivated phages for the photoreactivating enzyme in *E. coli* cells. *Intern. J. Rad. Biol.,* in press.

HARM, W. (1963). On the relationship between host-cell reactivation and ultraviolet-reactivation in UV-inactivated phages. *Zeit. fur Vererbungslehre,* **94,** 67–79.

HARRINGTON, W. F. and SCHELLMAN, J. A. (1956). Evidence for the instability of hydrogen-bonded peptide structures in water, based on studies on ribonuclease and oxidized ribonuclease. *C. R. trav. Lab. Carlsberg. Ser. chim.* **30,** 21.

HARRIS, D. T. (1926). Observations on the velocity of the photo-oxidation of proteins and amino acids. *Biochim. J.* **20,** 288.

HARRIS, L. and KAMINSKY, J. (1935). A precision actinometer for the ultraviolet region (including an exact test of the Einstein equivalence law). *J. Amer. Chem. Soc.* **57,** 1154.

HARRIS, L., KAMINSKY, J. and SIMARD, R. C. (1935). The absorption spectrum of malachite green leucocyanide and the mechanism of the dark reaction after photolysis. *J. Amer. Chem. Soc.* **57,** 1151.

HARRIS, R. J. C., ORR, S. F. D., ROE, E. M. F. and THOMAS, J. F. (1953). Investigation of polymorphism and isomerism in cytidine phosphate. *J. Chem. Soc.*, 489.

HARRISON, G. R. (1934). Simply constructed ultraviolet monochromator for large area illumination. *Rev. Sci. Inst.* **5,** 149.

HASSELSTROM, T., HENRY, M. C. and MURR, B. (1957). Synthesis of amino acids by β-radiation. *Science*, **125,** 350.

HATCHARD, C. G. and PARKER, C. A. (1956). A new sensitive chemical actinometer. II. Potassium ferrioxalate as a standard chemical actinometer. *Proc. Roy. Soc.* (*London*) *A.* **235,** 518.

HAUROWITZ, F. and ASTRUP, T. (1939). Ultraviolet absorption of genuine and hydrolysed proteins. *Nature*, **143,** 118.

HAUROWITZ, F. and TÜMER, A. (1949). The proteolytic cleavage of irradiated proteins, a contribution to the problem of quantum biology and the amplifier action of living cells. *Enzymologia*, **13,** 229.

HAUSSER, K. H. (1960). Detection and identification of long-lived radicals. *Rad. Res.*, Suppl. 2, 480.

HEIDT, L. J. and BOYLES, H. B. (1951). Influence of several variables encountered in photochemical work upon the intensity of wavelength 254 mμ produced by a quartz mercury vapor lamp of the low pressure type operating in water at 0–95°. *J. Amer. Chem. Soc.* **73,** 5728.

HELPRIN, J. J. and HIATT, C. W. (1959). Photosensitization of T2 coliphage with toluidine blue. *J. Bacteriol.* **77,** 502.

HENRI, V., WEIZMANN, C. and HIRSHBERG, Y. (1934). The action of ultraviolet light on glycine. *Compt. rend.* **198,** 168.

HENTZ, R. R. and BURTON, M. (1951). Studies in photochemistry and radiation chemistry of toluene, mesitylene and ethylbenzene. *J. Amer. Chem. Soc.* **73,** 532.

HEPPEL, L. A., ORTIZ, P. and OCHOA, S. (1957a). Studies on polynucleotides synthesized by polynucleotide phosphorylase. I. Structure of polynucleotides with one type of nucleotide unit. *J. Biol. Chem.* **229,** 679.

HEPPEL, L. A., ORTIZ, P. and OCHOA, S. (1957b). Studies on polynucleotides synthesized by polynucleotide phosphorylase. II. Structure of polymers containing a mixture of bases. *J. Biol. Chem.* **229,** 695.

HERMANS, J., DONOVAN, J. W. and SCHEREGA, H. A. (1960). Thermodynamic data from difference spectra. *J. Biol. Chem.* **235,** 91.

HERSHENSON, H. M. (1956). *Ultraviolet and Visible Absorption Spectra.* Academic Press, New York.

HERSCOVITS, T. S. (1963). Non-aqueous solutions of DNA: Denaturation by urea and its methyl derivatives. *Biochemistry*, **2,** 335–340.

HERSKOVITS, T. T. (1962). Non-aqueous solutions of DNA. II. *Arch. Biochem. Biophys.* **97,** 474.

HERSKOVITS, T. T., SINGER, S. J. and GEIDUSCHEK, E. P. (1961). Non-aqueous solutions of DNA. I. *Arch. Biochem. Biophys.* **94,** 99.

HERZBERG, G. and SHOOSMITH, J. (1956). Absorption spectrum of free CH_3 and CD_3 radicals. *Can. J. Phys.* **34,** 523.

HERZBERG, K. (1933). Weitere untersuchungen uber die zerstorbarkeit des variolavakzinevirus durch lichtenergie. *Z. Immunitats.* **80,** 507.

HESS, E. L., YASNOFF, D. S. and LANG, S. (1955). The sensitivity of ultraviolet radiation of a simple protein isolated from a lymphatic organ. *Arch. Biochem. Biophys.* **57,** 323.

HEYROTH, F. F. and LOUFBOUROW, J. R. (1931). Changes in the ultraviolet absorption spectrum of uracil and related compounds under the influence of radiations. *J. Amer. Chem. Soc.* **53,** 3441.

HIATT, C. W., KAUFMAN, E., HELPRIN, J. J. and BARON, S. (1960). Inactivation of viruses by the photodynamic action of toluidine blue. *J. Immunol.* **84,** 480.

HILDEBRAND, D. and KERSTEN, J. (1959). Amino acid analysis of wool keratin after UV-irradiation. *Biochem. Z.* **332,** 67.

HILL, R. F. (1958). Independent inactivation of bacteriophage T1 by X-rays and ultraviolet light. *Rad. Res.* **8,** 46.

HILL, R. F. and ROSSI, H. H. (1952). Absence of photoreactivation in T1 bacteriophage irradiated with ultraviolet in the dry state. *Science,* **116,** 424.

HILL, R. F. and ROSSI, H. (1954a). The ultraviolet sensitivity and photoreactivibility of T1 bacteriophage. I. Effect of irradiation conditions upon survival curves. *Radiation Research,* **1,** 282.

HILL, R. F. and ROSSI, H. (1954b). The ultraviolet sensitivity and photoreactivibility of T1 bacteriophage II. Interpretation of the survival curves. *Radiation Research,* **1,** 358.

HINKS, J. H. and JOHNSON, F. F. (1951). The stabilization of albumin during the UV irradiation of plasma. *J. Amer. Pharm. Assn.* **40,** 537.

HIRATA, M. (1958). Photolysis of arginine. *Nagasaki Igakkai Zasshi,* **33,** 1699. *Chem. Abs.* **53,** 6324b.

HIRS, C. H. W., STEIN, W. H. and MOORE, S. (1954). The amino acid composition of ribonuclease. *J. Biol. Chem.* **211,** 941.

HIRSHBERG, Y. (1956). Reversible formation and eradication of colours by irradiation at low temperature. A photochemical memory model. *J. Amer. Chem. Soc.* **78,** 2304.

HOCHSTRASSE, R. M. and PORTER, G. B. (1960). Primary processes in photo-oxidation. *Quart. Rev.* **14,** 146.

HOLIDAY, E. R. (1936). Spectrophotometry of proteins: I. Absorption spectra of tyrosine, tryptophane and their mixtures; II. Estimation of tyrosine and tryptophane in proteins. *Biochem. J.* **30,** 1795.

HOLIDAY, E. R. and BEAVEN, G. H. (1950). Some practical aspects of the effects of stray radiation in spectrophotometry. *Photoelectric Spectrometry Group Bulletin No.* **3,** Oct. 1950.

HOLLAENDER, A. and DUGGAR, B. M. (1936). Irradiation of plant viruses and of microorganisms with monochromatic light. III. Resistance of the virus of typical tobacco mosaic and *Escherichia Coli* to radiation from 3000 Å to 2250 Å. *Proc. Nat. Acad. Sci. U.S.* **22,** 19.

HOLLAENDER, A., GREENSTEIN, J. P. and JENRETTE, W. V. (1941). Effects of ultraviolet radiation on sodium thymonucleate. *J. Nat. Canc. Inst.* **2,** 23.

HOLTZ, P. (1936). Uber reduktions-und oxydationswirkungen bestrahlter zucher. *Arch. exptl. Pathol. Pharmakol.* **182,** 141.

HOLTZ, P. and BECKER, J. P. (1936). Uber die wirkung der rontgenstrahlen auf glucose. *Arch. exptl. Pathol. Pharmakol.* **182,** 160.

HOPKINS, G. R. and SINSHEIMER, R. L. (1955). Visible and ultraviolet light scattering by tobacco mosaic virus nucleic acid. *Biochim. Biophys. Acta,* **17,** 476.

HOUSTON, R. A. (1909). On the mechanism of the absorption spectra of solutions. *Proc. Roy. Soc. London,* **82A,** 606.

HOWARD-FLANDERS, P., BOYCE, R. P. and THERIOT, L. (1962). Mechanism of sensitization to ultraviolet light of T1 bacteriophage by the incorporation of 5-bromodeoxyuridine or by pre-irradiation of the host cell. *Nature,* **195,** 51–54.

HOWE, A. F., BALAZS, E. A. and LAURENT, T. C. (1958). Effect of irradiation on mucopolyssacharides. *IVth Intern. Congr. Biochem.,* Vienna, No. 1–51.

Hull, D. E. (1960). Thermodynamics and kinetics of spontaneous generation. *Nature*, **186,** 693.

Hunt, R. E. and Davis, W. (1947). Filters for the isolation of the 3130 Å spectral group of the mercury arc. *J. Amer. Chem. Soc.* **69,** 1415.

Hvidberg, E., Kvorning, S. A., Schmidt, A. and Schon, J. (1958). Reduction of viscosity of a potassium hyaluronate solution by ultraviolet irradiation. *Nature*, **181,** 1338.

Hvidberg, E., Kvorning, S. A., Schmidt, A. and Schon, J. (1959). Effect of ultraviolet irradiation on hyaluronic acid in vitro. *Acta Pharmacol. Toxicol.* **15,** 356.

Ide, I. (1960). Photolysis of glycine. *Nagasaki Igakkai Zasshi*, **35,** 99. *Chem. Abst.* **55,** 2251g, 1961.

Imahori, K. (1955). The ultraviolet inactivation of pepsin. *Biochim. Biophys. Acta*, **18,** 216.

Imahori, K. and Tanaka, J. (1959). Ultraviolet absorption spectra of poly-L-glutamic acid. *J. Mol. Biol.* **1,** 359.

Ingold, C. K. (1953). *Structure and Mechanism in Organic Chemistry.* Cornell University Press, Ithaca, New York.

Isaacs, A. and Burke, D. C. (1958). Interferon: A possible check to virus infections. *The New Scientist*, June 5, 109; *ibid.* October, 30, 1170.

Isaacs, A. and Burke, D. C. (1958). Mode of action of interferon. *Nature*, **182,** 1073.

Issacs, A., Klemperer, H. C., and Hitchcock, G. (1961). Studies on the mechanism of action of interferon. *Virology*, **13,** 191.

Jagger, J. (1958). Photoreactivation. *Bacteriol. Rev.* **22,** 99.

Jagger, J. (1960). Photoreactivation, in *Radiation Protection and Recovery* (Hollaender, A. ed.), Pergamon Press, London and New York, chap. 12.

Jagger, J. (1961). A small and inexpensive UV dose-rate meter useful in biological experiments. *Rad. Res.* **14,** 394.

Jagger, J. and Latarjet, R. (1956). Spectres d'action de la photo-restauration chez *E. Coli B/r. Ann. Inst. Pasteur*, **91,** 858.

Janion, C. and Shugar, D. (1960). Absorption spectra, structure, and behaviour towards enzymes of dihydropyrimidines and dihydrooligonucleotides. *Acta Biochim. Polon.* **7,** 309.

Jansen, E. F., Curl, A. L. and Balls, A. K. (1951). A crystalline, active oxidation product of *a*-chymotrypsin. *J. Biol. Chem.* **189,** 671.

Johns, H. E., Rapaport, S. A. and Delbruck, M. (1962). Photochemistry of thymine dimers. *J. Mol. Biol.* **4,** 104.

Johnson, M. B. (1961). Formation of an ommatin pigment from tyrosine by ultraviolet irradiation. *Nature*, **190,** 924.

Johnson, T. B. and Johns, C. O. (1914). Researches on pyrimidines. LXVIII. The structure of Ritthansen's divicine. *J. Amer. Chem. Soc.* **36,** 545.

Jordon, D. O. (1951). Physicochemical properties of the nucleic acids. *Prog. Biophys., Biophys. Chem.* **2,** 51.

Jordan, D. O. (1952). Nucleic acids, purines and pyrimidines. *Ann. Rev. Biochem.* **21,** 209.

Jordan, D. O. (1955). The physical properties of nucleic acids, in *The Nucleic Acids* (Chargaff, E. and Davidson, J. N. ed.), Academic Press, New York. vol. 1, chap. 13.

Kanazir, D. and Errera, M. (1955). Actions des rayons ultraviolets sur l'acide adenosine triphosphate des spermatozoides de grenouille. *Biochim. Biophys. Acta*, **16,** 198.

Kaplan, H. S. and Zavarine, R. (1962). *Biochem. Biophys. Res. Comm.* **8,** 432–436.

KAPLAN, J. G. and FRASER, M. J. (1952). The effect of ultraviolet radiation on monomolecular films of egg albumin. *Biochim. Biophys. Acta*, **9,** 585.

KAPLAN, J. G., ANDREWS, D. H. and FRASER, M. J. (1954). Ultraviolet irradiation of dilute solutions of ovalbumin. *J. Colloid Sci.* **9,** 203.

KAPLAN, J. G. and FRASER, M. J. (1954b). The expansion of monomolecular films of ovalbumin. *J. Biol. Chem.* **210,** 57.

KAPLAN, R. W. (1955a). Mutation und totung durch UV von bakterien verschiedenen wassergehelt. *Naturwiss.* **42,** 184.

KAPLAN, R. W. (1955b). Influence of the water content on the ultraviolet sensitivity of deoxyribonucleic acid and its bearing on biological effects. *Naturwiss.* **42,** 466.

KAPLAN, R. W. and GUNKEL, W. (1960). Reversion of mutation and reactivation by visible light, as well as by different salt solutions, after ultraviolet irradiation of Serratia. *Arch. Mikrobiol.* **35,** 63.

KAPLAN, R. W., WINKLER, U. and WOLF-ELLMAUER, H. (1960). Induction and reversion of C-mutations by irradiation of the extracellular *x*-phage of *Serratia. Nature,* **186,** 330.

KARREMAN, G., STEELE, R. H. and VON SZENT-GYÖRGYI, A. (1958). Resonance transfer of excitation energy between aromatic amino acids in proteins. *Proc. Nat. Acad. Sci. U.S.* **44,** 149.

KASHA, M. (1948). Transmission filters for the ultraviolet. *J. Opt. Soc. Amer.* **38,** 929.

KASHA, M. (1961). The nature and significance of $n \rightarrow \pi^*$ transitions, in *Light and life* (E. D. MCELROY and B. GLASS eds.), Johns Hopkins Press, Baltimore.

KATCHALSKI, E., BERGER, A., BIOCHOWSKY-SLOMNICKI, L. and KURTZ, J. (1955). Antibiotically active amino acid copolymers related to gramicidin S. *Nature,* **176,** 118.

KATCHMANN, B. (1955). Some observations on the photochemistry of soybean trypsin inhibitor. *Rad. Res.* **2,** 467.

KATRITZKY, A. R. and WARING, A. J. (1962). *Chem. and Ind.* (London), p. 695.

KAUFMAN, E. and HIATT, C. W. (1959). Photodynamic action of proflavine on T2 coliphage. *Virology,* **9,** 478.

KAUZMANN, W. (1954). Denaturation of proteins and enzymes, in *The Mechanism of Enzyme Action* (MCELROY, W. B. and GLASS, B. eds.), Johns Hopkins Press, Baltimore. pp. 70–110.

KAWADE, Y. (1957). Interaction between desoxypentose nucleic acid and histone in strong saline. *Biochim. Biophys. Acta,* **26,** 61.

KEILEEN, D. and HARTREE, E. F. (1958). Spectrophotometric study of suspensions of pigmented particles. *Biochim. Biophys. Acta,* **27,** 173.

KELLENBERGER, G. and WEIGLE, J. J. (1958). Etude au moyen des rayons ultraviolets de l'interaction entre bacteriophages temperes et bacterie hote. *Biochim. Biophys. Acta,* **30,** 112.

KELNER, A. (1949a). Effect of visible light on the recovery of *Streptomyces griseus* conidia from ultraviolet irradiation injury. *Proc. Nat. Acad. Sci. U.S.* **35,** 73.

KELNER, A. (1949b). Photoreactivation of ultraviolet irradiated *E. coli* with special reference to the dose-reduction principle and to ultraviolet-induced mutation. *J. Bacteriol.* **58,** 511.

KENDREW, J. C., BODO, G., DINTZIS, H. M., PARRISH, R. G. and WYCKOFF, H. (1958). A three-dimensional model of the myoglogin molecule obtained by X-ray analysis. *Nature,* **181,** 662.

KERR, G. P. (1947). New meters employing high sensitivity cells for the measurement of erythemal energy. *Rev. Sci. Inst.* **18,** 472.

KHENOKH, M. A. (1955). Action of γ-radiation of ^{60}Co on carbohydrates. *Dokl. Akad. Nauk U.S.S.R.,* **104,** 746.

KITA, H., MAEDA, H., HANAZAKI, B., SHIMIZU, F. and FUJITA, H. (1954). Studies in the deviations of absorption spectra caused by photo-reactivation. *Bull. Tokyo Med. Dent. Univ.*, No. 1, 37.

KLAMERTH, O. (1957). Uber die optische verhalten von desoxyribonukleinsaure-Histon kombination. *Z. Naturforsch.* **12b,** 186.

KLAND, M. J. and JOHNSON, L. A. (1957). A kinetic study of the ultraviolet decomposition of biochemical derivatives of nucleic acid. I. Purines. *J. Amer. Chem.* **79,** 6187.

KLECZKOWSKI, A. (1954a). Stability of chymotrypsin and tobacco mosaic virus decreased by ultraviolet radiation. *Biochem. J.* **56,** 345.

KLECZKOWSKI, A. (1954b). Inactivation of antibodies by ultraviolet radiation. *Brit. J. Experimental Pathology*, **35,** 402.

KLECZKOWSKI, A. (1957). Effects of non-ionizing radiations on viruses. *Adv. Virus Res.* **4,** 191.

KLECZKOWSKI, A. (1959). Aggregation of the protein of tobacco mosaic virus with and without combination with nucleic acid. *Virology*, **7,** 385.

KLECZKOWSKI, A. (1963). The inactivation of ribonucleic acid from tobacco mosaic virus by ultraviolet light at different wave lengths. *Photochem. Photobiol.* **2,** in press.

KLECZKOWSKI, A. and GOLD, A. H. (1962). Effect of ultraviolet radiation on antigenicity of horse serum albumin: formation of new determinants. *Photochem. Photobiol.* **1,** 299–304.

KLECZKOWSKI, J. and KLECZKOWSKI, A. (1953). The behavior of *Rhizobium* bacteriophages during and after exposure to UV radiation. *J. Gen. Microbiol.* **8,** 135.

KLISS, A. M. and MATTHEWS, C. N. (1962). HCN dimer and chemical evolution. *Proc. Nat. Acad. Sci.*, (U.S.), **48,** 1300.

KLOTZ, I. M. (1958). Protein hydration and behavior. *Science*, **128,** 815.

KLOTZ, I. M., AYERS, J., HO, J. Y. C., HOROWITZ, M. G. and HEINEY, R. E. (1958). Interaction of proteins with disulfide compounds: some implications for electron transport in proteins. *J. Amer. Chem. Soc.* **80,** 2132.

KLOUWEN, H. M., APPLEMAN, A. W. M. and BARENDSEN, G. W. (1962). *Nature*, **194,** 554–555.

KNIGHT, C. A. and WOODY, B. R. (1958). Phosphorus content of tobacco mosaic virus. *Arch. Biochem. Biophys.* **78,** 460.

KODA, T., HARADOUN, M. and KAMIYAMA, M. (1958). A band-pass filter using an organic solution for separating the mercury 2537 Å line. *Oyo Butsuri*, **27,** 176. *Chem. Abst.* **52,** 13418h.

KOFFLER, H. and MARKERT, I. L. (1951). Effect of photodynamic action on viscosity of deoxyribonucleic acid. *Proc. Soc. Exp. Biol. Med.* **76,** 90.

KOFMAN, E. F. (1950). Photochemical changes in solutions of egg albumin proceeding independently of photodenaturation. *Biokhimiya*, **15,** 534.

KOFMAN, E. B. (1957). On the photo-oxidation of sulphydryl groups of dissolved egg albumin. *Biophysics*, **2,** 212.

KOLLER, L. R. (1952). *Ultraviolet Radiation.* John Wiley, New York.

KONEV, S. V. (1959). Fluorescence reaction spectra of proteins. *Izvest. Akad. Nauk U.S.S.R., Ser. Fiz.* **23,** 90.

KOSHLAND, D. E., RAY, W. J. and ERWIN, M. J. (1958). Protein structure and enzyme action. *Fed. Proc.* **17,** 1145.

KOVACS, J. and NAGY (KOVACS), H. (1961). Polypeptide formation from asparagine under hypothetically primitive conditions. *Nature*, **190,** 531.

KRETCHMER, N. and TAYLOR, R. (1950). Effect of pH on the far ultraviolet absorption of tyrosine. *J. Amer. Chem. Soc.* **72,** 3291.

KRIEG, D. R. (1959). A study of gene action in ultraviolet-irradiation bacteriphage T4. *Virology*, **8,** 80.

KRIEG, D. R. (1960). Ultraviolet-induced mutation of extracellular bacteriophage T4. 8th Meeting Radiation Res. Soc., *Rad. Research*, **12,** 450.

KRISHNAN, V. G. and GOODMAN, L. (1961). Protonation effects on the $n \rightarrow \pi^*$ transitions in pyrimidine. *J. Amer. Chem. Soc.* **83,** 2042.

KUNITZ, M. (1946). A spectrophotometric method for the measurement of ribonuclease activity. *J. Biol. Chem.* **164,** 563.

KUNITZ, M. (1947). Isolation of a crystalline protein compound of trypsin and soybean trypsin-inhibitor. *J. Gen. Physiol.* **30,** 311.

KUNITZ, M. (1950). Crystalline desoxyribonuclease. I. Isolation and general properties. *J. Gen. Physiol.* **33,** 349.

KWIECINSKI, L., MAYER, J., and MARCHLEWSKI, L. (1928). The absorption of ultraviolet light by carbohydrates. *Z. physiol. Chem.* **176,** 292.

LABEYRIE, F. and SHUGAR, D. (1951). Spectre d'absorption, energie d'activation, et inactivation de l'aldolase par la lumiere ultra-violette. *J. Chim. Phys.* **48,** 1.

LADIK, J. (1960). Investigations of the electronic structure of DNA. *Acta. Phys. Acad. Scient. Hungar.* **11,** 239.

LALAND, S. G., LEE, W. A., OVEREND, W. G. and PEACOCKE, A. R. (1954). The ultraviolet absorption of some degraded desoxyribonucleic acids. *Biochim. Biophys. Acta*, **14,** 356.

LANHAM, U. N. (1952). Oparin's hypothesis and the evolution of nucleoproteins. *The American Naturalist*, **86,** 213.

LARIN, N. M. (1958). Effect of ultra-violet irradiation on electrophoretic properties of serum protein. *Nature*, **181,** 65.

LASKOWSKI, M. and LASKOWSKI, M., JR. (1954). Naturally occurring trypsin inhibitors. *Adv. Protein. Chem.* **9,** 203.

LASKOWSKI, M., WIDOM, J. M., MCFADDEN, M. L. and SCHERAGA, H. A. (1956). Differential ultraviolet spectra of insulin. *Biochim. Biophys. Acta*, **19,** 581.

LATARJET, R., EPHRUSSI-TAYLOR, H. and REBEYRETTE, N. (1959). On the target size of a transforming factor based on X-ray inactivation. *Rad. Res.* Suppl. 1, 417.

LATARJET, R. and MORENNE, P. (1951). Inactivation d'une bacteriophage par un rayonnement ultraviolet de tres faible intensite. *Ann. inst. Pasteur*, **80,** 220.

LATARJET, R., MORENNE, P. and BERGER, R. (1953). Un appareil simple pour le dosage des rayonnements ultraviolets emis par les lampes germicides *Ann. Inst. Past.* **85,** 174.

LATARJET, R. and WAHL, R. (1945). Precisions sur l'inactivation des bacteriophages par les rayons ultraviolets. *Ann. inst. Pasteur*, **71,** 336.

LATIMER, P. and EUBANKS, C. A. H. (1962). Absorption spectrophotometry of turbid suspensions—a method of correcting for large systematic distortions. *Arch. Biochem. Biophys.* **98,** 274.

LAUFFER, M. A., ANSEVIN, A. T., CARTWRIGHT, T. E. and BRINTON, C. C. (1958). Polymerization–depolymerization of TMV protein. *Nature*, **181,** 1338.

LAUFFER, M. A., TRKULA, D. and BUZZELL, A. (1956). Mechanism of inactivation of tobacco mosiac virus by X-rays. *Nature*, **177,** 890.

LAUNER, H. F. (1949). Exposure meter for precision light dosage. *Rev. Sci. Inst.* **20,** 103.

LAURENT, T. C. (1956). Effect of ultraviolet light on alkaline solutions of glucose and certain other sugars. *J. Amer. Chem. Soc.* **78,** 1875.

LAURENT, T. C. and WERTHEIM, E. (1952). Effect of ultraviolet light on the absorption spectra of carbohydrates. *Acta Chem. Scand.* **6,** 678.

LAWLEY, P. D. (1956). Interaction studies with DNA. III. The effect of changes in sodium ion concentration, pH and temperature on the ultraviolet absorption spectrum of sodium thymonucleate. *Biochim. Biophys. Acta*, **21,** 481.

LEACH, S. J. and SCHERAGA, H. A. (1960a). Effect of light scattering on ultraviolet difference spectra. *J. Amer. Chem. Soc.* **82,** 4790.

LEACH, S. J. and SCHERAGA, H. A. (1960b). Ultraviolet difference spectra and the internal structure of proteins. *J. Biol. Chem.* **235,** 2827.

LEE, H. H. and PECK, T. T. (1958). Single cell techniques in the study of ultraviolet irradiation of mammalian cells. *Intern. Cong. Radiation Research*, Burlington, Vt., Paper no. 150.

LEIGHTON, W. G. and FORBES, G. S. (1930). Precision actinometer with uranyl oxalate. *J. Amer. Chem. Soc.* **52,** 3139.

LEITNER, N. (1950). Thesis, Polytechnic Institute of Brooklyn.

LENNOX, F. G. (1959). Spectrophotometric measurement of color in wool fabrics *Textile Res. J.* **29,** 906.

LENNOX, E. S., LURIA, S. E. and BENZER, S. (1954). On the mechanism of photoreactivation of ultraviolet inactivated bacteriophage. *Biochim. Biophys. Acta,* **15,** 471.

LERMAN, L. S. and TOLMACH, L. J. (1959a). Genetic transformation. II. The significance of damage to the DNA molecule. *Biochim. Biophys. Acta,* **33,** 371.

LERMAN, L. S. and TOLMACH, L. J. (1959b). *Abstracts Biophys. Soc.* **110.**

LEVIN, G. and BROWN, G. B. (1962). Photodecomposition of adenine-N-oxide. *Fed. Proc.* **21,** (2), 372.

LEVINE, M. and COX, E. (1960). Increased ultraviolet sensitivity of frozen phage. *Radiation Research,* **12,** 452 (Communication No. 95 of 8th meeting of Radiation Research Society).

LEVINSON, A. B., MILZER, C. D., SHAUGHNESSY, O. M., NEAL, V. D. and OPPENHEIMER, F. (1944). Production of potent inactivated vaccines with ultraviolet irradiation. *J. Am. Med. Assoc.* **125,** 531.

LEVINSON, A. B., MILZER, C. D., SHAUGHNESSY, O. M., NEAL, V. D. and OPPENHEIMER, F. (1945). Production of potent inactivated vaccines with ultraviolet irradiation. *J. Immunol.* **50,** 317.

LEY, H. and ARENDS, B. (1932). Absorptionsmessungen im kurzwelligen ultraviolet. I. Karboxylsauren, amine und aminosauren. *Z. physik. Chem.* **B17,** 177.

LIEBEN, F. and GETREUER, V. (1933). Uber die aufspaltung von stickstoffhaltigen, zylischen verbindungen im lichte der quartz-lampe. *Biochem. Zeit.* **259,** 1.

LIENER, I. E. (1957). The essentiality of the disulfide linkages in trypsin. *J. Biol. Chem.* **225,** 1061.

LINDERSTRØM-LANG, K. U. (1952). *Proteins and Enzymes.* Stanford University Press, Stanford, California.

LIPSETT, M. N. (1960). Evidence for helical structure in poly-U. *Proc. Nat. Acad. Sci. (U.S.)* **46,** 445.

LITMAN, R. L. and EPHRUSSI-TAYLOR, H. (1959). Inactivation et mutation des facteurs genetiques de l'acide desoxyribonucleique du pneumocoque par l'ultraviolet et par l'acide nitreux. *Compt. Rendus,* **249,** 838.

LIVINGSTON, R. (1955). Photochemistry, in *Radiation Biology* (Hollaender, H. ed.), McGraw-Hill, New York, Vol. II, Chapt. 7.

LIVINGSTON, R. (1959). The potentialities and limitations of the paramagnetic resonance method in radiation research. *Rad. Res.* Suppl. 1, 463.

LOEWUS, M. W. and BRIGGS, D. R. (1952). The number of catalytically active sites present on the chymotrypsin molecule. *J. Biol. Chem.* **199,** 857.

LoGrippo, G. A. and Hartmann, F. W. (1958). Chemical and combined methods for plasma sterilization. *Bibliotheca Haematol.*, Suppl. *Acta Haematol.*, No. 7, 225 (*Chem. Abs.* **53,** 496h, 1959).

Loofbourow, J. R. (1948). The effects of ultraviolet radiation on cells. *Growth Symposium*, **12,** 75.

Loofbourow, J. R., Englert, M. A. and Dwyer, C. M. (1941). Increased yield of nucleic acid-like substances from irradiated yeast. *Nature*, **148,** 113.

Loofbourow, J. F. and Joyce, L. (1941). Increased ultraviolet absorption of cells following irradiation with ultraviolet light. *Nature*, **148,** 166.

Loofbourow, J. R., Oppenheim-Errera, S., Loofbourow, D. G. and Yeats, C. A. (1947). Intercellular hormones. 8. Release of nucleotides and nucleosides by damaged living cells. *Biochem. J.* **41,** 122.

Loofbourow, J. R. and Stimson, M. M. (1940). Ultraviolet spectra of nitrogenous heterocyclic compounds. Part I. Effect of pH and irradiation on the spectrum of adenine. *J. Chem. Soc.*, 844.

Luckiesh, M., ed. (1946). *Applications of Germicidal, Erythemal and Infra-red Energy*. Van Nostrand, New York.

Luckiesh, M. and Taylor, A. H. (1940). An ultraviolet meter for germicidal lamps. *Rev. Sci. Inst.* **11,** 110.

Luria, S. E. (1944). A growth-delaying effect of ultraviolet radiation on bacterial viruses. *Proc. Natl. Acad. Sci. U.S.* **30,** 393.

Luria, S. E. (1947). Reactivation of irradiated bacteriophage by transfer of self-reproducing units. *Proc. Nat. Acad. Sci. U.S.* **33,** 253.

Luria, S. E., ed. (1953). *General Virology*. John Wiley, New York.

Luria, S. E. (1955). Radiation and viruses, in *Radiation Biology* (Hollaender, A. ed.), McGraw-Hill, New York. vol. 2, pp. 333–364.

Luria, S. E. and Delbruck, M. (1942). Interference between bacterial viruses. II. Interference between inactivated bacterial virus and active virus of the same strain and of a different strain. *Arch. Biochem.* **1,** 207.

Luria, S. E. and Dulbecco, R. (1949). Genetic recombinations leading to production of active bacteriophage from ultraviolet inactivated bacteriophage particles. *Genetics*, **34,** 93.

Luse, R. A. (1961). The mechanism of enzyme inactivation by ultraviolet light. Ph. D. Dissertation, Univ. of Calif., Berkeley.

Luse, R. A. and McLaren, A. D. (1963). Mechanism of enzyme inactivation by ultraviolet light and the photochemistry of amino acids (at 2537 Å). *Photochem. Photobiol.* **2,** 343.

Lyons, W. E. (1948). Photolysis of organic disulfides. *Nature*, **162,** 1004.

Magasanik, B. (1955). Isolation and composition of the pentose nucleic acids and the corresponding nucleoproteins, in *The Nucleic Acids* (Chargaff, E. and Davidson, J. N. eds.), Academic Press, New York. vol. I, chap. II.

Magasanik, B. and Chargaff, E. (1951). Studies on the structure of ribonucleic acids. *Biochim. Biophys. Acta*, **7,** 396.

Magill, M. A., Steiger, R. E. and Allen, A. J. (1937). Amino acids, acyl amino acids, dipeptides, acyl dipeptides and derivatives of these compounds. I. *Biochem. J.* **31,** 188.

Mains, G. J., Roebber, J. L. and Rollefson, G. K. (1955). Photolysis with a high intensity spark. *J. Phys. Chem.* **59,** 733.

Mandl, I., Levy, B. and McLaren, A. D. (1950). Photolysis of the peptide bond at 2537 Å. *J. Amer. Chem. Soc.* **72,** 1790.

Mandl, I. and McLaren, A. D. (1949a). A quantum yield for the inactivation of ficin by ultraviolet light. *Arch. Biochem.* **21,** 408.

MANDL, I. and McLAREN, A. D. (1949b). Quantum yields for photolysis of the peptide bond at 2537 Å. *Nature*, **164,** 749.

MANDL, I. and McLAREN, A. D. (1951). A contribution to the photochemistry of amino acids. *J. Amer. Chem. Soc.* **73,** 1826, 5929.

MARCHLEWSKI, L. and URBANCZYK, W. (1933). Zur kenntnis reduzierende zucker III. *Biochem. Z.* **262,** 248.

MARKHAM, R., HILLS, G. and FREY, S. (1963). *Virology*, in press.

MARKHAM, R., MATTHEWS, R. E. and SMITH K. M. (1948). Specific crystalline protein and nucleoprotein from a plant virus having insect vectors. *Nature*, **162,** 88.

MARKHAM, R. and SMITH, J. D. (1952). The structure of RNA. 3. The end groups, the general structure and the nature of the core. *Biochem. J.* **52,** 565.

MARLMBERG, J. H. (1957). Millimicrosecond duration light source. *Rev. Sci. Inst.* **28,** 1027.

MARMUR, J., ANDERSON, W. F., MATTHEWS, L., BERNS, K., GAJEWSKA, E., LANE, D. and DOTY, P. (1961). The effects of ultraviolet light on the biological and physical chemical properties of deoxyribonucleic acids, in *Recovery of cells from Injury*, Research Conf. of Oak Ridge Nat. Lab., Gatlinburg, Tenn.

MARMUR, J. and DOTY, D. (1959). Heterogeneity in DNA's. I. Dependance on composition of the configuration stability of DNA's. *Nature*, **183,** 1427.

MARMUR, J. and FLUKE, D. J. (1955). Uniformity of ionizing radiation action on several transforming factors of Pneumococcus. *Arch. Biochem. Biophys.* **57,** 506.

MARMUR, J. and GROSSMAN, L. (1961). Ultraviolet light induced linking of DNA strands and its reversal by photoreactivating enzyme. *Proc. Nat. Acad. Sci. U.S.* **47,** 778.

MARSHALL, J. R. and WALKER, J. (1951). An experimental study of some potentially tautomeric 2- and 4(6)-substituted pyrimides. *J. Chem. Soc.*, 1004.

MASON, R. (1959). Charge transfer processes in biological systems. *Disc. Faraday Soc.*, 129.

MASON, S. F. (1959). The electronic spectra of *N*-heteroaromatic systems. Parts I–III. *J. Chem. Soc.*, 1240.

MASSON, C. R., BOEKELHEIDS, V. and NOYES, W. A. (1956). Photochemical reactions, in *Techniques of Organic Chemistry* (Weissberger, A., ed.), McGraw-Hill, New York, vol. 2.

MATSUDA, G. (1953). Photolysis of tryptophan. I and II. *Nagasaki Igakkai Zassi*, **28,** 438. *Chem. Abs.* **48,** 6842d.

MATSUDA, G., MAEKAWA, T., OHKAWA, I. and YAMADA, M. (1954). Photolysis of phenylalanine. *Nagasaki Igakkai Zassi*, **29,** 10. *Chem. Abs.* **48,** 7660c.

McCALL, K. B., GORDON, F. H., BLOOM, F. C., HYNDMAN, L. A., TAYLOR, H. L. and ANDERSON, H. D. (1957). Methods for the preparation and ultraviolet irradiation of human fibrinogen for intravenous use. *J. Am. Pharm. Ass., Sci. Ed.* **46,** 295.

McKINSTREY, D. and READING, E. H. (1944). The preparation and immugenic properties of an ultraviolet-inactivated poliomyelitis vaccine. *J. Franklin Inst.* **237,** 71.

McLAREN, A. D. (1947). Photochemistry of enzymes. *J. Polymer Sci.* **2,** 107.

McLAREN, A. D. (1949a). Concerning the supposed absorption of ultraviolet energy by the peptide linkage. *Acta Chem. Scand.* **3,** 648.

McLAREN, A. D. (1949b). Photochemistry of enzymes, proteins and viruses. *Ad. Enzymology*, **9,** 75.

McLAREN, A. D. (1957). Photochemistry of proteins. XVIII. Some observations with enzymes and nucleic acids and discussion. *Enzymologia*, **18,** 81.

McLaren, A. D. (1961). Ultraviolet effects on RNA and enzymes. *Conference on Molecular and Radiation Biology*, Pub. 823, Nat. Acad. Sci., Washington.

McLaren, A. D., Doering, A. and Philips, G. (1952). Inactivation of pepsin by high intensity ultraviolet light and the reciprocity law. *Biochem. Biophys. Acta*, **9,** 597.

McLaren, A. D. and Finkelstein, P. (1950). The influence of reagents and conditions on the quantum yield for the inactivation-denaturation of chymotrypsin. *J. Amer. Chem.* **72,** 5423.

McLaren, A. D., Gentile, P., Kirk, D. C. and Levin, N. A. (1953). Inactivation of enzymes with ultraviolet light and photolysis of the peptide bond. *J. Polymer Sci.* **10,** 333.

McLaren, A. D. and Lewis, C. (1950). Concerning the inactivation–denaturation of pepsin. *J. Polymer Sci.* **5,** 379.

McLaren, A. D. and Luse, R. A. (1961). Mechanism of inactivation of enzyme proteins by ultraviolet light. *Science*, **134,** 836, 1410.

McLaren, A. D. and Moring-Claesson, I. (1961). An action spectrum for the inactivation of infectious nucleic acid from tobacco mosaic virus by ultraviolet light. *3rd Intern. Congr. Photobiology*, Copenhagen, 1960; *Progress in Photobiology* (Christenson, B. and Buchmann, B., eds.), Elsevier, Amsterdam.

McLaren, A. D. and Pearson, S. (1949). Effect of pH and urea on ultraviolet light inactivation of crystalline pepsin. *J. Polymer Sci.* **4,** 45, 408

McLaren, A. D. and Peterson, G. H. (1961). Montmorillonite as a caliper for the size of protein molecule. *Nature*, **192,** 960.

McLaren, A. D., Peterson, G. H. and Barshad, I. (1958). The adsorption and reactions of enzymes and proteins on clay minerals. IV. *Soil Sci. Soc. Am. Proc.*, **22,** 239.

McLaren, A. D. and Rowen, J. W. (1951). Sorption of water vapor by proteins and polymers. *J. Polymer Sci.* **7,** 289.

McLaren, A. D., Sheppard, E. and Wagman, J. (1948). Study of the action of ultraviolet light on urease by means of the ultracentriguge. *Nature,* **162,** 370.

McLaren, A. D. and Takahashi, W. N. (1957). Inactivation of infectious nucleic acid from tobacco mosaic virus by ultraviolet light (2537 Å). *Rad. Res.* **6,** 532.

McLaren, A. D. and Takahashi, W. N. (1959). Scattering of light by tobacco mosaic virus and χ-protein from infected tobacco plants. *Biochim. Biophys. Acta*, **32,** 555.

McLaren, A. D. and Waldt, L. (1952). Photochemistry of proteins. XV. Some observations on the ultraviolet spectra of proteins. *Biochim. Biophys. Acta*, **9,** 240.

McLean, D. J. and Giese, A. C. (1950). Absorption spectra of proteins and amino acids after ultraviolet irradiation. *J. Biol. Chem.* **187,** 537.

McLean, I. W. and Taylor, A. R. (1958). Experiences in the production of poliovirus vaccines, in *Progress in Medical Virology* (Karger, S., ed.), vol. 1, 122, Basle.

Mellors, H. (1952). Ultraviolet irradiation of air supplies with special reference to ultraviolet intensity measurements. *J. Appl. Chem.* **2,** 681.

Merrified, R. B. and Woolley, D. W. (1952). The structure and microbiological activity of some dinucleotides isolate from yeast ribonucleic acid. *J. Biol. Chem.* **197,** 521.

Merrified, R. B. and Woolley, D. W. (1958). Synthesis of peptides with streptogenin activity. *J. Amer. Chem. Soc.* **80,** 6635.

Meselson, M. and Weigle, J. J. (1961). Chromosome breakage accompanying genetic recombination in bacteriophage. *Proc. Nat. Acad. Sci.* **47,** 857.

MICHELSON, A. M. (1958a). Hyperchromicity and nucleic acids. *Nature*, **182,** 1502.

MICHELSON, A. M. (1959a). Polynucleotides. Part I. Synthesis and properties of some polyribonucleotides. *J. Chem. Soc.*, 1371.

MICHELSON, A. M. (1959b). Polynucleotides. Part II. Homopolymers of cytidylic and pseudouridylic acids, copolymers with repeating subunits and the stepwise synthesis of polyribonucleotides. *J. Chem. Soc.*, 3655.

MICHELSON, A. M. (1959c). Some aspects of the chemistry of polynucleotides and nucleotide coenzymes. *Acta Biochim. Polon.* **6,** 335.

MICHELSON, A. M. and TODD, A. R. (1955). Deoxyribonucleosides and related compounds. V. Cyclothymidine and other thymidine derivatives. The configuration of the glycosidic centre in thymidine. *J. Chem. Soc.*, 816.

MIHALYI, E., BRADLEY, D. F. and KNOLLER, M. I. (1957). Some physical and chemical properties of the RNA contaminant of rabbit muscle myosin preparations. *J. Amer. Chem. Soc.* **79,** 6387.

MILLER, S. L. (1955). Production of some organic compounds under possible primitive earth conditions. *J. Amer. Soc.* **77,** 2351.

MILLER, S. L. (1957a). The mechanism of synthesis of amino acids by electric discharges. *Biochim. Biophys. Acta*, **23,** 480.

MILLER, S. L. (1957b). Formation of organic compounds on the primitive earth, in *1st Intern. Symp. on Origin of Life on the Earth*, Academy of Sciences, U.S.S.R., 1957; Pergamon Press, London, 1959).

MILLER, S. L. and UREY, H. C. (1959). Organic compounds synthesized on the primitive earth. *Science*, **130,** 245.

MILLICH, F. and OSTER, G. (1959). Photoreduction of acridine dyes. *J. Amer. Chem. Soc.* **81,** 1357.

MILZER, A. and LEVINSON, S. O. (1949). Active immunization of mice with ultraviolet inactivated lymphocytic choriomeningitis virus vaccine and results of immune serum therapy. *J. Infectious Diseases*, **85,** 251.

MILZER, A., OPPENHEIMER, F. and LEVINSON, S. O. (1944). Production of potent inactivated vaccines with ultraviolet irradiation. *J. Am. Med. Assoc.* **125,** 704.

MILZER, A., OPPENHEIMER, F. and LEVINSON, S. O. (1945). Production of potent inactivated vaccines with ultraviolet irradiation. *J. Immunol.* **50,** 331.

MISCHENKO, P. I. (1952). Action of ultraviolet light on the absorption spectra of amino acids. *Biokhimiya*, **17,** 82.

MISCHENKO, P. I. (1954). The effect of ultraviolet rays on the absorption spectra of amino acids irradiated under air, argon or nitrogen. *Biokhmija*, **19,** 268.

MITCHELL, H. K. (1955), cited by RAPPORT *et al.*, 1955.

MITCHELL, P. (1950). Spectrophotometric estimation of nucleic acids in bacterial samples. *J. Gen. Microbiol.* **4,** 399.

MITZ, M. A. and SCHLUETER, R. J. (1958). Direct spectrophotometric measurement of the peptide bond: application to the determination of acylase I. *Biochim. Biophys. Acta*, **27,** 168.

MOGILEVSKII, S. and LAUFER, A. L. (1957). Action of ultraviolet radiation on hyaluronic acid. *Dokl. Akad. Nauk S.S.R.* **76,** 239.

MONK, G. S. and EHRERT, C. F. (1956). Design and performance of a biological spectrograph. *Rad. Res.* **5,** 88.

MONTGOMERY, P. O. and BONNER, W. A. (1958). A new technique for ultraviolet microbeam irradiation of living cells. *AMA Arch. Pathol.* **66,** 418.

MONTGOMERY, P. O. and HUNDLEY, L. L. (1960). Effects of selective ultraviolet irradiation of cytoplasm of living cells. *Proc. Soc. Exp. Biol. Med.* **105,** 117.

MONTGOMERY, P. O., ROBERTS, F. and BONNER, W. (1956). The flying spot monochromatic ultraviolet television microscope. *Nature*, **177,** 1172.

MONTY, K. J. and DOUNCE, A. L. (1958). The properties and the enzymatic degradation of desoxyribonucleoprotein from liver cell nuclei. *J. Gen. Physiol.* **41,** 595.

MOORE, A. M. (1958). Ultraviolet irradiation of pyrimidine derivatives. II. Synthesis of the product of reversible photolysis of uracil. *Can. J. Chem.* **36,** 281.

MOORE, A. M. (1959). Ultraviolet irradiation of pyrimidine derivatives. III. Effect of HCN on the photolysis. *Can. J. Chem.* **37,** 1281.

MOORE, A. M. (1962). Anionic control in the photochemistry of aminopyrimidines. *Radiation Research,* **16,** 610; *Can. J. Chem.* **4,** 1937 (1963).

MOORE, A. M. and THOMSON, C. H. (1955). Ultraviolet irradiation of pyrimidine derivatives. *Science,* **122,** 594.

MOORE, A. M. and THOMSON, C. H. (1956). Photodecomposition of pyrimidine compounds. *Prog. in Radiobiology,* **4,** 75.

MOORE, A. M. and THOMSON, C. H. (1957). Ultraviolet irradiation of pyrimidine derivatives. I. 1,3-dimethyluracil. *Can. J. Chem.* **35,** 163.

MOORE, W. M., HAMMOND, G. S. and FOSS, R. P. (1960). Photochemical evidence for triplet state quenching by paramagnetic species. *J. Chem. Phys.* **32,** 1594.

MOROKUMA, T. Photolysis of lysine. *Nagasaki Igakkai Zassi,* **29,** 813. *Chem. Abst.* **49,** 4051g.

MOROWITZ, H. J. (1950). Absorption effects in volume irradiation of micro-organisms. *Science,* **111,** 229.

MUKAI, I. (1959). Photolysis of DL-ornithine. *Nagasaki Igakkai Zassi,* **34,** 577. *Chem. Abst.* **53,** 18138d.

MULLIKEN, R. S. (1953). See series of articles in *J. Chem. Phys.* **7,** 14, 20, 121.

MUSTAFA, A. (1952). Dimerization reactions in sunlight. *Chem. Rev.* **51,** 1.

NAKATARII, M. (1961). Studies on histidine residues in heme proteins related to their activities. V. Photo-oxidation of catalase in the presence of methylene blue. *J. Biochem.* (*Japan*) **49,** 98.

NATH, R., COHLY, M. A. and GIRI, K. V. (1955). A microelectrophoretic study of the complex formation of bovine plasma albumin and ribonucleic acid. *J. Indian Inst. Sci.* **37,** 163.

NELSON, L. S. (1956). Sapphire lamp for short wavelength photochemistry. *J. Opt. Soc. Amer.* **46,** 768.

NELSON, L. S. (1962). Intense rapid heating with flash discharge lamps. *Science.* **136,** 296.

NELSON, L. S. and RAMSAY, D. A. (1956). Flash photolysis experiments with a sapphire flash lamp. *J. Chem. Phys.* **25,** 372.

NESBET, R. K. (1962). Interaction of two ethylene molecules. *Molecular Physics,* **5,** 63–69.

NEUFELD, E. F., GINSBERG, V., PUTMAN, E. W., FANSHIER, D. and HASSID, W. Z. (1957). Formation and interconversion of sugar nucleotides by plant extracts. *Arch. Biochem. Biophys.* **69,** 602.

NEURATH, H. and DIXON, G. H. (1957). Structure and activation of trypsinogen and chymotrypsinogen. *Fed. Proc.* **16,** 791.

NEURATH, H., GREENSTEIN, J. P., PUTNAM, F. W., and ERICKSON, J. O. (1944). The chemistry of protein denaturation. *Chem. Rev.* **34,** 157.

NEURATH, H. and SCHWERT, G. W. (1950). The mode of action of the crystalline pancreatic proteolytic enzymes. *Chem. Rev.* **46,** 60.

NICHOLAS, J. W. and POLLAK, F. F. (1950). The isolation of the lines of the mercury arc by filters. *The Analyst,* **75,** 662.

NIKOLOV, T. K. and SAEV, G. K. (1957). The effect of temperature and substrate on amylase inactivation by ultraviolet radiation. *Biokhimiya,* **22,** 865.

NIRENBERG, M. W. and MATTHEI, J. H. (1961). *Proc. Nat. Acad. Sci.* **47,** 1588–1602.

NODA, L. H., KUBY, S. A. and LARDY, H. A. (1953). Properties of thiolesters: Kinetics of hydrolysis in dilute aqueous media. *J. Amer. Chem. Soc.* **75,** 913.

NORMAN, A. (1951). Inactivation of *Neurospora conidia* by ultraviolet radiation. *Exp. Cell. Res.* **2,** 454.

NORMAN, A. (1960). Ultraviolet inactivation of Poliovirus ribonucleic acid. *Virology,* **10,** 384.

NORRISH, R. G. W. and THRUSH, B. A. (1956). Flash photolysis and kinetic spectroscopy. *Quart. Rev.* **10,** 149.

NORTHROP, J. H. (1934). Crystalline pepsin, inactivation by Beta and Gamma rays from radium and by ultraviolet light. *J. Gen. Physiol.* **17,** 359.

NORTHROP, J. H., KUNITZ, M. and HERIOTT, R. M., eds. (1948). *Crystalline Enzymes.* Columbia University Press, New York.

NOVICK, A. and SZILARD, L. (1949). Experiments on light reactivation of ultraviolet-inactivated bacteria. *Proc. Nat. Acad. Sci. U.S.* **35,** 591.

NOYES, W. A. (1959). Oxygen effects in photochemical systems. *Rad. Res.*, Suppl. 1, 164.

NOYES, W. A. and LEIGHTON, P. A. (1941). *The Photochemistry of Gases.* Reinhold, New York.

NURNBERGER, J. I. (1955). Ultraviolet microscopy and microspectroscopy, in *Analytical Cytology* (Mellors, R. C. ed.), McGraw-Hill, New York, chapt. 4.

OBATA, Y. and KOSHIKA, M. (1959). Effect of ultraviolet light irradiation on the peptide bonds of protein. *Bull. Agr. Chem. Soc. Japan,* **23,** 107. *Chem. Abst.* **54,** 8930a (1960).

OBATA, Y. and SAKAMURA, S. (1958). Denaturation of casein with ultraviolet light. *Nippon Nogei-Kagaku Kaishi,* **32,** 251.

OPATOWSKI, I. (1950). On the interpretation of the dose-frequency curve of radiogenetics. *Genetics,* **35,** 56.

OPATOWSKI, I. and CHRISTIANSEN, A. M. (1950). On the single event hypothesis of radiogenetics. *Bull. Math. Biophys.* **12,** 19.

ORLANS, E. S. (1953). The effect of photo-oxidation on antibodies. *Brit. J. Exp. Pathol.* **34,** 164.

ORO, J. (1961a). Comets and the formation of biochemical compounds on the primitive earth. *Nature,* **190,** 398.

ORO, J. (1961b). Mechanism of synthesis of adenine from HCN under possible primitive earth conditions. *Nature,* **191,** 1193.

ORO, J. (1963). Ultraviolet absorbing compound(s) reported present in the Murray meteorite. *Nature,* **197,** 756–758.

ORO, J. and KIMBALL, A. P. (1960). Direct synthesis of purines from amino acids. *Fed. Proc.* **10,** (I), 314.

ORO, J. and KIMBALL, A. P. (1961). Synthesis of purines under possible primitive earth conditions. I. Adenine from HCN. *Arch. Biochem. Biophys.* **94,** 217–227.

ORO, J., KIMBALL, A. P., FRITZ, R. and MASTER, F. (1959). Amino acid synthesis from formaldehyde and hydroxylamine. *Arch. Biochem. Biophys.* **85,** 115.

OSTER, G. (1950). Two-phase formation in solutions of tobacco mosaic virus and the problem of long-range forces. *J. Gen. Physiol.* **33,** 445.

OSTER, G. (1953). The photochemistry of dyes in solution. *Photographic Engineering,* **4,** 173.

OSTER, G. (1955). Light scattering, in *Physical Techniques in Biological Research* (Oster, G. and Pollister, A. W., eds.), Academic Press, New York. vol. 1, chap.2.

OSTER, G. and ADELMAN, A. H. (1956). Long-lived states in photochemical reactions. I. Photoreduction of eosin. *J. Amer. Chem. Soc.* **78,** 913.

OSTER, G. and BELLIN, J. S. (1960). Regression of tumour growth in mice brought about by photodynamic action. *3rd Internat. Congr. Photobiol.*, Copenhagen.

OSTER, G., BELLIN, J. S. KIMBALL, R. W. and SCHRADER, M. E. (1959). Dye sensitized photo-oxidations. *J. Amer. Chem. Soc.* **81,** 5095.

OSTER, G. and MCLAREN, A. D. (1950). The ultraviolet light and photosensitized inactivation of tobacco mosaic virus. *J. Gen. Physiol.* **33,** 215.

OTEY, M. C. and GREENSTEIN, J. P. (1954). Apparent dissociation constants, and ultraviolet and infrared absorption spectra of isomeric cystinylcystine peptides. *J. Biol. Chem.* **53,** 501.

PAKULA, R., HULANICKA, E. and WALCZAK, W. (1959). Sensitivity to ultraviolet irradiation of the streptomycin resistance activity in transforming DNA of different bacteria. *Bull. Acad. Polon. Sci., Cl. II,* **7,** 217.

PAKULA, R., HULANICKA, E. and WALCZAK, W. (1960). Photoreactivation of UV damaged streptomycin-resistance marker in DNA of streptococci and pneumococci. *Bull. Acad. Polon. Sci., Cl. II,* **8,** 269.

PAL, A. B. (1948). Sensitive ultraviolet meters. *Rev. Sci. Inst.* **19,** 529.

PARKER, C. A. (1953). A new sensitive chemical actinometer. *Proc. Roy. Soc.* **A220,** 104.

PASCHKE, R., CHANG, R. W. H. and YOUNG, D. (1957). Importance of γ-rays for the origin of life. *Science,* **125,** 881.

PAULING, L. (1960). *The Nature of the Chemical Bond.* Cornell University Press, Ithaca, New York.

PAULING, L. and COREY, R. B. (1956). Specific hydrogen-bond formation between pyrimidines and purines in desoxyribonucleic acids. *Arch. Biochem. Biophys.* **65,** 164.

PAVLOVSKAYA, T. E. and PASSYNSKY, A. G. (1955). Changes caused by radiation in the ultraviolet and infrared spectra of proteins. *Colloid J., U.S.S.R.,* **17,** 305.

PAVLOVSKAYA, T. E. and PASSYNSKY, A. G. (1956). The protective action of some substances in the irradiation of protein solutions. *Colloid J., U.S.S.R.,* **18,** 583.

PAVLOVSKAYA, T. E. and PASSYNSKY, A. G. (1957). Primary formation of amino acids in ultraviolet rays and in electric discharge, in *1st Intern. Symp. on Origin of life on the Earth,* Academy of Sciences, U.S.S.R., Moscow, 1957; Pergamon Press, 1959.

PAVLOVSKAYA, T. E. and PASSYNSKY, A. G. (1958). Amino acid formation when exposing formaldehyde and ammonium salt solutions to ultraviolet radiation. *IVth Intern. Congr. Biochem., Vienna,* Sect. 1, Nr. 77, 12.

PAVLOWSKAYA, T. E., PASSYNSKY, A. G. and GREBENIKOVA, A. I. (1960). Formation of amino acids by the action of light on solutions of formaldehyde and ammonium salts in the presence of adsorbents. *Doklady Akad. Nauk S.S.R.,* **135,** 743.

PEACOCKE, A. R. (1954). Effect of drying at 110° on sodium desoxyribonucleate. *Biochim. Biophys. Acta,* **14,** 157.

PEEL, C. N. (1937). The operation and characteristics of quartz mercury vapour lamps for ultraviolet radiation. *Brit. J. Radiol.* **12,** 99.

PERLMANN, G. E. (1956). Observations on the effect of urea on crystalline pepsin. *Arch. Biochem. Biophys.* **65,** 210.

PERRY, R. P. (1957). Changes in ultraviolet absorption spectrum of parts of living cells following irradiation with an ultraviolet microbeam. *Exp. Cell Res.* **12,** 546.

PERUTZ, M. F., JOPE, E. M. and BARER, R. (1950). Observation of proteins in polarized ultraviolet light. *Dis. Faraday Soc.* **9,** 423.

PETUELY, F. and MEIXNER, N. (1953). U.V. spectroskopische unterschungen an zuckern: I Mitteil: Die einvirkung von natrium hydroxyd auf glucose; der nachweis der ediolatform der glucose. *Chem. Ber.* **86,** 1255.

Phillips, G. O. and Moody, G. J. (1958). Radiation chemistry of carbohydrates. II. Irradiation of aqueous solutions of dextran with γ-radiation. *J. Chem. Soc.*, 3534.

Phillips, G. O. and Moody, G. J. (1960a). Radiation chemistry of carbohydrates. III. The effect of γ-radiation on aqueous solutions of *d*-fucose. *J. Chem. Soc.*, 754.

Phillips, G. O. and Moody, G. J. (1960b). Radiation chemistry of carbohydrates. IV. The effect of γ-irradiation on aqueous solutions of sucrose. *J. Chem. Soc.*, 762.

Phillips, G. O., Moody, G. J. and Mattock, G. L. (1958). Radiation chemistry of carbohydrates. I. Action of ionizing radiations on aqueous solutions of *d*-glucose. *J. Chem. Soc.*, 3522.

Phillips, J. M., Braun, W. and Plescia, O. J. (1957). Antigencity of a bacterial desoxyribonucleic acid. *Nature*, **181,** 573.

Pichl, H. (1960). Writszellenfreie Photoreacktivierung UV-inaktivierter T2-phagen. *Naturwissenschaften*, **47,** 283.

Pierce, S. and Giese, A. C. (1957). Photoreversal of ultraviolet injury to frog and crab nerves. *J. Cellul. Comp. Physiol.* **49,** 303.

Pillemer, L., Ecker, E. E., Myers, V. C. and Muntwyler, E. (1938). Chemical and immunological studies of the effects of radiant energy and of oxidation on crystalline urease. *J. Biol. Chem.* **123,** 365.

Pitts, J. N., Margerum, J. D., Taylor, R. P. and Brin, W. (1955). Determination of light intensities with a highly sensitive uranyl oxalate actinometer. *J. Amer. Chem. Soc.* **77,** 5499.

Pollard, E. C. (1953). *The Physics of Viruses.* Academic Press, New York.

Ponzio, M. (1952). Fizykalische und biologische kriterien der aktinologische dosimetrie. *Strahlentherapie*, **88,** 544.

Porter, G. (1950). Flash photolysis and spectroscopy. A new method for the study of free-radical reactions. *Proc. Roy. Soc.* (*London*) **A200,** 284.

Porter, G. (1959). Application of flash photolysis in irradiation studies. *Rad. Res.*, Suppl. **1,** 479.

Porter, K. and Volman, D. H. (1962). Uranyl oxalate actinometer for microphotochemistry. *J. Am. Chem. Soc.* **84,** 2011.

Potter, J. L. and Dounce, A. L. (1956). Nucleotide-amino acid complexes in alkaline digests of ribonucleic acid. *J. Amer. Chem. Soc.* **78,** 3078.

Pouyet, J. (1952). Études de la viscosite des solutions diluees de desoxy-ribonucleate de sodium. *C. R. Acad. Sci.* **234,** 152.

Praus, R. (1955a). Photo-oxidation of blood proteins. II. Manometric study of the photo-oxidation of human globin in the presence of methylene blue. *Chem. Listy* (Czechoslovakia), **49,** 921.

Praus, R. (1955b). Photo-oxidation of blood proteins. III. Reaction of methylene blue with human globin. *Chem. Listy* (*Czechoslovakia*), **49,** 1064.

Praus, R. (1955c). Photo-oxidation of blood proteins. IV. Photo-oxidation of blood plasma and its reaction in presence of methylene blue. *Chem. Listy* (*Czechoslovakia*), **49,** 1069.

Preiss, J. W. and Setlow, R. (1956). Spectra of some amino acids, peptides, nucleic acids, and protein in the vacuum ultraviolet. *J. Chem. Phys.* **25,** 138.

Prusoff, W. H. (1963). Low temperature reversal of the ultraviolet photochemical reaction product of 2′-deoxyuridine. *Biochim. Biophys. Acta*, **68,** 302–310.

Pullman, A. and Pullman, B. (1958a). Structure electronique des pyrimidines entrants dans la constitution des acides nucleiques et des antimetabolites apparentes intervenant en chimiotherapie anticancereuse. *C. R. Acad. Sci.* **246,** 611.

Pullman, A. and Pullman, B. (1958b). Electronic structure of fundamental cellular constituents and their active compounds in anticancer chemotherapy. II. Fundamental purine bases. *Bull. Soc. Chim. France*, 766.

Pullman, A. and Pullman, B. (1959). The electronic structure of fundamental cellular constituents and of compounds active in cancer chemotherapy. VI. Basic natural pyrimidines. *Bull. Soc. Chim. France*, 594.

Pullman, A., Pullman, B. and Berthier, G. (1956). Recherche sur la structure electronique des composés actifs en chimiotherapie du cancer. *C. R. Acad. Sci.* **243,** 380.

Rajewsky, B., Bucker, H. and Pauli, H. (1959). The UV-action spectrum of the lysine decarboxylase induction in *Bacterium cadaveris*. *Arch. Biochem. Biophys.* **82,** 229.

Raman, C. V. (1922). *Molecular Diffraction of Light.* University of Calcutta, Calcutta.

Ramsay, D. A. (1957). Electronic spectra of polyatomic free radicals. *Ann. N.Y. Acid. Sci.* 67, 485.

Rapkine, L., Shugar, D. and Siminovitch, L. (1949). Inactivation de la triose phosphate dehydrogenase par la lumiere ultra-violette. *J. Chim. Phys.* **46,** 635.

Rapport, D. and Canzanelli, A. (1950). The photochemical action of ultraviolet light on the absorption spectra of nucleic acid and related substances. *Science*, **112,** 469.

Rapport, D., Canzanelli, A. and Sossen, R. (1955). Effect of ultraviolet irradiation on the capacity of uridine to maintain growth in a mutant of *Neurospora crassa*. *Fed. Proc.* **14,** 386.

Rebeyrotte, N. and Latarjet, R. (1960). Wirkungen nichtionisierender strahlung anfeine transformierende nucleinsaüre des pneumokokkus. *Strahlentherapie*, **111,** 85.

Reddi, K. K. (1957). Ultraviolet absorption studies on tobacco mosaic virus. *Biochim. Biophys. Acta*, **24,** 238.

Reddi, K. K. (1958). Ultraviolet absorption studies on TMV nucleic acid. *Biochim. Biophys. Acta*, **27,** 1.

Reid, C. (1957a). *Excited states in Chemistry and Biology.* Academic Press, New York.

Reid, C. (1957b). The relation between primitive and present-day photobiological processes, in *1st Intern. Symp. on Origin of Life on the Earth*, Academy of Sciences, U.S.S.R., Moscow, 1957; Pergamon Press, London, 619, (1959).

Rhodes, W. (1961). Energy transfer and hypochromicty in polynucleotides. *Abst. 139th meeting Amer. Chem. Soc.*, St. Louis; *J. Amer. Chem. Soc.* **82,** 6409.

Rice, E. W. (1952). The action of ultraviolet light on the pentose moiety of nucleic acids and related compounds. *Science*, **115,** 92.

Rich, A. (1958). Formation of 2- and 3-stranded helical molecules by polyinosinic acid and polyadenylic acid. *Nature*, **181,** 521.

Rich, A. (1959). Polynucleotide interactions and the nucleic acids, in *Structure and Functions of the Genetic Elements, Brookhaven Symp. in Biol.*, No. 12, 17.

Rich, A. and Kasha, M. (1960). A hybrid helix containing both deoxyribose and ribose polynucleotides and its relation to the transfer of information between nucleic acids. *Proc. Nat. Acad. Sci. (U.S.)*, **46,** 1044.

Richards, F. M. (1955). An assay for ribonuclease activity using a cyclic phosphate substrate. *C.R. trav. Lab. Carlsberg, Ser. chim.* **29,** 315.

Rideal, E. K. and Mitchell, J. S. (1937). Photochemical reactions in monolayers. I. *Proc. Roy. Soc. London*, **159A,** 206.

Rideal, E. K. and Roberts, R. (1951). The photochemistry of native proteins. *Proc. Roy. Soc.* **A205,** 391.

Riley, D. P. and Arndt, V. W. (1953). X-ray scattering evidence regarding the protein-nucleic acid combination in nucleoproteins. *Nature*, **172,** 294.

ROGER, M. and HOTCHKISS, R. D. (1961). Selective heat inactivation of pneumococcal transforming deoxyribonucleate. *Proc. Nat. Acad. Sci.* **47,** 653.

ROIZMAN, B., MAYER, M. M. and ROANE, P. R. (1959). Immunochemical studies of poliovirus. IV. Alteration of the immunological specificity of purified poliomyelitis virus by heat and ultraviolet light. *J. Immunol.* **82,** 19.

ROLLEFSON, G. K. and BURTON, M., eds. (1942). *Photochemistry and the Mechanism of Chemical Reaction.* Prentice-Hall, New York.

RORSCH, A., BEUKERS, R., IJLSTRA, J. and BERENDS, W. (1958). The effect of ultraviolet on some components of the nucleic acids. I. Uracil, thymine. *Rec. trav. chim.* **77,** 423.

ROSENKRANZ H. S. and BENDICH, A. (1958). On the nature of the deoxyribonucleic acid-methyl green reaction. *J. Biophys. Biochem. Cytology,* **4,** 663.

RUCH, F. (1957). Ultraviolet dichroism of nucleic acids and proteins in cell structures. *Acta histochim.* **4,** 193.

RUPERT, C. S. (1958). A biological test for ultraviolet damage and photoreversal in non-transforming DNA. *Fed. Proc.* **17,** 301.

RUPERT, C. S. (1960). Photoreactivation of transforming DNA by an enzyme from bakers' yeast. *J. Gen. Physiol.* **43,** 573.

RUPERT, C. S. (1962a). Photoenzymatic repair of UV damage in DNA. I. Kinetics of the reaction. *J. Gen. Physiol.* **45,** 703.

RUPERT, C. S. (1962b). Photoenzymatic repair of UV damage in DNA. II. Formation of an enzyme-substrate complex. *Ibid.,* **45,** 725.

RUPERT, C. S. (1963). In *Photobiology, action of light on living materials,* (A. C. GIESE ed.), Academic Press, N.Y.

RUPERT, C. S. and Boodgal, S. H. (1960). Shape of ultraviolet inactivation curves of transforming deoxyribonucleic acid. *Nature,* **185,** 556.

RUPERT, C. S., GOODGAL, S. H. and HERRIOT, R. M. (1958). Photoreactivation in vitro of UV inactivated *Hemophilus influenza* transforming factor. *J. Gen. Physiol.* **41,** 451.

RUPERT, C. S. and HERRIOT, R. M. (1958). Photoreactivation of ultraviolet damaged transforming factor and other deoxyribonucleic acids. *7th Intern. Congr. Microbiol., Stockholm, communication No. 4s.*

RUSHIZKY, G. W., KNIGHT, C. A. and MCLAREN, A. D. (1960). A comparison of the ultraviolet light inactivation of infectious ribonucleic acid preparations from tobacco mosaic virus with those of the native and reconstituted virus. *Virology,* **12,** 32.

RUSHIZKY, G. W. and PARDEE, A. B. (1962). Ultraviolet light induced spectral changes in nucleotides and nucleic acids. *Photochem. Photobiol.* **1,** 15.

RUSHIZKY, G. W., RILEY, M., PRESTIDGE, L. S. and PARDEE, A. B. (1960). Inactivation of enzyme formation by ultraviolet light. *Biochim. Biophys. Acta,* **45,** 70.

RUSSELL, G. (1958). Purity of gelatin as a protein. *Nature,* **181,** 102.

SAGAN, C. (1961). On the origin and planetary distribution of life. *Rad. Res.* **15,** 174–192.

SAIDEL, L. J. (1955a). Ultraviolet absorption spectra of peptides. I. Compounds containing a single monoalkyl-substituted amide link. *Arch. Biochem. Biophys.* **54,** 184.

SAIDEL, L. J. (1955b). Ultraviolet absorption of peptides. II. Glycine peptides containing more than one substituted amide link. *Arch. Biochem. Biophys.* **56,** 45.

SAIDEL, L. J. (1955c). Ultraviolet absorption spectra of peptides. III. N,N-dialkyl amides including polyvinylpyrrolidone. *J. Amer. Chem. Soc.* **77,** 3892.

SAIDEL, L. J., GOLDFARB, A. R. and KALT, W. B. (1951). False absorption bands in the region 200–230μ caused by stray radiation in the Beckman spectrophotometer. *Science*, **113**, 683.

SAIDEL, L. J., GOLDFARB, A. R. and WALDMAN, S. (1952). The absorption spectra of amino acids in the region 200–230μ. *J. Biol. Chem.* **197**, 285.

SAIDEL, L. J. and LIEBERMAN, H. (1958). Ultraviolet absoprtion spectra of peptides. IV. Alanine residues. *Arch. Biochem. Biophys.* **76**, 401.

SAKAMURA, S. and OBATA, Y. (1958). Denaturation of lactoglobulin by radiation with ultraviolet light. *Nippon Nogei Kagakn Kaishi*, **32**, 667.

SANTAMARIA, L. and ANGELOTTI, A. (1957). Misura dell'attivita fotodinamica di diverse sostanze mediante metodo batteriologico. *Atti della Soc. Italiana Patologia*, **5**, 463.

SAUERBIER, W. (1960). Effects of ultraviolet light on hydrogen bonds in deoxyribonucleic acid of phage T1. *Nature*, **188**, 329.

SAUTER, E. (1957). Die messung der strahlensterke der linie 1949 Å bei einer quecksilberdampf-Niederdrucklampe. *Z. angew. Phys.* **9**, 105.

SCHAUENSTEIN, E. (1949). Lichtabsorption von eiweisstoffen. *Monatsh.* **80**, 820.

SCHAUENSTEIN, E. (1957). Das UV-absorptionsspectrum der proteine—Grundlagen und Grenzen der ausvertung. *Acta histochim.* **4**, 208.

SCHAUENSTEIN, E., BAYZER, H. and KRONEGGER, K. (1955). Ein absorptions effekt im UV-spectrum hitzedenaturierter serumproteine. *Monatsh.* **86**, 637.

SCHAUENSTEIN, E. and TREIBER, E. (1950). Ultraviolet absorption spectra of actomyosin. *J. Polymer Sci.* **5**, 145.

SCHENCK, G. O. (1948). Photosensitized reactions with molecular oxygen. *Naturwiss.* **35**, 28.

SCHENCK, G. O. (1960). Selektivitat und typische reaktionsmechanismem in der strahlenchemie. *Z. Elektrochem.* **64**, 997.

SCHENCK, G. O. and KOCH, E. (1960). Intermediate reactions in photosensitized processes in solution. *Z. Elektrochem.* **64**, 170.

SCHERAGA, H. A. (1957). Tyrosyl-carboxylate ion hydrogen bonding in ribonuclease. *Biochim. Biophys. Acta*, **23**, 196.

SCHOCKEN, K. (1951). The effect of ultraviolet irradiation on certain amino acids. *Exper. Med. and Surgery*, **9**, 465.

SCHOCKEN, K. (1952). The effect of ultraviolet radiation on sulfhydryl-and disulfide-containing amino acids. *Science*, **116**, 544.

SCHOLES, G. and WEISS, J. (1959). Oxygen effects and formation of peroxides in aqueous solution. *Rad. Res.*, Suppl. 1, 177.

SCHOLES, G., WEISS, J. and WHEELER, C. M. (1956). Formation of hydroperoxides from nucleic acids by irradiation with X-rays in aqueous systems. *Nature*, **178**, 157.

SCHOMER, H. A. (1936). The effects of radiation on enzymes, in *Biological Effects of Radiation* (Duggar, B. M. ed.), McGraw-Hill, New York. pp. 1151–1165.

SCHONBERG, A. (1958). *Praparative Organisms Photochemie.* Springer-Verlag, Berlin.

SCHRAMM, G. and DANNENBERG, H. (1944). Uber die ultraviolettabsorption des tobakmosaikvirus. *Ber.* **77B**, 53.

SCHRAMM, G. and ENGLER, R. (1958). The latent period after infection with tobacco mosaic virus and virus nucleic acid. *Nature*, **181**, 916.

SCHULZ, P. (1947). Elektrische entladungen in edelgasen bei hohen druchen. *Ann. Physik.* **6** ser., 1, 95.

SCHUSTER, H. and SCHRAMM, G. (1958). Determination of the biologically active unit in ribonucleic acid of tobacco mosaic virus by chemical means. *Z. f. Naturfor.* **13**, 697.

SCOTT, J. F. (1955). Ultraviolet absorption spectrophotometry, in *Physical Techniques in Biological Research* (OSTER, C. and POLLISTER, A. W. eds.), Academic Press, New York, vol I, pp. 131–205.

SCOTT, J. F. and SINSHEIMER, R. L. (1955). Technique of study of biological effects of ultraviolet radiation, in *Radiation Biology* (HOLLAENDER, A. ed.), McGraw-Hill, New York. vol. 2, chap. 4, pp. 119–164.

SCOTT, J. F., SINSHEIMER, R. L. and LOOFBOUROW, J. R. (1948). Low-temperature spectroscopy of biological compounds. *Science*, **107**, 302.

SEEDS, W. E. and WILKINS, M. F. (1950). Ultraviolet microspectrographic studies of nucleoproteins and crystals of biological interest. *Dis. Faraday Soc.* **9**, 417.

SEIBERT, F. B. (1953). Tuberculin proteins. *Dis Faraday Soc.* **13**, 251.

SELA, M. and ANFINSEN, C. B. (1957). Some spectrophotometric and polarimetric experiments with ribonuclease. *Biochim. Biophys. Acta*, **24**, 229.

SELA, M. ANFINSEN, C. B. and HARRINGTON, W. F. (1957). The correlation of ribonuclease activity with specific aspects of tertiary structure. *Biochim. Biophys. Acta*, **26**, 502.

SERAYDARIAN, M. W. (1955). Inactivation of TPN by ultraviolet irradiation and the identification of decomposition products. *Amer. J. Physiol.* **181**, 291.

SERAYDARIAN, M. W., CANZANELLI, A. and RAPPORT, D. (1953). Correlation of changes in electrometric titration, absorption spectra and phosphate content of nucleic acid produced by ultraviolet radiation. *Amer. J. Physiol.* **172**, 42.

SERAYDARIAN, M. W., COHEN, A. I. and SABLE, H. Z. (1954). Inactivation of DPN by ultraviolet irradiation and identification of degradation products. *Amer. J. Physiol.* **177**, 150.

SETLOW, J. K. (1961). Ultraviolet inactivation of transforming DNA. *Rad. Res.* **14**, 500.

SETLOW, J. K. and SETLOW, R. B. (1960). Evidence for the existence of a single-stranded stage of T2 bacteriophage during replication. *Proc. Nat. Acad. Sci. U.S.* **46**, 791.

SETLOW, J. K. and SETLOW, R. B. (1961). Ultraviolet action spectra of single and double-stranded DNA. Biophys. Soc. meeting, Feb. 15–18, St. Louis, Miss.

SETLOW, J. K. and SETLOW, R. B. (1963). Nature of the photoreactiveable lesions in DNA. *Nature*, **197**, 560–562.

SETLOW, R. (1955a). A relation between cystine content and ultraviolet sensitivity of proteins. *Biochim. Biophys. Acta*, **16**, 444.

SETLOW, R. (1955b). Radiation studies of proteins and enzymes. *Ann. New York Acad. Sci.* **59**, 469.

SETLOW, R. (1957). Action spectroscopy. *Advances Med. Biol. Phys.* **5**, 37.

SETLOW, R. (1960). The use of action spectra to determine the physical state of DNA in vivo. *Biochim. Biophys. Acta*, **39**, 180.

SETLOW, R. (1961). The action spectrum for the reversal of the dimerization of thymine induced by ultraviolet light. *Biochim. Biophys. Acta*, **49**, 237.

SETLOW, R. and BOYCE, R. (1960). The ultraviolet light inactivation of ϕX174 bacteriophage at different wavelengths and pH's. *Biophys. J.* **7**, 29.

SETLOW, R. and CARRIER, W. L. (1963). Identification of ultraviolet-induced thymine dimers in DNA by absorbance measurements. *Photochem. Photobiol.* **2**, 49–57.

SETLOW, R. and DOYLE, B. (1953a). The effect of light of wavelength 2537 Å on proteins and nucleic acids at low temperatures. *Biochim. Biophys. Acta*, **12**, 508.

SETLOW, R. and DOYLE, B. (1953b). The combined effect of temperature and ultraviolet radiation on dry catalase. *Arch. Biochem. Biophys.* **46**, 31.

SETLOW, R. and DOYLE, B. (1954a). The effect of temperature on the ultraviolet light inactivation of trypsin. *Arch. Biochem. Biophys.* **48,** 441.

SETLOW, R. and DOYLE, B. (1954b). The action of radiation on dry DNA. I. *Biochim. Biophys. Acta,* **15,** 117.

SETLOW, R. and DOYLE, B. (1956). The action of ultraviolet light on paramecin and the chemical nature of paramecin. *Biochim. Biophys. Acta,* **22,** 15.

SETLOW, R. and DOYLE, B. (1957). The action of monochromatic ultraviolet light on proteins. *Biochim. Biophys.* **24,** 27.

SETLOW, R. and GUILD, W. R. (1951). The spectrum of the peptide bond and other substances below 230mμ. *Arch. Biochem. Biophys.* **34,** 223.

SETLOW, R. and MCNEAL, A. (1961). Physical state of intracellular DNA of phage ϕX174. Biophys. Soc. meeting, Feb. 15–18, St. Louis, Miss.

SETLOW, R., ROBBINS, S. and POLLARD, E. (1955). Action spectrum for latent period extension of T1 bacteriophage. *Rad. Res.* **2,** 262.

SETLOW, R., WATTS, G. and DOUGLAS, C. (1957). Inactivation of proteins by vacuum ultraviolet radiation. *Proc. 1st Nat Biophys. Conf., Columbus, 1957.* Yale Univ. Press, New Haven, 1959, pp. 174–183.

SHACK, J. and THOMPSETT, J. M. (1952). Studies of the anomalous titration of calf thymus des oxypentose nucleic acid and nucleohistone. *J. Biol. Chem.* **197,** 17.

SHAPIRO, H. S. and CHARGAFF, E. (1957). Studies on the nucleotide arrangement in desoxyribonucleic acids. I. The relationship between the production of pyrimidine nucleoside 3′,5′-diphosphates and specific features of nucleotide sequence. *Biochim. Biophys. Acta,* **26,** 596.

SHAUGHNESSY, H. J., WOLF, A. M., JANOTA, M., NEAL, J., OPPENHEIMER, F., MILZER, A., NAFTULIN, H. and MORRISSEY, R. A. (1957). *Proc. Soc. Exp. Biol. Med.* **95,** 251.

SHEPPARD, E. (1951). Thesis. Polytechnic Institute of Brooklyn.

SHIBATA, K. (1958). Spectrophotometry of intact biological materials. *J. Biochem. (Japan),* **45,** 599.

SHIBATA, K., BENSON, A. A. and CALVIN, M. (1954). The absorption spectra of suspensions of living organisms. *Biochim. Biophys. Acta,* **15,** 461.

SHIGEMOTO, T. (1958). Photochemical decomposition of some nucleotides. *Osaka Daigaku Igaku Zasshi,* **10,** 513. *Chem. Abst.* **52,** 13832i.

SHOOTER, K. V., DAVISON, P. F. and BUTLER, J. A. V. (1954). Physical properties of thymus nucleoprotein. *Biochim. Biophys. Acta,* **13,** 192.

SHORE, V. G. and PARDEE, A. B., (1956a). Fluorescence of some proteins, nucleic acids and related compounds. *Arch. Biochem. Biophys.* **60,** 100.

SHORE, V. G. and PARDEE, A. B. (1956b). Energy transfer in conjugated proteins and nucleic acids. *Arch. Biochem. Biophys.* **62,** 355.

SHUGAR, D. (1951a). Ultra-violet irradiation of triosephosphate dehydrogenase. *Biochim. Biophys. Acta,* **6,** 548.

SHUGAR, D. (1951b). Photoreactivation in the near ultraviolet of *d*-glyceraldehyde-3-phosphate dehydrogenase. *Experientia,* **7,** 26.

SHUGAR, D. (1952a). The ultraviolet absorption spectrum of ribonuclease. *Biochem. J.* **52,** 142.

SHUGAR, D. (1952b). The measurement of lysozyme activity and the ultraviolet inactivation of lysozyme. *Biochim. Biophys. Acta,* **8,** 302.

SHUGAR, D. (1953). The binding of basic dyes by ribonucleic acid and the measurement of ribonuclease activity. *Bull. Acad. Polon. Sci., Cl. II,* **1,** 39.

SHUGAR, D. (1954). The riboflavin photo-sensitized oxidation of some 3-substituted indole derivatives. *Acta Biochim. Polon.* **1,** 3.

SHUGAR, D. (1955). Enzymes hydrolyzing nucleic acids (ribonuclease and desoxyribonuclease). *Postepy Biochim.* **1,** 269.

SHUGAR, D. (1958). Applications of radioactive indicators to quantitative staining reactions. *Proc. 2nd Intern. Conf. Peaceful Uses Atomic Energy, Geneva,* **24,** 263.

SHUGAR, D. (1960a). Photochemistry of nucleic acids and their constituents, in *The Nucleic Acids* (CHARGAFF, E., DAVIDSON, J. N. eds.), Academic Press, New York. vol. 3, chap. 31, pp. 39–104.

SHUGAR, D. (1960b). Photoreactivation and reversible photolysis of nucleic acids. *3rd Intern. Cong. Photobiol.* Copenhagen.

SHUGAR, D. (1961). Quantitative staining in histo- and cytochemistry, in *Progress in Biophysics* (BUTLER, J. A. V., ed.), Pergamon Press, London, vol. 12.

SHUGAR, D., ADAMIEC, A. and SZTUMPF, E. (1959). Role of peptide bond absorption in protein photochemistry. *Acta Biochem. Polon.* **6,** 417.

SHUGAR, D. and BARANOWSKA, J. (1957). Quantitative staining with radioactive indicators. ^{131}I in gram staining of micro-organisms. *Intern. Conf. Radioisotopes in scientific research,* UNESCO, Paris, Paper No. 149, Pergamon Press, London, 1958.

SHUGAR, D. and BARANOWSKA, J. (1960). Aggregation and staining behavior of UV-irradiated films of nucleic acids. *Nature,* **185,** 33.

SHUGAR, D. and FOX, J. J. (1952). Spectrophotometric studies of nucleic acid derivatives and related compounds as a function of pH. I. Pyrimidines. *Biochim. Biophys. Acta,* **9,** 199.

SHUGAR, D. and RZENDOWSKA, F. (1956). The ultraviolet photolysis of ribonuclease. *Bull. Acad. Polon. Sci., Cl. II,* **4,** 293.

SHUGAR, D. and RZENDOWSKA, F. (1959a). Phenolic dissociation in, and the photochemistry and photo-sensitized inactivation of desoxyribonuclease. Unpublished results.

SHUGAR, D. and SZER, W. (1962). Secondary structure in poly-ribothymidylic acid. *J. Mol. Biol.* **5,** 580–582.

SHUGAR, D. and WIERZCHOWSKI, K. (1957). Reversible photolysis of pyrimidine derivatives and trials with nucleic acids. *Biochim. Biophys. Acta,* **23,** 657.

SHUGAR, D. and WIERZCHOWSKI, K. L. (1958a). Structure and photochemical behaviour of nucleic acids and related components. *J. Polymer Sci.* **31,** 269.

SHUGAR, D. and WIERZCHOWSKI, K. L. (1958b). Photochemistry of nucleic acids, nucleic acid derivatives and related compounds. *Post. Biochemii,* **4,** Suppl., 243.

SHUGAR, D. and WIERZCHOWSKI, K. L. (1958c). Preparation in quantitative yield of nucleoside-2′:3′-phosphates and some ribonucleotide isomers. *Bull. Acad. Polon. Sci., Cl. II,* **6,** 283.

SHUGAR, D. and WIERZCHOWSKI, K. L. (1960a). Photochemistry of pyrimidine derivatives in oligonucleotides and in single and double stranded poly-nucleotides. *4th Inter. Cong. Photobiol.,* Copenhagen.

SHUGAR, D. and WIERZCHOWSKI, K. L. (1960). Photochemistry of pyrimidine residues in oligonulceotides and in single and double stranded polynucleotides. *3rd Internat. Cong. Photobiol.,* Copenhagen.

SIEGEL, A. (1957). Ultraviolet irradiation of dry tobacco mosaic virus. *Nature,* **180,** 1430.

SIEGEL, A., GINOZA, W. and WILDMAN, S. G. (1957). The early events of infection with tobacco mosaic virus nucleic acid. *Virology,* **3,** 554.

SIEGEL, A. and NORMAN, A. (1958). Action spectrum for two strains of TMV. *Virology,* **6,** 725.

SIEGEL, A. and WILLIAMS, S. G. (1954). Some natural relationships among strains of tobacco mosaic virus. *Phytopatology,* **44,** 277.

SIEGEL, A. and WILDMAN, S. G. (1956). The inactivation of the infectious centers of tobacco mosaic virus by ultraviolet light. *Virology*, **2,** 69.

SIEGEL, A., WILDMAN, S. G. and GINOZA, W. (1956). Sensitivity to ultraviolet light of infectious TMV nucleic acid. *Nature*, **178,** 1117.

SIEGEL, J. M., MONTGOMERY, G. A. and BOCK, R. M. (1959). Ultraviolet absorption spectra of DPN and analogs of DPN. *Arch. Biochem. Biophys.* **82,** 288.

SIMON, M. I. and VAN VUNAKIS, H. (1962). The photodynamic reaction of methylene blue with DNA. *J. Mol. Biol.*, **4,** 488–499.

SINGER, M. F., HEPPEL, L. A., RUSHIZKY, G. W. and SOBER, H. A. (1963). Spectral properties of adenine oligoribonucleotides. *Biochim. Biophys. Acta*, **61,** 474–477.

SINGER, S. J. (1962). Properties of proteins in non-aqueous solvents. *Adv. Protein Chem.*, **17,** 1–60.

SINSHEIMER, R. L. (1954a). The photochemistry of uridylic acid. *Rad. Res.* **1,** 505.

SINSHEIMER, R. L. (1954b). The action of pancreatic DNase. I. Isolation of mono- and dinucleotides. *J. Biol. Chem.* **208,** 445.

SINSHEIMER, R. L. (1955). Ultraviolet absorption spectra, a review, *Radiation Biology* (Hollaender, A. ed.), McGraw-Hill, New York. vol. 2, pp. 165–201.

SINSHEIMER, R. L. (1957). The photochemistry of cytidylic acid. *Rad. Res.* **6,** 121.

SINSHEIMER, R. L. (1957). Some properties of bacteriophage ϕX174. *Fed. Proc.* **16,** 250.

SINSHEIMER, R. L. (1959). Is the nucleic acid message in a two-symbol code? *J. Mol. Biol.* **1,** 218.

SINSHEIMER, R. L. and Hastings, R. (1949). A reversible photochemical alteration of uracil and uridine. *Science*, **110,** 525.

SINSHEIMER, R. L. and KOERNER, J. F. (1952). Di-desoxyribonucleotides. *J. Amer. Chem. Soc.* **74,** 283.

SIZER, I. W. (1945). Essential groups of crystalline chymostrypsin. *J. Biol. Chem.* **160,** 547.

SKREB, Y. and ERRERA, M. (1957). Action des rayons UV sur des fragments nuclees et enuclees d'amibes. *Exp. Cell Res.* **12,** 649.

SMIETANOWSKA, A. and SHUGAR, D. (1961). Photochemistry of model oligo and poly-nucleotides. V. Properties of photodimers of uracil and thymine. *Bull. Acad. Polon. Sci., Cl. II.*

SMITH, D. B. and BUTLER, G. C. (1951). On the mechanism of action of ionizing radiations on sodium thymonucleate. *J. Amer. Chem. Soc.* **73,** 258.

SMITH, E. L., BROWN, D. M. and LASKOWSKI, M. (1951). Molecular weights of chymotrypsinogen and chymotrypsin α and β. *J. Biol. Chem.* **191,** 639.

SMITH, K. C. (1963). Photochemistry of the nucleic acids. In *Photophysiology* (GIESE, A. C. ed.), Academic Press, N.Y.

SMITH, K. C. and ALLEN, F. W. (1953). The liberation of polynucleotides by the alkaline hydrolysis of ribonucleic acid from yeast. *J. Amer. Chem. Soc.* **75,** 2131.

SOGO, P., JOST, M. and CALVIN, M. (1959). Evidence for free-radical production in photosynthesizing systems. *Rad. Res., Suppl.* **1,** 511.

SPECTOR, W. S. (1956). *Handbook of Biological Data*. W. B. Saunders, Philadelphia, Pa. p. 31.

SPIEGEL-ADOLF, M. (1958). Photosensitization of nucleic acids and pyrimidines. *Radiol. Clin.* **27,** 39.

STACEY, K. A. (1956). *Light-Scattering in Physical Chemistry*. Butterworth, London.

STADLER, L. J. and UBER, F. M. (1942). Genetic effects of ultraviolet radiation in maize. IV. Comparison of monochromatic radiation. *Genetics*, **27,** 84.

STAHL, F. W. (1959). Radiobiology of bacteriophage, *The Viruses* (BURNETT, F. M. and STANLEY, W. M. eds.), vol. 2, Chap. 10, Academic Press, New York.

STAHL, F. W. CRASEMANN, J. M., OHUN, L., FOX, E. and LAIRD, C. (1961). Radiation sensitivity of bacteriophages containing 5-bromodeoxyridine. *Virology*, **13,** 98.

STANLEY, W. M. (1936). The inactivation of crystalline tobacco mosiac virus protein. *Science*, **83,** 626.

STEELE, R. H. and SZENT-GYORGI, A. (1958). Studies on the excited states of proteins. *Proc. Nat. Acad. Sci.* **44,** 540.

STEIN, G. (1954). Etats metastables dans les systemes biologiques irradies. *J. Chim. Phys.* **51,** 133.

STEIN, W. and LASKOWSKI, W. (1956). Verleich der inaktivierugen bei *Escherichia coli* B durch organische peroxide and UV-bestrahlung unter besonderer beruck-sichtigung der warme- und photoreaktivierbarkeit. *Z. Naturforsch.* **11b,** 643.

STEIN, W. and MEUTZNER, I. (1950). Reaktivierung von UV-inaktivierten *Bakterium coli* durch warme. *Naturwiss.* **37,** 167.

STEINER, R. F. and EDELHOCH, H. (1961). Influence of pH and urea on the ultra-violet fluorescence of several globular proteins. *Nature*, **192,** 873.

STEINER, R. F. and EDELHOCH, H. (1962). Fluorescent protein conjugates. *Chemical Rev.*, **62,** 457–483.

STENT, G. S. and FUERST, C. R. (1960). Genetic and physiological effects of the decay of incorporated radioactive phosporus in bacterial viruses and bacteria. *Advances Biol. Med. Phys.* **7,** in press.

STIMSON, M. M. (1942). Ultraviolet absorption spectra of nitrogenous heterocycles. IV. Effect of pH and irradiation on the spectra of isoguanine and 2-hydroxy-6, 8-diaminopurine. *J. Amer. Chem. Soc.* **64,** 1604.

STOCKX, J., Vandendriessche, L. and PARIJS, R. (1960). A new method for the spectrophotometric determination of ribonuclease activites. *Arch. intern. physiol. biochim.* **68,** 417.

STRAIT, L. A., GOYAN, F. M. and KUMLER, W. D. (1956). Semi-solid nickel sulfate filter. *J. Opt. Soc. Amer.* **46,** 1038.

STREISINGER, G. (1956). The genetic control of ultraviolet sensitivity levels in bacteriophage T2 and T4. *Virology*, **2,** 1.

STRONG, J., ed. (1939). *Procedures in Experimental Physics.* Prentice-Hall, New York.

STRONG, R. L., CHIEN, J. C. W., GRAF, P. and WILLARD, J. E. (1957). Studies of iodine atom and bromine atom recombination following flash photolysis of gaseous I_2 and Br_2. *J. Chem. Phys.* **26,** 1287.

STROTHER, G. K. and WOLKEN, J. J. (1959). A simplified microspectrophotometer. *Science*, **130,** 1084.

STRYER, L. (1960). Energy transfer in proteins and polypeptides. *Rad. Res., Suppl.* **2,** 432.

STUY, J. H. (1961). Radiation inactivation of intracellular transforming DNA. Thesis, University of Utrecht, Netherlands.

SUMMER, J. B., GRALEN, N. and ERICKSON-QUENSEL, I. B. (1938). The molecular weight of urease. *J. Biol. Chem.* **125,** 37.

SUZATANI, E. (1958a). Photolysis of tyrosine. I. Photolysis velocity. *Nagasaki Igakkai Zasshi*, **33,** 326. *Chem. Abst.* **52,** 11590c).

SUZATANI, E. (1958b). Photolysis of tyrosine. II. Photolysis products of L-tyrosine. *Nagasaki Igakkai Zaschi*, **33,** 507. *Chem. Abst.* **52,** 12030h.

SUZUKI, K. and ONO, J. (1960). Biochemical heterogeneity of the RNA synthesized by *E. coli B* after irradiation with UV light. *Nature*, **183,** 395.

SWENSON, P. A. (1950). Action spectrum of the inhibition of galactozymase produced by ultraviolet light. *Proc. Nat. Acad. Sci.* **36,** 699.

SWENSON, P. A. and SETLOW, R. B. (1963). *Abs. Biophys. Soc.*, (U.S.).

SYMONDS, M. and MCCLOY, E. W. (1958). The irradiation of phage-infected bacteria: Its bearing on the relationship between functional and genetic radiation damage. *Virology*, **6,** 649.

SZENDRO, P. (1931). Photochemical transformations of imidazole compounds. *S. Arch. des Physiol. Pfugers*, **228,** 743.

SZENDRO, P., LAMPERT, U. and WREDE, F. (1933). Formation of cysteine from cystine by radiation. *Hoppe-Seyler's Zeit. f. physiol. Chemie*, **222,** 16.

SZER, W. and SHUGAR, D. (1961). The preparation and properties of high molecular weight polymers of N-methyluridylic acid. *Acta Biochim. Polon.* **8,** 235.

SZER, W. and SHUGAR, D. (1963). Preparation of poly-5-fluorouridylic acid and the properties of halogenated poly-uridylic acids and their complexes with poly-adenylic acids. *Acta Biochim. Polon.* **10,** 219–231.

SZER, W., SWIERKOWSKI, M. and SHUGAR, D. (1963). Secondary structure of poly-uridylic and poly-ribothymidylic acids, their *N*-methylated analogues and their 1:1 complexes with poly-A. *Acta Biochim. Polon.* **10**, 87–106.

SZTUMPF, E. and SHUGAR, D. (1962a). Photochemistry of model oligo- and poly-nucleotides. VI. Photodimerization and its reversal in thymine dinucleotide analogues. *Biochim. Biophys. Acta.* **61,** 555; (1962b). Microscope cover glasses as cut-off filters in the near ultraviolet. *Photochem. Photobiol.* **1,** 337.

TAFT, E. and APKER, L. (1953). Photo-emission from caesium and rubidium tellurides. *J. Opt. Soc. Amer.* **43,** 81.

TAMM, I. and FLUKE, D. (1950). The effect of monochromatic ultraviolet radiation on the infectivity and hemaglutinating ability of influenza virus, type A strain PR-8. *J. Bact.* **59,** 449.

TANFORD, C. and HAUENSTEIN, J. D. (1956). Hydrogen-ion equilibrium of ribonuclease. *J. Amer. Chem. Soc.* **78,** 5287.

TANFORD, C. and ROBERTS, G. L. (1952). Phenolic hydroxyl ionization in proteins. I. Bovine serum albumin. *J. Amer. Chem. Soc.* **74,** 2509.

TAYLOR, A. R. (1960). Effects of non-ionizing radiations on animal viruses. *Ann. N.Y. Acad. Soc.* **83,** 670.

TAYLOR, A. R. and HAYNES, H. (1947). New meters for germicidal energy. *Gen. Elect. Rev.* **50,** 27.

TAYLOR, A. R., KAY, W. W., MCLEAN, I. W., OPPENHEIMER, F. and STUIRPERT, F. D. (1957a). Effect of ultraviolet light on poliomyelitis virus. *J. Immunol.* **78,** 45.

TAYLOR, A. R., KAY, W. W., TIMM, E. A. HOOK, A. E. and MCLEAN, I. W. (1957). Inactivation of poliomyelitis virus for the preparation of vaccines. I. Formalin and ultraviolet irradiation. *J. Immunology*, **79,** 265.

TAYLOR, A. R., SHARP, D. G., BEARD, D., FINKELSTEIN, H. and BEARD, J. W. (1941). Influence of ultraviolet light on equine encephalomyelitis virus protein (Eastern strain). *J. Infectious Diseases*, **69,** 224.

TAYLOR, E. C. and KAN, R. O. (1963). Photochemical dimerization of 2-amino-pyridines and 2-pyridones. *J. Am. Chem. Soc.* **85,** 776–784.

TAYLOR, E. C., PAUDLER, W. W. and KUNTZ, I. (1961a). A photochemically-induced, valence-bond tautomerism in a six-membered aromatic system. *J. Am. Chem. Soc.* **83,** 2967–2968.

TAYLOR, E. C., KAN, R. O. and PAULDER, W. W. (1961b). The photolysis of 2-aminopyridines and 2-pyridones. *J. Am. Chem. Soc.* **83,** 4484–4485.

TAYLOR, J. F., VELICK, S. F., CORI, G. T., CORI, C. F. and SLEIN, M. W. (1948). The prosthetic group of crystalline *d*-glyceraldehyde-3-phosphate dehydrogenase. *J. Biol. Chem.* **173,** 619.

TAYLOR, L. W. and JONES, L. C. (1956). Modifications of a Beckman model DU quartz spectrophotometer for measurements to 192 mμ. *Anal. Chem.* **28,** 1706.

TEALE, F. W. J. (1960). The ultraviolet fluorescence of proteins in neutral solution. *Biochem. J.* **76,** 381.

TEALE, F. W. J. and WEBER, G. (1957). Ultraviolet fluorescence of the aromatic amino acids. *Biochem. J.* **65,** 476.

TERENIN, A. M. (1957). Photosynthesis in the shortest ultraviolet, *1st Intern. Symp. on Origin of Life on the Earth,* Academy of Sciences, U.S.S.R., Moscow, 1957; Pergamon Press, London, 1959, pp. 136–139.

TESSMAN, E. S. and OZAKI, T. (1960). The interaction of phage S13 with ultraviolet-irradiated host cells and properties of the ultraviolet-irradiated phage. *Virology,* **12,** 431.

TESSMAN, I. and OZAKI, T. (1957). Multiplicity reactivation of bacteriophage T1. *Virology,* **4,** 315.

THOMAS, R. (1954). Recherches sur la dénaturation des acides desoxyribonucléiques. *Biochim. Biophys. Acta,* **14,** 231.

THORELL, B. and RUCH, F. (1951). Molecular orientation and light absorption. *Nature,* **167,** 815.

THRUSH, B. A. (1959). Kinetic spectroscopy in the far ultraviolet: the flash photolysis of ethyl compounds. *Proc. Roy. Soc.* (*London*) **A243,** 555.

TIMOFEEF-ROSSOVSKY, N. K. and ZIMMER, K. G. (1947). *Das trefferprinzip in der biologie.* Hirzel, Leipzig.

TINOCO, I. (1960). Hypochromism in polynucleotides. *J. Amer. Chem. Soc.* **82,** 4785.

TOLLIN, G., SOGO, P. B. and CALVIN, M. (1958). Energy transfer in ordered and unordered photochemical systems. *Ann. New York Acad. Sci.* **74,** 310.

TOLMACH, L. J. and LERMAN, L. S. (1957). Selective effects of ultraviolet radiation and other agents on genetic transformation. *Rad. Res.* **7,** 454.

TOMBS, M. P. and MACLAGAN, N. F. (1961). The effect of hydrolysis on the ultraviolet absorption of proteins near 210 mu. *Biochem. J.* **81,** 41P.

TOMBS, M. P., SONTER, F. and MACLAGAN, N. F. (1959). The spectrophotometric determination of protein at 210 mu. *Biochem. J.* **73,** 167.

TOMINAGA, F. (1958). Products of photolysis of L-phenylalanine. *Nagasaki Igakkai Zassi,* **33,** 753. *Chem. Abs.* **52,** 18591a.

TOMOHIRO, Y. (1953). Photolysis of histidine. *Nagasaki Igakkai Zassi,* **28,** 1117. *Chem. Abst.* **48,** 6842g.

TRAMER, Z. and SHUGAR, D. (1959). Studies on phenolic hydrogen bonding in proteins. *Acta Biochim. Polon.* **6,** 235.

TRAUB, F. B., FRIEDEMANN, U., BRASCH, A. and HUBER, W. (1951). High intensity electrons as a tool for preparation of vaccines. *J. Immunology,* **67,** 379.

TRAUBE, W. (1898). Ueber die Einwirkung des Dicyans auf Acetessigester und Acetylacetone. *Ber.* **31,** 2938–2946.

TREIBER, E. and SCHAUENSTEIN, E. (1949). Einfluss und bericksichtigung der tyndall-streuung bei der bestimmung der konsumptiven lichtabsorption. Messungen an actomyosin, F- and B-actin, seidenfibroin und gelatine. *Z. Naturforsch.* **4b,** 252.

TRISTRAM, G. R. (1953). The amino acid composition of proteins, in *The Proteins* (NEURATH, H. and BAILEY, K. eds.), Academic Press, New York, pp. 181–233.

TSCHACOTIN, S. (1912). Die mikroskopische strahlenstichmethode, eine zelloperationsmethode. *Biol. Zentralbl.* **32,** 623.

TSCHACHOTIN, S. (1938), in *Handbuch der Biologishcen Arbeitsmethoden* (Abderhalden, G., ed.), Urban u. Schwarzenberg, Berlin.

TSUBOI, K. K. (1950). Mouse liver nucleic acids. II. Ultraviolet absorption studies. *Biochim. Biophys. Acta*, **6,** 202.

UBER, F. M. (1941). A quantum yield for the inactivation of tobacco mosaic virus protein. *Nature*, **147,** 148.

UBER, F. M. and MCLAREN, A. D. (1941). A photochemical yield for the inactivation of crystalline trypsin. *J. Biol. Chem.* **141,** 231.

UBER, F. M. and VERBRUGGE, F. (1940). Photochemistry and absorption spectrum of the pyrimidine component of vitamin B_1. *J. Biol. Chem.* **134,** 273.

URETZ, R. E. (1954). Irradiation of parts of individual cells. II. Effects of an ultraviolet microbeam focused on parts of chromosomes. *Science*, **120,** 197.

UREY, H. C. (1962). Lifelike forms in meteorites. *Science*, **137,** 623–628; (see also *Nature*, **193,** 1119–1133).

URRY, W. H., STACEY, F. W., HUYSER, E. S. and JUVELAND, O. O. (1954). The peroxide and light-induced additions of alcohols to olefins. *J. Amer. Chem. Soc.* **76,** 450.

VAN TAMELEN, E. E. and PAPPAS, S. P. (1962). Chemistry of Dewar benzene. 1,2,5-tri-t-butylbicyclo(2,2,0)hexa-2,5-diene. *J. Am. Chem. Soc.* **84,** 3789–3791.

VELICK, S. F. and RONZONI, E. (1948). The amino acid composition of aldolase and *d*-glyceraldehyde phosphate dehydrogenase. *J. Biol. Chem.* **173,** 627.

VENDRELY, R., KNOBLOCH, A. and MATSUDAIRA, W. (1958). A comparative biochemical study of nucleohistones from different vertebrates. *Nature*, **181,** 343.

VENDRELY, R. and VENDRELY, C. (1953). Arginine and DNA content of erythrocyte nuclei and sperms of some fishes. *Nature*, **172,** 30.

VERBRUGGE, F. (1942). Thesis, University of Missouri.

VERBRUGGE, F. (1943). The inactivation of trypsin by ultraviolet radiation. *J. Biol. Chem.* **149,** 405.

VERLINDE, J. D. and KLARENBECK, A. (1952). Studies on Immunity in Poliomyelitis. I. *Arch. ges. Virusforsch.* **5,** 27.

VLADIMIROV, I. A. (1959). Fluorescence of aromatic amino acids in solutions, crystals and proteins. *Izvest. Akad. Nauk S.S.S.R., Ser. Fiz.* **23,** 86.

VLADIMIROV, I. A. and KONEV, S. V. (1957). On the possiblity of energy transfer in protein molecules. *Biophysics*, **2,** 1.

VLADIMIROV, I. A. and LITVIN, F. F. (1960). Long-lived phosphorescence of aromatic amino acids and proteins at low temperatures. *Biophysics*, **5,** 151.

VODRAZKA, Z. (1959). Photo-oxidation of proteins. *Chem. Listy*, **53,** 829.

VODRAZKA, Z. and PRISTONPILOVA, K. (1957). Photo-oxidation of blood proteins. VII. Changes in amino acid composition. *Chem. Listy*, **51,** 1657.

VODRAZKA, Z. and SPONAR, J. (1957). Photo-oxidation of blood proteins. VI. Comparison of the effects of various sensitizers. *Chem. Listy*, **51,** 1649.

VOIGT, J. (1958). Uber den photochemischen abbau von ungesattigen polyesterharzen. Stadium der UV-absorptionsspektrum. *Makromol. Chem.* **27,** 80.

VOLKIN, E. and COHN, W. E. (1954). Estimation of nucleic acids, in *Methods of Biochemical Analysis*, Interscience, New York, vol. I.

VOLMAN, D. H. (1941). The photochemistry of acetamide in water solution. *J. Amer. Chem. Soc.* **63,** 2000.

VON BORSTEL, R. C. and WOLFF, S. (1955). Photoreactivation experiments on the nucleus and cytoplasm of the Habrobracon egg. *Proc. Nat. Acad. Sci. U.S.* **41,** 1004.

WACKER, A. (1963). In *Progress in Nucleic Acid Research* (J. N. DAVIDSON and W. E. COHN eds.), Academic Press, N.Y.

WACKER, A., DELLWEG, H. and LODEMANN, E. (1961a). Strahlenchemisches veranderung der nucleinsauren *in vivo* und *in vitro*. 2 Mitteil. *Angewandt. Chem.* **73,** 64.

WACKER, A., DELLWEG, H. and WEINBLUM, D. (1960). Strahlenschemische veranderung der nucleinsauren *in vivo* und *in vitro*. I Mitteil. *Naturwiss.* **47,** 477.

WACKER, A., MENNIGMANN, H. D. and SZYBALSKI, W. (1962). Effects of "visible" light on 5-bromouracil-labelled DNA. *Nature,* **196,** 685–686.

WACKER, A., TRAGER, L. and WEINBLUM, D. (1961b). Strahlenschemische veranderung der nucleinsauren *in vivo* und *vitro*. 3 Mitteil. Untersuchungen an pyrimidin-ribosiden und pyrimidinen. *Angewandt. Chem.* **73,** 65.

WANG, S. Y. (1958). Photochemistry of nucleic acids and related compounds. I. The first step in the ultraviolet irradiation of 1,3-dimethyluracil. *J. Amer. Chem. Soc.* **80,** 6196.

WANG, S. Y. (1958). Ultraviolet irradiation of 1,3-dimethyluracil. *Intern. Congr. Rad. Res.*, Burlington, Vt., Paper 304.

WANG, S. Y. (1959). Phototautomerization of cytosine derivatives by ultraviolet irradiation. *Nature,* **184,** suppl. 4, 184.

WANG, S. Y. (1961). Photochemical reactions in frozen solution. *Nature,* **190,** 690.

WANG, S. Y., APICELLA, M. and STONE, B. R. (1956). Ultraviolet irradiation of 1,3-dimethyluracil. *J. Amer. Chem. Soc.* **78,** 4180.

WARBURG, O. (1949). *Heavy Metal Prosthetic Groups and Enzyme Action.* Oxford University Press, London. pp. 99–101.

WARBURG, O. and CHRISTIAN, W. (1935). Zerstorung der wasserstoffubertragender Co-ferments durch ultraviolettes licht. *Biochem. Zeit.* **282,** 221.

WARBURG, O. and SCHOCKEN, V. (1949). A manometric actinometer for the visible spectrum. *Arch. Biochem.* **21,** 363.

WARNER, R. C. (1957). Studies on polynucleotides synthesized by polynucleotide phosphorylase. III. Interaction and ultraviolet absorption. *J. Biol. Chem.* **229,** 711.

WARNER, R. C. (1958). Formation and dissociation of polynucleotide complexes. *IVth Intern. Congr. Biochem.*, Vienna.

WATANABE, J. (1960). Photolysis of *dl*-alanine. *Nagasaki Igakkai Zasshi,* **35,** 22.

WEALE, R. (1949a). Absorption spectra, molecular weights, and visual purple. *Nature,* **163,** 163.

WEALE, R. (1949b). Absorption spectra, molecular weights, and visual purple. *Nature,* **164,** 959.

WEATHERWAX, R. S. and LANDMAN, O. E. (1960). Ultraviolet light-induced mutation and DNA synthesis in *E. coli. J. Bacteriol.* **80,** 528.

WEBER, G. (1958). Transfert d'energie dans la dihydro-diphosphopyridine nucleotide. *J. Chim. Phys.* **55,** 878.

WEBER, G. (1960). Fluorescence-polarization spectrum and electronic-energy transfer in proteins. *Biochem. J.* **75,** 345.

WEIGLE, J. J. and DELBRUCK, M. (1951). Mutual exclusion between an infecting phage and a carrier phage. *J. Bacteriol.* **62,** 301.

WEIL, L., BUCHERT, A. R. and MAHER, J. (1952). Photo-oxidation of crystalline lysozyme in the presence of methylene blue and its relation to enzymatic activity. *Arch. Biochem. Biophys.* **40,** 245.

WEIL, L., GORDON, W. G. and BUCHERT, A. R. (1951). Photo-oxidation of amino acids in the presence of methylene blue. *Arch. Biochem. Biophys.* **33,** 90.

WEIL, L., JAMES, S. and BUCHERT, A. R. (1953). Photo-oxidation of crystalline chymotrypsin in the presence of methylene blue. *Arch. Biochem. Biophys.* **46,** 266.

WEIL, L. and SEIBLES, T. S. (1955). Photo-oxidation of crystalline ribonuclease in the presence of methylene blue. *Arch. Biochem. Biophys.* **54,** 368.

WEIL, L., SEIBLES, T. S., SPERO, L. and SCHANTZ, E. J. (1957). Photo-oxidation of crystalline *Clostridium botulinus* type A toxin in the presence of methylene blue. *Arch. Biochem. Biophys.* **68,** 308.

WEISS, J. (1949). The mechanism of photosensitized reactions in solution. *J. Soc. Dyers and Colourists,* **65,** Symposium, Harrogate.

WEISMANN, S. I. (1942). Intramolecular energy transfer. *J. Chem. Phys.* **10,** 214.

WEIZMANN, C., BERGMANN, E. and HIRSCHBERG, Y. (1936). The photochemical deamination of amino acids in water solution. I. *J. Amer. Chem. Soc.* **58,** 1675.

WEIZMANN, C., BERGMANN, F. and HIRSHBERG, Y. (1939). Hydrolytic deamination of amino acids. *Nature,* **143,** 723.

WEIZMANN, C., HIRSHBERG, Y. and BERGMANN, E. (1938). The photochemical deamination of amino acids in water solution. II. *J. Amer. Chem. Soc.* **60,** 1799.

WELLS, P. H. and JOHNSON, H. (1953). Action of ultraviolet and visible light on ATP. *Anat. Record.* **117,** 644.

WELSH, J. N. and ADAMS, M. H. (1954). Photodynamic inactivation of bacteriophage. *J. Bacteriol.* **68,** 122.

WERZEL, R., ESCHKE, W. and ECKART, W. (1960). Uber den UV-inaktivierungs-mechanismus der Bakteriophagen T1 bei verschiedenen wellenlange. *Monatsber. Deut. Akad. Wissen. (Berlin)* **2,** 697.

WEST, W. (1956). Fluorescence and phosphorescence, in *Chemical Applications of Spectroscopy* (Weissberger, A. ed.), Interscience, New York. vol. 9, chap. 6.

WEST, W. and MILLER, W. E. (1940). Photosensitization and fluorescence by aromatic hydrocarbons. *J. Chem. Phys.* **8,** 849.

WETLAUFFER, D. B. (1962). Ultraviolet spectra of proteins and amino acids. *Adv. Protein Chem.* **17,** 303–390.

WETLAUFFER, D. B., EDSALL, J. T. and HOLLINGWORTH, B. R. (1958). Difference spectra in proteins and amino acids. *J. Biol. Chem.* **233,** 421.

WHITAKER, J. R. and JANDORF, B. J. (1956). Specific reactions of dinitrofluoro-benzene with groups of chymotrypsin. *J. Biol Chem.* **223,** 751.

WHITBY, L. G. (1953). New methods for the preparation of flavin adenine nucleotide. *Biochem. J.* **54,** 437.

WIBERG, K. B. (1955). The deuterium isotope effect. *Chem. Rev.* **55,** 713.

WIERZCHOWSKI, K. L. and SHUGAR, D. (1956). Die photolyse und die struktur der nukleinsaurederivaten. Annual Congress *Chemische Gesellschaft DDR,* Leipzig.

WIERZCHOWSKI, K. L. and SHUGAR, D. (1957). Photochemistry of cytosine nucleosides and nucleotides. *Biochim. Biophys. Acta,* **25,** 355.

WIERZCHOWSKI, K. L. and SHUGAR, D. (1959). Studies of reversible photolysis in oligo- and poly-uridylic acids. *Acta Biochim. Polon.* **6,** 313.

WIERZCHOWSKI, K. L. and SHUGAR, D. (1960a). Further studies on the photochemistry of pyrimidines with special reference to 5-substituted derivatives in relation to photoreactivation in the T-even bacteriophages. *Acta Biochim. Polon.* **7,** 63.

WIERZCHOWSKI, K. L. and SHUGAR, D. (1960b). Photochemistry of model oligo- and poly-nucleotides. II. Homopolymers of Ap, Gp and Cp and several heteropolymers. *Acta Biochim. Polon.* **7,** 377.

WIERZCHOWSKI, K. L. and SHUGAR, D. (1961a). Photochemistry of cytosine nucleosides and nucleotides. *Acta Biochim. Polon.* **8,**

WIERZCHOWSKI, K. L. and SHUGAR, D. (1961b). Photochemistry of model oligo- and polynucleotides. IV. Hetero-oligonucleotides and high molecular weight single and double stranded polymer chains. *J. Photochem. Photobiol.* **1,** 21–36.

WIERZCHOWSKI, K. L. and SHUGAR, D. (1961c). Photochemistry of 4-aminopyrimidines. *Proc. 3rd Intern. Congr. Photobiol.*, Copenhagen; *Progress in Photobiology* (CHRISTENSEN, B. C. and BUCHMANN, B. eds.).

WIERZCHOWSKI, K. L. and SHUGAR, D. (1963). Photochemistry of 4-aminopyrimidines: 2,6-dimethyl-4-aminopyrimidine. *Photochem.* and *Photobiol.* **2,** 377.

WIERZCHOWSKI, K. L., SHUGAR, D. and KATRITZKY, A. R. (1963). Primary photoproduct of 2,6-dimethyl-4-aminopyrimidine. *J. Am. Chem. Soc.*, **85,** 827–828.

WILKINSON, P. G. and TANAKA, Y. (1955). New xenon light sources for the vacuum ultraviolet. *J. Opt. Soc. Amer.* **45,** 344.

WITHROW, R. B. and WITHROW, A. P. (1955). Generation, control and measurement of visible and near visible radiant energy, in *Radiation Biology* (HOLLAENDER, A. ed.), McGraw-Hill, New York, vol. 3, chap. 3.

WITKIN, E. M. (1956). Time, temperature and protein synthesis: A study of ultraviolet-induced mutation in bacteria. *Proc. Cold Spring Harb. Symp. Quant. Biol.* **21,** 123.

WITKIN, E. M. and THIELE, E. C. (1960). The effect of posttreatments with chloramphenical on various ultraviolet-induced mutations in *E. coli. Proc. Nat. Acad. Sci. U.S.* **46,** 226.

WOLFROM, M. L., BINKLEY, W. W., MCCABE, L. J., SHENHAN, T. M. and MICHELALIS, A. M. (1959). The effect of ionizing radiations on carbohydrates. *Rad. Res.* **10,** 37.

WOOD, H. N. and BALLS, A. K. (1955). Enzymatic oxidation of α-chymotrypsin. *J. Biol. Chem.* **213,** 297.

WOOD, R. W., (1934). *Physical Optics.* Macmillan, New York.

WULFF, D. L. and FRAENKEL, G. (1961). On the nature of thymine photoproduct. *Biochim. Biophys. Acta,* **51,** 332.

WULFF, D. L. and RUPERT, C. S. (1962). Disappearance of thymine photodimer in ultraviolet irradiated DNA upon treatment with a photoreactivating enzyme from baker's yeast. *Biochem. Biophys. Res. Comm.* **7,** 237–240.

YAMADA, M. (1956). Photolysis of L-proline. *Nagasaki Igakkai Zasshi,* **31,** 999. *Chem. Abst.* **51,** 6359b.

YAMAMOTO, N. (1956). Photodynamic action on bacteriophage. I. The relation between chemical structure and the photodynamic activity of various dyes. *Virus,* **6,** 510.

YAMAMOTO, N. (1958). Photodynamic inactivation of bacteriophage and its inhibition. *J. Bacteriol.* **75,** 443.

YANARI, S. and BOVEY, F. A. (1960). Interpretation of the ultraviolet spectral changes of proteins. *J. Biol. Chem.* **235,** 2818.

YOSHINAGA, K. (1960). Photolysis of L-aspartic acid. *Nagasaki Igakkai Zasshi,* **35,** 103. *Chem. Abst.* **55,** 2251f (1961).

ZAHN, R. K., DOCTER, A., HEICKE, B., KUKLEMANN, H., OCHS, G. and TOVZ, E. (1958). A biological action of injected DNA. *Nature,* **182,** 1679.

ZAMENHOF, S. (1956). Biology and biophysical properties of transforming principles. *Prog. Biophys. Biophysical Chem.* **6,** 86.

ZAMENHOF, S. (1957). Properties of the transforming principles, in *Symposium on Chemical Basis of Heredity* (McElroy, W. D. and Glass, B., eds.), Johns Hopkins Press. Baltimore.

ZAMENHOF, S., LEIDY, G., GREER, S. and HAHN, E. (1957). Differential stabilities of individual heredity determinants in transforming principle. *J. Bact.* **74,** 194.

ZAMENHOF, S., LEIDY, G., HAHN, E. and ALEXANDER, H. E. (1956). Inactivation and unstabilization of the transforming principle by mutagenic agents. *J. Bacterol.* **72,** 1.

ZECH, H. (1961). Cytochemische unter suchungen zur reproduktion dis Tabakmosaik virus. *Zeit. für Naturforschung.* **16b,** 520.

ZEELEN, F. J. (1956). Stearoyl-aminozuren synthese, spreiding en photochemie. Thesis, Leiden.

ZELLE, M. R. and HOLLAENDER, A. (1954). Monochromatic ultraviolet action spectra and quantum yields and inactivation of T1 and T2 *E. coli* bacteriophages. *J. Bacteriol.* **68,** 210.

ZELLE, M. R. and HOLLAENDER, A. (1955). Effects of radiations on bacteria, in *Radiation Biology* (Hollaender, A. ed.), McGraw-Hill, New York, vol. **II,** pp. 365–430.

ZELLE, M. R., OGG, J. E. and HOLLAENDER, A. (1958). Photoreactivation of induced mutations and inactivation of *E. coli* exposed to various wavelengths of monchromatic ultraviolet irradiation. *J. Bacteriol.* **75,** 190.

ZILL, L. P. (1955). A monometric actinometer for the ultraviolet spectrum. *Arch. Biochem. Biophys.* **59,** 473.

ZILL, L. P. (1957). Ultraviolet inhibition of sucrose synthesis in plants. *Fed. Proc.* **16,** 276.

ZILL, L. P. and TOLBERT, N. E. (1958). The effect of ionizing and ultraviolet radiations on photosynthesis. *Arch. Biochem. Biophys.* **76,** 196.

ZINDER, N. D. (1959). Virology. *American Scientist,* **48,** 608.

ZIRKLE, R. E. (1957). Partial cell irradiation. *Adv. Biol. Med. Phys.* **5,** 103.

ZIRKLE, R. E. and URETZ, R. B. (1960; 1963). Energy relations in chromosome paling by ultraviolet microbeams. *Science,* **131,** 1316 (abstract); *Proc. Nat. Acad. Sci. U.S.* **49,** 45.

ZIRKLE, R. E. and URETZ, R. B. (1963). Action spectrum for paling (decrease in *u*) of UV-irradiated chromosome segments. *Proc. Nat. Acad. Sci. U.S.* **49,** 45–52.

ZWORYKIN, V. K. (1957). Television techniques in biology and medicine. *Adv. Biol. Phys.* **5,** 243.

ZUBAY, G. and DOTY, P. (1959). The isolation and properties of deoxyribonucleoprotein particles containing single nucleic acid molecules. *J. Mol. Biol.* **1,** 1.

Index

Absorption
band 8, 9, 10
Beer–Lambert equation 4
cross-section 17
extraneous, of proteins 46
pigment 46
spectra 4, 322
of adsorbed gases 351
amide bond 39
of fibrous proteins 43
of films 228
and hyperchromicity 69
influence of condensed state 10
influence of urea 38
of metastable states 57
of nucleic acids 222
of proteins 29, 46, 297, 324
solute–solvent interaction 10
and temperature 11
theory of 4
of viruses 287, 289
weighting effect 36
Acetabularia 386
Acetamide 89, 90
Acetic acid
filter 89
photolysis 90
Acetylalanine
photolysis 91, 95
spectrum 91
Acetylcholine esterase 149
Acetylglycine 39
Acetyl-L-methionine 49
Acetyltryptophan 40
Acridine
3, 6-diamino- 194
-orange 278, 315, 316
Acriflavin 276, 315
Acrylonitrile 123, 126
Actinometry 146, 378
ferrioxalate 381
malachite green 380
miscellaneous 383
procedures 382
uranyl oxalate 378, 380
Action spectrum (*see also* Individual substances) 29, 87, 110, 325
of aldolase 327
assumptions 320
of bacteriophage 297, 317
in biology 320
and chromosome paling 334
definition 18, 321
of enzymes 113, 121, 140, 142, 149, 154, 155, 327
of gramicidin 327
hemagglutination 292
and inducible enzyme formation 329
infectivity 292
of microorganisms 328, 330
of nucleic acids 327
of paramecin 328
and photoreactivation 333, 340
of proteins 327
ribonuclease 46
of RNA and DNA in bacteria 330
of RNA and DNA in cytoplasm 324
of single and double stranded DNA in phage 330, 332, 333
TMV 285
TMV-RNA 285, 289, 327
viruses 285, 286, 292
Active center or site
in enzymes 110, 137, 138, 158
in trypsin 138
Actomyosin 44
Adenine 75, 165, 166, 168, 176, 225
methyl- 75
-1-*N*-oxide 169
oligoribunocleotides 73
photosynthesis 356
Adenosine 165, 166, 171
Adenylic acid (A) 163
heteropolymers 268
spectrum 69
ADP 171
Adsorption of amino acids on clays 351, 354
Aggregation, of proteins 117, 118, 122, 123, 125, 131, 158

Air 105, 125, 180, 190
Albumin
  serum 161
    photolysis 121, 123
    spectrum 40, 48, 49, 116
Alcohol 148, 327
Aldolase 115, 133, 149, 155, 327
Alkylhalide, photolysis 22
Allantoic acid 170
Allantoin 170
Allele 310, 311, 312
Alumina, adsorption of gases by 351
Amides 94, 354
Amines
  photolysis 90
  spectrum 90
Amino-acids (*see also* Individual amino-acids)
  fluorescence spectra 52
  liberation from peptides 96
  photolysis 97, 98
  photosensitized destruction 107, 156
  photosynthesis of 349, 353, 354, 355
Amino groups 118, 175
Aminopyrimidine 60
Ammonia 175, 216, 347, 351, 354
  from amino-acids 92
  from purines and pyrimidines 168, 175
Amoeba 386
AMP 171
Amylase 133, 136, 156
Anthracene 190
Antibodies 129, 143, 150, 157
Antigenic activity of virus 233, 243, 281, 292
Apurinic acid (APA) 121, 228, 247, 260
Aromatic groups
  energy transfer 149, 257
  enzyme inactivation 142, 154
  photolysis 97, 120
  spectra 97, 122
Arsenophosphotungstic acid 174
Ascites tumor 277
Ascorbic acid 164
Aspartic acid 103
Atmosphere
  fall-out in 350
  primitive 348
ATP 163, 171

Bact. megaterium 297
Bacteriophage 224, 228, 240, 244, 252, 270, 295
  action spectra 297, 302
  of *Bact. megaterium*, M5 297
  C16 296, 317
  cross reactivation (CR) 306, 309
  DNA content 296
  dried state 300
  gene-controlled reactivation 310
  heat denaturation 343
  inactivation and 5-bromouracil 318
  marker rescue 309
  multiplicity reactivation (MR) 306
  mutagenesis 335
  P22 296, 303
  phage-host interactions 301
  phage lambda 301
  ϕX174 302, 326, 331, 333
  photochemistry 232, 295
  photodynamic inactivation 314, 316
  photoreactivable sector 305
  photoreactivation 300, 310, 311, 312
  S13 296, 317
  survival curves 299, 304, 311
  synthetase activity 302
  T1 297, 298, 299, 301, 302, 309, 326, 343
  T2 84, 295, 298, 299, 302, 303, 307, 308, 311, 312, 314, 318, 324, 331
  T2r 313
  T3 299, 303, 309, 315
  T4 264, 275, 295, 299, 303, 307, 308, 309, 311, 312, 318
  T4u 304
  T5 299
  T6 299, 303
  T6r 314
  T7 299, 303, 309
  T-even 212
  Vi-phage II 307
  vulnerable centers 307
  X-phage 335, 339, 343, 346
Bacterium cadaveris 329
Barbital 62
  dimethyl- 63
Barbituric acid 62, 176
  1, 3-dimethyl- 218
Beer–Bouger–Lambert equation, derivation 15
Benzene
  1,2,4-tri-t-butyl- 214

Benzoylalanine
  photolysis 95
  spectrum 95
Biological effects of radiation 221, 223, 225, 229, 231, 232, 237, 296, 328
Biological inactivation 232, 240, 242, 245, 250, 251, 261, 262, 276, 285, 345
  reversal 270, 272, 336, 342, 334
Birefringence, streaming 222
Blood plasma 294
Bond
  dissociation of 3
  thermochemical strengths of 2
Bromine 178
Buffer
  phosphate compared with acetate 146, 300

Caffeine 166
Caprylate 117
Carbohydrates 172, 204
  photolysis 163, 173
  spectrum 163
Carbon dioxide 354, 380
Carbon monoxide 110, 119, 349, 380
Carboxyl groups 118
  spectra 89
Carboxypeptidase 144
Carcinogens 276
Catalase 143, 160, 223
Catalysts as adsorbents 352
Cell nuclei 227, 334, 336, 386
Cell physiology and action of light 163, 221, 226, 311, 329, 333, 335
Cells 250
  incorporation of T-DNA 245
  partial cell irradiation 384
  red 165
Ceric sulfate 379
Cesium chloride gradient 250
Chloroplasts 25
Chromophore
  definition 10
  extraneous absorption 46
  in nucleic acids 274, 275
  in nylon photochemistry 93
  in proteins 88, 113, 123, 149
  peptenol 44
Chromosomes 303, 308
  paling 334
Chymotrypsin 129, 137, 327
  fluorescence 54, 112
  inactivation 115, 118, 120, 124, 127, 133, 136, 137, 146, 157
  spectrum 38, 40, 50, 116
Clays
  adsorption of amino-acids on 351
  adsorption of gases by 351
Clupein
  photolysis 93, 121
  spectrum 42
Collagen 44
Concentration effect and UV-sensitivity (*see* Quantum yield)
Cophosphorylase 165
Copper 151
Cross linkage
  between protein and DNA induced by UV light 250, 297
  in DNA 229, 250, 326
  in polynucleotide chains 184, 229, 246
Cross reactivation 306
Cyanoacetylacetone 213
Cyanuric acid 170
Cyclobutane ring 184, 344
Cyclohexadiene 27
Cysteine 104
  spectrum 30, 31
Cystine 104, 105, 327, 353
  photochemistry (*see* disulfide linkages)
  spectrum 30, 31
  in wool 117
Cytidine 203
  -5′-dimethylphosphate 205
  phosphate 61, 65, 197, 207, 263
    and reverse reactions 205
  photoproducts 205
  spectrum 61
Cytidylic acid (C) 163, 198, 203, 225
  deoxy- 260
  oligonucleotides of 263
  poly-, spectrum 70, 74, 263
  spectrum 69
Cytoplasm 324, 386
Cytosine 168, 196, 200, 302, 348
  alkylamino- 202
  analogues 61, 200
  carbohydrates, spectra 64
  derivatives and quantum yields 203
  dihydro- 62, 198
  glycosides 197, 198, 201

Cytosine (*contd.*)—
hydration 241, 264, 266
hydroxymethyl- 252, 264, 332
2 methoxy- 212, 215
1-methyl 196, 200, 202, 203, 210, 213
5-methyl- 251
1-methyl-monomethylamino- 203
nucleosides 211
nucleotides 199, 211
reversal reactions 206
spectrum 59
synthesis of 348

D. pneumoniae 271
Dark reactions, including thermal reversibility 133, 174, 176, 179, 194, 197, 255, 256, 263, 266, 268, 290, 301, 325, 344
and reversile of ultraviolet effects 342, 343
Degradation
of DNA 222
of RNA 230, 242
Dehydration 207, 272
Denaturation
of DNA 75, 239, 250
of enzymes 137
of nucleoprotein 81
of proteins 118, 120, 121, 122, 155, 158
spectral changes 38, 39, 51
and fluorescence 54
of viruses 281, 285, 290
Deoxyribonucleohistone 231
Desoxyribonuclease 143, 148, 160
Desoxyribonucleic acid (DNA)
in bacteriophage 296
cross-linkage 326
dimerization in 184, 249, 262
films 77, 326
halogen containing 211, 319
and histones 83
hyperchromicity 67, 72, 75, 76
hypochromicity 77, 78
irradiation *in vivo* and *in vitro* 226, 231
melting 75, 76
of ϕX174 phage 333
photolysis 163, 222
photosensitized inactivation 276
"primer" 272
protein complexes 82

Desoxyribonucleic acid (DNA) (*contd.*)—
spectrum 68, 72
transforming 224, 231, 234, 241, 245, 251, 261, 270, 276, 338, 341, 346
Watson–Crick structure 68, 74, 77
Deoxycytidine phosphate 203
Depolymerization
of nucleic acids 222
of urease 127
Detergent 117, 291
Deuterium 187, 200, 208
Diazine 215
Dichroism 322
Dimerization 190, 199
of chymotrypsin 134, 158
in nucleic acids (*see also* Individual acids) 250, 272, 325
in oligonucleotides 191, 253, 259, 260, 261, 264
and photodissociation 261, 262, 272
and photoreversile 340
in polynucleotides 191, 261, 344
of pyrimidines 229
of thymidine 187
in thymine (*see* Thymine)
Dinucleotides, spectra 69
Disulfide linkages
and enzyme inactivation 137, 138, 140, 146, 149, 150, 154
inner filter 136
photolysis 97, 110, 118, 119, 121
spectrum 30, 43, 327
Double bonds 182
DPN 163, 164, 165, 171, 173, 226
Dyes 28, 107, 156, 157, 160, 229, 242, 276, 294, 313, 315, 316, 317

E. coli 309, 311, 329, 332
photosensitized killing 160
and reactivation 270, 272, 337, 340, 342
ribonucleoprotein from 81, 227
Earth
atmosphere 347
primitive 347, 356
sources of radiant energy 348
Egg nuclei 334
Einstein, definition 1
Electrical discharge 349, 352, 356
Electron paramagnetic spin resonance 24

Electron transfer 110
Electrophoresis
of nucleic acids 226
of plasma proteins 294
of proteins 123
End groups
photochemistry of 89
spectra of 89
End groups, infection by phage 302, 304, 305
Energy of activation 134, 205
Energy levels 6
Energy transfer
in bacteriophage 331
in benzene 22
in nucleic acids 244, 257, 267, 270, 288
in peptides 95
in polynucleotides 332
in proteins 54, 111, 112, 113, 149, 288
Enzyme induction 330
Enzymes 322
photosensitized inactivation 156
quantum yields 111, 141, 142, 228
sulfhydryl 149
Eosin 27, 156, 276, 314
Ergosterol 23
Erythromycin 239
Ethane 353
Ethanol 182, 183, 213, 300, 301
Ethylbenzene 96
Excitation energy
electronic 9
resonance transfer 11
transfer mechanisms 11, 13, 88, 111
Excited state 3, 147, 187, 194, 207, 223, 242, 341
of amino-acids 56
half life 19, 277
of proteins 56, 57
Extinction coefficient
definition 4, 15
upper value 4
Extraneous absorption by proteins 46

Ferrioxalate actinometry 381
Ferrous salts 216, 348
Feulgen reaction 227, 334
Fibers of DNA and APA 228, 229, 231
Fibrinogen 160
Ficin 149
Films
of apurinic acid 228
and dimer formation 190
of phage 300
poly-A 78
of nucleic acids 227, 228, 249, 326, 345
and spectra 228
of viruses 293
Filters, optical 352, 369
acetic acid 89
band pass 373, 374
chemical 371, 372, 374, 375
cut-off 371
solid 369
Flash photolysis 25, 119, 162, 195, 366, 367
Flash-spectroscopy 367
Flavin derivatives 172
Fluorescein 160, 275
Fluorescence 8, 12, 26, 27, 96, 148, 163
of amino-acids 53, 111
depolarization of 56
of nucleic acid 111
of proteins 12, 46, 53, 54, 111, 112, 150, 152
quenching 112
spectra 51, 53
Formaldehyde 20, 293, 294, 302, 349, 354, 355
Formamide 250, 256
Formhydroxamic acid 355
Formic acid 354, 356
Formol titration 120
Franck–Condon principle 7, 20
Franck–Rabinowitch cage effect 90, 146, 153, 242
Free radical formation 23, 24, 25, 123, 124, 148, 163, 164, 191, 216, 275, 354
life time 26
recombination 124
test for 123, 124
Freezing 294
Fumaric acid 23
Furfural 163

Galactosidase 329
Gases, dissociation of 349, 351
Gel formation
in irradiated DNA 227

Gelatin
  fluorescence 53
  spectra 39
Gene 310
Genetic markers 235, 237
Genome 245, 301, 303, 308, 310, 311
Glass state 342
Globin, human 161
Globulin 120
γ-Globulin 161
Glucose 163, 164, 223
Glutamic acid 354
Glutamylalanylglycine 105
Glutathione 105, 223
Glycerol 242, 323
Glycine
  photolysis 99, 103, 350
  photosynthesis 350, 353, 354, 356
  quantum yields for 99, 350
  spectrum 39
Glycosidic linkage 164, 171, 174, 196, 251, 253, 260, 269, 318, 332
Glycylglycine 49
  photosensitization 108
  spectrum 89
Glycyltryrosine 105, 106
Gramicidin 149, 155, 327
Guanine 165, 225, 277, 302, 356
Guanylic acid (G) 171

Haem 110, 112
Haemophilus influenza 234, 238, 241, 270, 272, 333
Heat sensitivity of irradiated viruses and enzymes 129, 244, 281
Helix
  formation 73
  separation 238, 241, 250
  structure 253
Hemagglutination 292
Hemipeptides
  photolysis 91, 95
  spectrum 89
Hemocyanin 126
Hemoglobin 119, 137
Hemolytic streptococci
  streptolysin S 242
  transforming DNA 234
Hexamethylenetetramine 354
Hexane 213
Hexose 164
Histidine 104, 143, 157, 158, 159, 354
Histone 231
Homocystine 105
Hyaluronic acid 163
Hydration (*see* Uracil, thymine, etc.)
Hydrogen 195, 348
Hydrogen bonds 39, 69, 73, 345
  base-pair in DNA 77, 252
  and melting of DNA 76, 241
  in nucleic acids 222, 228, 241, 246, 251, 267
  in nucleotides 266, 269
  in phage 303
  in polycaprolactam 131
  in proteins 153
  in pyrimidines 213
  in TMV 112
Hydrogen cyanide 183, 356
Hydrogen isotope effect 199, 207, 209
Hydrogen peroxide 117, 278
Hydrogen sulfide 104
Hydroquinone 123
  and free radicals 124, 125
Hydroxyacids 354, 355
Hydroxylamine 355
Hyperchromicity 47, 67
  of DNA 73
  and nucleic acids 67, 222, 249, 251, 284
  of oligonucleotides 70, 71, 257, 266
  of poly-A-poly-U 222, 263, 268
  of poly-C 263
  of poly-U 73
  polyglutamic acid 51
  of polynucleotides 71, 78
Hyperconjugation and spectra 51
Hypochromicity
  of homopolymers of uridylic and cytidylic acids 69, 70, 253
  of nucleic acids 67
Hypoxanthine 168

Ice 179, 184, 242, 300
Imidazole 356
Immune serum 129
Inactivation cross-section
  definition 17
  trypsin 151, 154
Indirect irradiation effects 275
Indoleacetic acid 156, 160
Induction period 167

Infectious RNA (*see* Ribonucleic acid)
Influenza transforming DNA 237
Inosinic acid
  poly-, spectrum 74
Insulin
  fluorescence 54
  spectrum 37, 40
Intensity of irradiation 2, 3, 16, 125, 167, 280
Interferon 294
Internal conversion 23, 341
Ionic strength 146, 201, 205, 206
Ionization by irradiation at short wave lengths
  chymotrypsin 154
  trypsin 154
Ionizing radiation 164, 191, 231, 237, 239, 243, 245, 251, 252, 276, 278, 295, 329, 334, 335, 349, 355, 356
Isocytidine 211
Isoelectric point
  enzymes 145
  virus 280
Isoguanine 166, 169
Isosbestic point
  ribonuclease 33, 35
  tyrosine 32, 35
Isotope effect 200, 207, 254, 256
Isotopes, heavy 303
Isouracil 175

Kinetics
  of amino-acid decomposition 91
  of cytidylic acid photoreaction 263
  of decomposition of photoproducts 105, 114, 120, 135, 196
  of enzyme inactivation 115, 133, 134, 151
  of hetero-oligonucleotide reactions 266
  of inactivation by photosensitization 159
  of inactivation of bacteriophage 299, 303, 317
  of inactivation of dry enzymes 151
  of inactivation of TMV-RNA 288
  of inactivation of transforming DNA 237, 242
  of photolysis of uridylic acid polymers 253
  of photoreversile 272
Kinetics (*contd.*)—
  of protein denaturation 15, 114, 124
  of purine decomposition 167
  of reverse reactions 204
  of uracil transformations 191
  van't Hoff difference method 193
  of virus inactivation 280

Lactoblobulin, photolysis 118
Lamps
  Carbon 359
  daylight 318
  flash discharge 365
  for inactivation studies 364
  Krypton 352
  Mercury 358
    arc 353
    germicidal 360
    high pressure 324
    low pressure, resonance 328, 360, 362, 364
  Oxygen 366
  for preparing photochemistry 363
  Xenon 292, 352, 359
Life, origin of 346, 356
  and adsorption of organic matter on clay 351
  synthesis of glycine 350
Light intensity (*see* Reciprocity law) 146, 366
  measurement of 376
  visible 107, 171, 172, 334
Light scatter 15, 16, 29, 51, 322, 323, 350
  bacteriophage 84, 298
  clupein 42
  constant 46
  nucleic acids 223
  proteins 45, 115, 117, 287
  of suspensions 323
  theory 45
  viruses 84, 280, 285, 323
  X-protein 287
Luminal test 124
Luminescence 23, 51
Lysine decarboxylase 329
Lysogenic bacteria 295
Lysozyme
  and DNA complexes 82
  fluorescence 54, 112
  inactivation 142, 156, 157, 158
  photolysis 118

Lysozyme (*contd.*)
spectrum 38

M. pyogenes 323
Malachite green 381
Malonamide
N,N'-dimethyl- 218
N,N'-dimethylmethyl- 219
Marker rescue 309
Melting point
of DNA 76, 238, 240, 250, 345
of oligonucleotides 269
of poly-U 73
Metastable states 225, 341
Meteorites 348
Methane 348
Methionine 105
Methyl green 227, 228
Methylene blue 107, 157, 158, 160, 161, 276, 277, 313, 314, 317
Microbeam irradiation 333, 384
Micro-organisms 174, 196, 206, 210, 243, 249, 250, 251, 294, 295, 321, 323, 324, 328, 336, 342, 343, 346
and humidity 228
Microspectrophotometry 226, 323, 386
Microsurgery 333
Moisture
and films of DNA 228
and micro-organisms 228
Molecular rearrangements 23
Molecular weight
influence of irradiation 125, 280
and irradiation of DNA 223
and quantum yields of proteins 143, 149
and RNA from TMV 282
Molecules
diatomic 5
excited 8
molecular states 7
polyatomic 10
selection rules 7
Monochromatic light 334
Monochromator 265, 324
Monolayers, spread
enzymes 148
peptides 94
proteins 119, 129, 131
Multi-component inactivation curves 236
"Multi-hit process" 234, 243, 245, 291, 292, 299, 314, 318
Multiplicity 7
Multiplicity reactivation 306, 309, 311
Mutagenesis in RNA and DNA 335
Mutation 228, 243, 275, 285, 335, 338, 343, 345, 346
Myoglobin 110
energy transfer in 14, 110

Neurospora crassa 179
Nicotinamide 165, 171
Nitriles 354
Nitrogen 115, 123, 158, 167, 195, 196, 223, 228, 259, 281, 348
Nuclear magnetic resonance 186, 203, 213
Nucleic acids (*see also* RNA, DNA, oligonucleotides, and polynucleotides)
absorption spectra 58, 66, 67
free radical formation 24
metabolism of 210
number of residues affected during inactivation 274
photochemistry 222, 246
photoreactivation 336, 339
photosynthesis 356
potential reversible transformations 343
summary of primary photochemical effects 273
Nucleohistone 232
Nucleophilic addition 183, 198, 200
Nucleoprotamines 232
Nucleoproteins
composition 81
photolysis 231
and photoreactivation 336
spectra 79, 81, 82
synthetic 82
Nucleosides, spectra 64
Nucleotides, spectra 64
Nylon
photolysis 93, 131
spectrum 93

Ocean, synthesis in 354
Oligonucleotides 190, 209, 233, 242
of cytidylic acid 263

Oligonucleotides (*contd.*)—
dimerization in 261, 298
hetero- 264
natural 251
photochemistry of 205, 296
spectra of 66, 68, 70
of thymidylic acid 260
or uridylic acid 253
"One-hit" process 16, 126, 127, 134, 243, 280, 283
Optical density definition 15
Optical rotation
and irradiation 280
of proteins 122
Optochin resistance 236
Orcinol 165
Organic molecules, photosynthesis of 353
Orotic acid 60, 195, 196, 356
dihydro- 356
Osmotic pressure, changes with irradiation 125
Ovalbumin
excited state 57
monolayers 131
phosphorescence 55
photolysis 117, 119, 121, 136
spectrum 37, 57
Oxamide 217
Oxazine dyes 316
Oxidase 151
Oxygen 20, 26, 58, 93, 104, 105, 106, 108, 115, 117, 119, 123, 125, 143, 148, 156, 160, 161, 164, 167, 168, 190, 191, 194, 196, 216, 223, 259, 276, 281, 315, 366
Ozone 104, 106, 117

Paramagnetism 195
Parabanic acid 217
Paramecin 327, 328
Partial cell irradiation 333
Pepsin 136, 142, 146, 148
absorption spectrum 38
fluorescence 54
Pentose 164
Peptide bond
absorption above 2400 Å 39
absorption below 2400 Å 48, 50, 154
and energy transfer 111
and fluorescence 53
Peptide bond (*contd.*)
photolysis 90, 102, 119, 120, 123, 157
photolysis and wave length 149
spectra 41, 42, 44
Peptides
helix formation 73
hyperchromicity 47
photolysis 90, 95, 96, 118
photosynthesis 354, 355
spectra 12, 31, 39, 40, 42, 89
Peroxides 131, 173, 190, 223, 224
pH
and action spectra 331
and dimer formation 188
influence of
on enzyme inactivation 134, 142, 145, 146
on photochemistry of DNA 225
on photochemistry of proteins 145
on photolysis of amino-acids 106
on photolysis of nucleic acids 164
on purines and pyrimidines 167, 177, 179, 196, 203, 204, 205
on reverse reactions 205
Phenol 56, 117
Phenolic groups 118, 145
Phenyl groups 93
photolysis 95, 96
spectra 94
Phenylalanine (*see* amino-acids)
photolysis 106
and protein synthesis 262
spectrum 30, 48
Phosphodiesterase 250
Phosphoglucomutase 158
Phosphorescence 8, 51
of amino-acids 55
of proteins 55, 153
Phosphorylation, oxidative 295
Phosphotungstic acid 175
Photo-autoxidation 277
Photochemistry, laws of 1, 2
Photochromism 342
Photodissociation of dimers (*see* Dimerization)
Photodynamic action 156, 276, 294, 314, 316
Photographic emulsions 341
Photolysis
acetylalanine 18
aldehydes 18, 20
formaldehyde 20

Photolysis (*contd.*)—
  ketones 18
Photo-oxidation 122, 123, 148, 156
Photoreactivating enzyme 271
Photoreactivation (PR) 334, 335
  and action spectra 333, 336
  of bacteriophage 295, 299, 304, 305
  biological 173, 174, 221, 275
  characteristics 336
  of nucleic acids 250, 261, 270, 283
  physical models for 340
  of plant viruses 291
  radiation receptors in 339
  of seed germination 340
  of TMV-RNA 233
  of triosephosphate dehydrogenase 150
  viruses 336
Photoreactivation sector 305
Photorestoration (*see* Photoreactivation)
Photosensitization 3, 22, 23, 25, 110, 230
  and enzyme inactivation 156
  and oxygen 26
  and photolysis 96
  and viruses 317
Photoxidation 315
Pigment
  absorption 46
  photosensitization by 156, 317, 318
Planets, atmosphere 347
Plant viruses 291
Pneumococcol transforming DNA 235, 241
Polyadenylic acid (poly-A) 73, 78, 228, 253
Poly-A-Poly-U 269, 270, 345
Polycytidylic acid (poly-C) 70
Polyesters 230
Polyguanine 228
Polyguanylic acid (poly-G) 253
Polyinosinic acid (poly-I) 74
Poly-I-Poly-C 269, 270
Polymerase, calf thymus 261
Polymerization 134, 135, 230
Polymethylacrylic acid 276
Polynucleotide 196
  dimerization in 184, 261
  mechanism of inactivation of DNA 241, 332
  photosynthesis of 356
Polynucleotides 190
  hydration in 251
Polynucleotides (*contd.*)—
  photochemical reactions 191, 211, 251, 344
  spectra 66, 75
  twin stranded 267, 331
Poly-ribosethymine 269
Polythymine (poly-T) 259, 325
Polyuridylic acid (poly-U) 70, 73, 78, 228, 258, 283, 325
  and protein synthesis 262
Potential energy 9
Predissociation 215
Preparative photochemistry 362
Primary process 3
Propiolactone 294
Prosthetic group
  diphosphopyridine dinucleotide 36
  heme 12
  in phycocyanin 12
Protection
  of enzymes from inactivation 124, 136, 150
  inner filter 136
Proteins
  absorbing groups 29, 34, 36
  bound phenolic groups in 37, 38
  excited states 51
  free radical formation 24
  physical changes 130
  spectra 106
    changes in proteins 114
    contribution of peptide bonds 43
    fibrous proteins 44
Purine (*see* Individual compounds)
  derivatives 166
  oligonucleotides 253
  photosynthesis 356
Purines 165, 166, 176
  chromophore 60
  spectra 59, 60, 63, 65
Purines and quantum yields (*see* Individual compounds)
Pyranosylcytosines, spectra 65
Pyrimidine
  4-amino- 176, 215, 264
  2-chloro-4-amino- 176
  derivatives 166
  2,4-diethoxy- 181
  dihydro- 356
  2,4-diketo-dihydro- 188
  2,6-dimethyl-4-amino- 213

Pyrimidine (*contd.*)—
  2,6-dimethyl-4-amino-photoproduct 213
  2-methyl-5-ethoxy methyl-4-amino- 176, 215
  photosynthesis 356
Pyrimidines 165, 167, 174
  amino- 356
  4-amino- 213
    photochemical reactions Table 6–9
  chromophore 60
  degradation products 216
  dimerizations 191
  and excited states 194
  nucleosides 65
  and quantum yields 180
  spectra 59, 60, 63, 65
Pyridone 214
Pyruvic acid
  from acetylalanine 92
  from amino acids 98
  from pyrimidines 216

Quantum, definition 1
Quantum numbers 6
Quantum yield
  and aggregation 125
  of apurinic acid 248
  of bacteriophage 298, 326, 331
  and biological systems 321
  calculation of 16
  and chain length 95, 258
  and concentration 193, 194, 233, 236, 240, 254, 257, 280, 283
  and cystine content 138
  of cytosine compounds 203, 263, 264, 266
  definition 3
  determination 15
  of dimer formation 188, 190, 344
  for dissociation of dimers 188, 261
  for dissociation of gases 349
  of DPN co-enzyme activity 173
  for enzymes 113, 121
  and fiber formation of DNA 228
  of fluorescence 51, 52, 53
  in heavy water 200
  hemocyanin 127
  for inactivation of RNA 223, 283, 288, 289, 326
    and kinetics of uridin-phosphate transformation 193

Quantum yield (*contd.*)—
  and molecular weight 143
  and multi-hit inactivation 234, 243
  of nucleic acids 284, 289, 290, 324
  and phosphate groups 206
  and presence of oxygen 116
  of protein monolayers 132
  of proteins 122
  of poly-U 258
  of pyrimidines 180
  streptolysin 242
  of tobacco mosaic virus 285, 289, 290
  of transforming-DNA 237, 241
  of uracil 176
  of uridine derivatives 254, 258
  of uridylic acid photolysis 193, 258
  of viruses 285, 288
  and viscosity of DNA and RNA 222, 223
  and water 148
  and wave length 148, 149, 173, 285, 298, 324, 326
Quenching 195
Quinones 117

Radiation receptors in biology 320
Reactivation
  of bacteriophage 299
  cross 295
  multiplicity 295
Reciprocity law 2, 3, 16, 125, 146, 147, 167, 245, 280, 300, 322, 330
Refractive index 45, 323
Resonance 13
Reversible reactions 176, 179, 196, 201, 246, 258, 262, 264, 342
Riboflavin 156, 157, 160, 172, 313, 314
Ribonuclease 234, 258, 268, 314
  action spectrum 46
  fluorescence 54, 112
  photosensitized inactivation 156, 160
  quantum yield 46, 136, 142, 148
  spectrum 33, 38, 46, 47
Ribose 163, 165, 172
Ribonucleic acid (RNA)
  core 242
  depolymerization of 227
  infectious
    from polio virus 231

Ribonucleic acid (RNA) (*contd.*)—
infectious (*contd.*)—
from TMV 231, 232, 233, 242, 262, 275, 336
in irradiated yeast 226
melting 72
number of residues affected by inactivation 275
photolysis 226, 230
photoreversile of inactive 262, 337
spectra 72, 247
S-RNA 251, 329, 330
synthesis of 227
transfer–RNA 262
yeast 163, 242, 246
Ribosidic linkage 171, 216
5-Ribosyluracil 251
Rose bengal 276

S. cerevisae 342
Safranin 231
Salmonella typhi 307
Sedimentation
and inactivation of chymotrypsin 127
of proteins 123, 125
of trypsin and wavelength of inactivation 149
of urease 127
Seeds, germination 340
Selection rules 10
Serine, and protein synthesis 262
Serineformamide carbonate 356
Serological properties 280, 291, 313, 314, 316
Serratia marcesens 228, 335, 339, 343, 346
Serum
immune 129
oxidase 151
proteins 160
rabbit 129
Silicates as adsorbents 351, 352
Single-hit theory (*see* "One-hit" process)
Skin 322
Snake venom 157
Sodium caprylate 117
Solid state (*see also* Ice) 148, 340
Solubility changes
of irradiated chymotrypsin 128, 136, 157
of proteins 121, 122, 134
of virus 281
Solutions, photochemical reactions in 15, 18
Solvent effects 146, 173, 201, 213, 215, 227, 231, 288, 300, 320, Table 6-9
Space
micro-organisms in 346
satellites 346
vehicles 346
Spectra
absorption band 9
action 18
aldehydes and ketones 19
change with irradiation 107, 114, 116, 275
and chelation 213
difference, of X-protein 287
flash 367
flashphotolysis 25
of hemipeptides 94
and hyperconjugation 51
influence of temperature on 11, 151
of nucleic acids and irradiation 222, 247, 329
of peptides 89
of phenyl compounds 94
and photosensitized reactions 275
of proteins 29, 113
and quantum yields 180
reproducibility among laboratories 139
residue 51
of thymine 11
triplet 26
of uracil compounds 69, 177, 178
weighting effect 36
Spindle in newt cells 334
Spiropyran 342
Staining, and irradiation of DNA 227, 228, 231
Stationary state 321
Stearylamide 94
Stokes' rule 8
Streptomyces griseus 340
Streptomycin, -resistance 234, 239, 272
Sugars 355
Survival curves (*see* Individual entities)
theory 243, 244, 299
Synthesis of organic molecules and UV light 347, 352, 353
Synthetase activity 302

Target theory 17, 240, 280, 329
Temperature
and excited state 147
and inactivation of DNA 238, 240
influence on aggregation 126
influence on enzymes 23, 133, 134
influence on inactivation cross section 152, 154
influence on cross-linkage in DNA 326
influence on photochemical action 23, 121, 150, 164, 197, 203, 205, 241, 242, 290, 317
influence on photoreactivation 338
influence on spectra 11, 151
and irradiation of phage 310, 316
and reactivation of DNA 270, 283
Thermal reactions, *see* dark reactions
Thermal reactivation (TR) or thermal restoration 342, 343
Thiamine 175
Thiazine 313
Thiazine dyes 314, 316
Thiocytosine 181
Thionine 277
Thioranthone 313
Thiouracil 181
Thymidine 186, 187
Thymidylic acid 186, 225, 248
Thymidylic acid oligonucleotides 260
Thymine 75, 302, 318
addition reactions 183
6-amino- 210
dimerization 184, 189, 191, 194, 241, 259, 312, 325
1,3-dimethyl- 219
1,2-dimethyl-cyclobutane-*cis*-di- 189
6-hydroperoxy-5-hydroxy- 190, 191
methyl- 75
N,N'-dimethyl- 186
peroxide 224
products of irradiation 216
spectrum 11
Thymocytes 226
Tobacco mosaic virus 143
action spectrum 285
course of infection by 291
fluorescence 112
hydration 281
inactivation 114, 224, 244, 280, 315
infectious RNA 233, 236, 242, 316
photoreactivation 336
Tobacco mosaic virus (*contd.*)—
reconstitution 231, 233, 282, 285
spectrum 83, 85, 86, 281
X-protein 283
Toluenediamine 277
Toluidine blue 294, 314, 316
Toxin, botulinus 157
TPN 164, 172, 226
Transforming DNA (*see* desoxyribonucleic)
Triglycine 49
Triosephosphate dehydrogenase 163
inactivation 146, 148, 150
spectrum 36
Triphenyl methane dyes 317
Triplet state 8, 26, 195, 215, 316
Tritium 184, 187, 249
Triuret 169
Tropomyosin, spectrum 79
Trypsin 120, 129, 137
action spectrum 113, 121, 140, 155
cross-section 152
inactivation 137
monolayers 131
spectrum 151, 154
and wave length 149
Trypsin inhibitor 107, 122, 136, 146
Tryptophan
and enzyme inactivation 142, 149, 156
fluorescence 54
in hemocyanin 127
photolysis 106, 157
spectrum 30, 33, 34, 48, 49, 106
Tryptophanase 329
Turbidity (*see* light scatter)
Twin strand DNA 229, 241
polynucleotides 268, 269, 345
Tyrosinase 156
Tyrosine
bound 37, 131, 145
fluorescence 54
influence on aggregation of protein 123
influence on quantum yield 142
o-methyl- 35
peptides 12
photolysis 105, 149
spectrum 12, 30, 32, 34, 35, 49, 57, 122
Tyrosyltyrosine 12
Tyrosyl-tyrosyltyrosine 12

Ultrasonic irradiation 171
of nucleic acid 68, 237
Ultraviolet light and synthesis of bio-organic molecules 347
Uracil 174, 176, 183, 356
additional of ethanol 183
addition of HCN 183
6-amino- 210
analogues 200
analogues and dimer formation 187, 191
5-bromo- 211, 312, 318
derivatives 209
dihydro- 61
dimerization 188, 191, 259, 262
dimethyl- 191, 194, 196, 218
1,3-dimethyl- 177, 182, 210, 216, 356
dimethylhydro- 178
1,3-dimethyl-6-hydroxy-5-hydro- 178, 356
2-ethoxy- 181, 182, 209
halogen substituted 210, 319
hydration 187, 262, 325
5-hydroxy-1,3-dimethyl-hydro- 218
6-hydroxy-1,3-dimethyl-hydro- 218
1-methyl- 177, 200
3-methyl- 356
6-methyl- 210
methyl-dihydro- 62
photolysis of 164, 168, 178, 181
photoproducts 196, 216, 219
spectrum 61
1,3,6-trimethyl- 210
uracil copolymers 268
Uranyl-oxalate 146, 378
Urea 112, 136, 168, 216, 356
Urease
inactivation and antibody formation 150
inactivation and molecular weight 127
inactivation and wavelength 146
inactivation by photosensitization 156, 157
Ureide group 168, 216
Ureidosuccinic acid 356
Uric acid 166, 167, 168
Uridine 194
additions to 183
derivatives 178, 179
-diphosphate glucose (UDPG) 163
isopropylidene- 193, 194
Uridine (*contd.*)—
phosphate 66, 179, 192, 322, 325
Uridylic acid (U) 178
kinetics of transformation 191, 193, 254
oligonucleotides 253, 261
photochemistry 255
poly-, spectrum 70
polynucleotides 261
spectrum 69, 254

Vaccines 244, 291, 293, 294
eastern equine encephalmyelitis 294
ectromelia 293
lymphocytic choriomeningitis 293
poliomyelitis 293, 294
SK-mosaic poliomyelitis 293
vaccinia 293
Vacuum irradiation 228
Van't Hoff difference method 193, 199
Virus interference 294
Viruses 196, 210, 239, 273, 279, 322
adenovirus 315
Asian 294
coxsackie B2 315
cucumber mosaic and tobacco 291
ECHO-1 315
equine encephalomyelitis 115
hepatitis 294
inactivation of 279
influenza 84, 292, 295
jaundice 294
nucleic acid (*see* nucleic acids, TMV, etc.) 315
photochemistry 232, 244
photosensitized inactivation 276, 313
plant 280
poliovirus 285, 293, 314, 315
potato X, bushy stunt, tobacco necrosis 291
rabies 293
Rift valley 294
ring spot 291
St. Louis encephalitis 293
spectrum 83
T3 coliphage 315
tobacco mosaic (TMV), spectrum 85, 86
vaccinia 313, 315
vesicular exanthemo 315
yellow fever 294

Viscosity
  of nucleic acids 222, 223, 225, 228, 230, 237, 240, 276, 277
  of proteins 157, 158
  of TMV-RNA 283
  of virus 281
Vision 321
Vitamin B_1 175, 212, 215
Vulnerable centers 307

Water 173, 228
  elimination from 5,6 bond 207
  and energy transfer 111
  heavy 200, 208, 209, 256
  and hydrogen bonds in phage 303
  influence on quantum yield 148, 149, 300
  photolysis 349, 351
Water (*contd.*)—
  uptake by double bonds 182, 194, 198, 219, 246
  uptake by irradiated uracil 178, 182, 210
Watson–Crick structure of DNA 227, 251
Wavelength
  and photodissociation 190
  and photosensitization 156
  and quantum yields 154, 173, 227, 289
Wheat germ DNA 251
Wool 93, 117, 121, 122

Xanthine 167
X-protein 283, 287

Yeast 226, 242